Optimal Management Strategies in Small and Medium Enterprises

Milan B. Vemić
Higher School of Academic Studies "DOSITEJ", Serbia

A volume in the Advances in Logistics, Operations, and Management Science (ALOMS) Book Series

IGI GLOBAL
DISSEMINATOR OF KNOWLEDGE
www.igi-global.com

Published in the United States of America by
IGI Global
Business Science Reference (an imprint of IGI Global)
701 E. Chocolate Avenue
Hershey PA, USA 17033
Tel: 717-533-8845
Fax: 717-533-8661
E-mail: cust@igi-global.com
Web site: http://www.igi-global.com

Library of Congress Cataloging-in-Publication Data

Names: Vemic, Milan, 1962- editor.
Title: Optimal management strategies in small and medium enterprises / Milan Vemic, editor.
Description: Hershey : Business Science Reference, [2017]
Identifiers: LCCN 2016048352| ISBN 9781522519492 (hardcover) | ISBN 9781522519508 (ebook)
Subjects: LCSH: Small business--European Union countries--Case studies. | Small business--Management--Case studies. | Sustainable development--European Union countries. | Technological innovations--Economic aspects--European Union countries. | Knowledge management--European Union countries.
Classification: LCC HD2346.E85 O68 2017 | DDC 658.4/012094--dc23 LC record available at https://lccn.loc.gov/2016048352

This book is published in the IGI Global book series Advances in Logistics, Operations, and Management Science (ALOMS) (ISSN: 2327-350X; eISSN: 2327-3518)

British Cataloguing in Publication Data
A Cataloguing in Publication record for this book is available from the British Library.

For electronic access to this publication, please contact: eresources@igi-global.com.

Advances in Logistics, Operations, and Management Science (ALOMS) Book Series

John Wang
Montclair State University, USA

ISSN:2327-350X
EISSN:2327-3518

Mission

Operations research and management science continue to influence business processes, administration, and management information systems, particularly in covering the application methods for decision-making processes. New case studies and applications on management science, operations management, social sciences, and other behavioral sciences have been incorporated into business and organizations real-world objectives.

The **Advances in Logistics, Operations, and Management Science** (ALOMS) Book Series provides a collection of reference publications on the current trends, applications, theories, and practices in the management science field. Providing relevant and current research, this series and its individual publications would be useful for academics, researchers, scholars, and practitioners interested in improving decision making models and business functions.

Coverage

- Networks
- Operations Management
- Political Science
- Services management
- Computing and information technologies
- Marketing engineering
- Decision analysis and decision support
- Organizational Behavior
- Production management
- Information Management

IGI Global is currently accepting manuscripts for publication within this series. To submit a proposal for a volume in this series, please contact our Acquisition Editors at Acquisitions@igi-global.com or visit: http://www.igi-global.com/publish/.

The Advances in Logistics, Operations, and Management Science (ALOMS) Book Series (ISSN 2327-350X) is published by IGI Global, 701 E. Chocolate Avenue, Hershey, PA 17033-1240, USA, www.igi-global.com. This series is composed of titles available for purchase individually; each title is edited to be contextually exclusive from any other title within the series. For pricing and ordering information please visit http://www.igi-global.com/book-series/advances-logistics-operations-management-science/37170. Postmaster: Send all address changes to above address.

Titles in this Series

For a list of additional titles in this series, please visit: www.igi-global.com

Supply Chain Management in the Big Data Era
Hing Kai Chan (University of Nottingham Ningbo, China) Nachiappan Subramanian (University of Sussex, UK) and Muhammad Dan-Asabe Abdulrahman (University of Nottingham Ningbo, China)
Business Science Reference • copyright 2017 • 299pp • H/C (ISBN: 9781522509561) • US $195.00 (our price)

Knowledge Management Initiatives and Strategies in Small and Medium Enterprises
Andrea Bencsik (Széchenyi István University, Hungary & J. Selye University, Slovakia)
Business Science Reference • copyright 2017 • 442pp • H/C (ISBN: 9781522516422) • US $200.00 (our price)

Handbook of Research on Information Management for Effective Logistics and Supply Chains
George Leal Jamil (InescTec, Portugal) António Lucas Soares (University of Porto, Portugal) and Cláudio Roberto Magalhães Pessoa (FUMEC University, Brazil)
Business Science Reference • copyright 2017 • 554pp • H/C (ISBN: 9781522509738) • US $310.00 (our price)

Optimum Decision Making in Asset Management
María Carmen Carnero (University of Castilla – La Mancha, Spain) and Vicente González-Prida (University of Seville, Spain)
Business Science Reference • copyright 2017 • 523pp • H/C (ISBN: 9781522506515) • US $215.00 (our price)

Green Supply Chain Management for Sustainable Business Practice
Mehmood Khan (Abu Dhabi University, UAE) Matloub Hussain (Abu Dhabi University, UAE) and Mian M. Ajmal (Abu Dhabi University, UAE)
Business Science Reference • copyright 2017 • 330pp • H/C (ISBN: 9781522506355) • US $185.00 (our price)

Multinational Enterprise Management Strategies in Developing Countries
Mohammad Ayub Khan (Tecnológico de Monterrey, Mexico)
Business Science Reference • copyright 2016 • 490pp • H/C (ISBN: 9781522502760) • US $200.00 (our price)

Strategic Management of Sustainable Manufacturing Operations
Rameshwar Dubey (Symbiosis International University, India) and Angappa Gunasekaran (University of Massachuesetts, Dartmouth, USA)
Business Science Reference • copyright 2016 • 419pp • H/C (ISBN: 9781522503507) • US $215.00 (our price)

Handbook of Research on Global Fashion Management and Merchandising
Alessandra Vecchi (London College of Fashion, U.K.) and Chitra Buckley (London College of Fashion, U.K.)
Business Science Reference • copyright 2016 • 862pp • H/C (ISBN: 9781522501107) • US $335.00 (our price)

IGI GLOBAL
DISSEMINATOR OF KNOWLEDGE
www.igi-global.com

701 E. Chocolate Ave., Hershey, PA 17033
Order online at www.igi-global.com or call 717-533-8845 x100
To place a standing order for titles released in this series, contact: cust@igi-global.com
Mon-Fri 8:00 am - 5:00 pm (est) or fax 24 hours a day 717-533-8661

Editorial Advisory Board

Table of Contents

Detailed Table of Contents

as fundamental attribute perceptions, together with the importance of an agreeable "taste". The study stresses the strategic importance for small and medium Italian enterprises offering traditional Italian food products to adopt more appropriate marketing techniques and practices which take into account the indications emerging from the survey for developing the Chinese market.

The aim of this research is to shed light on establishing a successful integration of environmental practices among small and medium enterprises (SMEs). The purpose is to develop a better understanding of what drives SME participation into potential environmental practices within a context in Finland. Specific objectives are to: 1) assess the willingness to participate (WTP) among Finnish SMEs based on four drivers - financial performance, internal readiness, supplier pressure and governmental regulations; 2) separate respondents into demographical subgroups based on their WTP; and 3) determine features that characterize Finnish SMEs which are WTP based from the studied sample.

The chapter presents the essence of online/virtual travel communities and examines their current significance for SMEs. It presents types and characteristics as well as the most popular virtual travel SME communities and discusses ways of optimizing the development of SMEs by applying a strategy of virtual communities by selected hospitality SMEs. An eight step strategy to implement virtual travel communities by SMEs is elaborated. One of the main findings of the author is that virtual travel communities can be very useful for optimizing the development of SMEs by supporting all the stages in the decision-making process, focusing on pre-planning, making the decision and after the tourist and business visit stage. In order to understand how virtual travel community can optimize the development of SMEs, the author discusses tools for measuring the effectiveness of social media. Both future research directions and a conclusion are outlined.

The European Union (EU) and the Republic of Serbia have recognized the importance of SMEs and are developing accordingly the institutional framework of this sector. In the process of EU accession, Serbia has to fulfil political, economic and legal requirements by implementing systemic reforms of its legal system in order to align it with the EU acquis. Some of these requirements have a direct or indirect impact on SME regulations, to which our chapter has been dedicated within a series of our papers on

This chapter addresses several approaches in investigation of the SME life cycles, particularly in respect to the structure, functions and behavior of these businesses. Once SMEs have been established they start to change both their structure and behavior. In fact, it could be stated that SMEs operate in a permanently changing environment. There are a lot of factors which influence these changes and optimizing their management is a great challenge for most SMEs. As a result of these factors and suboptimal SME management practices survival rate of new established SMEs is low. For researchers and practitioners it is therefore crucial to discover how to increase long-term survival chances of SMEs through relevant structural change. In order to contribute to this goal the chapter will consider the conceptual and research framework of the structure and practical management behavior of SMEs in change management processes. The chapter also provides suggestions on how to use practical management tools such as the PEST, SWOT and TOWS matrices and ways of their analysis.

In recent years, the idea of integration of sustainability into project management has been put forth, which requires a more comprehensive and holistic approach to project management. Integration of sustainability increases the complexity of project management. Therefore, project management organizations require a framework capable of opening up the traditional organization of the project business and providing competitive advantages. This study argues that the open innovation model presents the requisite capability to approach this issue. This research, through a structured review of the literature and an overview of the open innovation and project management knowledge areas, explains how open innovation can provide a fruitful contribution to the integration of sustainable development into project portfolio management and why it should be favoured. The findings show that open innovation can facilitate the environmental, social and economic sustainability of projects, while rendering the project organization more agile.

Foreword

IMPORTANCE OF THIS BOOK

Optimal Management Strategies in Small and Medium Enterprises is an important addition to the literature. Unlike most books that deal with optimization and typically only implicitly deal with the milieu of small and medium enterprises, this book explicitly deals with decision contexts (small and medium enterprises) in which risk is magnified because of a relative lack of slack resources, a relative small margin of error, or a relative lack of management science expertise at the ownership or executive level. I use the term "relative" simply because while "size matters" size can often be a detrimental to optimization, especially if a firm wittingly or unwittingly substitutes satisficing for optimizing.

Optimization is ultimately about making the most realistic choice given a set of constraints and desired outcomes. But what constitutes a realistic choice for a large firm is often unrealistic for a medium enterprise, much less a small enterprise.

A realistic choice presumes that a firm has a tenable sense of its external and internal environments and their permeable boundary points such that known and unknown states of nature are perceived and respected and that known and unknown alternative courses of action are perceived and respected. *Optimal Management Strategies in Small and Medium Enterprises* presents theoretical paradigms and concomitant praxis that enable small and medium firms to identify and to address perception versus reality as regards states of nature and alternative courses of action.

EXPERTISE DRAWN UPON IN THIS BOOK

The book draws upon the expertise of a wide array of academicians and practitioners who represent a rich tapestry of disciplines, technical expertise, institutions, and cultural contexts. The depth and scope are compelling and provide a filter and a prism through which to see the universal efficacy of optimization methodology.

I use the term "methodology" because optimization is an approach to management as opposed to a mere set of techniques, technologies, or procedures. That is why the expertise drawn upon in *Optimal Management Strategies in Small and Medium Enterprises* will be profoundly beneficial to readers.

BENEFITS TO BE GAINED FROM THIS BOOK

It has been my great fortune to have had the opportunity to manage at all levels in a variety of settings including academe, small companies, the military, Fortune 100 firms, and philanthropic entities. I am convinced that the modest successes I enjoyed would have been greatly enhanced if *Optimal Management Strategies in Small and Medium Enterprises* had been available to me as a philosophical treatise on the practice of management as well as a practical guide for decision making. So it is with a bit of envy but even greater joy that I am able to commend this book which will (if you allow it) "feed your soul" and "fill your manager's toolkit." *Optimal Management Strategies in Small and Medium Enterprises* meets and exceeds its promised benefits as stated in this book's preface and as enumerated in this book's table of contents.

Keith Harman
Oklahoma Baptist University, USA

Preface

This book is a collection of chapters representing the development of SME thinking of the authors. A preface will perhaps be useful in order to indicate what the principal questions that are to be discussed are, and how they are connected. Therefore, the preface provides an abbreviated overview of the essential content of this book in a very condensed form.

THE CHALLENGE

The original objective of this book was to create a unique SME Toolkit to help small businesses expand globally by providing detailed elaboration of optimizing all main management functions such as working capital and marketing, with guidance on how to implement sustainable business management practice. The book is multidisciplinary in nature and considers a wide range of topics, each of which is related to *Optimal Management Strategies in Small and Medium Enterprises*. Therefore the publication broadens and expands on existing business optimization knowledge and assists SMEs with theoretic and practical examples on how to achieve and sustain business performance by implementing optimal and holistic management models allowing real time monitoring and corrective actions alike. Academically scholars will also be able to use the book with Hands-on learning, more formally known as Experiential Education, reflecting a teaching philosophy that promotes learning by doing. The strategy allows students to practice guided tactile learning in which they absorb knowledge of management not only by listening, but by experiencing. Experiential learning is becoming praised as a top teaching method by higher educational institutions, especially those in the fields of economic science and management.

As frequently happens when academic and business terms confront, there is necessary discussion about the precise meaning of the term optimal management strategy. I did not attempt to restrain this reflection, as it would be early and unnecessary to expect full agreement in this emerging area. Several chapter authors offered possible meanings for the term so I have no intention of forcing a single definition; however, a review of the proposed definitions suggests that optimization demonstrates several key features. Having said that, I provide the following definition stemming from contributions of chapter authors:

An optimal management strategy in Small and Medium Enterprises is an alternative or approach that best fits a business situation, makes use of all available information, employs assets in the most efficient and effective manner which yields the highest possible return under the circumstances and creates sustained value for shareholders.

This book is particularly unique in several distinct ways. The target audience of this book includes SME managers, professionals and researchers working in the field of SME development in various disciplines, e.g. management, economics, education, adult education, library, information and communication sciences, administrative sciences and sociology, computer science, and information technology. Moreover, the book provides insights and supports executives concerned with the management of expertise, knowledge, information and organizational development in different types of work communities and environments directly and indirectly related with SMEs.

The book is organized into 17 chapters based on three sections of optimal management strategies.

All chapters relate to practical optimal management in SMEs. In the first section, the six authors have examined various tools, techniques and technologies illustrating how marketing, market orientation, regulatory and policy issues contribute to optimal management. The next four chapters focus on the challenges associated with the role of pricing, finance and insurance in optimal SME management. The third final section involves seven chapters dealing with strategic aspects of optimal management such as knowledge management, structural change, process management and open innovation. Together these three sections provide an interesting overview at how SME managers, professionals and researchers may use opportunities and threats to accomplish their mission, vision and strategy. Obviously, like with any emerging field of science, there will be controversies; however, wise SME leaders will exploit these dilemmas to create real and best value. Ultimately, the authors hope that academics will benefit from the research and future lines of enquiry.

SECTION 1: SME OPTIMIZATION TOOLS, TECHNIQUES, AND TECHNOLOGIES

In Chapter 1, Marcin Soniewicki establishes the significance of the role of market orientation in creating competitive advantage of small and medium companies and uses the data from an empirical quantitative study conducted by the author in which more than 1,274 companies from Poland were examined. In the first section the chapter provides and elaborates on relevant literature with thorough presentation of the concept of market orientation. This is followed by explanation of its importance for small and medium companies. The second section presents methodology of this empirical study. The third part presented the market research results and finally brought are conclusions and practical advice for contemporary small and medium enterprises on the importance of market orientation concept.

In Chapter 2, Maria Giovanna Tongiani and Sheng Zaho discuss the significance of the traditional Italian food products in the Chinese Market and offer the best practice for small and medium Italian enterprises. The chapter argues that despite the importance of these products, a lot of confusion still exists among consumers with regard to this concept. This is why two surveys of 350 Chinese consumers were conducted by the authors who conclude that traditional products have acquired an important role in daily consumption, becoming a topic of study and research for numerous scholars. A recent study conducted in six European countries also demonstrated that consumers perceive a strong link between traditional products and their territory of origin in historical and cultural terms. Therefore, traditional products are considered all typical agri-food products, the result of which is characterized by a collective and localized historical process of accumulation of contextual knowledge based on a combination of specific territorial resources of a physical and cultural nature that creates a strong, unique and inimitable link with the territory of origin.

In Chapter 3, Nurul Aida Abdul Malek and Josu Takala illustrate the application of regression analysis for green practices participation among 170 small and medium sized companies in Finland. The authors discover that mostly large firms emphasize this area while SMEs remain underexposed. Their research explored how the SMEs reacted toward the driven factors of their behavior to participate in sustainable practices. The implications of the study can be beneficial by SMEs and large-sized companies as well as governments who always strive to improve the environmental performance along the value chains. Based from the research results, the influential factor of willingness to participate in green supply chain among SMEs is significantly associated by SMEs readiness. Although also correlated, financial performance, buyer practices and government involvement have not heavily determined the SMEs motivation to participate in green supply chain. The study found that companies with high level of awareness related to environmental concern, have human resources to deal with environmental issues and are implementing environmental management systems have a significant motivation to take part in green practices.

In Chapter 4, Silvena Yordanova considers how optimizing virtual communities in tourism facilitates development of small and medium-sized enterprises and argues that the Internet has revolutionized the whole world including the way SMEs operate in this sector of the economy. The author defines virtual travel communities (VTC), further examines their current significance for SMEs in the available literature, presents type and characteristics of the most popular virtual travel SME communities and discusses strategies of applying virtual communities by selected hospitality SMEs. It is expected that in the future many more SMEs will use the economic benefits of sustainable travel by applying virtual communities in their business models particularly for smart travel tech for SMEs business trips, optimal travel management and online booking strategies for SMEs, SME contract for car rental, airline ticket, hotel and ticker pricing strategies. Finally, this chapter analyzes and recommends for further consideration the use of VTC for SME travel agency optimal marketing strategy.

In Chapter 5, Olgica Milošević provides a critical review of SME regulation optimization in Serbia with reflection on harmonization with European Union law. The central part of her chapter discusses whether and how company law enables or hinders SME growth and development. Key principles addressed involve the essentiality of the intention to demarcate SMEs as companies from the so-called institutions, the principle of equality of companies in easier market access, the principle of tripartite interest position of the majority / minority owner, creditors / employees, the principle of compatibility of legal forms of companies, the principle of strengthening the legal institute of liability of members of the management body, controlling members and shareholders, the principle of transparency of companies' operations, the principle of modernity and sophistication of legal institutes and the principle of liberal enactments. The author also analyzes important facilitating and restrictive factors of corporations of interest to SME development and contemplates new legal forms of business organization for SMEs.

In Chapter 6, Peter Malega presents Small and Medium Enterprises in the Slovak republic with their status and competitiveness of SME´s in the global markets with elaboration of possibilities of optimization. It can be noted that SME development is one of the priorities of economic development of Slovak economy. The most important factor of this development is the creation of an appropriate business environment, which can benefit from simplification and clarification of legislation, reducing administrative, social contributions and the tax burden, reinforcement of infrastructure and improvement of access to capital. Different other factors also participate in the development of entrepreneurship such as state support, investments related to costs and operation of the businesses. In the Slovak Republic small and medium enterprises have an irreplaceable role particularly in job creation and regional development.

SECTION 2: OPTIMAL PRICING, FINANCING, AND INSURANCE MANAGEMENT STRATEGIES FOR SMEs

In Chapter 7, Mariusz Maciejczak and Adrian Słodki elaborate initial price strategies of Polish micro and small enterprises conducting research on a sample of 153 Polish micro and small enterprises. An application of game theory for industrial organization of SMEs sector is then discussed. It is argued thus, that entrepreneurs can have different visions about their company, buy they usually have the same goal: to generate profit and try to eliminate competitors. In achieving that goal, the game theory seems to be a reasonable tool to use in establishing initial price strategy. For solving the model there was used maxmin strategy, which maximizes the worst-case scenario from the game. As a final conclusion one can stress that there is a real chance that, if entrepreneurs would analyze their initial price situation with respect of game theory, their strategies would be more accurate and provide better outcomes.

In Chapter 8, Milan Vemić examines working capital optimization in medium-sized enterprises and tries to quantitatively - qualitatively synthesize and upgrade existing theories and practices of this field followed by practical optimization examples for financial decision making under uncertainty. The theory and practical examples in this chapter are supported using several scientific approaches including the Laplace, Wald, Hurwicz criteria, the Savage minimax regret criterion and Bayes' criterion. The author further develops a structural model with a mathematical notation, elaborates risk and uncertainty in stock exchanges for medium-sized enterprises. After discussing specific development constraints and opportunities he provides a codification of working capital combinations and desired changes in order to attain sustainable development of these business entities.

In Chapter 9, Katarina Ivančević elaborates insurance as an optimization tool for risk management in business operations of small and medium-sized enterprises. The author argues that insurance business has a number of important social functions in the market all over the world. Through the transfer of an individual risk to insurers it contributes to the security and stability of economic entities and business operations. Risk management in the context of SMEs can be well set and carried out but it is still not a total guarantee against occurrence of an SME business crisis and possible detrimental effects. Once an SME manager obtains relevant information from brokers about the level and extent of the risks to which it is exposed, and who is eligible for insurance, he or she will be able to use such information for strategy coordination and optimization. It is realistic to assume that growth of insurance coverage in SMEs would contribute to the stability of their business, Further, it is expected that the conclusions and recommendations of this study will indicate possible directions for further development of risk management optimization in SMEs.

In Chapter 10, Milan Vemić approaches financial innovation in medium-sized enterprises in developing their growth opportunities for gravitation toeards formal stock exchanges. In fact, significant amount of attention is given to the steps by which exchanges could get involved in this process. Practical and theoretical SME finance issues are elaborated in such a way as to point out the interconnection between medium sized firms and exchanges in order to demonstrate that each new source of finance is either based on growth or innovation or else follows logically from the previous stage of development in the firm life cycle. Financial innovation is defined both in the narrow sense of the word and within a broader approach of the author which is followed by a codification and typology of modifications and changes aimed at solving the problems of financing.

SECTION 3: OPTIMAL STRATEGIC MANAGEMENT FOR SMEs

In Chapter 11, Binod Timilsina elaborates on the current practice of strategic planning, with an online survey of 36 Finnish SMEs. From practical view point potential solutions in overcoming the identified barriers have been proposed. For strategy the author particularly emphasizes the resource based view (RBV) and industry organization theory (IO) which have been widely recognized not only as a tool to differentiate the performance difference between competing firms but also as a foundation for a firm's strategy. The chapter also classifies the competencies of a good strategic planner offering practical examples how a strategic leader or manager can enhance their practice of strategic management. Conclusions of the research argue that results and findings of this study will assist managers of SMEs and strategic planners in improving the practice of strategic management, hence in improving organizational performance in a changing business environment.

In Chapter 12, Aleksandar Damnjanović explains how significant knowledge management optimization becomes through IT and e-business utilization. The author conducted a research on a sample of 137 Serbian SMEs. Statistically tested and confirmed were the hypothesis that optimization of SMEs business through IT and E-business had a positive impact on competitiveness in the observed research period. The author concluded that competences of the employed and of managers in the field of ICT, their knowledge and their computer skills are key factors for the enhancement of companies' competitiveness generally and especially for small and medium-sized enterprises. It is conclude that the greatest positive effect of investments into innovations and new information technologies in relation to the consumers will also be obtained if the discussed approach is pursued.

In Chapter 13, Julia Stefanova and Zachary Wenner discuss financing socially innovative enterprises in various EU approaches and instruments. Authors explain that the EU has increasingly prioritized three forms of economic expansion: "smart growth," which favors scalable expansion, "trans-border" innovation and economic diversification, "sustainable and inclusive growth," which promote environmental sustainability and diffuse economic access using a variety of financial engineering instruments" for SMEs. Also discussed is EU VC Program which seeks to increase capital access for SMEs in environments of highly constrained bank lending with limited availability to alternative finance. In conclusion it is stressed that a system that links capital access to the materialization of social outcomes will advance the goals of the EU only to the extent that it bolsters the social impact, financial sustainability and scale potential of SMEs. As ad hoc and exploratory development strategies give way to more systematic approaches, there exists an invaluable opportunity to highlight the financial engineering instruments and SME types that will drive the development of a smart, sustainable and inclusive economy.

In Chapter 14, Lidia Sanchez-Ruiz and Beatriz Blanco analyze and compare process management in SMEs focusing on barriers, enablers and benefits with empirical evidence from 68 companies in Cantabria, Spain. In recent decades, globalization has led to changes in market conditions. Thus, now companies have to deal with more dynamic environments characterized by a fierce competition, better informed and more demanding customers, greater importance of the product/service quality, or greater awareness on the environmental and social issues they face. According to the authors process management could be seen as a mechanism that arises to meet the new business environment requirements, giving the company the flexibility to develop their business in today's competitive environment, and leading the company towards business excellence.

In Chapter 15, Olja Arsenijević, Drago Orčić and Edita Kastratović discuss intellectual capital which has become a subject of serious study, in the last twenty years. Values began to be noticed that were not enumerated in the classic financial statements. Science and practice were seriously researching the concept of intellectual capital management, to increase its value. The authors examined the methods for measuring intellectual capital in the companies, in order to establish measurable targets for stakeholders. A newly created tool is provided for optimizing intellectual potential in SMEs, verified through pilot research in the examined Serbian economy reflecting manufacturing and services activities. This contributes to the finding that intellectual capital is present, measurable, and can also be optimized in SMEs in Serbia.

In Chapter 16, Robert Molnar elaborates Optimizing Structural Change Management in Sustainable SME Development. One approach in investigation of the SMEs offered by Molnar is life cycles in respect to their structure, functions and behavior. Once SMEs have been established they start to change both their own structure and own behavior. So, it could be stated that SMEs operate in permanent changes which originate both internally and externally. The author argues that there are a lot of factors which influence these changes and optimizing their management is a great challenge for most SMEs. As a result, and strongly dependent on success of SMEs management, numerous statistical evidence discussed by the author show that less than a half of established SMEs survive more than five years.

In Chapter 17, Hosein Daneshpour reviews integrating sustainable development into portfolio management through application of open innovation. The author explains that the idea of integration of sustainability into project management has been coined and requires a more comprehensive and holistic approach in project management the complexity of which increases. Therefore, organization of project management needs a capable framework that opens up the traditional organization of project business and provides competitive advantages. This study argues that, open innovation model presents a proper capability to cope with this issue. This research, coupled with a sound review of literature, explains why and how open innovation can provide a fruitful contribution for the integration of sustainable development into project portfolio management.

Obviously, all authors have carried the theory and practice far enough to show that we can explain the essential features of *Optimal Management Strategies in Small and Medium Enterprises*.

Such a theory is now capable of having a new empirical content, especially in connection with the domain of main topics in this book. Moreover, we have seen that this type of theory opens up new possibilities for further development. Of course, the SME theory as elaborated here is far from complete. All of this will be published later and elsewhere. This leads to an indication of some lines of research along which it may be possible to meet the urgent challenge to develop a more comprehensive SME strategy theory.

Nevertheless, even in its present incomplete form, the theory of *Optimal Management Strategies in Small and Medium Enterprises* does answer the basic criticisms of those who regarded such a theory as impossible, or who felt that it could never concern itself with any real empirical problems in real economic and business life. SME questions and challenges are, of course, enormous and could in any case probably never be resolved ultimately and completely. At the very least, the book does seem to throw some light on a number of such researched problems.

Finally, it is hoped that the presentation of research and concepts in these chapters may help to convey to the reader how the subject *of Optimal Management Strategies in Small and Medium Enterprises* itself has actually developed, so that the form of the book is, as it were, a demonstration of what may be meant by the content.

Milan B. Vemić
Higher Education Institution "DOSITEJ", Serbia

Acknowledgment

An SME book like this results only from the contributions of many talented people. I would like to acknowledge and thank all of those involved in the development of this book. The whole team at IGI Global was very abetting and accessible. Their support was fascinating, timely useful and appreciated. In particular, I wish to thank Austin M. DeMarco, Managing Editor – Book Development at IGI Global.

The team of reviewers deserves special gratitude as they profusely contributed their resources and knowledge to enable an excellent review process. Frequently the work associated with chapter review is not appreciated enough – thanks to all of you for your excellent work.

Thank you to the Advisory Board members, all of whom supported the editor in developing this publication project and sharing valuable ideas about small and medium-sized enterprises. Each of the Board members is a very busy professional in their own field of expertise and I undoubtedly appreciate their great contribution to the book.

Lastly, I would like to thank the chapter authors that make up this book for their clear and informative explanations of the powerful SME concepts and optimization tools. In this era of technology and the Internet, while most of the underlying concepts only seem to remain fixed, SME applications are changing dynamically, compelling the authors to recommit themselves to their professions. Without their acumen, this book would naturally not be conceivable.

Milan B. Vemić
Higher Education Institution "DOSITEJ", Serbia

Section 1

SME Optimization Tools, Techniques and Technologies

Chapter 1
The Importance of Market Orientation in Creating a Competitive Advantage of Micro, Small, and Medium-Sized Companies in the Internationalization Process

Marcin Soniewicki
Poznan University of Economics and Business, Poland

ABSTRACT

The chapter concentrates on the concept of market orientation. It uses data from an empirical quantitative study conducted by the author in which more than 1,200 companies were examined. Market orientation intensity has been measured with the use of the MKTOR scale, which over the years has been tested by many researchers and is widely accepted in literature. The chapter consists of four parts. The first one presents internationalization theory and contains a thorough presentation of the concept of market orientation, as well as an explanation of its importance to micro, small and medium-sized enterprises. The following parts contain the methodology and results of the empirical study. Its analyses compare companies of various sizes that operate only locally with those undergoing an internationalization process in terms of market orientation intensity and competitiveness. At the end of the chapter, there are conclusions and practical advice concerning market orientation for contemporary micro, small and medium-sized companies.

INTRODUCTION

The key issue in this chapter is explaining why market orientation (MO) is important for micro, small and medium-sized enterprises (SMEs). One of the main characteristics of today's Knowledge Based Economy (KBE) is continuous volatility and uncertainty (Soniewicki, 2015). This is noticeable especially

DOI: 10.4018/978-1-5225-1949-2.ch001

for companies operating in foreign markets, but not only. Because of globalization, even companies that do not operate abroad often need to cooperate with foreign or international companies. Gorynia (2000) calls this passive internationalization. As a result, KBE and globalization have made today's market more demanding for all kinds of companies, which is felt especially by micro, small and medium-sized companies that usually have much more limited resources than large companies do. Another problem is that in many micro, small, and even some medium-sized companies, marketing is often associated only with promotion and advertising. Moreover, in some micro companies one can see the attitude of owners who think that they do not need market research, which is usually expensive. They often believe that if they need some product or service, then everyone needs it. However, in today's globalized world, starting and keeping a business without understanding the contemporary dynamic market is very difficult and even dangerous. The successes of companies that are not in possession of current market knowledge are, in many cases, just a stroke of luck. Nevertheless, in a constantly changing market, even such firms cannot last long. Lack of market orientation and the popular association of marketing with promotion only are probably some of the main reasons why many businesses are unsuccessful. Managing any company without at least some sort of market research (primary or secondary) can be associated, to some extent, with playing roulette.

The market is changing fast, so market orientation is more important for companies than ever before, as market knowledge is quickly becoming out of date. Micro, small and medium-sized companies tend to be the most vulnerable entities in these conditions. For example, in the past there were many small convenience stores in Poland. Nevertheless, in the last 10 years, they have been gradually disappearing and have been replaced by more effective and profitable shops belonging to large chains similar to 7-Eleven, which is popular in the United States. One of the causes of such chains' higher competitiveness might be a better understanding of the market and more intensive market orientation, which is reflected by a product offer that is better tailored to customers' changing needs.

Another unique aspect taken into account in this chapter is the inclusion of the issue of internationalization in the analysis. Companies in the internationalization process operate in a less favorable environment which is often culturally more or less distant. Their customers and competitors are usually more diversified, as they are more geographically dispersed. This may require more effort and a more intensive market orientation from these companies if they intend to achieve a competitive advantage.

These issues need to be thoroughly investigated with the use of detailed empirical data, which is what the present chapter presents. Its main goal is to verify the following hypothesis:

Micro, small and medium-sized companies with a more intensive market orientation are more competitive than companies of the same size but less market oriented.

Moreover, this chapter is going to check if there are any differences in market orientation intensity and the strength of its influence on competitiveness between micro, small, medium-sized and large companies. The final aspect which will be investigated here are differences in market orientation intensity, as well as its influence on company competitiveness among businesses of various sizes operating only locally or undergoing the internationalization process.

THE CONCEPT OF MARKET ORIENTATION

In the opinion of Noble, Sinha and Kumar (2002, p. 25), and Pazio (2007, p. 14), a company's orientation reflects its methods of functioning. In the past, firms were mostly concentrated on their interior, and then they gradually focused on their exterior, which is reflected in various orientations: production, product, sales and buyer (marketing) (Niestrój, 2003, pp. 18-21; Pazio, 2007, pp. 14-17; Fonfara, 2014, pp. 15-25; Soniewicki, 2015).

Currently, the most popular orientation among companies is the buyer (or marketing) orientation, which is related to the marketing concept (Fonfara, 2014, pp. 17, 18, 22).

There are many definitions of marketing in literature. Their most important elements are, in particular, satisfying customers' needs (Armstrong & Kotler, 2011, p. 33), the company taking the customer's perspective (Tajeddini, Trueman & Larsen, 2006, p. 533), or a specific way of thinking which should be present in the entire entity (Altkorn, 2001, p. 3). Unfortunately, the marketing concept has a serious disadvantage, as it concentrates practically on buyers only. Fonfara (2014, p. 22) underlines that today, to be effective, companies need to concentrate not only on the buyer but also on many other external entities. This is why the concept of market orientation has been introduced.

Judging by an increased number of its citations, the term "market orientation" became popular in literature at the end of the 1980s (Soniewicki, 2015). Although many publications have appeared on the topic since then, they are mostly based on several key works. One of the most popular articles in this area is that published by Shapiro (1988). In his publication, the author presents the concept of market orientation using the example of the manufacturing company *Wolverine*. He demonstrates that two of the most important steps towards the competitiveness of any business are to understand how customers use their products and to use this knowledge effectively. Another very important element is observing key competitors, the steps they have decided to make, and our customers in order to effectively notice market opportunities (Shapiro, 1988, p. 3).

The authors of other important articles in this area were Kohli and Jaworski (1990), Narver and Slater (1990), and Kohli, Jaworski and Kumar (1993). In the first one, the authors start from a general concept of marketing and Kotler's considerations by distinguishing three marketing pillars which are customer orientation, profitability and coordinated marketing. In their opinion, market orientation should be regarded by the entire organization as market intelligence or knowledge generation, which is related to both current and future customer needs. Kohli and Jaworski (1990) underline that generation of market knowledge is more than just exploring the needs of customers: it is identifying the factors that influence their preferences, factors which could include government regulatory moves or competitor actions. The authors also mention other entities, such as distributors. Such market knowledge should be constantly disseminated among all company departments. To do this, all kinds of methods should be used, including formal or informal ones. The final, and probably the most important operation is the firm's continuous reaction to this knowledge, which should take place not just in the marketing department but also in the entire entity (Kohli & Jaworski, 1990, pp. 3-6; Soniewicki, 2015).

Narver and Slater (1990) distinguish three elements of market orientation: customer orientation, competitor orientation and interfunctional coordination (Narver & Slater, 1990, pp. 22, 29). The first element, customer orientation, means delivering the highest possible value to the company's customers. To do this, the firm needs to learn who are its current and future customers (Narver & Slater, 1990, p. 21). In contrast to Kohli and Jaworski (1990), Narver and Slater (1990) give more attention to a company's competitors by calling competitor orientation one of the three key elements of market orientation.

These are defined by authors not only as current competitors but also those entities which may become competitors in the future (Narver & Slater 1990, pp. 21-22). The third market orientation element distinguished by Narver and Slater (1990) is interfunctional coordination. One can expect it to be related to the issue of barriers existing between departments in companies, described earlier by Shapiro (1988) and resembling "islands" (Shapiro, 1988). Narver and Slater (1990) compare a company to an orchestra in which, to achieve a desired result, each of its instruments has to play its part well (Soniewicki, 2015).

All the concepts of market orientation presented here have their advantages and disadvantages. The article by Shapiro (1988) is some sort of classic in this area, although it is rather general and less practical than other, also very important, studies by Kohli and Jaworski (1990), and by Narver and Slater (1990). Both of them distinguish three elements of marketing orientation, and in both the first element is customer orientation. Nevertheless, the latter concentrates on just two types of entity – customers and competitors – while Kohli and Jaworski (1990) underline the importance of other entities, such as distributors. However, it seems that these authors do include another, more important aspect: cooperation between departments, which is, on the other hand, included in the Narver and Slater (1990) concept. It is an important factor, to which Shapiro (1988) devotes considerable attention. In general, all of the three concepts described are valuable and useful for companies. This is because each of them brings something new and important that might be helpful in creating the firm's competitive advantage (Soniewicki, 2015).

The aforementioned groups of researchers created two questionnaires whereby one can examine the intensity of a company's market orientation: MKTOR (Narver & Slater, 1990) and MARKOR (Kohli et al., 1993). Both of them are still popular, and are frequently used in their original form in contemporary research (Kaur, Sharma & Seli, 2013, p. 52).

INTERNATIONALIZATION PROCESS

There are many attitudes to the concept of internationalization in literature. One can find numerous definitions which refer to numerous aspects of companies' operation (Andersen, 1997, p. 28; Daszkiewicz, 2004, p. 15; Duliniec, 2009, p. 15; Kozioł-Nadolna, 2013, p. 66).

First of all, internationalization might be defined statically as a result, a point in time, or dynamically as a process (Stępień, 2011, p. 99). Nevertheless, in the opinion of Fonfara (2009, p. 12), the latter of the two is more popular in literature. Another important aspect of internationalization is its gradation, which may manifest itself, for example, as the number of markets in which a company operates or as international turnover (Stępień, 2011, p. 100). Finally, a company's internationalization does not always need to mean that an enterprise operates aboard. According to some authors, this may also mean that the company cooperates with foreign enterprises in the local market through, for example, a franchise or license agreement, or through the imports of goods from abroad. Such activities are usually referred to in literature as passive internationalization. On the other hand, when a company itself undertakes operations in foreign markets, it is often called active internationalization (Gorynia, 2000, p. 16; Stępień, 2011, p. 99; Fonfara, 2012, p. 23).

A very interesting definition of internationalization has been proposed by Rymarczyk. He defines it as "any type of a company's business activity which is undertaken abroad" (Rymarczyk, 2004, p. 19). Nevertheless, in the author's opinion, taking into account the aforementioned company cooperation with foreign enterprises is also important. Consequently, for the purpose of the survey, the author decided to define internationalization very widely as: *any type of business activity which a company undertakes abroad or any cooperation with foreign enterprises in the local market.* In this chapter, no gradation of companies' internationalization is necessary, as there are only two types of enterprises distinguished in this matter: those in the internationalization process and those operating only locally.

METHODOLOGY OF THE EMPIRICAL STUDY

The data used in this chapter have been obtained in a quantitative study conducted in Poland in 2012 and in early 2013 as part of a research project financed by Poland's National Science Centre, Preludium grant, decision no. DEC-2011/03/N/HS4/00429. The sampling frame used was the *Kompass Poland database*. The research questionnaire was created in two forms: traditional (sent by post) and electronic. In order to conduct an effective Internet based study, a custom-made surveying system was created by the present author with the help of a computer scientist. It consisted of invitations to take part in the research which were sent by email together with custom links to a web-based questionnaire. Such a system allowed this author to send reminders only to those respondents that had not filled in the questionnaire by a certain deadline.

The research was conducted among companies of any size, operating in all industries. Almost 1,300 completed questionnaires were received. Some of them were rejected due to partially missing data. Ultimately, 1,274 companies were analyzed. The structure of the examined sample, taking into account the size of the sample, is presented in Table 1.

As Table 1 demonstrates, the majority of the investigated companies are micro, small and medium-sized companies, namely 1,182 entities. The largest part of this group consists of small companies: 547 entities; the number of micro and medium-sized companies is similar: around 300. Large companies are the smallest element of the sample – 92 entities – although it is sufficient for examining their characteristics and comparing them with smaller firms.

Table 1. The number of companies of a particular size in the investigated sample

Type of Companies	***No. of Companies in the Sample***
Micro companies (under 10 employees)	304
Small companies (10-49 employees)	547
Medium-sized companies (50-250 employees)	331
Large companies (over 250 employees)	92
Total	*1,274*

Source: own study.

Another important aspect of the analyzed sample structure is presented in Table 2. It shows how many companies of each size operate only locally, and how many of them are in the internationalization process; additionally, the share of the latter is shown. This table, as well as all the other analyses which take into account the internationalization process, does not include several companies because of the absence of relevant data.

As Table 2 shows, most of the companies in the sample (718) are undergoing the internationalization process. This is also true for enterprises of all sizes, except from micro companies. Among them, there are a little more entities operating only locally than those in the internationalization process. The general principle that can be formulated on the basis of the table under discussion is that the share of companies in the internationalization process increases together with firms' size.

An important part of the quantitative research on which this chapter is based was examining the intensity of companies' market orientation. This was done with the use of a series of questions developed by Narver and Slater (1990) – MKTOR. It consists of three parts: customer orientation, competitor orientation and interfunctional coordination. The MKTOR scale consists of 14 questions. Six of them apply to customer orientation, four to competitor orientation and the last four to interfunctional coordination. Each item was assessed on the five-point Likert scale, where 1 was the lowest rating and 5 the highest one.

The choice of the MKTOR scale, instead of MARKOR, was driven by several factors. First of them was that MKTOR had been previously tested in Poland by Hooley et al. (2000). This is important as, for example, Farell and Oczkowski tried to use both MKTOR and MARKOR scales for studies conducted in Australia and found that both scales do not fit into the "Australian business culture" (Farrell & Oczkowski, 1997, pp. 6-7). Another important argument is that the statistical reliability of MKTOR is higher than that of MARKOR (Pelham & Wilson, 1996, p. 33). Moreover, one can find opinions in literature against the MARKOR, claiming that it actually measures the level of marketing rather than the implementation of market orientation in a company (Wrenn, 1997, p. 33; Hooley et al., 2000, p. 274).

Table 2. The number of companies of a particular size, operating only locally and in the internationalization process, in the investigated sample

Type of Companies	***No. of Companies in the Sample***		***Share of Companies in the Process of Internationalization among all Entities of a Particular Size***
	Operating only Locally	***In the Process of Internationalization***	
Micro companies (under 10 employees)	153	148	49.2%
Small companies (10-49 employees)	246	297	54.7%
Medium-sized companies (50-250 employees)	127	203	61.5%
Large companies (over 250 employees)	21	70	76.9%
All companies	*547*	*718*	*56.8%*

Source: Own study.

The competitiveness of companies investigated was assessed with the use of the Competitiveness Index developed by Fonfara (2009, 2012). Its idea is to compare the examined companies with their closest competitors in four areas: profit, return on investment (ROI), value of sales and market share. The index also uses the five-point Likert scale to assess the position of the examined company in each of the areas mentioned. As part of this tool, each response should be interpreted as follows: 1 – much worse (than the company's closest competitor), 2 – worse, 3 – more or less the same, 4 – better, 5 – much better.

The results of company competitiveness analyses presented in the tables below are in most cases shown for the following intensities of market orientation (MO): "<=3", ">3", ">3.5", ">4", ">4.5". In one case, when the entire sample was taken into account, competitiveness was also shown separately for companies with MO intensity "=5", but later on, in more detailed analyses, such entities were spread among different groups, so in the majority of collections there were not enough entities under analysis with MO=5 to present reliable results for this intensity of market orientation.

EMPIRICAL ANALYSES OF MARKET ORIENTATION INTENSITY IN COMPANIES OF VARIOUS SIZE

This section consists of two parts. In the first one, companies will be analyzed only from the perspective of their size. In the second part, another criterion – internationalization process - will be added. Companies will be divided into two groups: those operating only locally and those undergoing the internationalization process.

The initial, most basic analysis in Table 3, presents the intensity of market orientation and its components in companies of various size.

The average intensity of market orientation in companies of different size is similar. The exception is large companies in which the value is slightly higher. The most important thing is that in all companies that are considered to be SMEs market orientation intensity is similar.

A slightly different picture is revealed when at the constituent elements of market orientation are looked at more closely. In their case, there are visible differences between companies of a particular size. The biggest are differences between SMEs and large companies in terms of competitor orientation. There

Table 3. Market orientation intensity in companies of various size

Type of Companies	*Market Orientation Intensity*	*Customer Orientation*	*Competitor Orientation*	*Interfunctional Coordination*
Micro companies (under 10 employees)	3.61	3.71	3.54	3.54
Small companies (10-49 employees)	3.60	3.69	3.53	3.53
Medium-sized companies (50-250 employees)	3.61	3.72	3.52	3.54
Large companies (over 250 employees)	3.66	3.76	3.64	3.52

Source: Own study.

are also differences, but smaller ones, in customer orientation. In terms of interfunctional coordination intensity, companies of all sizes are almost equal.

For companies, market orientation is not so much an objective in itself as a tool to increase their profits and to outperform competitors. This is why the crucial and most interesting analyses are those that confront market orientation intensity with firms' competitiveness. Table 4 shows companies' competitiveness divided into groups with particular levels of market orientation intensity.

The most important aspect to notice in Table 4 is that companies in which market orientation intensity is equal to or lower than 3 are more competitive than their closest competitors (Competitiveness Index = 2.70). Those firms in which market orientation value is over 3 are much more competitive (Competitiveness Index = 3.22), which means that they are more competitive than their closest competitors. A very important conclusion is that companies' competitiveness level always rises in parallel with their market orientation intensity. This means that this is a very significant factor influencing companies' competitiveness and as such is always worth increasing.

Another interesting conclusion that comes from Table 4 is that most of the companies analyzed (84.2%) are characterized by a market orientation of more than 3, and only 15.8% of the firms have a market orientation equal to or lower than 3. To some extent, one can see Pareto's 80/20 principle at work here. Even more interesting is the fact that only 26% of companies with a market orientation intensity higher than 3 have a market orientation higher than 4. Nevertheless, when we look at higher market orientation levels (>3.5, >4, >4.5, =5), the share of companies with such high levels is quickly decreasing. The most competitive elite is made up of companies with market orientation >4, >4,5 and =5. The final conclusion that can be drawn from Table 4 is as follows: the majority of companies with MO>3 are a little more competitive than their closest competitors, but to become exceptionally competitive enterprises in the market, they need to increase their market orientation much more, to the level of 4.5 or even close to 5.

Up to this point, the present chapter has proved that companies' competitiveness increases as market orientation intensity rises. However, the analyses have been performed among all kinds of firms without any specific features. There may be differences in this respect between particular types of firms, just

Table 4. Competitiveness of all the examined companies with varying intensity of market orientation

Market Orientation (MO) Intensity	***Competitiveness of Companies with a Particular Intensity of MO***	***Share of Entities with a Particular Intensity of MO***	***No. of Companies***
<=3	2.70	15.8%	201
>3	3.22	84.2%	1073
>3.5	3.34	57.5%	732
>4	3.53	22.4%	285
>4.5	3.71	6.7%	85
=5	4.23	0.9%	12

Source: Own study.

as, for example, micro companies differ from medium-sized or large ones. In order to perform a more detailed analysis of the data presented in Table 4, the companies under analysis were divided into four groups, depending on their size. The results are shown in Table 5.

As Table 5 reveals, competitiveness of all kinds of the featured companies rises with growth in market orientation intensity. Nevertheless, there are some important differences between them. In particular,

Table 5. Competitiveness of companies of different size with varying market orientation intensity

Market Orientation (MO) Intensity	***Competitiveness of Companies with a Particular Intensity of MO***	***Share of Entities with a Particular Intensity of MO***	***No. of Companies***
Micro Companies (fewer than 10 employees)			
<=3	2.63	17.8%	54
>3	3.03	82.2%	250
>3.5	3.16	59.5%	181
>4	3.29	21.1%	64
>4.5	3.34	7.9%	24
Small Companies (10-49 employees)			
<=3	2.63	15.2%	83
>3	3.19	84.8%	464
>3.5	3.30	56.1%	307
>4	3.55	21.0%	115
>4.5	3.75	6.8%	37
Medium-Sized Companies (50-250 employees)			
<=3	2.85	15.1%	50
>3	3.33	84.9%	281
>3.5	3.44	58.3%	193
>4	3.52	23.6%	78
>4.5	3.71	4.2%	14
Large Companies (over 250 employees)			
<=3	2.88	15.2%	14
>3	3.58	84.8%	78
>3.5	3.84	55.4%	51
>4	4.04	30.4%	28
>4.5	4.48	10.9%	10

Source: Own study.

we could note that the average competitiveness of micro companies (fewer than 10 employees) is lower than that of other kinds of companies. For firms with MO<3, the value of Competitiveness Index is quite low: 2.63. It is increased with a rise in market orientation, but more slowly than for larger entities. Moreover, the proportion of entities with a low MO intensity (<=3) is a little larger (17.8%) than of other kinds of companies (around 15.2%).

Small companies (10-49 employees) and medium-sized companies (50-250 employees) are the two biggest groups of companies examined here. In their case, the increase in the Competitiveness Index value in parallel with growth in market orientation is greater than that among the previously discussed micro companies. It is worth noting that companies with a high (>3.5, >4) – and especially those with a very high (>4.5) – market orientation intensity are very competitive entities. This means that the market orientation level is especially important for these companies.

Large companies (over 250 employees) are on average the most competitive entities. Moreover, growth in their Competitiveness Index value with an increase in market orientation intensity is much higher than among SMEs. There are also more large companies with a very high MO (>4.5) than among smaller companies. This means that such enterprises are very difficult competitors for small and medium-sized companies.

In general, Table 5 demonstrates that there are significant differences between SMEs and large companies, in terms of both market orientation intensity and competitiveness level. Large companies are much more competitive entities, with a higher average level of market orientation.

In the subsequent analyses, another factor will be added: companies will be divided into two groups – those that operate only locally and those in the internationalization process. Nevertheless, the level of employment will remain the most important factor taken into account. Companies in which it could not be determined whether they operate just locally or internationally were not taken into consideration in the following analyses. This slightly decreases the overall number of analyzed companies. The first analysis, which takes into account the issue of companies' internationalization, is presented in Table 6.

Table 6. Market orientation intensity in companies of a particular size, operating only locally and in the internationalization process

Type of Companies	***Market Orientation Intensity***	
	Operating only Locally	***In the Process of Internationalization***
Micro companies (under 10 employees)	3.54	3.69
Small companies (10-49 employees)	3.50	3.67
Medium-sized companies (50-250 employees)	3.52	3.67
Large companies (over 250 employees)	3.67	3.64
All companies	3.52	3.67

Source: Own study.

Table 6 shows the intensity of market orientation in companies operating only locally and in those in the internationalization process. As we can see, the latter are characterized by a higher level of market orientation intensity than the former. The single exception is large companies in which the reverse is true. Nevertheless, the difference in their case is rather small in comparison to SMEs, where the differences under study are considerable. This issue requires a more detailed breakdown showing market orientation component factors for enterprises of each size. Such an analysis is presented in Table 7.

As Table 7 demonstrates, the biggest differences between companies operating locally and those in the internationalization process can be found in two market orientation components: customer orientation and competitor orientation. Nevertheless, there exists a specificity of these differences which varies depending on the companies' size. In the case of micro companies (fewer than 10 employees), the largest difference is in customer orientation and competitor orientation, but there is practically no difference in interfunctional coordination. The latter is probably due to the fact that the size of these companies facili-

Table 7. The intensity of market orientation and its constituent elements in companies of different size, operating only locally and in the internationalization process

Type of Companies	***Market Orientation Intensity***	***Customer Orientation***	***Competitor Orientation***	***Interfunctional Coordination***
Micro Companies (fewer than 10 employees)				
Operating only locally	3.54	3.58	3.44	3.52
In the process of internationalization	3.69	3.83	3.63	3.53
Difference (int.-local.)	*0.15*	*0.25*	*0.18*	*0.01*
Small Companies (10-49 employees)				
Operating only locally	3.50	3.61	3.40	3.44
In the process of internationalization	3.67	3.75	3.64	3.60
Difference (int.-local.)	0.17	*0.13*	*0.24*	*0.15*
Medium-Sized Companies (50-250 employees)				
Operating only locally	3.52	3.62	3.40	3.48
In the process of internationalization	3.67	3.78	3.60	3.57
Difference (int.-local.)	0.15	*0.17*	*0.21*	*0.09*
Large Companies (over 250 employees)				
Operating only locally	3.67	3.79	3.68	3.50
In the process of internationalization	3.64	3.75	3.62	3.51
Difference (int.-local.)	-0.03	-0.04	-0.06	0.01

Source: Own study.

tates this kind of processes. Among small companies (10-49 employees), a bigger difference than in micro companies is in competitor orientation. There is also a visible difference in interfunctional coordination. In the case of medium-sized companies (50-250 employees), there are considerable differences among all three components. The market orientation of large companies (over 250 employees) differs from the previously described smaller companies. Firms of this size, operating only locally, are characterized by higher market orientation intensity than those in the internationalization process. Nevertheless, this difference is small, and the level of market orientation of both is rather high, comparable with that among smaller companies in the internationalization process. Quite a high market orientation level is maintained in the case of large companies, both operating only locally and in the internationalization process, and the difference which has been noticed is the result of the imperfection of the research tool applied.

The subsequent step will be to analyze the competitiveness of micro, small, medium-sized and large companies with various levels of market orientation intensity, either operating only locally or in the internationalization process. The results are presented in Tables 8, 9, 10 and 11.

Table 8 reveals that micro companies operating only locally with low market orientation intensity are much less competitive than their closest competitors: their Competitiveness Index is 2.52. Entities with the same area of operation which are characterized by an above-average market orientation (>3) are much more competitive (Competitiveness Index – 3.00), but this still means that they are more or less equal to their closest competitors. As for micro companies in the internationalization process, there are fewer companies with a low intensity of market orientation (<3), and even these are more competitive than their counterparts operating only locally (Competitiveness Index – 2.83). Among micro companies in the internationalization process, there are also greater proportions of companies with a high (>3.5 and >4) and a very high (>4.5) market orientation intensity.

Small companies operating only locally with a low market orientation intensity (<=3) are almost as uncompetitive as the previously-discussed micro companies. In the case of their counterparts in the

Table 8. Competitiveness of micro companies (fewer than 10 employees) with varying intensity of market orientation, operating locally and in the internationalization process

Market Orientation (MO) Intensity	Operating only Locally			In the Process of Internationalization		
	Competitiveness of Companies with a Particular Intensity of MO	Share of Entities with a Particular Intensity of MO	No. of Companies	Competitiveness of Companies with a Particular Intensity of MO	Share of Entities with a Particular Intensity of MO	No. of Companies
<=3	2.52	20.3%	31	2.83	14.9%	22
>3	3.00	79.7%	122	3.07	85.1%	126
>3.5	3.17	56.9%	87	3.17	62.2%	92
>4	3.28	17.0%	26	3.32	25.0%	37
>4.5	3.41	7.2%	11	3.29	8.8%	13

Source: Own study.

Table 9. Competitiveness of small companies (10-49 employees) with varying intensity of market orientation, operating locally and in the internationalization process

Market Orientation (MO) Intensity	Operating only Locally			In the Process of Internationalization		
	Competitiveness of Companies with a Particular Intensity of MO	Share of Entities with a Particular Intensity of MO	No. of Companies	Competitiveness of Companies with a Particular Intensity of MO	Share of Entities with a Particular Intensity of MO	No. of Companies
<=3	2.54	19.5%	48	2.76	11.8%	35
>3	3.13	80.5%	198	3.24	88.2%	262
>3.5	3.22	52.4%	129	3.36	59.3%	176
>4	3.48	17.9%	44	3.60	23.6%	70
>4.5	3.62	5.3%	13	3.82	8.1%	24

Source: Own study.

internationalization process, competitiveness is higher (Competitiveness Index – 2.76) but lower than among micro companies with similar market orientation intensity. Nevertheless, it is important to note that competitiveness among both types of small entities – those operating only locally and those in the internationalization process – competitiveness quickly increases with growth in market orientation intensity. This progress is much faster than among micro companies.

Table 10. Competitiveness of medium-sized companies (50-250 employees) with varying intensity of market orientation, operating locally and in the internationalization process

Market Orientation (MO) Intensity	Operating only Locally			In the Process of Internationalization		
	Competitiveness of companies With a Particular Intensity of MO	Share of Entities with a Particular Intensity of MO	No. of Companies	Competitiveness of Companies with a Particular Intensity of MO	Share of Entities with a Particular Intensity of MO	No. of Companies
<=3	2.93	16.5%	21	2.79	14.3%	29
>3	3.23	83.5%	106	3.38	85.7%	174
>3.5	3.31	52.8%	67	3.50	61.6%	125
>4	3.35	15.7%	20	3.58	28.6%	58
>4.5	3.30	3.9%	5	3.94	4.4%	9

Source: Own study.

Among medium-sized companies, we can see quite a large gap in competitiveness and companies' proportion in particular categories between entities operating locally and those in the internationalization process among firms with a higher-than-average (>3), high (>3.5, >4) and very high (>4.5) market orientation intensity. Medium-sized companies in the internationalization process with low market orientation intensity are not very competitive (Competitiveness Index – 2.79). Nevertheless, with an increase in market orientation intensity, the Competitiveness Index rises very quickly, unlike among firms of the same size that operate only locally. Among medium-sized companies in the internationalization process, there are also greater proportions of entities with an above-average (>3), high (>3.5, >4) and very high (>4.5) market orientation intensity. It seems that, among companies of this size and in the internationalization process, the role of market orientation intensity is very important for their competitiveness[1].

Large companies are in many aspects unusual in comparison to other types of companies. Such entities with an above-average market orientation intensity (>3), both operating only locally and in the internationalization process, are much more competitive than other smaller companies. Moreover, an increase in competitiveness together with growth in MO is the highest among the examined companies, even higher than among medium-sized companies in the internationalization process. Especially high competitiveness can be seen among companies with a very high level of market orientation intensity (>4.5): the Competitiveness Index reaches 4.33 among firms operating only locally and 4.54 among entities in the internationalization process. What is interesting is that a greater proportion of large companies operating only locally than those in the internationalization process are characterized by an above-average (>3) market orientation intensity. This may be due to the fact that not so many large companies took part in this survey. Nevertheless, the most important conclusion in this chapter is that large companies are very difficult and dangerous competitors for SMEs. Their MO intensity is mostly higher, and the companies themselves are more competitive.

Table 11. Competitiveness of large companies (over 250 employees) with varying intensity of market orientation, operating locally and in the internationalization process

Market Orientation (MO) Intensity	Operating only Locally			In the Process of Internationalization		
	Competitiveness of Companies with a Particular Intensity of MO	Share of Entities with a Particular Intensity of MO	No. of Companies	Competitiveness of Companies with a Particular Intensity of MO	Share of Entities with a Particular Intensity of MO	No. of Companies
<=3	2.63	9.5%	2	2.92	17.1%	12
>3	3.54	90.5%	19	3.60	82.9%	58
>3.5	3.80	47.6%	10	3.86	57.1%	40
>4	4.17	28.6%	6	4.04	30.0%	21
>4.5	4.33	14.3%	3	4.54	10.0%	7

Source: Own study.

Tables 12 and 13 present a different perspective on the previously described results by making it possible to compare the competitiveness levels of different-size companies with various market orientation levels – those operating only locally and those in the internationalization process.

Tables 12 and 13 show that the competitiveness of companies with an above-average market orientation (>3) intensity is always higher than of those with an MO intensity equal to or below the average (<=3) for both companies operating only locally and those in the internationalization process. Another important fact is that among companies operating only locally, an increase in competitiveness resulting from growth in MO is visible, although in most cases it is not as high as in firms of the same size but in the internationalization process. The results also demonstrate that an increase in competitiveness accom-

Table 12. Competitiveness of companies operating only locally with a particular intensity of market orientation

Market Orientation (MO) Intensity	Competitiveness of Companies Operating only Locally with a Particular Intensity of MO			
	Micro Companies (under 10 employees)	Small Companies (10-49 employees)	Medium-Sized Companies (50-250 employees)	Large Companies (over 250 employees)
<=3	2.52	2.54	2.93	2.63
>3	3.00	3.13	3.23	3.54
>3.5	3.17	3.22	3.31	3.80
>4	3.28	3.48	3.35	4.17
>4.5	3.41	3.62	3.30	4.33

Source: Own study.

Table 13. Competitiveness of companies in the internationalization process with a particular intensity of market orientation

Market Orientation (MO) Intensity	Competitiveness of Companies in the Internationalization Process with Particular Intensity of MO			
	Micro Companies (under 10 employees)	Small Companies (10-49 employees)	Medium-Sized Companies (50-250 employees)	Large Companies (over 250 employees)
<=3	2.83	2.76	2.79	2.92
>3	3.07	3.24	3.38	3.60
>3.5	3.17	3.36	3.50	3.86
>4	3.32	3.60	3.58	4.04
>4.5	3.29	3.82	3.94	4.54

Source: Own study.

panying growth in market orientation intensity is, in general, faster among larger companies. Nevertheless, there is an exception, namely medium-sized enterprises operating only locally. Another interesting conclusion is that, on average, market orientation intensity is more important for the competitiveness of companies in the internationalization process than for those operating only locally. However, in the majority of cases, companies in the internationalization process[2] are more competitive than enterprises operating only locally with a similar market orientation intensity, including even those with an MO below average (MO<=3). The final crucial conclusion drawn from Tables 12 and 13 is that, in most cases, every increase in an average company's market orientation intensity pays off owing to a high probability of its growth in competitiveness.

PRACTICAL ADVICE AND RECOMMENDATIONS FOR CONTEMPORARY MICRO, SMALL AND MEDIUM-SIZED ENTERPRISES

The above research results concerning the role of market orientation in the competitiveness of companies of various sizes may be a source of important practical recommendations. The most important of them is the one mentioned in the previous section: that it is beneficial for SMEs to enhance their market orientation intensity, as it practically always increases their competitiveness. Moreover, the results of the research have shown that among micro, small and medium-sized companies usually just over 80% of entities have a market orientation intensity higher than average (>3). Nevertheless, SMEs that have a very high (>4.5) market orientation constitute only between 3.9% and 8.8% of their groups and, apart from very few exceptions, they are the most competitive firms. This means that increasing market orientation to a very high level is the simplest recipe for increasing SMEs' competitiveness among firms operating only locally and those in the internationalization process. This can be achieved quite quickly by simply carrying out the activities described by Narver and Slater (1990) in their article and MKTOR questionnaire.

It should be underlined that in the process of increasing market orientation intensity SMEs should concentrate more on their customers and their competitors, but especially on the latter. An interesting practical example is Nokia. It has lost its lead in the mobile technology because it did not observe the market thoroughly enough. It missed Apple's activity in this area and the new trend of smart phones. On the other hand, Samsung has become one of the leaders in this area by closely observing new market trends and leaders, Apple in particular. This demonstrates the importance of market orientation, especially competitor orientation, and of observing and quickly discovering new trends that emerge, as well as meeting customers' new needs. Currently, markets are changing faster than ever before, so intensive market orientation becomes increasingly important for companies' performance and sometimes, as the case of Nokia demonstrates, even for their survival.

FUTURE RESEARCH DIRECTIONS

Future studies on the importance of market orientation for the competitiveness of small and medium-sized companies should focus on examining this type of firms in greater detail. This could include dividing them into different groups of similar enterprises using various criteria, for example the industry in which they operate. Clearly, interesting conclusions could be derived from more advanced statistical analyses of quantitative-research results. Another interesting direction is using a different, qualitative, type of study, which would be complementary to quantitative research. Moreover, this could help to identify factors that have an influence on market orientation intensity among particular types of enterprises.

CONCLUSION

The empirical analyses conducted in this chapter prove the expected importance of an intensive market orientation for the competitiveness of micro, small and medium-sized companies. The analyses confirm the hypothesis presented in the introduction that *micro, small and medium-sized companies with a more intensive market orientation are more competitive than companies of the same size but less market oriented.*

Market orientation intensity does not vary considerably among small and medium-sized enterprises. Nevertheless, it is a little higher among large companies. Moreover, these entities are often more competitive than smaller enterprises. Large firms are strong and dangerous competitors for SMEs, but this usually applies only to entities with at least an above-average market orientation intensity (MO>3). Unfortunately, in most cases the larger the company is, the greater is the effect of an increase in market orientation intensity for the firm's competitiveness. Despite this, the research results also show that, for firms of all sizes, it always pays off to increase the company's market orientation, as practically each time the investigated firms increased their market orientation, their competitiveness rose, too.

The study has also shown that, on average, market orientation is higher among SMEs undergoing internationalization processes than among SMEs operating locally. However, among large companies the opposite is the case, although the market orientation intensity of both kinds of companies is actually similar. More detailed results show that SMEs operating only locally and those in the internationalization process differ especially in terms of two market orientation components – competitor orientation and customer orientation, which are more intensive in the latter companies. However, with regard to the competitiveness of SMEs operating locally and those in the internationalization process, there are no considerable differences in terms of the importance of market orientation for their competitiveness. Among SMEs operating locally and those undergoing the internationalization process, competitiveness almost always grows with an increase in market orientation intensity. Nevertheless, the volume of these changes varies, yet the direction, apart from several exceptions, is the same. This means that market orientation is important for both companies operating only locally and those in the internationalization process.

REFERENCES

Altkorn, J. (2001). Orientacje Marketingowe Polskich Przedsiębiorstw. In *VII Kongres Ekonomistów Polskich, Sesja IV*. Warszawa: Polskie Towarzystwo Ekonomiczne.

Andersen, O. (1997). Internationalization and Market Entry Mode: A Review of Theories and Conceptual Frameworks. *Management International Review, 37*, 27-42.

Armstrong, G., & Kotler, Ph. (2011). *Marketing. An Introduction*. Upper Saddle River, NJ: Pearson.

Daszkiewicz, N. (2004). *Internacjonalizacja Małych i Średnich Przedsiębiorstw we Współczesnej Gospodarce*. Gdańsk: Scientific Publishing Group.

Duliniec, E. (2009). *Marketing międzynarodowy*. Warszawa: Polskie Wydawnictwo Ekonomiczne.

Farrell, M. A., & Oczkowski, E. (1997). An analysis of the MKTOR and MARKOR measures of market orientation: An Australian perspective. *Marketing Bulletin*, *8*, 30–40.

Fonfara, K. (2009). Istota i zakres procesu internacjonalizacji firmy. In K. Fonfara (Ed.), Zachowanie przedsiębiorstwa w procesie internacjonalizacji – podejście sieciowe (pp. 11-25). Warszawa: Polskie Wydawnictwo Ekonomiczne.

Fonfara, K. (2012). *The Development of Business Networks in the Company Internationalization Process*. Poznań: Poznań University of Economics Press.

Fonfara, K. (2014). Marketing partnerski na rynku przedsiębiorstw (3rd ed.). Warszawa: Polskie Wydawnictwo Ekonomiczne.

Gorynia, M. (2000). Luka konkurencyjna w przedsiębiorstwach a przystąpienie Polski do Unii Europejskiej. *Gospodarka Narodowa*, *10*, 48–67.

Hooley, G., Cox, T., Fahy, J., Shipley, D., Beracs, J., Fonfara, K., & Snoj, B. (2000). Market Orientation in the Transition Economies of Central Europe. *Journal of Business Research*, *50*(3), 273–285. doi:10.1016/S0148-2963(99)00105-8

Kaur, G., Sharma, R. D., & Seli, N. (2013). A Balanced Approach towards Market Orientation. *Vikalpa*, *38*(3), 51–65.

Kohli, A. K., & Jaworski, B. J. (1990). Market Orientation: The Construct, Research Propositions, and Managerial Implications. *Journal of Marketing*, *54*(April), 1–18. doi:10.2307/1251866

Kohli, A. K., Jaworski, B. J., & Kumar, A. (1993). MARKOR: A Measure of Market Orientation. *JMR, Journal of Marketing Research*, *30*(November), 467–477. doi:10.2307/3172691

Kozioł-Nadolna, K. (2013). *Internacjonalizacja Działalności Badawczo-rozwojowej w Kształtowaniu Procesów Innowacyjnych Przedsiębiorstw w Polsce*. Warszawa: CeDeWu.

Narver, J. C., & Slater, S. F. (1990). The Effect of a Market Orientation on Business Profitability. *Journal of Marketing*, *54*(4), 20–35. doi:10.2307/1251757

Niestrój, R. (2003). Istota i znaczenie marketingu. In J. Altkorn (Ed.), Podstawy marketingu (pp. 13-44). Kraków: Instytut Marketingu.

Noble, Ch. H., Sinha, R. K., & Kumar, A. (2002). Market Orientation and Alternative Strategic Orientations: A Longitudinal Assessment of Performance Implications. *Journal of Marketing*, *66*(4), 25–39. doi:10.1509/jmkg.66.4.25.18513

Pazio, N. M. (2007). *Podstawy marketingu*. Warszawa: Oficyna Wydawnicza Politechniki Warszawskiej.

Pelham, A. M., & Wilson, D. T. (1996). A Longitudinal Study of the Impact of Market Structure, Firm Structure, Strategy, and Market Orientation Culture on Dimensions of Small-Firm Performance. *Journal of the Academy of Marketing Science*, *24*(1), 27–43. doi:10.1007/BF02893935

Rymarczyk, J. (2004). *Internacjonalizacja i Globalizacja Przedsiębiorstwa*. Warszawa: Polskie Wydawnictwo Ekonomiczne.

Shapiro, B. P. (1988). What the Hell is 'Market Oriented'? *Harvard Business Review*, *66*, 119–125.

Soniewicki, M. (2015). *The company's international competitive advantage – the role of knowledge*. Warsaw: Difin.

Stępień, B. (2011). Międzynarodowa kooperacja gospodarcza – przyczyny i uwarunkowania instytucjonalne i kulturowe. In B. Stępień (Ed.), Międzynarodowa kooperacja gospodarcza z polskiej perspektywy (pp. 90-130). Warszawa: Polskie Wydawnictwo Ekonomiczne.

Tajeddini, K., Trueman, M., & Larsen, G. (2006). Examining the Effect of Market Orientation on Innovativeness. *Journal of Marketing Management*, *22*(5-6), 529–551. doi:10.1362/026725706777978640

Wrenn, B. (1997). The Market Orientation Construct: Measurement and Scaling Issues. *Journal of Marketing Theory and Practice*, *5*(3), 31-54.

ADDITIONAL READING

Baker, W. E., & Sinkula, J. M. (1999). The Synergistic Effect of Market Orientation and Learning Orientation on Organizational Performance. *Journal of the Academy of Marketing Science*, *27*(4), 411–427. doi:10.1177/0092070399274002

Becker, J., & Homburg, C. (1999). Market-Oriented Management: A Systems-Based Perspective. *Journal of Market-Focused Management*, *4*(1), 17–41. doi:10.1023/A:1009804011447

Darroch, J., & McNaughton, R. (2003). Beyond Market Orientation. Knowledge Management and the Innovativeness of New Zealand Firms. *European Journal of Marketing*, *37*(3/4), 572–593. doi:10.1108/03090560310459096

Farell, M. A. (2000). Developing a Market-Oriented Learning Organisation. *Australian Journal of Management*, *25*(2), 201–222. doi:10.1177/031289620002500205

Greenley, G. E. (1995). Forms of Market Orientation in UK Companies. *Journal of Management Studies*, *32*(1), 47–66. doi:10.1111/j.1467-6486.1995.tb00645.x

Helfert, G., Ritter, T., & Walter, A. (2002). Redefining market orientation from a relationship perspective. Theoretical considerations and empirical results. *European Journal of Marketing*, *36*(9/10), 1119–1139. doi:10.1108/03090560210437361

Hooley, G., Fahy, J., Greenley, G., Beracs, J., Fonfara, K., & Snoj, B. (2003). Market orientation in the service sector of transition economies in Central Europe. *European Journal of Marketing*, *37*(1/2), 86–106. doi:10.1108/03090560310453975

Johanson, J., & Vahlne, J. E. (2009). The Uppsala internationalization process model revisited: From liability of foreignness to liability of outsidership. *Journal of International Business Studies*, *40*(9), 1411–1431. doi:10.1057/jibs.2009.24

Kumar, V., Jones, E., Venkatesan, R., & Leone, R. P. (2011). Is Market Orientation a Source of Sustainable Competitive Advantage or Simply the Cost of Competing? *Journal of Marketing*, *75*(January), 16–30. doi:10.1509/jmkg.75.1.16

Lafferty, B. A., & Hult, G. T. M. (2001). A synthesis of contemporary market orientation perspectives. *European Journal of Marketing*, *35*(1/2), 92–109. doi:10.1108/03090560110363364

Laine, A., & Kock, S. (2000). *A Process Model of Internationalization – New Times Demands New Patterns*. Paper presented at 16th IMP-conference in Bath, United Kingdom, Retrieved May 5, from http://impgroup.org/uploads/papers/83.pdf

Martin, J. H., & Grbac, B. (2003). Using supply chain management to leverage a firms market orientation. *Industrial Marketing Management*, *32*(1), 25–38. doi:10.1016/S0019-8501(01)00192-4

Soniewicki, M. (2015). The use of external knowledge sources in manufacturing companies. *Kwartalnik Nauk o Przedsiębiorstwie, 4*(37), 89-97.

Soniewicki, M. (2016). Synergy between market orientation and knowledge management in building competitive advantage of trading companies. *Marketing i Rynek*, *1*, 9–17.

Soniewicki, M., & Wawrowski, Ł. (2015). The use of external knowledge sources by Polish private-owned and state-owned enterprises in the internationalization process. *Journal of Economics and Management, 22*, 75-96. Retrieved May 5, from http://www.ue.katowice.pl/en/units/journal-of-economics-and-management/archive-issues/archive/journal-of-economics-and-management-no-2215.html

Wang, C. L., Hult, G. T. M., Ketchen, D. J. Jr, & Ahmed, P. K. (2009). Knowledge management orientation, market orientation, and firm performance: An integration and empirical examination. *Journal of Strategic Marketing*, *17*(2), 99–122. doi:10.1080/09652540902879326

KEY TERMS AND DEFINITIONS

Competitiveness Index: A method of measuring the level of company competitiveness devised by Fonfara. It consists of four elements: profit, ROI (return on investment), value of sales and market share, which are estimated by respondents on the 5-point Likert scale and compared with the company's closest competitors.

Competitor Orientation: One of the three key elements of market orientation, as defined by Narver and Slater. It refers to a company's current interest in a competitor's actions. It is measured on Narver and Slater's MKTOR scale as the arithmetic average of responses to four questions.

Customer Orientation: One of the three key elements of market orientation, as defined by Narver and Slater. It refers to a company's current interest in customer needs. It is measured on Narver and Slater's MKTOR scale as the arithmetic average of responses to six questions.

Interfunctional Coordination: One of the three key elements of market orientation, as defined by Narver and Slater. It refers to a company's current effectiveness of cooperation and level of commitment to satisfying customer needs. It is measured on Narver and Slater's MKTOR scale as the arithmetic average of responses to six questions.

Internationalization: Any operation undertaken abroad by a company or cooperation with a foreign company in the local market.

Market Orientation: The concept of a company's orientation, one level higher than marketing orientation, which became popular around 1990 mainly thanks to Shapiro, Kohli and Jaworski, and Narver and Slater. According to this concept, the entire company should concentrate not only on the customer but also on many elements of the market. All departments of such an enterprise should cooperate as one team, for example to discover market trends, observe competitors' actions and effectively satisfy customers' needs.

Micro Companies: Entities employing fewer than 10 people.

MKTOR: One of the most popular methods of measuring companies' market orientation intensity, invented by Narver and Slater. It consists of 14 questions divided into three sections: customer orientation (six questions), competitor orientation (four questions), and interfunctional coordination (six questions). Companies' responses are measured on the Likert scale.

SMEs: In this chapter, the term SMEs (Small and Medium-Sized Enterprises) is used to refer to micro (fewer than 10 employees), small (10-49 employees) and medium-sized (50-249 employees) enterprises.

ENDNOTES

1 Editors' note: This finding corresponds to the conclusion from chapter 12, on knowledge management optimization, that entities of this size increase competitiveness through responsible management system, communication with customers and developing user-friendly web-sites.

2 Editors' note: For a comparative point of view on internationalization issues see chapter 6 on Slovak SMEs.

Chapter 2
Traditional Italian Food Products on the Chinese Market:
Best Practices for Italian Small and Medium Enterprises

Maria Giovanna Tongiani
University of Pisa, Italy

Sheng Zhao
University of Pisa, Italy

ABSTRACT

This work focuses attention on the area of a specific sector consisting of Italian small and medium enterprises that market traditional Italian products. A questionnaire submitted to a group of Chinese consumers provides their definition of a traditional product and, in particular, gives their positive opinion on traditional Italian food products, whose "reliability" and "health" content they acknowledge as fundamental attribute perceptions, together with the importance of an agreeable "taste". The study stresses the strategic importance for small and medium Italian enterprises offering traditional Italian food products to adopt more appropriate marketing techniques and practices which take into account the indications emerging from the survey for developing the Chinese market.

INTRODUCTION

SMEs play an important role in an economic system because of their impact in numerical terms (share in the total number of firms) and their contribution to the number of employed. Operations in SMEs consist of a combined set of implementation methodologies for the marketing function.

DOI: 10.4018/978-1-5225-1949-2.ch002

Over the past few years, traditional products have acquired an important role in daily consumption, and as highlighted by several authors, the consumers seem to pay more attention to the quality of the product, evaluating its safety and wholesomeness, rather than considering the links existing between the products and specific geographical areas (Gellynck, Kühne, Banterle, Carraresi, & Stranieri, 2012). On the international scene, Italy is one of the nations with the most plentiful and variegated agri-food resources and at European Union (EU) level, it is the country with the greatest number of brands of food products of a particularly traditional nature covered by EU legal protection. As observed by some authors, the Food and Beverage industrial sector in Europe is characterised by the presence of SMEs (Banterle, Cavaliere, & De Marchi, 2016), and in Italy this sector consists almost entirely of SMEs, each characterized by a limited assortment and with scanty economic, managerial and financial resources. Exports currently play a fundamental role in the development of the food sector. China's economic development promotes the improvement of the per capita income, a direct consequence of which is also the increase and expansion of the level of consumption on the market. China represented the third most important export destination market, while Italy holds fourth place among EU countries for both imports and exports with China (Camera di Commercio Italo-Cinese). The Association of Italian Food Industries forecasts that by 2018 China will be the world's largest consumer of imported goods. The bestseller Italian food products in China are wine and *spumante* (Italian *sparkling wine*), pasta, fruit and vegetables, extra virgin olive oil, coffee and cheese (COLDIRETTI, 2015).

Recent studies (Gellynck et al., 2012; Banterle et al., 2016) show that the Italian SMEs operating in the traditional food sector are very limited in size being for the main part micro enterprises. Therefore, in this sector, according to surveys in literature (Gellynck et al., 2012; Banterle et al., 2016), despite the presence of a certain number of enterprises with an adequate marketing orientation which are consequently equipped with marketing capabilities in line with the market demands, the great majority of companies show shortcomings and in some cases, a total lack of marketing orientation. According to these studies (Banterle, Carraresi, & Stranieri, 2010; Gellynck et al., 2012), this marketing insufficiency implies a deficiency in terms of marketing capabilities, which thus entails a poor ability to operate properly on international markets for the majority of these companies, especially the micro enterprises. In the light of these elements, this chapter focuses on a specific sector consisting of Italian SMEs that market traditional Italian products, and addresses identification of more appropriate marketing practices for developing the Chinese market.

In the background section, the chapter clearly formulates its two main objectives: that of making an academic review of the characteristics of SMEs compared to large enterprises and their marketing function; and that of characterising the traditional Italian food products on the Chinese market with highlighting the significance of the country of origin. This study focuses on the marketing activities of Italian SMEs that offer traditional food products, gathering information by means of a questionnaire filled out directly by a sample of Chinese consumers in two separate research phases. An initial phase entailed the collection of preliminary information by means of a convenience sampling survey in July 2015. The second phase tested various elements which emerged during previous data collection in March 2016 by means of online probability sampling of 200 Chinese purchasers. The empirical research conducted on the Chinese market and the practical results obtained regarding the perception of traditional Italian products revealed factors that influence purchasing decisions and communication channels used by consumers to search for and obtain information about traditional Italian products. Based on the survey results, the analysis allowed the authors to identify *optimal* solutions for marketing activities that can be implemented by the Italian SMEs due to accurate and in-depth knowledge of the reference Chinese customer, in order to gain a competitive advantage on the Chinese market. Finally, the study proposes future research avenues for monitoring changes in the purchasing habits of Chinese consumers and any possible threats to traditional Italian products.

BACKGROUND WITH LITERATURE REVIEW ON SME CHARACTERISTICS AND THEIR MARKETING OPTIMIZATION FUNCTION

In this section authors examine contributions made by literature to these three different areas of interest:

1. Contributions made by literature to SME characteristics and the specificity of their marketing function;
2. Definition of the traditional food products and the role played by the *Country of Origin*;
3. Analysis of marketing capabilities of Italian SMEs in the traditional food sector and the importance of the word of mouth.

SMEs play an important role in the economic system because of their impact in numerical terms and their contribution to the number of employed. The distinguishing characteristics of these firms are attributable to various aspects:

1. Their limited size in terms of operators;
2. Their low sales volume;
3. The limited resources available;
4. Lastly – and an extremely important aspect – the presence, in most cases, of an entrepreneur-manager who makes, and is responsible for all the firm's strategic and operational decisions.

The above elements imply that large enterprises have a different and considerably more favourable potential and capacity to operate than SMEs (Resnick, Cheng, Simpson, & Lourenço, 2016; Franco, de Fàtima Santos, Ramalho, & Nunes, 2014).

The possibility of SMEs to respond and have an impact on the environment is virtually rather limited, unlike large enterprises which, conversely, have competences and resources that are able to influence and induce change in those contexts where they are present (Atanassova & Clark, 2015; Resnick et al., 2016).

It must however be pointed out that the specific aspects of SMEs counter these non-positive circumstances and help them to act in an extremely favourable situation with regard to flexibility; speed of adaptation to change; creativity (Qureshi & Kratzer, 2011; Atanassova & Clark, 2015).

The differences between large and SMEs are therefore reflected in the marketing activities. In fact, larger enterprises adopt traditional marketing strategies and formal planning via the acquisition of information on reference markets through market research, advertising with huge investments of financial resources, and planning with suitable marketing mix policies (Resnick et al., 2016).

Marketing assumes an even more important role in SMEs because their target market is numerically very limited and it is therefore vital to implement marketing activities in order to maintain a strong relationship with customers and avoid even a minimum loss of clients (Reijonen & Komppula, 2010; Franco et al., 2014). This function which is implemented by SMEs without any formalization and planning takes a basically tactical view; a very different method from that adopted by large enterprises but which, however, brings certain advantages deriving from simplified procedures and prompt feedback, together with a simple organization structure and natural customer orientation (Reijonen & Komppula, 2010; Franco et al., 2014). Marketing activity is the result of a combination of competences, resources and individual skills determined by the presence of an entrepreneur-manager. The latter, in his role as general manager, defines the business and since he is closely acquainted with its characteristics, impresses his own style based on informal relationships and the creation of contacts and dealings. The entrepreneur-manager, posing as the market interface, promotes the business with interpersonal relations and support which materialize into marketing characterized by the enterprise's individual situations (Harris & Deacon, 2011; Resnick et al., 2016). Marketing for SMEs therefore assumes specific connotations, partly due to the lack of available financial resources, the difference in size, and a scant knowledge of complex specialist experiences, which makes it impossible to achieve traditional marketing operations. On the other hand, this characteristic is the consequence of the presence of an intuitive, down-to-earth entrepreneur-manager with a capacity for intra-organizational communication and an application of informal procedures. All of which go to support the development of an emotional involvement with employees and the formation of relationships with customers that are much closer to the relative business reality than in larger enterprises (Reijonen & Komppula, 2010; Resnick et al., 2016). Therefore, operations in SMEs consist of a combined set of implementation methodologies for the marketing function (Harris & Deacon, 2011).

Marketing interfaces where business efficiency derives from the central role of an entrepreneur-manager who is extremely competent in business notions and concepts and gains benefits from advance reading of trends with consequent prompt adjustment to market conditions. Albeit in the absence of planning and control, these results develop flexible responses, identifying opportunities and user requirements, together with the ability to create informal networks of privileged relations with customers and the competitive environment (Harris & Deacon, 2011; Atanassova & Clark, 2015).

Administrative marketing is an approach in which activities concerning managerial processes envisage a structure based on traditional rules and regulations, with particular reference to promotional tools. Many SMEs take part in fairs and exhibitions and set up websites with personalized features. This classical type of operation, however, fails to define specific objectives. The use and management of internet websites is carried out without fully understanding the application and potential of digital channels. It also takes place without caring about or troubling to evaluate results. No attention is paid to the method of processing or selecting the large amount of information transmitted. The presence in virtual space, even if implemented informally and intuitively, allows SMEs to extend their contacts efficiently and with reduced communication costs (Harris & Deacon, 2011; Atanassova & Clark, 2015).

Marketing authenticity refers to the entrepreneur's awareness in creating an offer based on profound values related to a culture and a deep-seated conviction which, albeit associated with subjective interpretations, is communicated and shared. In SMEs this result is closely linked to the entrepreneur's specific features, unlike in large enterprises where it may even not be real and is to some extent "invented" with complex, sophisticated and costly marketing strategies. The presence of the owner-manager figure whose knowledge and all-round value-based competence are acknowledged, creates an opportunity for and the achievement of an advantage based on the perception of real authenticity which the subject who is responsible for and legitimated to conduct the business formulates through the dissemination of information and applications in his/her network of relationships with the environment and customers (Harris & Deacon, 2011).

The combined presence of different types of marketing in the activities of smaller enterprises thus materializes into *marketing in context* (Harris & Deacon, 2011) with differentiated results in different environmental and business situations, and consequently with varying levels of efficiency on the market.

DEFINITION OF TRADITIONAL FOOD PRODUCTS AND THE ROLE OF THE COUNTRY OF ORIGIN

Over the past few years, traditional products have acquired an important role in daily consumption, becoming a topic of study and research for numerous scholars (Veale, Quester, & Karunaratna, 2006; Vanhonacker et al., 2010). Despite their importance, a lot of confusion still exists among consumers with regard to their concept.

A recent study conducted in six European countries demonstrated that consumers perceive a strong link between traditional products and their territory of origin in historical and cultural terms. According to this study, traditional products are "products frequently consumed or associated with specific celebrations and/or seasons, passed down from one generation to the next, produced in a specific manner in accordance with the gastronomic heritage, naturally processed, recognised and known for their sensorial properties, and associated with a specific local area, region or country" (Vanhonacker et al., 2010, p.472).

The term '*traditional*' highlights the close link between the product and the past, handed down from one generation to the next, that has existed and been consumed for a long time, and which has always formed a part of the consumer's life (Vanhonacker et al., 2010).

Therefore, traditional products are considered all typical agri-food products, the result of which is characterised by a collective and localised historical process of accumulation of contextual knowledge based on a combination of specific territorial resources of a physical and cultural nature that gives rise to a strong, unique and inimitable link with the territory of origin (Arfini, Belletti, & Marescotti, 2010; Belletti, Brunori, Marescotti, Pacciani, & Rossi, 2010).

There are four European consumer segments, identified in studies and literature based on the importance given to the items listed above that are associated with traditional food, and which include the multi-concept, usual/family, typical/authentic and hereditary sectors (Vanhonacker et al., 2010).

From this segmentation it can be observed how none of the definitions of the concept of traditional food products are linked to the nationality of the interviewees, but rather, every segment consists of individual consumers from different countries and the national samples are distributed over the various segments. This demonstrates a transversal aspect among the various countries regarding the definitions and concepts linked to the traditional food products (Vanhonacker et al., 2010).

The "*Country of origin*" is understood as territory, area, and region to which the planning, creation, manufacture, and design of a product or service are associated according to the aforementioned definition. It represents an essential element for traditional food products. In literature it is pointed out how the country of origin enters among those evaluation factors that customers refer to when making a final decision, as it strongly influences the consumers' perception and increases their initial trust. Initial trust is a condition in which consumers, despite not knowing a product or country, place trust in the goods purchased. Literature in this sector considers that this has a higher value because it is based on the good faith of the parties, irrespective of past experience. This fact assumes a relevant role for companies launching themselves on the market for the first time and the product that manages to foster this type of relationship with the reference market will gain a competitive advantage (Michaelis, Woisetschlager, Backaus, & Ahlert, 2008).

In fact, one study reveals that the consumers' willingness to pay may vary depending on the image of the country of origin, because a favourable image of a country will positively influence consumers who are encouraged to pay more in order to obtain a specific product. Instead, if the country image is negative, the consumers will not be willing to pay more than necessary to purchase a product or a service (Koschate-Fischer, Diamantopoulos & Oldenkotte, 2012).

ITALIAN SMEs IN THE TRADITIONAL FOOD SECTOR: MARKETING CAPABILITIES AND THE WORD-OF-MOUTH TOOL

Recent studies have investigated the traditional food sector both in Europe and Italy. The analysis revealed that the structure of enterprises in terms of employees is characterised by a limited number. This situation is particularly prevalent in Italy where, according to data collected in the 2011 census, about 99% of the companies employ less than 250 workers, while approximately 98% employ less than 50 people, and finally, more than 87% of companies have less than 10 employees, a size structure that has also been confirmed by the Eurostat data for the following years (Gellynck et al., 2012; Banterle et al., 2016). The researchers analysed the marketing settings implemented by means of investigations on samples of companies in the traditional food sector. The results have made it possible to identify four groups of enterprises with shared similarities in their marketing orientation and marketing capabilities (Banterle et al., 2010; Gellynck et al., 2012). Therefore, in this sector, according to literature surveys, (Banterle et al., 2010; Gellynck et al., 2012) clusters indicate that despite the presence of a certain number of enterprises with an adequate marketing orientation and consequently endowed with marketing capabilities in line with the market demands, the great majority of them show shortcomings, and in some cases a total lack of marketing orientation. These studies (Gellynck et al., 2012; Banterle et al., 2016) indicate that this marketing failure implies a shortcoming in terms of marketing capabilities and most of these companies, especially the micro enterprises, have very little chance of operating adequately on international markets. As regards the shortcomings in terms of marketing capabilities of the SMEs operating in the traditional food sector, the studies shed light on how certain marketing activities such as care in adapting the offer to the customers' requirements and specific situations, the propensity for product differentiation compared to competitors, and efforts to improve the goods offered, are still present in the business practices while the operability of the traditional promotional and advertising activities are lacking in the marketing mix (Banterle et al., 2014).

Emerging elements confirm other research which involved in-depth analyses of the SMEs in the food sector. Highlighted are the attitudes of direct and ongoing relationships with customers and the lack of any propensity towards the use of impersonal and distant communication tools such as advertising (Resnick et al., 2016). In fact, the entrepreneurs-managers tend to consider this communication means insincere and are not able to maintain an appropriate authenticity in line with the transparency and consistency demanded by customers. As a result, word of mouth is the most valid communication means for both consumers and entrepreneurs (Harris & Deacon, 2011). This instrument stimulates the potential for positive persuasion of new customers and increases the loyalty of the existing clientele (Berger, 2014; Herold, Tarkiainen & Sundqvist, 2016). Indeed, the opinions expressed directly by customers are considered sincere and devoid of risks associated with economic interests. They are consequently accepted with confidence by possible purchasers and give rise to a potentially greater attraction capacity of the companies and the products offered (Stephen & Lehman, 2016).

ELABORATION OF TRADITIONAL FOOD PRODUCTS ON THE CHINESE MARKET

Over the past few years, traditional products have acquired an important role in daily consumption, and as highlighted by several authors, the consumers seem to pay more attention to the quality of the product, evaluating its safety and wholesomeness, rather than considering the links existing between the products and specific geographical areas (Gellynck et al., 2012). On the international scene Italy is one of the nations with the most plentiful and variegated agri-food resources and at EU level it is the country with the greatest number of brands of food products of a particularly traditional nature covered by EU legal protection. As observed by some authors, the Food and Beverage industrial sector in Europe is characterised by the presence of SMEs (Banterle et al., 2016), and in Italy this sector consists almost entirely of SMEs, each characterized by a limited assortment and with scanty economic, managerial and financial resources. In fact, as stressed by the same authors, it must be noted that 87.5% of the Italian food industry is represented by micro enterprises with less than ten employees, and for these the process of internationalisation and exports of the products is more difficult (Banterle et al., 2016). Furthermore, Italy has about 25% of the total protected designation of origin and typical geographical designation certificates issued in the EU which protect food (fruit and vegetables, cheese, extra virgin olive oil and cured meat products), (Eurostat, 2016).

China's economic development promotes the improvement of the per capita income, a direct consequence of which is also the increase and expansion of the level of consumption on the market. The size of population and strong momentum of economic growth makes China a market in which many enterprises are trying to expand their presence. In 2015, exports from the EU to China equalled to 9.5 percent of total EU exports (Eurostat, 2016). China represented the third most important export destination market, while Italy holds fourth place among EU countries for both imports and exports with China (Camera di Commercio Italo-Cinese, 2014).

Recent studies (Gellynck et al., 2012; Banterle et al., 2016) show that the Italian SMEs operating in the traditional food sector are very limited in size being for the main part micro enterprises. Therefore, in this sector, according to surveys in literature (Gellynck et al. 2012; Banterle et al., 2016), despite the presence of a certain number of enterprises with an adequate marketing orientation which are consequently equipped with marketing capabilities in line with the market demands, the great majority of companies show shortcomings and in some cases, a total lack of marketing orientation. According to these studies (Banterle et al., 2010; Gellynck et al., 2012), this marketing insufficiency implies a deficiency in terms of marketing capabilities, which thus entails a poor ability to operate properly on international markets for the majority of these companies, especially the micro enterprises. In the light of these elements, this chapter highlights attention on the area of a specific sector consisting of Italian SMEs that market traditional Italian products, as well as the identification of more appropriate marketing practices for developing the Chinese market.

This study focuses on the marketing activities of Italian SMEs that offer traditional food products. The authors gathered information by means of a questionnaire filled out directly by a sample of Chinese consumers in two separate research phases. An initial phase entailed the collection of preliminary information by means of a convenience sampling survey in July 2015. The sample consisted of 150 subjects residing in various cities characterised by free trade and exposed to many Western goods and services. The majority of these were women, representing approximately 70% of the total sample. The interviewees were almost exclusively from the younger population with almost 90% aged 35 or younger, and at least 60% aged between 18 and 25 years.

In the first phase of the survey, questionnaires were submitted to a convenience sample of consumers in order to discover how the Chinese consider traditional products and how they perceive the attributes that define them.

The following aim of this chapter is to interpret the data collected, by evaluating the responses given in the questionnaires.

To discover their perception of traditional products, respondents were first asked to express their level of agreement with various definitions and characteristics of traditional products in general, taking into consideration only scores over 4 which correspond to the response options "agree" and "strongly agree". The results which are shown in Figure 1 indicate that consumers consider a traditional product one which is steeped in tradition (89% of respondents) and made from raw materials that come from certain geographical areas (70%). Chinese consumers therefore echo the same judgement of what is also acknowledged as traditional by Europeans (Vanhonacker et al., 2010). They appreciate the close tie between the product and the past to which it belongs, corresponding to the results highlighted in literature regarding European consumers (Belletti et al., 2010).

For Europeans, in fact, a study carried out in six European countries ascertained that consumers perceive the strong tie between a traditional product and the territory to which it belongs from a historical and cultural point of view (Vanhonacker et al., 2010).

While for the Chinese the lowest percentage concerns the perception of high quality, only 24% of respondents consider a product to be of high quality. To be noted, over half the respondents consider a product prestigious, (therefore, for Chinese consumers, prestige does not correspond to high quality), and comparing the results of the survey with the results of the survey on the perception of Europeans, the majority of Europeans confirm they acknowledge the high quality of traditional products and perceive that a traditional product is flavoursome and tasty (Almli, Verbeke, Vanhonacker, Næs, & Hersleth, 2011).

The Chinese perception of traditional products is shown in Figure 1.

Figure 1. The Chinese perception of traditional products in general
Source: *Own presentation of author.*

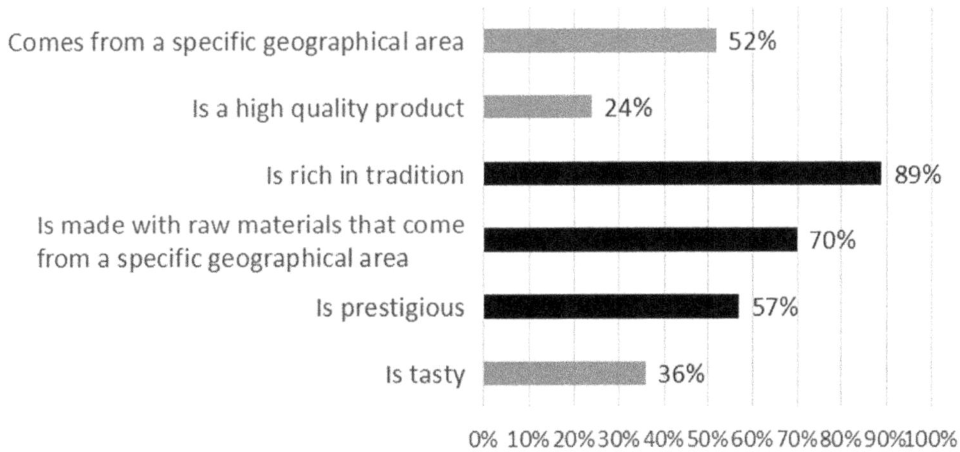

The different perception of the two groups is attributable to the situation in the Chinese food sector where consumers do not trust traditional Chinese products. The counterfeiting of traditional Chinese products (for example, the replacement of a raw material with another of poor quality or harmful in order to reduce costs) leaves traditional Chinese products with unsavoury tastes (not as good as before), because in China there are no precise regulations to control the genuineness of products just as there are no suitable procedures for following and monitoring production activities. Many producers manufacture traditional products, each declaring their traditional origin, but no-one checks up on this, particularly in smaller towns. Consequently when Chinese consumers were asked to express an opinion on traditional products in general, their negative judgement embraced all products.

HIGHLIGHTS OF TYPES OF TRADITIONAL ITALIAN FOOD PRODUCTS ON THE CHINESE MARKET

A second survey was conducted on an online sample of 200 Chinese consumers and the data were collected on the field in March 2016 by a Chinese market research agency. It was clear that there was equal gender distribution among the respondents and the sample had a high level of education, with the age of Chinese consumers ranging between 26 and 55 years.

The main objectives of this survey were to investigate the following:

1. Chinese perception of the traditional Italian products;
2. The elements taken into consideration at the time of purchasing or consuming traditional Italian products and whether consumers are prepared to pay a premium price;
3. The information sources used by consumers to search for and obtain information on traditional Italian products.

To find out how Chinese consumers perceive traditional Italian products the responses concerning the characteristics of a traditional Italian products were analysed, always taking into consideration responses with a value equal to or higher than 4, corresponding to "agree" or "strongly agree". The results of the survey are shown in Figure 2.

Figure 2. The Chinese perception of traditional Italian products
Source: *Own presentation of author.*

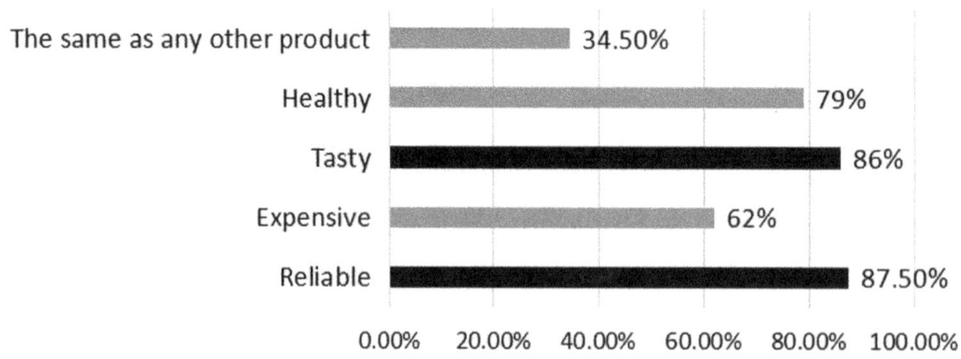

It can be observed how 87.5% of respondents believe that reassuring factors are the most distinguishing properties of traditional Italian products, as in the study conducted in Shanghai and Xi'an (Wang, Gellynck, & Verbeke, 2015). "*Reliability*" is one of the most important factors associated with a positive image of European food products on the Asian market. An analysis in literature of the data collected from European consumers also confirms that the safety factor is an important characteristic of traditional products for almost all EU countries, with the exception of Norway and France. For these purchasers, reliability is also the most distinguishing quality of a traditional product (Almli et al., 2011).

The Chinese pay particular attention to health and safety because, as previously mentioned, there have been various scandals over food safety in China and Chinese consumers no longer trust either the quality or safety of national food products, while those from the west are often considered safer, reinforcing the Chinese consumers' trust in traditional Italian products. The Chinese do not consider traditional Chinese products reliable, whereas traditional Italian products are viewed positively, particularly with regard to their reliability. Therefore, even though the Chinese are not well acquainted with Italian food products, they trust them, and 79% of the sample considers them to be healthy and 86% consider them tasty (see Figure 2).

The positive perception of traditional Italian products and their reputation can be observed in the response to the question on their willingness to give a traditional Italian product as a gift. Overwhelming majority of 94% of all respondents said that they were prepared to give it as a gift. The most frequently-given reasons were:

1. It is a foreign product
2. Italian food is famous or tasty.

The positive significance of this response for the purpose of developing the Chinese market for traditional Italian products can be traced back to the Chinese culture. In relation to the importance of giving gifts, the Chinese culture differs considerably from the western world. When introduced to someone, it is of the utmost importance to give a prestigious gift, which for the Chinese represents respect and a sound financial position; a gift is the basis for establishing social relations and friendship (Luo, 2009).

In order to better understand this concept it is necessary to analyse the two Chinese terms, "*Mianzi*" and "*Guanxi*". The first means self-image and the second social relations, so it very common to hear the expression "to have mianzi". This means that people expect others to be under the impression that they are rich, dignified and have good taste, etc. (Luo, 2009). In fact, for the majority of Chinese, having a good relationship with others is a requirement for achieving success and in the national culture an effective way of maintaining good long-term relations is to give "gifts" (Luo, 2009). Therefore, the fact of considering traditional Italian products suitable for being given as presents sheds light on the high distinction and elevated image these hold in the minds of Chinese consumers. Indeed, a large number of those interviewed (more than 60%) judged the traditional Italian reference products different from other products (see Figure 2).

The positive evaluation also emerges with regard to the preferred countries for purchasing traditional products, not counting China, because the Chinese consumers are not sufficiently used to western tastes and even though not ethnocentric (Parker, Haytko, & Hermans, 2011), they are still influenced by a sort of pride in and fondness for their nation. Italy ranks first among the western European countries and as many as 87.5% of those interviewed consider products from Italy the most pleasing. Italy is known and acclaimed for its food products in almost every part of the world, and its reputation continues to increase.

Figure 3. Country of origin of most highly appreciated products (the sum of points 4 and 5)
Source: *Own presentation of author.*

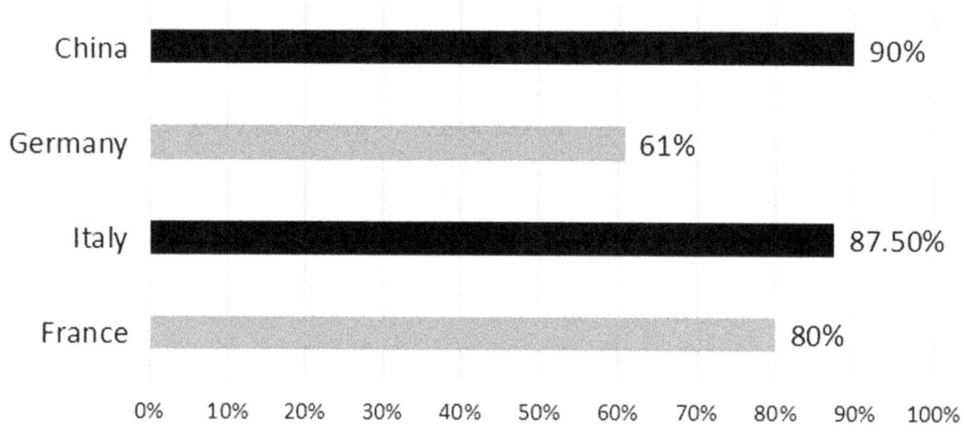

As revealed in literature, when the "country image" and the product category are highly congruent, consumers place more trust in the country of origin (Hu, Li, Xie, & Zhou, 2008). The graph in Figure 3 shows the level of appreciation in the various countries.

Continuing the research, the authors investigated the favourite places for purchasing traditional Italian products. According to 87.5% of the interviewees, the favourite place is a specialised shop, followed by 77% who prefer the traditional Italian restaurants, and 76% who choose the supermarket. The restaurants and specialised shop are undoubtedly the favourite places due to their specialities, however, the supermarket cannot be overlooked and even though imported products are now rarely found there, they still offer the possibility of success for European exporters. The supermarket is a place that appeals to a large cross section of the population as they are widespread throughout the city and easy to reach; the interviewees' preference showed their potential for developing this distribution structure (Peach International, 2012).

Only 16% expressed a preference for traditional stores. This choice is due to the fact that in the past the less prestigious goods found in traditional stores were considered to be of low quality and without any assurance of quality, consequently the Chinese population is not used to indicating these as their favourite places. At the time of making a purchase, consumers are also interested in the importance of the factors that guide their choice. They are accustomed to perceiving the products and services as a set of attributes and characteristics they themselves use as indicators for forming opinions and making judgments (Veale et al., 2006).

Analysis of the authors also took into account the sum of the percentages of those who declared to consider each element important and very important at the time of making a purchase. In this study, the most important intrinsic factor is the flavour, and 95% gave a high value for this. The quality perceived associated with the intrinsic factors of the food product (flavour, aroma, taste, natural materials, etc.) has a positive and significant influence on the consumers' intention to purchase (Botonaki, Polymeros, Tsakiridou, & Mattas, 2006).

However, consumers usually lack sufficient knowledge about foreign products and they do not have any great opportunities to sample them, therefore the extrinsic cues such as product origin (83%) and origin of the ingredients (81%) are important factors in their purchase decisions. When their experience with purchasing is insufficient, the use of information like country of origin helps reduce the time of deciding what to buy (Veale et al., 2006).

Chinese consumers attribute a superior value to traditional Italian products through their willingness to pay the premium price. The survey conducted in China by the authors shows how as many as 98.5% of the interviewees are willing to pay the premium price for Italian products compared to those coming from other countries and more than half of them are willing to pay a 10% higher price; this demonstrates that if Chinese consumers believe that the price and the quality of a product are linked together, then they usually have a positive image of Italian food products. Finally, a previous study maintains that the country of origin has a positive influence on the consumers' willingness to pay for products from countries that have a positive image (Koschate-Fischer et al., 2012).

Chinese consumers are undoubtedly of the type extremely concerned for quality of products due to the fact that the current situation of food is not as good in China as in western countries. Authors have also found that compared to industrialised countries, the consumers in developing countries are more willing to pay higher prices in exchange for healthier sustenance. Online retail sales have generally been on the rise over the last ten years and this derives from the fact that both the penetration of internet and the absolute size of the population using internet have increased. The online ordering and delivery of food and beverages by services which supply European products is now beginning to appear on the market. In order to evaluate the possibility of using a virtual distribution channel, the willingness to purchase traditional products through internet has been investigated and there is a high tendency (94.5%) of the interviewees to purchase these reference products online. In addition, the willingness to pay a higher price also persists for this purchase channel which in nearly 70% of cases exceeds 10%.

To discover the means of communication most widely used by consumers searching for useful information before purchasing, respondents were asked by authors to express on a scale from 1 to 5 the level of interest in various mass media. Even though the sum of percentages of respondents who gave values of 4 and 5 are considered, Figure 4 shows the preference for information means.

Figure 4 shows that first and foremost, respondents trust their friends' advice (83%), while with almost the same intensity purchasers turn to the social network (81%) and the blogs and websites of chefs and experts (81%) to find information before purchasing traditional Italian products. This indicates that the

Figure 4. Preference for information means (the sum of points 4 and 5)(1=not at all interested,5=very interested)
Source: *Own presentation of author.*

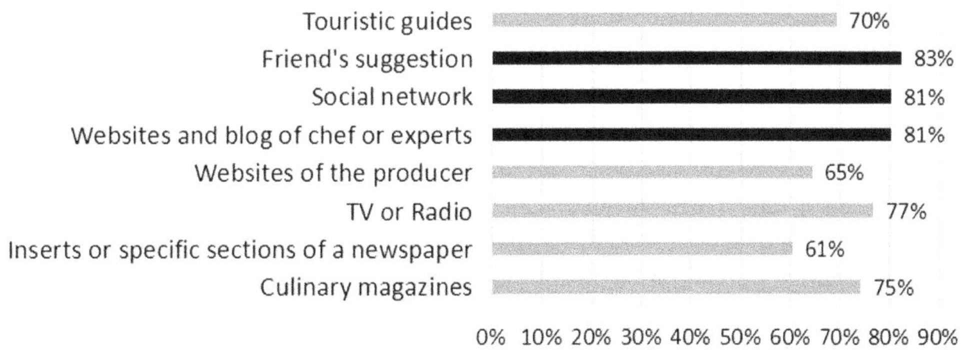

consumer is becoming a spectator and a protagonist at the same time (Rodriguez, Peterson, & Krishnan, 2012). Another important aspect to be analysed is the evolution of the web platform which has led to an exponential increase in the use of word of mouth. From mere verbal communication, "*grapevine communication*" has been reinforced by the web through blogs, forums, emails, chats and social networks. These web tools enable opinions to be efficiently and instantly conveyed by one person to many others (Rodriguez et al., 2012).

Word of mouth and social network responses netted the highest scores, and even though, as referred above, there are numerous ways of obtaining information, Chinese customers are more willing to accept information from their acquaintances, in other words, through the grapevine[1].

This depends on the characteristic of *collectivism* which plays a very important role because the Chinese consider it important to be accepted by others (Luo, 2009). The social network is also important. In China there is a famous microblog, widely followed by young people, where information is spread rapidly and effectively; the Chinese have great faith in this medium and many enterprises use it to advertise information about their products and to establish a good image. However, the more traditional mass media, like television and radio, are still important for providing information for Chinese consumers, confirming what other studies have already observed and our findings continue to suggest, i.e. the adoption of a balance between old and new communication channels (Chao, Corus, & Li, 2012).

Lastly, authors conducted a correlation and significance analysis to verify whether the willingness of Chinese customers to pay a premium price is associated with the information channels used or the characteristics defining traditional Italian products. (Authors refer to Kendall's tau-b index that measures the intensity of the association and also its direction (Bora & Di Ciccia, 2008).

Further analyses were conducted to investigate the association between premium price and:

1. Opinions on traditional Italian food products;
2. Preferences for the mass media adopted.

Authors used Kendall's tau-b index and if the index sign is positive, the two data sets are concordant. If the sign is negative, the two data sets are discordant (Bora & Di Ciccia, 2008)[2]. Figure 5 illustrates the correlation between willingness to pay the premium price and the perception of the attributes of traditional Italian products.

Figure 5 reveals a positive association between the question regarding the willingness to pay a premium price for an Italian product as opposed to a product from another country and consumers consider Italian food healthy (α=0.039), significant at level 0.05, but with a weak association (0.144). If the sign is positive, the two data sets are concordant: in other words, the more subjects consider it to be healthy food the more they are willing to pay a premium price. The analysis does not however appear to pinpoint any single decisive factor for explaining the willingness to pay a premium price for traditional Italian products.

Figure 6 illustrates the relationship between the question regarding willingness to pay the premium price and preference for information means.

Figure 6. reveals the positive correlation to pay a premium price for an Italian product as opposed to a product from another country and considering "TV and radio" the preferred method of acquiring information before purchasing a traditional Italian product (α=0.015) significant at level 0.05 (positive sign, the two data sets are concordant): in other words, the more subjects consider TV or radio to be the

Figure 5. Association between willingness to pay the premium price and the perception of the attributes of traditional Italian products
Source: *Own presentation of author.*

Correlations

			premium price	Reliable	Expensive	Tasty	Healthy	The same as any other product
Kendall's tau_b	premium price	Correlation Coefficient	1.000	.055	.003	.084	.144*	-.049
		Sig. (2-tailed)	.	.432	.960	.232	.039	.465
		N	200	200	200	200	200	200
	Reliable	Correlation Coefficient	.055	1.000	-.049	.365**	.198**	-.081
		Sig. (2-tailed)	.432	.	.479	.000	.005	.225
		N	200	200	200	200	200	200
	Expensive	Correlation Coefficient	.003	-.049	1.000	.009	-.021	-.075
		Sig. (2-tailed)	.960	.479	.	.897	.760	.247
		N	200	200	200	200	200	200
	Tasty	Correlation Coefficient	.084	.365**	.009	1.000	.326**	-.079
		Sig. (2-tailed)	.232	.000	.897	.	.000	.235
		N	200	200	200	200	200	200
	Healthy	Correlation Coefficient	.144*	.198**	-.021	.326**	1.000	-.007
		Sig. (2-tailed)	.039	.005	.760	.000	.	.914
		N	200	200	200	200	200	200
	The same as any other product	Correlation Coefficient	-.049	-.081	-.075	-.079	-.007	1.000
		Sig. (2-tailed)	.465	.225	.247	.235	.914	.
		N	200	200	200	200	200	200

*. Correlation is significant at the 0.05 level (2-tailed).

**. Correlation is significant at the 0.01 level (2-tailed).

preferred means, the more they are willing to pay a premium price, but the association is weak (0.169). Again in this case, this result does not pinpoint any single decisive means for correlating the consumers' willingness to pay a higher price for purchasing a traditional Italian product.

OPPORTUNITIES AND ADVANTAGES FOR ITALIAN SMEs ON THE CHINESE MARKET

Research to date has given authors direct indications of the behaviour of Chinese consumers regarding traditional Italian products and this highlights two important elements for Italian SMEs selling these products on the Chinese market. Firstly, it highlights the opportunity of interesting spaces on the market, and secondly, it offers valuable information about the possible marketing activities that SMEs operating in the traditional food product sector could put into practice.

From the empirical analyses conducted, the traditional food products are generally defined by the Chinese interviewees as products rich in tradition, and either made with raw materials that come from a specific geographical area or made from raw materials coming from a given territory. Chinese consumers confirm their opinion of the traditional aspect of the products, that is, the close link between the product and the past, where the finished products are made with local raw materials according to traditional customs, and a strong connection exists between a traditional product and the territory it belongs to. There is a common denominator among these aspects concerning the perception of traditional food products for both European and Chinese consumers (Almli et al., 2011).

Figure 6. Association between willingness to pay the premium price and preference for information means
Source: *Own presentation of author.*

Correlations

			Tv or radio
Kendall's tau_b	premium price	Correlation Coefficient	.169*
		Sig. (2-tailed)	.015
		N	200
	Culinary magazine	Correlation Coefficient	.219**
		Sig. (2-tailed)	.001
		N	200
	Inserts or specific sections of a newspaper	Correlation Coefficient	.201**
		Sig. (2-tailed)	.003
		N	200
	Tv or radio	Correlation Coefficient	1.000
		Sig. (2-tailed)	.
		N	200
	Websites of the producer	Correlation Coefficient	.179**
		Sig. (2-tailed)	.009
		N	200
	Websites and blog of chef or experts	Correlation Coefficient	.309**
		Sig. (2-tailed)	.000
		N	200
	Social networks(weibo. wechat..)	Correlation Coefficient	.188**
		Sig. (2-tailed)	.006
		N	200
	Friend's suggestion	Correlation Coefficient	-.029
		Sig. (2-tailed)	.680
		N	200
	Touristic guides	Correlation Coefficient	.192**
		Sig. (2-tailed)	.005
		N	200

*. Correlation is significant at the 0.05 level (2-tailed).

**. Correlation is significant at the 0.01 level (2-tailed).

The empirical study authors conducted revealed the positive opinion of the interviewees and their appreciation of traditional Italian food products, the local environmental conditions, and the cultural and natural factors that distinguish the high quality of these products. These elements suggest that in order to obtain an advantage on the Chinese market, it is recommended that Italian SMEs adopt *optimal marketing solutions* consisting of the following four strategies:

1. Make the Italian provenance of every element of the product distinguishable by means of the various methods accessible at reduced costs, such as the use of colours or signs on the packaging that clearly indicate that the product is Italian, the use of labels providing information about the components and materials used and any national and/or international certifications, the supplying of cards and/or brochures attached to the packaging that illustrate the characteristics and safe provenance of the ingredients and raw materials used;
2. Make the associated tradition and history known in reference to the product marketed that guarantees authenticity and customs of the territory of provenance of the traditional Italian food products. Illustrate the ties and relationships of the entrepreneur - manager with the history, tradition and

customs of the product marketed in order to adopt a marketing authenticity that brings Chinese customers closer to the authentic values of the territories of provenance, also because the consumers, as indicated in several other studies, never purchase a product bearing an Italian brand that is manufactured in another country since there would be no assurance of the quality of the ingredients and manufacturing processes (Paciolla & Mai, 2011). In view of the lack of resources distinguishing the SMEs, the communication instruments that could be used for an *optimal marketing solutions* are indicated by the survey where it is evident that the means used by Chinese consumers before buying a traditional Italian food product are first and foremost the advice of friends and family (83%), followed by the social networks (81%). The preferences collected directly from the purchasers indicate that the main means of communication is the social marketing media. The adoption of this marketing channel is also possible for the smaller sized enterprises due to the lower costs and ease of access. In addition, it also allows for direct contact between the company and the customers, thus expanding the relational capacities of the enterprise (Hansson, Wrangmo, & Solein, 2013). Finally, it allows participants to interact directly one with the other, encouraging a group spirit, and what is of even greater interest to the Chinese, it makes it possible to conform to the group and share familiarity with others (Luo, 2009);

3. Adopt as an *optimal price policy*, the choice of a medium-high price level using this marketing solution also as a sign of value and quality while taking into account the fact that among all interviewees there was greater willingness to pay a premium price for traditional food products from Italy rather than from other countries. In fact, if consumers think that the price and quality of a product are linked together, by paying a lower price it would mean accepting a lower quality, and vice-versa, in order to obtain top quality a higher price should be paid (Veale et al., 2006). Lastly, out of all the western European countries, Italy is the most highly appreciated in China with regard to traditional food products, and according to one study, the positive image of the country of origin has a favourable influence on consumers who find it an added incentive to pay more to obtain those particular products (Koschate-Fischer et al., 2012).
4. Adopt the choice of an online channel as a more effective distribution policy, since it has emerged from the authors' survey that the use of internet to go shopping online is normal, just as it is to make other decisions, and the majority of consumers would be willing to purchase traditional Italian food products directly online. In fact, the outsourcing chain is becoming increasingly more developed and the consumers' habits are changing. "*Alibaba*" and "*Jingdong*" are two of the main online sales platforms in China and there are specific channels on the websites for imported products with the result that the online purchasing of imported food products is gradually turning into a trend. Even though convenience is what principally prevails when purchasing online, this seems to have little impact on the willingness to pay a premium price when purchasing Italian food products. This element, associated with a greater assurance of the quality and security of shopping online, would make more consumers willing to access this channel (Luo, 2009; Peach International, 2012). Moreover, for the organisation of the online activities and appropriate logistics, in view of the need for SMEs to be able to access this function with limited resources, there is an excellent marketing solution entailing the choice of a specific opportunity that is being configured on the Chinese market. In addition to the highly developed online sales channels, there are also operators who act as external partners of the enterprise and are able to ensure necessary management of coordination of the network of distribution of information and logistics activities. These situations offer an exceptional solution for this marketing activity which would not otherwise be implementable

internally by SMEs due to lack of resources. Indeed, the existence of a coordinator company in the supply chain could help improve the processes, and above all, the relations and the network with the customers and suppliers of activities and services. Furthermore, the services of these operators could also be exploited for managing the activities on the social media so that the company would directly control the information to convey and be conveyed on these important relational platforms, thus *optimising web marketing* (Jüttner et al., 2010). Table 1 contains a list of some of the operators who offer coordination services on the Chinese market, reviews the functions and activities they carry out.

FUTURE RESEARCH DIRECTIONS

This chapter focuses on marketing activities of the Italian SMEs that offer traditional food products. Information was gathered by means of a questionnaire filled out directly by a sample of Chinese consumers. The choice of an empirical study was made in order to acquire opinions on the most significant attributes and expectations expressed by the Chinese with regard to these products. The analysis allowed authors to identify *optimal solutions for marketing activities* that can be implemented by Italian SMEs in virtue of accurate and in-depth knowledge of the reference Chinese customer. The solutions should in any case be subjected to careful monitoring for verifying their effectiveness more clearly, also in relation to the characteristics of limited resources and the lack of formality typical of SMEs (Harris & Deacon, 2011).

Additional analyses are recommended in relation to the opinions of the Chinese consumers for monitoring any changes occurring in their purchasing habits, and more particularly in their methods of consumption. These would also allow researchers and practitioners to better understand the impact of internet marketing and the word-of-mouth marketing, which for customers and entrepreneurs alike is the most valid communication tool in future marketing, and also whether the change in consumer methods is influencing the marketing methods (Stephen & Lehman, 2016). In fact, an interesting chain of future research could concern the propensity to purchase ready-to-eat foods or those which only require very little preparation at the time of consumption as this could configure not only as an opportunity, but also as a possible threat to traditional Italian food products. And since the Chinese language and culture are

Table 1. Coordination operators and the functions carried out

Company name	Analysis of consumer behaviour	Assistance to find a logistics company and the right platform for products	Homepage design on the B2C platforms	Assistance in managing the financial part of the enterprise	Design of a strategic marketing campaign with B2C platform to enlarge brand impact
Chamago		O	O	O	O
Hangzhou Netops Technology	O		O	O	O
Leqee	O		O	O	O
Joywell	O	O	O	O	O

Source: Data collected from authors' survey

so different to those of European countries, there are numerous Italian products which are purchased for one specific use; for example olive oil is only used to dress salads, very few use it to cook, consequently there are limited purchases of this product and future research must explore whether Chinese consumers would buy more if the products also included information in Chinese. If we look at the world situation, future research could also consider the competition of other countries selling similar products, by investigating which type of Italian product is more popular with Chinese consumers compared to other countries, and not only in general terms of traditional products but also divided by category.

In the opinion of authors, the adoption of the marketing configurations suggested in their model would lead to full exploitation of the valid positioning held by the products of Italian SMEs in the traditional food sector on the Chinese market, while also giving rise to an increase in turnover and added value of exports of traditional Italian products in China. In this perspective, interesting future research could entail an in-depth analysis of several cases of Italian SMEs equipped with websites that offer traditional products, in order not only to identify and suggest government policies for supporting and encouraging these Italian enterprises to set up websites that are also visible in China, but also to understand the *optimal marketing solutions* based on the importance of the country of origin.

CONCLUSION

This research has highlighted the possibility for Italian SMEs in the traditional food sector to implement *optimal marketing solutions* that would enable them to fully exploit the potential that the smaller dimensions offer for achieving an edge on the market. In particular, the presence of the entrepreneur – manager who ensures a close relationship with the territory of origin, incorporates in-depth knowledge of the customs and traditions represented by the traditional food products in the local history that become key elements in a marketing authenticity allowing reliable relations with customers. This situation in the Chinese context where the consumers are vitally interested in the reliability of the food products and their components and ingredients, creates a solution which is extremely favorable for having an adequate perception of the attributes of the product by the Chinese consumers, and provides a chance to exploit *optimal marketing solutions* that are also accessible by smaller sized enterprises with limited resources. These excellent solutions are also determined by the context of the Chinese market. Indeed, the accelerated development not only on the economic and income front, but also of the digital infrastructures and the high propensity of the Chinese population to use the digital marketing channels, makes it possible for Italian SMEs to implement *optimal marketing solutions*. The accessibility through the marketing social media enables the setting up of relations and direct networks with customers even on markets that are so far away, thus permitting the maintaining of a customer-orientation which is the fundamental requirement of the marketing capability of the Italian SMEs.

REFERENCES

Almli, V., Verbeke, W., Vanhonacker, F., Naes, T., & Hersleth, M. (2011). General image and attribute perceptions of traditional food in six European countries. *Food Quality and Preference*, *22*(1), 129–138. doi:10.1016/j.foodqual.2010.08.008

Atanassova, I., & Clark, L. (2015). Social media practices in SME marketing activities: A theoretical framework and research agenda. *Journal of Customer Behaviour, 14*(2), 163-183.

Banterle, A., Carraresi, L., & Stranieri, S. (2010). Small Business Marketing Capability in the Food Sector:The Cases of Belgium, Hungary and Italy. *International Journal on Food System Dynamics, 2*, 94–102.

Banterle, A., Cavalieri, A., Carraresi, L., & Stranieri, S. (2014). Food SMEs Face Increasing Competition in the EU Market: Marketing Management Capability Is a Tool for Becoming a Price Maker. *Agribusiness, 30*(2), 113–131. doi:10.1002/agr.21354

Banterle, A., Cavaliere, A., De Marchi, E. (2016). The Italian food industry in the era of the TTIP negotiate. *British Food Journal, 118*(8), 1930 – 1945.

Berger, J. (2014). Word of mouth and interpersonal communication: A review and directions for future research. *Journal of Consumer Psychology, 24*(4), 586–607. doi:10.1016/j.jcps.2014.05.002

Belletti, G., Brunori, G., Marescotti, A., Pacciani, A., & Rossi, A. (2010). *Guide to the Valorisation of Typical Agri-food Products. Concepts, Methods and Tools*. Regional Agency for Development and Innovation in Agriculture and Forestry.

Bora, S., & Di Ciccia, A. (2008). *Statistica: metodologie per le scienze economiche e sociali*. Milan: Italy McGraw-Hill Education.

Botonaki, A., Polymeros, K., Tsakiridou, E., & Mattas, K. (2006). The role of food quality certification on consumers food choices. *British Food Journal, 108*(2), 77–90. doi:10.1108/00070700610644906

Camera Di Commercio Italo-Cinese. (2014). *Commercio estero: i trend cinese del prime semestre 2014*. Retrieved August 16th from Camera Di Commercio Italo-Cinese web site: http://www.chinaitaly.com/it/commercio-estero-i-trend-cinesi-del-primo-semestre-2014

Chao, C. N., Corus, C., & Li, T. (2012). Balancing traditional media and online advertising strategy. *International Journal of Business, Marketing, and Decision Sciences, 5*(1), 12–24.

Coldiretti. (2015). *Expo: Coldiretti, + 300% export cibo Made in Italy in Cina da inizio crisi*. Retrieved August 16th from Coldiretti web site: http://www.coldiretti.it/News/Pagine/384---9-Giugno-2015.aspx

Eurostat. (2016). *Scambi internazionali di merci*. Retrieved August 16th from Eurostat web site: http://ec.europa.eu/eurostat/statistics-explained/index.php/International_trade_in_goods/it

Franco, M., Santos, M. D. F., Ramalho, I., & Nunes, C. (2014). An exploratory study of entrepreneurial marketing in SMEs. *Journal of Small Business and Enterprise Development, 21*(2), 265–283. doi:10.1108/JSBED-10-2012-0112

Gellynck, X., Banterle, A., Kühne, B., Carraresi, L., & Stranieri, S. (2010). Market orientation and marketing management of traditional food producers in the EU. *British Food Journal, 114*(4), 481–499. doi:10.1108/00070701211219513

Harris, J., & Deacon, J. (2011). Marketing in context – the marketing authenticity of owner/entrepreneurs of small firms: Case evidence from Welsh (UK) SME food and drink producers and retailers. *Small Enterprise Research*, *18*(1), 33–50. doi:10.5172/ser.18.1.33

Hansson, L., Wrangmo, A., & Søilen, K. S. (2013). Optimal ways for companies to use Facebook as a marketing channel. *Journal of Information. Communication and Ethics in Society*, *11*(2), 112–126. doi:10.1108/JICES-12-2012-0024

Herold, K., Tarkiainen, A., & Sundqvist, S. (2016). How the source of word-of-mouth influences information processing in the formation of brand attitudes. *Journal of Marketing for Higher Education*, *26*(1), 64–85. doi:10.1080/08841241.2016.1146387

Hu, X., Li, L., Xie, C., & Zhou, J. (2008). The effects of country-of-origin on Chinese consumers wine purchasing behavior. *Journal of Technology Management in China*, *3*(3), 292–306. doi:10.1108/17468770810916195

Jüttner, U., Christopher, M., & Godsell, J. (2010). A strategic framework for integrating marketing and supply chain strategies. *The International Journal of Logistics Management*, *21*(1), 104–126. doi:10.1108/09574091011042205

Koschate-Fischer, N., Diamantopoulos, A., & Oldenkotte, K. (2012). Are Consumers Really Willing to Pay More for a Favorable Country Image? A Study of Country-of-Origin Effects on Willingness to Pay. *Journal of International Marketing, American Marketing Association*, *20*(1), 19–41.

Luo,Y. (2009). Analysis of Culture and Buyer Behavior in Chinese Market. *Asian Culture and History*, *1*(1), 25-30.

Michaelis, M., Woisetschlager, D. M., Backhaus, C., & Ahlert, D. (2008). The effects of country of origin and corporate reputation on initial trust. *International Marketing Review*, *25*(4), 404–422. doi:10.1108/02651330810887468

Parker, S. R., Haytko, D. L., & Hermans, C. M. (2011). Ethnocentrism and Its Effect on the Chinese Consumer: A Threat to Foreign Goods? *Journal of Global Marketing*, *24*(1), 4–17. doi:10.1080/08911762.2011.545716

Paciolla, R., & Mai, L. W. (2011). The Impact of Italianate on Consumers' Brand Perceptions of Luxury Brands. *European Advances in Consumer Research*, *9*, 360–366.

Peach International. (2012). *The Chinese food market*. Beijing: Author.

Qureshi, S., & Kratzer, J. (2011). An Investigation of Antecedents and Outcomes of Marketing Capabilities in Entrepreneurial Firms: An Empirical Study of Small Technology-Based Firms in Germany. *Journal of Small Business and Entrepreneurship*, *24*(1), 49–66. doi:10.1080/08276331.2011.10593525

Resnick, S. M., Cheng, R., Simpson, M., & Lourenço, F. (2016). Marketing in SMEs: A 4Ps self-branding model. *International Journal of Entrepreneurial Behavior & Research*, *22*(1), 1–20. doi:10.1108/IJEBR-07-2014-0139

Reijonen, H., & Komppula, R. (2010). The adoption of market orientation in SMEs: Required capabilities and relation to success. *Journal of Strategic Marketing*, *18*(1), 19–37. doi:10.1080/09652540903511282

Rodriguez, M., Peterson, R. M., & Krishnan, V. (2012). Social Medias Influence on Business-to-Business Sales Performance. *Journal of Personal Selling & Sales Management, 32*(3), 365–378. doi:10.2753/PSS0885-3134320306

Stephen, A. T., & Lehman, D. R. (2016). *How Word-Of-Mouth Transmission Encouragement Affect Consumers' Transmission Decisions, Receiver Selection, and Diffusion Speed.* Oxford, UK: Saïd Business School.

Vanhonacker, F., Verbeke, W., Guerrero, L., Claret, A., Sulmont-Rossé, C., & Raude, J. (2010). How European consumers define the concept of traditional food: Evidence from a survey in six countries. *Agribusiness, 26*(4), 453–476. doi:10.1002/agr.20241

Veale, R., Quester, P., & Karunaratna, A. (2006). *The role of intrinsic (sensory) cues and the extrinsic cues of country of origin and price on food product evaluation.* Paper presented at the 3rd International Wine Business & Marketing Research Conference, Montpellier.

Wang, O., Gellynck, X., & Verbeke, W. (2015). General image of and beliefs about European food in two mainland Chinese cities. *British Food Journal, 117*(5), 1581–1595. doi:10.1108/BFJ-07-2014-0232

ADDITIONAL READING

Banterle, A., Carraresi, L., & Cavaliere, A. (2011). What is the role of marketing capability to be a price maker? An empirical analysis in Italian food SMEs. *Economia E Diritto Agroalimentare*, 16(2), 245-261.

Cho, H., & Tansuhaj, P. S. (2013). Becoming a Global SME: Determinants of SMEs Decision to Use E-Intermediaries in Export Marketing. *Thunderbird International Business Review, 55*(5), 513–530. doi:10.1002/tie.21567

Eid, R., & El-Gohary, H. (2013). The impact of E-marketing use on small business enterprises marketin success. *Service Industries Journal, 33*(1), 31–50. doi:10.1080/02642069.2011.594878

Guido, G., Marcati, A., & Peluso, A. M. (2011). Nature and antecedents of a marketing approach according to Italian SME entrepreneurs. *International Journal of Entrepreneurial Behaviour & Research, 17*(4), 342–360. doi:10.1108/13552551111139610

Liu, J., Guan, Z., & Xie, X. (2015). B2C e-commerce logistic channel structure in china.*WHICEB 2015 Proceedings,*Paper 5.

Mazzarol, T. (2015). SMEs engagement with e-commerce, e-business and e-marketing. *Small Enterprise Research, 22*(1), 79–90. doi:10.1080/13215906.2015.1018400

Newbert, S. L. (2012). Marketing amid the uncertainty of the social sector: Do social entrepreneurs follow best marketing practices? *Journal of Public Policy & Marketing, 31*(31), 75–90. doi:10.1509/jppm.11.038

Spielmann, N. (2015). Anything but Typical: How Consumers Evaluate Origin Products Based on Their Cues. *International Journal of Wine Business Research, 27*(1), 23–39. doi:10.1108/IJWBR-07-2014-0031

Vanhonacker, F., Lengard, V., Hersleth, M., & Verbeke, W. (2010). Profiling European traditional food consumers. *British Food Journal*, *112*(8), 871–886. doi:10.1108/00070701011067479

Wirtz, J., Tuzovic, S., & Kuppelwieser, V. G. (2014). The role of marketing in todays enterprises. *Journal of Service Management*, *25*(2), 171–194. doi:10.1108/JOSM-01-2014-0037

KEY TERMS AND DEFINITIONS

Country Image: The general perception of a country.

Food Safety: The quality of food is guaranted and tested so that no harm to human health is produced.

Marketing Authenticity: Entrepreneur's awareness of creating an offer based on profound values related to a culture and a deep-seated conviction.

Premium Price: Willing to pay more for a certain product in respect to another product.

Territory of Origin: An area where the product or material of the product comes from.

Traditional Food Product: Products passed down from one generation to the next, produced in a specific manner, associated with a specific local area, region or country.

Traditional Italian Food Products: Products passed down from one generation to the next, produced in a specific manner, associated with a specific Italian Territory or Italian region.

Word of Mouth: The characteristic of customers to share knowledge and discoveries regarding market offers by means of specific conversations about their impressions of the products.

ENDNOTES

1 Editors' note: A similar research result is obtained in chapter 4, on virtual travel communities.

2 Editors' note: Correlation analysis as a tool for determining the direction and strength of a relationship between two variables is also used in chapters 3 and 14.

Chapter 3
Regression Analysis for Environmental Practices:
Participation among Finnish SMEs

Nurul Aida Abdul Malek
University of Vaasa, Finland

Josu A. Takala
University of Vaasa, Finland

ABSTRACT

The aim of this research is to shed light on establishing a successful integration of environmental practices among small and medium enterprises (SMEs). The purpose is to develop a better understanding of what drives SME participation into potential environmental practices within a context in Finland. Specific objectives are to: 1) assess the willingness to participate (WTP) among Finnish SMEs based on four drivers - financial performance, internal readiness, supplier pressure and governmental regulations; 2) separate respondents into demographical subgroups based on their WTP; and 3) determine features that characterize Finnish SMEs which are WTP based from the studied sample.

INTRODUCTION

There has been increasing pressure from public institutions, consumers as well as non-governmental agencies regarding environmental performance of business industries due to their association as being the most visible contributor to environmental pollution. In such conditions, firms have begun to enhance competitiveness through environmental performance by minimizing environmental impact in their business activities. Despite emerging literature on various justifications on factors influencing firms in green participation, the emphasis is mostly on large firms while SMEs remain relatively unknown. SMEs are a source of job creation and introducing innovation and competition to the market; however, it is estimated that SMEs account for around 60 percent of carbon dioxide emissions (Revell & Blackburn, 2007). In addition, they also contribute 70 percent of the total global pollution (Aragón-Correa & Sharma, 2008). According to previous research, little attention was paid to SMEs which represent suppliers within the

DOI: 10.4018/978-1-5225-1949-2.ch003

supply chain. As a result, SME suppliers can be treated as missing links between SME environmental management and green issues in the supply chain management.

Most Finnish SMEs in the manufacturing sector are part of the value chain of domestic large companies. In order to compete globally, large companies have to fulfill the growing needs of environmental practices in their operations. In order to remain in the value chain, SMEs have to incorporate green elements in their operations as well. The government has played their role in providing a number of incentives to assist in the installation of environmental practices among the SMEs. However, it is interesting to note that a relatively high proportion of firms receiving public support have yet to be translated into a higher proportion of SMEs offering these products or investing in resource efficiency (DG Enterprise and Industry, 2014).

Therefore, the aim of this study is to develop a better understanding of drivers that can influence potential environmental practices in the context of Finnish SMEs because each country has their own particular characteristics such as policies, natural assets and technological capabilities. The objectives are to:

1. Assess the willingness to participate (WTP) among Finnish SMEs based on four drivers - financial performance, internal readiness, supplier pressure and governmental regulations which were highlighted in the literature
2. Separate respondents into demographical subgroups based on their WTP
3. Determine features that characterize SMEs that are WTP based from the studied sample.

The study addressed two important gaps in literature from the perspective of the practitioner and academia. It is essential for practitioners to recognize the driving factors of green participation for the benefit of future strategy formulation as well as to extend knowledge from a new point of view in the academia based on a sample from Finnish SMEs.

This study gave rise to a current discussion out of existing studies with an extension on the influence of financial performance on Finnish SMEs toward their WTP in environmental management. The conceptual model was partly based on a framework by Lee (2008) with an additional variable focusing on the influence of financial performance. The study attempted to find out the possibility of financial performance to become an antecedent to WTP in environmental management.

BACKGROUND

The Small Business Act (SBA) for Europe conducts a review of the act annually to provide a better understanding of recent trends and national policies related to SMEs in Europe. According to the DG GROW (2015), the estimated number of Finnish SMEs was 219,584 which covered 63.7 percent of the employment rate in Finland and did not include enterprises within agriculture, forestry, fisheries, education and health industry. This implies their importance as a job creator and contributor to national value added industries. Finland being one of the Europe Union (EU) member states is not excluded from being monitored in policy measures which incorporate ten principles including environmental factors. In a nutshell, the 2013 review indicated that the environmental performance of Finnish SMEs were all above average and ranked at the top of the list of all the EU member states. Finnish SMEs scored remarkably well in resource efficiency practices along with the development of green products and services. Therefore, this study is relevant to be conducted within the Finnish SME context which could provide an indicator to other SME populations with regards to environmental focus. Moreover, Finnish SMEs are recovering from the economic crisis in 2008 and expecting future growth especially in internalization. Consequently, this study could enhance understanding on how SMEs are stimulated to strengthen their environmental performance.

Previous empirical literature has used several theoretical perspectives to explore the internal driver(s) of a firm to respond to the demands of environmental concerns. Based on strategy resource based logic, both financial and non-financial resources are crucial to support any firm to continue to develop certain capabilities in order to enhance overall competitiveness (Wernerfelt, 1984, 1995). The availability of internal resources enables for the type of strategy that they intend to formulate and compete in. The major difference between small and large firms is the amount of resources that are available. Small firms are often associated with an impression that they are very limited in resources and less capable to invest for a new particular strategy that can bring them more advantages in market positioning. In conjunction with environmental practices, most SMEs are financially incapable of availing their firm to be more environmentally responsible oriented (Pacheco, Dean & Payne, 2010; Roberts, Lawson & Nicholls, 2006 and Simpson, Taylor & Barker, 2004). Moreover, SMEs also lack the competencies that it takes to bring them to become more environmentally active. Limitations of knowledgeable staff on environmental regulations and the ability to meet environmental performance requirements suppress the interest to participate.

Financial resource is a key driver to stimulate the adoption of environmental practices (Ghisetti, 2015). In order to develop green strategies, companies require huge amounts of investment which increase the role of public and private actors to support the initiatives. Financial constraints always reduce the ability of firms to innovate (Savignac, 2008). In view of these needs, Collector Bank Ab (a private credit company, formerly known as Collector Finland Oy) is interested to seek a specific role of finance as a driver to sustainability growth within Finnish SMEs. This study could promote their understanding in decision-making on credit granting processes and indirectly remove the barrier of environmental practices adoption. Therefore, this study separates the dimensions of finance from the previous framework into a new variable named financial performance. In this study, financial performance is the current financial measures of each SME as indicated in Table 1. On the other hand, readiness is referring to internal characteristics of SMEs which permits environmental practices adoption. Consequently, this study is interested to examine whether the internal factors of Finnish SMEs stimulate them to participate in green practices.

It is common for global players to have suppliers from all over the world to take advantage of the lowest cost source possible in order to minimize the total cost of end products. These type of suppliers are mostly SMEs (Walton, Handfield & Melnyk, 1998). However, this approach could end up with far more costly consequences if the products are rejected due to their non-compliance with environmentally oriented markets. In such situations, large firms have to assure their suppliers to integrate environmental standards required by their targeted markets. Consequently, SMEs are pressured to integrate environmental considerations into their business processes. In this respect, Asian SMEs that supply goods to foreign markets have to consider getting accreditation such as ISO 14001 to fulfill the standards of their buyers to secure their position as suppliers (Lee & Klassen, 2008; Rao, 2007; Zhu, Geng, Fujita & Hashimoto, 2010 and Zhu, Sarkis & Geng, 2005).

On the other hand, literature has highlighted the importance of regulations as a significant driver of environmental participation among firms. Regulatory pressures are viewed from penalties or fines issued to industrial organizations that are found to violate the environment. SMEs are always associated with being reluctant to adopt environmental practices in their business operations and therefore, regulation is an important driver for them to initiate the transition with at least minimal participation. Moreover, government plays an important role to stimulate awareness among SMEs by providing appropriate supports to promote environmental practices adoption (Lee, 2008). Therefore, buyer's practices and government involvement are chosen as external factors which stimulate the WTP in environmental practices among Finnish SMEs. Buyer's practices refer to the buyer's practices to improve environmental performance of their suppliers which includes supplier assessment criteria, the environmental performance of incoming supplies and environmental focus supporting activities. Meanwhile, government' involvement includes environmental efforts related to technical and financial incentives, economic benefit such as tax reduction and infrastructure supports. Following are the hypotheses of the study.

Hypothesis$_1$: Financial performance has a significant influence on the willingness to participate in environmental practices among Finnish SMEs.

Hypothesis$_2$: Firm readiness has a significant influence on the willingness to participate in environmental practices among Finnish SMEs.

Hypothesis$_3$: Buyer pressure has a significant influence on the willingness to participate in environmental practices among Finnish SMEs.

Hypothesis$_4$: Governmental involvement has a significant influence on the willingness to participate in environmental practices among Finnish SMEs.

Table 1 lists all the items measured under each variable.

METHOD

This study used primarily a quantitative approach with a survey method. The conceptual models of this study is partly based on the framework developed by Lee S. S.-Y. (2008) with an extension of additional variable that is focusing on the influence of financial performance toward WTP in green practices among Finnish SMEs. Hubbard and Armstrong, (1994) stated that a successful replication leads to reliability of results and a replication with extension provides an enhancement generalization of the findings beyond the prior context.

According to DG GROW (2015), the number of Finnish SMEs is 219,584. Therefore, the research samples of this study totaled 383 in accordance to Krejcie and Morgan (1970) who indicated that it is sufficient and valid to represent the population. The survey was conducted online for the duration of two months from May to April 2015. The total response compiled was 170 sets which was equivalent to 44.3 percent response rate.

The survey questionnaire contained 32 items; 7 related to the demographic and 25 related to the conceptual framework and measured on a seven-point Likert scale from strongly disagree to strongly agree. Cronbach's alpha value for all constructs was 0.769 which provided sufficient value for the data to be used in this study.

Table 1. Conceptual framework

<table>
<tr><th colspan="2">Internal drivers</th></tr>
<tr><th>Readiness</th><th>Financial performance</th></tr>
<tr><td>• Managers are aware of the importance of environmental issues
• Environmental issues are well communicated in our company
• Implementing environmental management system
• Will implement environmental practices if we can have financial support
• Has knowledge related to emerging environmental issues
• Has human resources to deal with emerging environmental issues</td><td>• Company's turnover category (million euro)
• Has financial reserves to invest in environmental technology
• Firm's current financial situation
• Firm's financial situation over the last few years
• Firm's financial situation for the coming few years</td></tr>
<tr><th colspan="2">External drivers</th></tr>
<tr><th>Buyer's influence</th><th>Government' involvement</th></tr>
<tr><td>• Incorporate environmental considerations in selecting their supplies and suppliers.
• Firm's financial situation for the coming few years
• Have interest in greening the supply chain.
• Environmental management system (e.g. ISO 14001).
• Provide us with environmental training, education or technical assistance.</td><td>• Coordinate the environmental practices initiatives
• Increase funds for the environmental practices initiatives
• Provide guidelines to SMEs in environmental management practices.
• Spread knowledge of environmental management.
• Build infrastructure for facilitating environmental practices initiatives.</td></tr>
<tr><th colspan="2">Company Willingness to Participate</th></tr>
<tr><td colspan="2">• Aware of the environmental management practices initiatives.
• Willing to participate in the environmental management practices initiatives.
• Have managers who have interest in the environmental management practices initiatives.
• Expects environmental and economic benefits from the environmental management practices initiatives</td></tr>
</table>

Source: Own presentation of authors.

The researchers performed analysis of the data and exploration of relationships determined through descriptive and inferential statistics. All research objectives and hypotheses were addressed accordingly. First, the dispersion of respondents were described and analyzed. Next the correlation results were evaluated to determine the relationship between the drivers, company profiles and their participation in green practices. Lastly the regression results were tested to evaluate the association of the drivers in terms of the participation of Finnish SMEs in green practices[1].

ANALYSIS AND FINDINGS FROM HYPOTHESES TESTING

Demographic Profile of Respondents

Table 2 describes the dispersion of respondents with respect to their demographic variables. 170 SMEs took part in the survey and the majority of them were from metal, building, wholesale and retail trade industries. 66 percent of the respondents belonged to entrepreneur job positions. A cumulative percentage of 94 percent generated up to 10 million euro of sales turnover within a year. Two-thirds of the total number of respondents comprised family-oriented companies with more than 20 years of operating experience. A combined percentage of 80 percent of the respondents belonged to SMEs with less than 19 employees. More than half of the respondents did not export their products or services internationally.

Table 2. Frequencies of demographic data by types of industries

	Metal	Electric / Electronic	Building / Construction	Rubber/ Plastic	Chemical	ICT	Forest Based Products	Food	Energy	Wholesale and Retail Trade	Others	Total	Percentage
N	20	8	29	3	4	14	8	6	6	15	57	170	100
Job Position													
Entrepreneur	13	3	20	3	2	9	7	3	5	10	37	112	65,9
Shareholder	2	2	5	0	1	2	1	2	0	2	9	26	15,3
Chairman	1	1	1	0	0	3	0	1	0	0	2	9	5,3
Others	4	2	3	0	1	0	0	0	1	3	9	23	13,5
Company's Turnover													
0-1	4	4	11	1	0	7	2	0	0	5	18	52	30,6
1-2	4	0	7	0	1	2	1	2	3	2	16	38	22,4
2-10	11	3	9	2	2	5	4	4	3	6	21	70	41,2
10-20	1	1	1	0	0	0	1	0	0	1	0	5	2,9
20-100	0	0	1	0	0	0	0	0	0	1	2	4	2,4
100	0	0	0	0	1	0	0	0	0	0	0	1	0,6
Company's Employees													
1-4	4	2	8	0	0	5	4	1	0	2	16	42	24,7
5-9	7	3	10	2	1	1	1	1	1	3	14	44	25,9
10-19	2	2	8	1	1	4	1	2	3	8	19	51	30,0
20-49	6	0	0	0	1	4	1	1	2	0	6	21	12,4
50-99	1	1	2	0	0	0	0	1	0	1	1	7	4,1
100	0	0	1	0	1	0	1	0	0	1	1	5	2,9
Company orientation													
Family	14	7	18	1	4	11	7	4	3	10	33	112	65,9
Non-family	6	1	11	2	0	3	1	2	3	5	24	58	34,1
Age of the company													
0-5	1	0	2	0	1	1	0	1	0	3	5	14	8,2
6-10	4	1	5	1	0	2	0	0	3	2	13	31	18,2
11-15	1	0	4	0	0	3	1	0	1	2	6	18	10,6
16-20	0	0	3	0	1	3	0	0	0	0	5	12	7,1
>20	14	7	15	2	2	5	7	5	2	8	28	95	55,9
Percentage of sales for export													
0	10	4	17	1	0	9	3	2	3	9	33	91	53,5
1-19	5	2	6	0	4	1	5	3	1	4	13	44	25,9
20-39	0	1	2	1	0	1	0	0	1	1	6	13	7,6
40-59	2	0	3	1	0	1	0	0	1	0	0	8	4,7
60-79	1	0	0	0	0	1	0	0	0	0	1	3	1,8
80-100	2	1	1	0	0	1	0	1	0	1	4	11	6,5

Source: Own presentation of authors.

Percentage Distribution of Drivers

Figure 1 is the percentage distribution of agreement for each dimension under readiness. At first glance, 50 to 60 percent of the respondents agreed with all the measurements and up to 80 percent were aware of environmental issues. The percentage of Finnish SMEs which have already implemented environmental management systems exceeded 60 percent. This reflected a considerable level of participation with regards to environmental practices. In terms of human resources, half of the respondents already have the knowledge and the employees related to environmental concerns. Their low priority on environmental practices investment is signified by varying responses on issues of implementation with financial support (see Figure 1).

As indicated in Figure 2, even though 50 percent of their buyers incorporated environmental elements in choosing their supplies and suppliers, other indicators suggested weak buyer influence within the respondents' value chain pertaining to environmental considerations. Only 10 percent of their buyers had initiatives to offer environmental knowledge or skills enhancement. Their buyers' slight interest of greening the supply chain with low request on environmental concerns indicated their relatively minor role in stimulating respondent participation.

Meanwhile, Figure 3 illustrates the distribution of percentage for government drivers. The percentage of respondents who chose to be neutral was just as many as those who disagreed to government linked measures. It seemed that only a modest 20 percent were receptive to governmental support in promoting environmental concerns in their company.

Correlation Analysis

The strength of prediction of one variable to another is indicated by a correlation coefficient which ranges from +1.00 to -1.00. The value of +1.00 suggests a strong positive relationship, whilst -1.00 shows a strong negative relationship and 0.00 indicates no relationship. Table 3 shows the correlation coefficient values between the dependent variable in this study, which is the willingness of Finnish SMEs

Figure 1. Percentage level of agreement on dimensions under readiness
Source: *Own presentation of authors.*

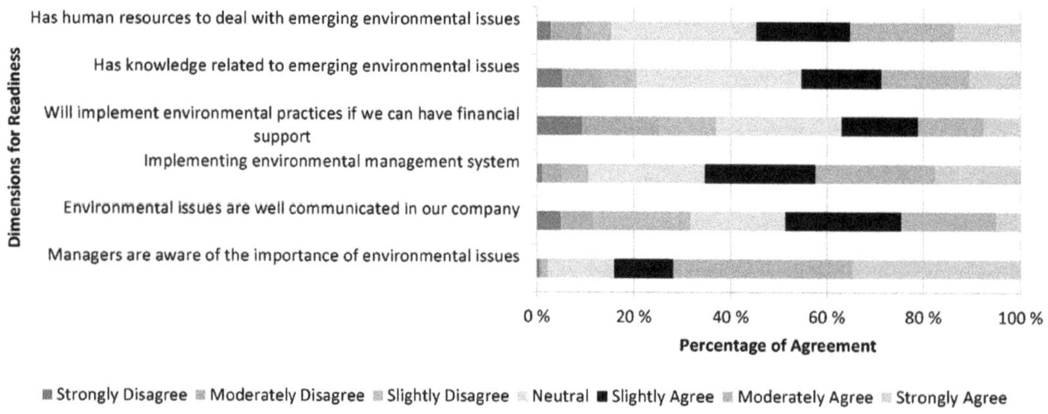

Figure 2. Percentage level of agreement on dimensions under buyer driver
Source: *Own presentation of authors.*

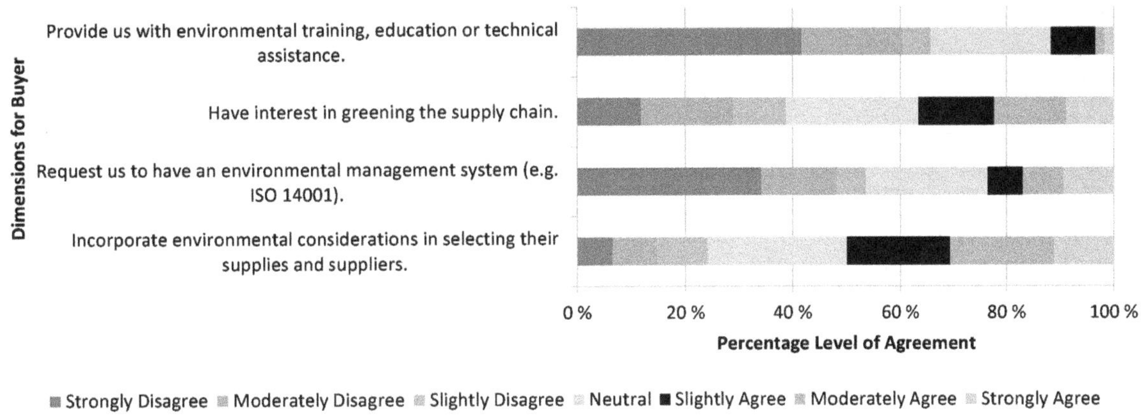

Figure 3. Percentage level of agreement on dimensions under governmental drivers
Source: *Own presentation of authors.*

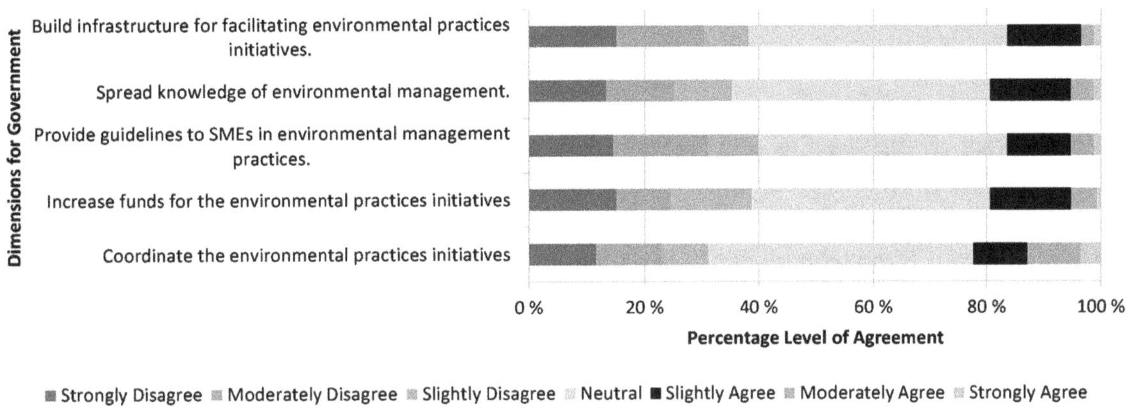

to participate in environmental practices (WTP) and their respective independent variables: readiness, buyer pressure, governmental influence and financial performance.

As illustrated in Table 3, all relationships are significant toward WTP. Company readiness with organizational resources associated with environmental concerns has the highest correlation value of 0.619. This is followed by buyer pressure, 0.529 and governmental influence, 0.439. Although significantly correlated, financial performance holds the lowest correlation value toward WTP with 0.365.

The researchers were interested to find out the relationship of the respondents' profile with their WTP. Table 4 portrays the correlation coefficient values according to demographical categories. Post, company turnover as well as company size appeared to have a significant relationship with WTP. Even though the strength of the relationships was moderate, it provided potential insights on certain groups within these variables which could diversify the level of WTP.

After identifying the significant demographic variables, the results of the correlation analysis between the drivers and WTP were divided according to the respective groups. The reason they were divided was to trace groups that were more driven toward WTP by comparing the strength of their correlation values

Table 3. Pearson correlations between drivers and willingness to participate

	1	2	3	4	5
1. Readiness	1				
2. Buyer	.475**	1			
3. Government	.284**	.486**	1		
4. Financial	.336**	.256**	.271**	1	
5. Willingness	.619**	.529**	.439**	.365**	1

**. Correlation is significant at the 0.01 level (2-tailed)
Source: Calculation of authors.

Table 4. Spearman correlations between demographic variables and willingness to participate

	Job position	Company turnover (million euro)	Company size	Type of industry	Ownership of the company	Age of the company	Sales exporting
Spearman Coefficient	,187*	,204**	,179*	,027	-,021	,031	,123

Source: Calculation of authors.

with those in Table 3. The status of each relationship is indicated in the status column. As illustrated in Table 5, there are 15 relationships with higher correlation coefficient values. The relationship which was most significant was within a group with a company size of more than 100 employees with governmental involvement and WTP with the r value of 0.960 which suggested a very strong relationship. Moreover, a group of company turnover with one to two million also represented improved relationships in terms of their readiness, buyer pressure and governmental involvement toward WTP.

Regression Analysis

Regression analysis was performed to investigate the degree of company readiness, buyer pressure, governmental pressure and financial performance of a company to predict the participation of Finnish SMEs in environmentally sustainable practices (WTP). The result of the regression is summarized in Table 6 which provides all the parameters and if the p value < 0.05, it would indicate that the driver would have a significant impact on the WTP. The normality test of the residuals was also performed and the requirement for normality was met which suggests linear relationship between the dependent and independent variables. A significant regression equation was found ($F(4,165)=40.940$, $p<.000$) with an R square of 0.498. The predicted WTP is equal to -2.396 with + 0.418 readiness, + 0.185 buyer pressure, + 0.153 governmental pressure, and + 0.196 financial performance. The value of R square suggests that the drivers investigated provided 49.8 percent variation in determining Finnish SME participation. Based on the coefficient table, all factors were significant and supported the participation with readiness as the most influential factor.

Table 5. Pearson correlation between drivers and willingness to participate for each sub-group

Demographic Variables		Readiness	Status	Buyer	Status	Government	Status	Financial	Status
Job Positions	Entrepreneur	.598**	x	,584**	Improve	,557**	Improve	.384**	Improve
	Shareholders	.668**	Improve	.462*	x	.369	x	.299	x
	Chairman	-.250	x	-.013	x	-.241	x	-.183	x
	Others	.736**	Improve	.467*	x	-.048	x	.413	x
Company's Turnover (millions)	0-1	.359**	x	.542**	Improve	.456**	Improve	.254	x
	1-2	.794**	Improve	.611**	Improve	.564**	Improve	.278	x
	2-10	.644**	Improve	.404**	x	.270*	x	.329**	x
	10-20	.748	x	.736	x	.919*	x	.359	x
	20-100	-.796	x	.108	x	.483	x	.584	x
Company Size	1-4	.448**	x	.643**	Improve	.427**	x	.438**	Improve
	5-9	.598**	x	.463**	x	.324*	x	.113	x
	10-19	.744**	Improve	.519**	x	.391**	x	.238	x
	20-49	.633**	Improve	.206	x	.617**	Improve	.379	x
	50-99	-.196	x	.245	x	.677	x	.728	x
	100	.799	x	.771	x	.960**	Improve	.145	x

Source: Calculation of authors.

Table 6. Multiple regression models for willingness of SMEs to participate in green practices

Model Summary				ANOVA		Coefficients		
R	R^2	Adjust R^2	SEE	F	Sig.	Independent Variables	B	Sig.
.706a	.498	.486	3.81804	40.940	.000b	Constant	-2.396	0.175
						Readiness	0.418	0.000
						Buyer	0.185	0.004
						Government	0.153	0.004
						Financial	0.196	0.050

b. Predictors: (Constant), Government, Readiness, Buyer, Financial
Source: Calculation of authors.

DISCUSSION

This research explored how Finnish SMEs reacted toward driving factors of their behavior to participate in environmental practices. The implications of the study can be useful for SMEs and large-sized companies as well as governments which continuously strive to improve environmental performance along the value chain within nations. Based on the findings, it can be concluded that all hypotheses proposed at the beginning of the study were supported.

Discussion of Hypothesis 1: Readiness has Significant Influence on the Willingness to Participate in Environmental Practices among Finnish SMEs

Regarding the extent of readiness as a driver toward WTP, the findings suggested that the majority of Finnish SMEs were stimulated to some extent. There exists a significant association between organizational resources and capabilities with environmental concerns towards WTP given a high correlation coefficient. The findings supported the notion embedded within resource based views which claimed that companies could sustain competitive advantage based on internal firm characteristics (Barney, 2001a, 2001b and Wernerfelt, 1984, 1995). Interestingly, the results suggested that more than 60 percent of Finnish SMEs are already involved in environmental practices by engaging environmental management systems within their operational systems. This is consistent with the findings of TNS Political & Social (March 2012b) which reported almost 93 percent of EU SMEs are using at least an approach to become more resource efficient. More than 80 percent of the companies are aware of emerging environmental issues although it is moderately communicated internally. In Finland, SMEs are more likely to be conscious of the rising cost of non-renewable energy and most of them are supporting efforts to look for alternative renewable energy solutions which are actively developed within the country. Human capital plays a vital role to ensure the successful growth of a company. Half of the respondents claimed that they have the knowledge and skills in environmental concerns. Universities in Finland provide assistance initiatives through regular attempts to engage their SMEs in training with an environmental focus (DG Enterprise and Industry, 2014). Dynamic capabilities development through employee empowerment could be achieved through these opportunities. SMEs should be mindful that green practices lead to long term cost efficiency as well as promoting synergy among their current and potential buyers. They should put more effort in gradually improving their internal readiness capabilities through training to develop skills and competencies among their human resources and also exploiting the wide-ranging potential for enhancing energy efficiency in their operations.

This highlights the importance of flexibility development within a company for future adaptability to the dynamics of a market. In order to sustain a competitive advantage, companies have to develop inbuilt flexibility within their value chain to face the uncertainties of future global environmental markets.

Discussion of Hypothesis 2: Buyer Practices have Significant Influence on the Willingness to Participate in Environmental Practices among Finnish SMEs

This study also indicated an important influence on buyer practices and governmental involvement. Although more than half of the respondents did not export their products or services internationally, 50 percent of the respondents' buyers strongly incorporated environmental requirements in conducting business transactions. This could explain the reason that more than half of them have already implemented environmental management systems. Even though this does not associate largely with international environmental regulations and requirements, effects of second tier regulations are seemingly delivered through the value chain. Furthermore, Finnish SMEs are largely attached to the value chains of large domestic companies. It can be seen that medium-sized companies with more than a 100 employees are very strongly stimulated by buyer pressure as compared to smaller-sized companies2.

Only 10 percent of buyers include knowledge and skills enhancement assistance which suggests less urgency of environmental needs in supplies. The buyers could be relying on government support to facilitate the SMEs with regards to environmental concerns. The interest to promote environmental practices in the value chain has drawn a variety of responses which implies that they have yet to see it as a main competitive advantage. In this case, the external network which is the buyer provides only moderate pressure to Finnish SMEs to mobilize their internal resources and capabilities for environmental performance.

Discussion of Hypothesis 3: Governmental Involvement has Significant Influence on the Willingness to Participate in Environmental Practices among Finnish SMEs

Based on the results, most of the respondents are sitting on the fence with regards to governmental involvement in implementing green initiatives in the value chain. Essentially, in Finland there is SITRA (The Finnish National Fund for Research and Development), an environmental program to support the creation of start-up companies with an environmental focus as well as to promote specific markets. Furthermore, the government provides support to two of the main industries in Finland i.e., metal and construction. For the construction industry, enterprises as well as research institutes with the government's assistance have developed products and processes which are embedded into construction projects such as heating and cooling systems. Likewise, awareness campaigns are also conducted within the metal industry. The government has also been providing resource-efficiency improvement support to assist companies to enhance their environmental performance (DG Enterprise and Industry, 2014). Finnish SMEs also ranked second amongst EU SMEs for their interest in bidding for a public procurement which require them to include environmental practices (TNS Political & Social, March 2012a). This suggests that Finnish government has managed to stimulate them to some extent. However, based on the measures of this study, these efforts have not yet reached the Finnish SMEs as a whole. It is recommended that the government provides indirect assistance such as by promoting collaborative opportunities between SMEs and large companies.

Discussion of Hypothesis 4: Financial Performance has Significant Influence on the Willingness to Participate in Environmental Practices among Finnish SMEs

Internal source of financing is not a strong determinant for Finnish SMEs to participate in environmental practices based on a weak relationship coefficient. It has been argued that the SMEs' lack of participation is mainly due to their financial constraints to fund the adoption. Apparently, this research suggests that financial performance led slightly for Finnish SMEs to actively increase their environmental performance. Although SMEs' economic development in Finland has recovered since the financial crisis in 2008, the study has shown that they could possibly have other concerns other than being environmentally oriented (DG Enterprise and Industry, 2014). Varying responses on the issue of 'environmental implementation with financial support' reflects the priority of Finnish SMEs toward environmental practices adoption which is perceived more like an option rather than a need.

Nevertheless, the demographical variable on 'sales turnover' has a significant relationship with WTP. It appears that rather than higher sales turnover, Finnish SMEs with a 'one to two million euro sales turnover' and 'smaller-sized companies' have a higher probability to participate. It may be that companies with small capacities are more flexible to accommodate their resources and engage in environmental practices (Shukla, Deshmukh & Kanda, 2010).

The correlational relationship of financial performance and WTP also implies that both variables can be performed in a similar manner. Significant results suggest that Finnish SMEs which choose to participate in environmental practices adoption also have a high tendency to perform well financially.

LIMITATIONS AND FUTURE RESEARCH DIRECTIONS

The sample of this study was obtained through Collector Oy's customer directory. Therefore, generalizations from the results may not comprehensively represent all Finnish SMEs. Furthermore, there is potential existence of social desirability bias where respondents may have a tendency to respond in a favorable manner towards environmental issues. However, the number of responses was sufficient and the respondents also came from various industries in Finland which increased the presence of heterogeneity. In addition, each item was measured in a seven-point scale to minimize the presence of social desirability bias.

In order to achieve an in depth view of Finnish SMEs based on the model of this study, it is suggested that a qualitative approach be adopted in gathering and evaluating responses. Three entities of the model could be incorporated in future research i.e. companies, suppliers and government sectors within a study to determine how these variables may be able to drive and complement one another to promote environmental practices adoption nationally.

CONCLUSION

The managerial implication of this study is that this analysis enhances the awareness to improve the governance of Finnish SMEs through an understanding of the drivers that influence them in environmental practices adoption. The study has statistically analyzed 170 responses and demonstrated that Finnish SMEs are aware to some extent that their companies need to play a role in achieving environmental sustainability. Internal and external resources play a critical role to steer the growth of Finnish SMEs towards environmental improvement. The buyers of Finnish SMEs provide both direct and indirect pressure along the value chain. Governmental involvement should be more comprehensive within the SME context. Universities as well as the public sector are encouraged to equip the SMEs on the significance of environmental practices and assist them to execute the transition.

The addition of 'financial performance' in the previous conceptual framework developed by Lee S. S.-Y. (2008) has improved the *r squared value* of the multiple regression analysis. This implies that the revised model in this study contributed to a higher variation in determining the participation of Finnish SMEs. Furthermore, this model also provided a stronger significance level in all drivers toward the willingness to participate in environmental practices rather than just two out of three drivers as in previous models.

From the financial perspective, this study provided some insights as to how the current financial performance of Finnish SMEs was associated with environmental concerns. This study can assist in promoting public and private funder understanding in decision-making of credit granting processes and indirectly remove barriers in environmental practices adoption.

REFERENCES

Aragón-Correa, J. A., & Sharma, S. (2008). The influence of stakeholders on the environmental strategy of service firms: The moderating effects of complexity uncertainty and munificence. *British Journal of Management, 19*(2), 185–203. doi:10.1111/j.1467-8551.2007.00538.x

Barney, J. (2001a). Is the resource-based "view" a useful perspective for strategic management research? Yes. *Academy of Management Review, 26*(1), 41–56.

Barney, J. (2001b). Resource-based theories of competitive advantage: A ten-year retrospective on the resource-based view. *Journal of Management, 27*(6), 643–650. doi:10.1177/014920630102700602

DG Enterprise and Industry. (2014). *SBA Fact Sheet: Finland.* Retrieved August 22nd 2016 from EU web site: http://ec.europa.eu/DocsRoom/documents/16121/attachments/11/translations/en/renditions/native

DG GROW. (2015). *SBA Fact Sheet: Finland.* Retrieved August 22nd 2016 from EU web site: http://ec.europa.eu/DocsRoom/documents/16344/attachments/10/translations/en/renditions/native

Ghisetti, C., Mazzanti, M., Mancinelli, S., & Zoli, M. (2015). *Do financial constraints make the environment worse off? Understanding the effects of financial barriers on environmental innovations.* SEEDS Working Paper Series.

Hubbard, R., & Armstrong, J. S. (1994). Replications and extensions in marketing: Rarely published but quite contrary. *International Journal of Research in Marketing, 11*(3), 233–248. doi:10.1016/0167-8116(94)90003-5

Krejcie, R., & Morgan, D. (1970). Determining sample size for research activities. *Educational and Psychological Measurement.*

Lee, S. S.-Y. (2008). Drivers for the participation of small and medium-sized suppliers in green supply chain initiatives. *Supply Chain Management: An International Journal, 13*(3), 185–198. doi:10.1108/13598540810871235

Lee, S.-Y., & Klassen, R. D. (2008). Drivers and Enablers That Foster Environmental Management Capabilities in Small- and Medium-Sized Suppliers in Supply Chains. *Production and Operations Management, 17*(6), 573–586. doi:10.3401/poms.1080.0063

Pacheco, D., Dean, T., & Payne, D. (2010). Escaping the green prison: Entrepreneurship and the creation of opportunities for sustainable development. *Journal of Business Venturing, 25*(5), 464–480. doi:10.1016/j.jbusvent.2009.07.006

Rao, P. (2007). Greening of the supply chain: An empirical study for SMES in the Philippine context. *Journal of Asia Business Studies.*

Revell, A., & Blackburn, R. (2007). The business case for sustainability? An examination of small firms in the UKs construction and restaurant sectors. *Business Strategy and the Environment, 16*(6), 404–420. doi:10.1002/bse.499

Roberts, S., Lawson, R., & Nicholls, J. (2006). Generating regional-scale improvements in SME corporate responsibility performance: Lessons from responsibility Northwest. *Journal of Business Ethics, 67*(3), 275–286. doi:10.1007/s10551-006-9184-4

Savignac, F. (2008). *Impact of financial constraints on innovation: what can be learned from a direct measure?* Econ. Innov. New Techn.

Shukla, A. C., Deshmukh, S. G., & Kanda, A. (2010). Flexibility and Sustainability of supply chains: Are they together? *Global Journal of Flexible Systems Management, 11*(1/2), 25.

Simpson, M., Taylor, N., & Barker, K. (2004). Environmental responsibility in SMEs: Does it deliver competitive advantage? *Business Strategy and the Environment, 13*(3), 156–171. doi:10.1002/bse.398

TNS Political & Social. (2012a). *SMEs, Resource Efficiency and Green Markets*. Retrieved August 22nd 2016 from EU web site: http://ec.europa.eu/public_opinion/flash/fl_342_sum_en.pdf

TNS Political & Social. (2012b). *SMEs, Resource Efficiency and Green Markets*. Retrieved August 22nd 2016 from EU web site: http://ec.europa.eu/public_opinion/flash/fl_342_en.pdf

Walton, S. V., Handfield, R. B., & Melnyk, S. A. (1998). The green supply chain: integrating suppliers into environmental management processes. *International Journal of Purchasing and Materials Management, 34*(1), 2-11.

Wernerfelt, B. (1984). A resource-based view of the firm. *Strategic Management Journal, 5*(2), 171–180. doi:10.1002/smj.4250050207

Wernerfelt, B. (1995). The resource-based view of the firm: Ten years after. *Strategic Management Journal, 16*(3), 171–174. doi:10.1002/smj.4250160303

Zhu, Q., Geng, Y., Fujita, T., & Hashimoto, S. (2010). Green supply chain management in leading manufacturers. *Management Research Review, 33*(4), 380–392. doi:10.1108/01409171011030471

Zhu, Q., Sarkis, J., & Geng, Y. (2005). Green supply chain management in China: Pressures, practices and performance. *International Journal of Operations & Production Management, 25*(5), 449–468. doi:10.1108/01443570510593148

ADDITIONAL READING

European Environment Agency. (2014). Resource-efficient green economy and EU policies. (EEA Report No 2/2014). Retrieved from EU web site: http://www.eea.europa.eu/publications/resourceefficient-green-economy-and-eu

Hottenrott, H., & Peters, B. (2012). Innovative capability and financing constraints for innovation: More money, more innovation? *The Review of Economics and Statistics, 94*(4), 1126–1142. doi:10.1162/REST_a_00227

KEY TERMS AND DEFINITIONS

Buyer's Practices: Buyer's practices to improve environmental performance of their suppliers which includes supplier assessment criteria, the environmental performance of incoming supplies and environmental focus supporting activities.

Environmental Practices: Operational activities that reduce the amount of damaging impact to the environment.

Financial Performance: Current financial measures of the Finnish SMEs.

Government' Involvement: Finnish government' environmental efforts related to technical and financial incentives, economic benefit such as tax reduction and infrastructure supports.

Readiness: Internal characteristics of the Finnish SMEs.

Willingness: The willingness of Finnish SMEs to participate in environmental practices.

ENDNOTES

1. Editors' note: Regression analysis as a mathematical method of modeling the relationships among three or more variables is also used in chapter 12.
2. Editors' note: Research results from chapters 1, 6, 7, 8, 12 and 14, also inductively suggest this notion of the unique significance of medium-sized enterprises in the whole set of ensembles of criteria.

Chapter 4
Optimizing Virtual Communities in Tourism to Facilitate Development of Small and Medium-Sized Enterprises

Silvena Dencheva Yordanova
Varna University of Management, Bulgaria

ABSTRACT

The chapter presents the essence of online/virtual travel communities and examines their current significance for SMEs. It presents types and characteristics as well as the most popular virtual travel SME communities and discusses ways of optimizing the development of SMEs by applying a strategy of virtual communities by selected hospitality SMEs. An eight step strategy to implement virtual travel communities by SMEs is elaborated. One of the main findings of the author is that virtual travel communities can be very useful for optimizing the development of SMEs by supporting all the stages in the decision-making process, focusing on pre-planning, making the decision and after the tourist and business visit stage. In order to understand how virtual travel community can optimize the development of SMEs, the author discusses tools for measuring the effectiveness of social media. Both future research directions and a conclusion are outlined.

INTRODUCTION

Virtual travel community has many definitions and advantages for small and medium sized enterprises (SMEs). One of those benefits is the internet which has fundamentally changed the way SME customers make their decisions. Nowadays, they are more affected to make decision by travelers from travel virtual communities, rather than their friends. In this vein, virtual travel communities support consumers in the pre-planning and planning process, during the trip and after it as well (Gulbahara & Yilidirima, 2015). Therefore, guest experience has become a dominant factor in the selection of a hotel, so review

DOI: 10.4018/978-1-5225-1949-2.ch004

sites continue to be the fore-front when consumers are planning a hotel room purchase (Gulbahara & Yilidirima, 2015).

Consequently, travel online communities provide unbiased information and customers can easily without any effort find out what the others recommend about visiting a certain destination, which hotels are recommended by travelers who had already visited the place (Casaló, Flavián & Guinalíu, 2011). On the other hand, social media have become a very important part of the marketing strategy of SMEs as they provide an opportunity for engaging with customers and generating brand advocates (Oztamur & Karakadilar, 2014). In this regard, social media allow destinations to contact visitors at relatively low cost and higher levels of efficiency that can be achieved with more traditional communication tools (Kiralova & Pavliceka, 2015). In other words, travel online communities help SMEs in building their image, reputation and creating long-term relationships with customers by engaging them in constant communication by the blog or forum they provide to their web-site. Moreover, *web-based platforms* allow for real-time recording and sharing of tourist experiences and thereby add to advances of personal virtual identities (Munar & Jacobsen, 2014).

This chapter has three main aims:

1. *To present the meaning of online/virtual travel communities* (VTC) and examine their current significance for SMEs in the available literature;
2. To present VTC types and characteristics;
3. *To provide ways of optimizing the development of SMEs* by applying a strategy of virtual communities by a selected hospitality SME.

Following the above described structure, the VTC discussion will proceed with future research directions. Finally, the chapter is completed with a conclusion.

BACKGROUND

Internet affected the business by changing the business models. In this regard, SMEs have in fact started taking advantage of social media, and it has become quite a popular tool (Oztamur & Karakadilar, 2014). SMEs can benefit in many ways from applying virtual communities as part of their business. In this regard, SMEs now have the potential to access increased opportunities to innovate on an unprecedented scale (Jones, Alford & Wolfenden, 2015). SMEs define target markets, as digital marketing offers an unparalleled access to information at a granular scale, which can redefine the ways in which customer segments are created using free-to-use web-based tools (Jones, Alford, & Wolfenden, 2015). This allows the potential user to enter any person or brand's Twitter handle and analyze their followers.

The benefits of using a travel online community can be summed up, as follows:

- The traveler receives useful information about the trip planned;
- There exists easy access to information from other travelers.

In addition, companies and SMEs benefit from using online communities in the following ways:

- Through deep and constant relationship with customers.
- Through better customer service.
- Through personalization of the message sent to the customer, according to his/her preferences and needs.
- By increasing brand awareness, brand engagement, word-of-mouth marketing, friends liking, trust and social validation (Kiralova & Pavliceka, 2015).

HIGHLIGHTS OF VIRTUAL TRAVEL COMMUNITIES IN SME DEVELOPMENT

The main focus of this chapter section is to present the essence of virtual travel communities, its types and characteristics. This will be followed by development of a strategy that will optimize the development of SMEs.

First of all, different types of virtual travel communities are presented and discussed. Based on the type of information exchanged between parties these are the main types of virtual interaction between travelers:

- **Blogs:** Short recordings that are chronologically ordered and contain media materials; Blogs are very useful for SMEs as they provide opportunities for feedback with customers. If the SMEs have a blog in their web-site, they can easily get constant feedback and understand the weak points in the service provided.
- **Chats:** Real-time communication through instant messaging programmes; Some customers prefer online assistant or living chat in making decision to buy from the web-site. They can improve optimization of SMEs by developing close contact to customer and assisting him during his decision.
- **Forums:** Web-sites for online discussions on specific topics, give more communication opportunities that chats or blogs. Most of the customers visit forums in their pre-research on certain topic. They are very useful and help increase competitiveness of SMEs by helping the customer to get acquainted with the information he is looking for. The more useful the web-site is, the more visited and preferred by customers it will be.
- **Wiki-Sites:** The web-sites aimed to deliver collective knowledge.
- **Social Networks:** A virtual network aimed to establish relations among its users, e.g. Facebook, MySpace, Vkontakte, etc (Liu & Park, 2015). They are very popular in the hospitality as SMEs use them for many reasons. They are the most used as virtual community as promotion tool, providing special promotions, etc.
- **Online Travel Reviews:** They are very useful tool for gathering information of certain destination as well as making the decision what to visit and why. They are very important information source that allow consumers to search for detailed and reliable information by sharing past consumption experiences (Liu &Park, 2015). The extensive travel information available online by using social media enables people to save time and efforts (Liu & Park, 2015). The researchers in online consumer reviews have focused on two main questions: the consumer decision making as a process and how it is affected by trust of people. Online reviews are useful for business as well as for consumers. On the business side, the hoteliers can understand how customers perceive the quality of the service provided such as accommodation, restaurant facilities. On the other hand, the potential tourist is affected to a greater extent by the marking given to a certain destination by a traveler

who has been already there (Nguyen & Coudounaris, 2015). In other words, the tourist image of a certain destination can be formed through online reviews, involving:

- Specific aspect of the service offering reviewed;
- The overall valence of the set of reviews;
- Valence of the information presented first;
- Whether or not easy to process graphical information is present (Browning & Sparks, 2011).

The content of review-staff or core values affects the willingness to make online booking and perceptions of trust in a hotel. Then the overall valence of the set of reviews is held constant, a series of hotel reviews that is framed with negative reviews results will lead to lower willingness to book the hotel (Browning & Sparks, 2011). The virtual travel communities with this form of communication optimize the contact with a customer by converting it to a long term relationship.

Characteristics of Online Communities

There are several characteristics of online communities. Some of them are succinctly defined by Preece, (2001):

1. **Purpose (Content of Interaction):** Virtual communities are built around a common purpose, but there can be many different common interests among the members such as travel, golf etc. The purpose of the community defines the goals and members it has as well as its content.
2. **Place (Extent of Technology Mediation of Interaction):** It concerns the technologies, protocols and norms of the online community inside the community's space.
3. **Platform (Design of Interaction):** In virtual communities, interaction can be either synchronous (e.g. chat rooms) or asynchronous (bulletin boards, newsgroups). It concerns the response time of the participants. The faster the time for response is the more reliable is the community for its participants.
4. **Population Interaction Structure (Pattern of Interaction):** Interaction pattern in the online community is related to the group structure and type of social ties.
5. **Profit Model:** It is related how profit is created by using this business model.
6. **Sense of Belonging:** The sense of belonging plays a prominent role in understanding the realities of the community as a social universe (Qu &Lee, 2011). It is a crucial for the existence of a travel community. It can be formed on the net though travel members' active communication around shared travel interests. There is a relationship between community identification and positive member behavior. If managers can understand how community affects its members, then the community manager can use this advantage by exerting influence on individual travel members through applying a more specific management strategy (Qu &Lee, 2011). In this regard, community manager can provide rewards. The community rewards are often given to a member to recognize its efforts, consistent behavior and contributions this member makes to the virtual travel community (Wang et al., 2015). In other words, being a member of certain community is rewarded and is valued highly by its members.

Main Components of a Virtual Community

In this regard, following Wang, Shuang & Li (2015) the author defines the following *main components* that compose a virtual community of interest for SME development:

1. *The users* who interact online with a mutual interest to satisfy their needs, which is where SMEs directly come in;
2. A *purpose* for a certain information exchange. In the author's view this is a very important component as it defines why the community exists;
3. Guiding *principles* that compile a communication *policy* which concern users and their access.
4. *Network facilities* that work as an intermediary in the communication process, whereby the facilities can be managed by SMEs;
5. *Relationship building* in online communities is a very important component of the online community. Relationships can lead to offline transactions. For example, people look for someone known in Facebook (Facebook, n.d.) and add them as friends. It is interesting how online relationships can lead to offline. Good online community reputation coupled with good online communication behavior leads to higher travel community member's willingness to participate in an offline relationship;
6. *Sustainability of community* is another characteristic of the virtual community (Wang et al., 2015). Sustainability of the community is its longevity of over an extended period of time. It can be measured by the number of participants the community has and the number of posts and its quality. Participant loyalty and trustworthiness are important for making a community sustainable. This component is also of utmost importance as it defines the quality of the relationship that users have. That is the main idea to form online communities, i.e. to gather people having similar interests and preferences at one virtual place. In online community its members are strangers and their aim is to share impressions and experiences about trips and discuss specific information about destinations, so it helps future tourists to avoid wasting extra money or missing important attractions in an unfamiliar place (Wang et al., 2015). Tourists and SMEs can share their experience and thus support potential new clients or help them choose which destination to visit.

The Role of the Word-of-Mouth

In relationships the electronic word of mouth (EWOM) is a very powerful tool affecting the communication in travel virtual communities. It is connected with the opportunities provided by online communities with constant interaction between them. In this regard, it can be said that *Internet* provides a platform for immediate interaction and communication in real time. EWOM has important implications for the design and development of online environments and interactions with consumers (Standing, Holzweber &Mattson, 2016). It plays an important role in influencing customers to use online recommendation sources and has influence on their product choice. According to Standing et al., (2016) EWOM is crucial, especially when people are socially independent or the level of commitment in decision-making is low. In this regard, it involves building trust, attracting traffic, creating awareness and most of all turning browsers or surfers into buyers. One of the main functions of EWOM is to transmit emotions of its participants and as such it is very important for tourism. According to the study, women are more likely to use the positive comments. For example, one negative review does not affect the next so much, whereas several negative emotions expressed influence product value evaluations (Standing et al., 2016). The intention to visit certain destination is directly affected by the online word of mouth. Since, the decision to visit a destination is interpreted as a rational calculation of the costs and benefits of a site of alternative destinations, which were derived from external information sources, including word of mouth and traveler's blogs (Abubakar & Ilkan, 2015).

Word-of–mouth (WOM) has a strong influence in the tourism industry, primarily due to the intangible nature of tourism services. According to the reports, more than hundreds of millions of potential tourists visit online reviews and consult with them prior to making any decision to travel. In this regard, WOM has the potential to reduce uncertainty, risk and has a significant impact on the tourists' destination image. The income, destination trust and WOM affect positively to the intention to travel to certain medical destination (Abubakar & Ilkan, 2015). In optimizing the development of SMEs, it is crucial for managers to know what the key components of the travel online community are.

Main Customer Online Intentions and Traveler Motivation

It is necessary to understand what provokes customers and SMEs to be involved in a virtual community. What is their motivation? Few studies focus on the real reasons provoking customer participation in virtual communities, and even less so for SMEs. In this regard, Amaro & Duarte (2015) discuss the *main intention* of the customers to purchase online and mention the following:

- *Intention to purchase online* which is the main predictor of customer behavior;
- *Attitude towards online travel shopping*, because positive attitudes clearly lead to purchase.
- *Online travel shopping self-efficacy* defined by the authors as consumers' self-assessment of their own capabilities.
- *Controllability* which is defined as individual judgments about the availability of resources and opportunities.
- *Perceived relative advantage of online travel* that increases trust in online travel shopping.

Several internal motivations for customers to participate in virtual community such as self-discovery, maintaining interpersonal interactivity, social enhancement and entertainment value are revealed by through research (Wang et al., 2015). These motivations are categorized as motivations for interdependence and motivations for independence. Interdependence focuses on building and developing relationships and affiliation with others. Independence is related to internal traits. It is very important for marketing managers of SMEs in hospitality to know the motivation of customers which is not only to attract them but also to keep them as loyal to the online community. For example, SMEs in hospitality in Korea experienced some difficulties in application of electronic commerce (Kim, 2006). Therefore, knowing real customer motives can be crucial as it will not only attract customers to the community but will keep them as loyal. Therefore, it is a factor leading to an optimization of the whole virtual travel community involving customers and SMEs alike.

One of the observed motives for joining a community discussed above is a sense of loneliness (Qu & Lee, 2011). By providing connectedness the travel online communities permit individuals to reduce their sense of loneliness among users (Qu & Lee, 2011). In this regard, online interaction provides individuals with an opportunity to share experience and exchange ideas with people who are physically distant form them. In this regard, *Internet* provides an opportunity for some customers to overcome their shyness and share emotions with unknown people, having similar interests or hobbies. Lonely individuals are likely to utilize online travel communities to uncover their emotions within respect to their travel needs and desires through effective peer interactions because online communities serve as an important platform for supporting community organized trips (Qu & Lee, 2011). The participation of the customer is rewarded by providing certain motivators, such as rewards. Several types of rewards that can be given to community members, such as a member-ranking list, virtual medals and granting access to special information which others cannot access (Wang et al., 2015).

Furthermore, *the motivation* of travelers to participate in a travel virtual community also includes the following identified factors:

- **Knowledge Sharing:** It is the travel member's willingness to share their expertise or experiences with other members (Qu & Lee, 2011). For example, one member may be an information seeker, while the other prefers to be more actively engaged. Based on these criteria, there are several types of customers. It is very important that community manager knows what needs the customer has in order to send a personalized content, based on his interest;
- **Self- Construal:** It is the extent to which an individual interacts with others and views the self as an independent identity or in relation to others (Wang et al., 2015).
- **Behavioral Changes:** The participation in virtual travel community leads to changes in behavior. This is a result of having a sense of belonging to a certain VTC. The more members categorize themselves as part of a VTC, the more likely are they to accept and purchase products according to their members' suggestions (Qu & Lee, 2011).

Significance of Online Travel Communities for SMEs

The online travel communities are useful in the three stages of the traveler and SME behavior: the pre-travel phase (a destinations research, booking, and purchase), the during-travel phase (interactive forums, blogs) and the post-travel stage (feedbacks, reviews on hotels and destinations, shared visuals). In this regard, online travel communities should be included in the web-site of the hotel as they support the customer during his relationship with the company. Obviously a hotel can be an SME. The travel online community can help the customer with the choice of a hotel, during the booking and after the visit of the destination. It is important to know not only the motives of consumers in joining virtual travel community, but also to recognize different characteristics of the online community.

In sharing of ideas to people with common interests who meet online together and exchange ideas, social media may in fact create online community which spreads a sense of belonging to its members. The main aim of the organization is to cultivate sense of belonging to a group in the real or the virtual world, which engages consumers and introduces them to a sense of belonging (Kavoura & Stavrianea, 2011). Clearly, if the SMEs can create a sense of belonging for its customers, it is a prerequisite for the development of the business organization.

Travelers trust more the comments made in the social networks where other travelers place their point of view, share information and their experiences, they are influenced by the information provided and these networks consists of socialization and interaction among like-minded people with similar interests (Kavoura & Stavrianea, 2011).

There are many examples of travel virtual communities of significance for SMEs. A classification made on the criteria type of shared knowledge is used by the author. In this regard, the professional oriented online communities involve *Learning and Expert networks*, where SMEs are typically service providers (Dencheva, 2014). Also, the commercially-oriented online communities have the main purpose of generating revenue, and as such are most important for SMEs involving two types:

1. Business-to-Business or SME2SME: it supports the supply chain, better organization and communication between business SME organizations. Trading is involved between SMEs.
2. Business-to-Consumer or SME2Customer: communities are created by SME organizations to support and market a product to the end customers.

AN OUTLINE OF STRATEGY TO IMPLEMENT TRAVEL ONLINE COMMUNITIES BY SMEs

The advent of digital technologies affected all spheres of life, as well as SME businesses. The customer has become the most powerful factor of SME development due to the information provided by *Internet*. It is expected that in the future many more SMEs will use the economic benefits of sustainable travel by applying virtual communities in their business models particularly for smart travel tech for SMEs Business trips, optimal travel management and online booking strategies for SMEs, SME contract for car rental, airline ticket, hotel and ticker pricing strategies (Amaro & Duarte, 2015).

In the next section the chapter analyzes and recommends for further consideration the use of VTC for SME travel agency optimal marketing strategy[1]. The strategy of the SMEs follows certain steps. The most important is to set the goals of the SME business organizations. They are quite specific for each business, but the main idea is to understand what the aim of the *Internet* presence for the company is. The company might exist in the real world, not just as an Internet company and use *Internet* only as another communication channel to boost its sales. On the other hand, it may be real virtual company which aims to conduct business through Internet. The right definition of goals means defining who the target customers are. It is easier targeting on *Internet*, as customers are self-targeting. They decide which websites to visit, from which to buy. In this regard, it is very important to choose the right business model.

Strategic Step 1: Choice of SME Internet Business Model

The first step in the strategy is definition of the right internet business model responding to the main goals of the SME and defining its value proposition. According to Timmers (1998) definition of a business model is: "An architecture for product, service and information flows, including a description of the various business actors and their roles, a description of the potential benefits for the various business actors and a description of the sources of revenue" (p.2).

The business model is focused on understanding the customer, his need and how to meet his expectations. The most common types of e-business models on a worldwide basis are B2B (business-to-business), B2C (business-to-consumer), B2E (business-to-employee) and C2C (customer-to customer) all of which are applicable for SMEs.

The Business-to-business model is focused mainly on conducting sales relationships between two companies or developing relationship between company and its suppliers.

For example, in the Bulgarian hospitality industry, the Business-To-Business e-business model represents in most of the cases, relationship between a travel agency and a hotel. Other possible business-to-business relationships can be relationship between a hotel and an airline. Bulgaria has 1000 travel agencies. Based on this, it can be said that the travel industry is quite developed in the country and this represents a huge opportunity for the development of B2B e-business model. Based on the facts that, through almost every single tour agent website, the online users can afford whole packages, including accommodation, flight tickets, it can be emphasized that the B2B e-business model in this country is very developed.

Tourist Service Providers (business) such as airlines and hotels sell their products to other businesses such as tour operators like (Wizzair, n.d.). To promote tourism, this model works with the collaboration of different businesses (i.e. Airlines, Hotels, Tour Operators, Agents etc.) and can provide dynamic packages to the tourists. On the *Internet*, services are often realized by huge networks and at the onset it is not always clear who profits and how much. This is especially true in the field of E-Tourism where many different companies co-operate in order to produce the final customers' service.

With the advent of Internet technologies, new forms of business models have appeared. One of the most popular remains the Business to Customer Model (B2C). What is specific for this model is the direct contact of the tourist provider with tourists. Tourists have an opportunity to make reservations on their own directly from the web-site of the hotel without the assistance of a tour agency. B2C business model in tourism is an advantage for both sides. Customers have access to information and can contact directly the hotel. On the other hand the hoteliers can receive fast feedback about the level of service they provide to the tourists.

The C2C e-business model represents online business relationships between with a seller (negotiator) and a buyer. In this model, a tourist can communicate with other tourists through the *Internet* (forums, blogs and email groups etc.) and can develop a tourist community.

The choice of a business model is affected by the usability and design of the web-site. In this regard, visual aesthetics is defined by Sanchez –Franco & Cataluna (2010) as"an artistically beautiful or pleasing appearance" (p.3.). Visual aesthetics is the extent to which customers believe a virtual community is pleasing to the eye. If customers find virtual communities pleasing, it is likely that both their state of mind and subsequent evaluations will be favorably enhanced and it will influence their emotions and their inclination to conduct business (Sanchez-Franco & Cataluna, 2010). The next feature that marketing managers should take into account when they develop their web-site as well as choose the business model to be applied is the usability of the virtual community. The usability of the web-site is connected with the ease that the customer can experience in what he looks for and make it fast (Dencheva, 2014). Therefore, usability is the ease with which the customer is able to learn to use the system and memorize its basic routine operations (Sanchez-Franco & Cataluna, 2010). Ultimately, the usability and aesthetics have a positive influence on customer satisfaction in virtual travel community.

Strategic Step 2: Analysis of the Competition

Second step of the strategy is conducting analysis of the competition. By analysis of the competition, the author means comparing the best practices, used by competitive web-sites and taking into account how the competition uses social media.

Strategic Step 3: Choice of Key Words

This is a very important tool, leading to a competitiveness of the company to be among the first ten results displayed in a product category. If the potential customers are interested in trips, and the company offers packages for a certain destination, marketing managers must think how to reach them. In this regard, Italian SMEs hospitality firms are promoting their destinations by using social media. According to Mariani, Felice, & Mura (2016) the type of posts made by regional domestic destination has an impact on the engagement, especially the visual types of posts, such as photos are having positive impact on the engagement of tourists. According to the study, evening and weekend posts have a positive impact on the engagement of tourists. The engagement rate is more important for marketing managers. Clearly, it is more important how people react to a Facebook post in terms of sentiment, rather than using just the numbers (Mariani et al., 2016).

Social media marketing initiatives cannot be measured commercially, as they mean to what extent their strategies on Facebook (Facebook, n.d.) and other social media are increasing the number of tourist packages and services purchased at the destination. In other words, by using social media SMEs in hospitality, at least on the example of Italian hospitality enterprises, cannot only reach potential travelers but also keep the loyal ones. The other opportunity is to create a forum, where the travelers can discuss their experiences with others. If the business organization is focusing on increasing the traffic to its web-site, it is a very useful tool. In this way, the number of visitors will be increased as well as the conversion rate which is the number of customers who will actually become buyers and loyal customers.

It also can be pointed out that customers trust in greater extent to their friends in making decision to visit a certain destination. At the same time, the number of people who trust a traveler is higher too (Dencheva, 2014). Virtual communities are an excellent tool to achieve the goals of the company, the trust of customers and develop SME brand recognition. In this regard, it is important to know what the real reasons are why people join the virtual travel community. There are different types of consumers on *Internet*, having different motives. Some are entertainment-seekers, while others are information-seekers.

Strategic Step 4: Analysis of Customers

Analysis of the customers is needed to understand what type of customers represent the SME audience, what interests and motives they have on *Internet*. In this regard, there are several types of customers:

- **Content Creators:** These are people publishing information in blogs who can be offered to upload or download their own video, audio.
- **Conversationalists:** These are people who visit social media, as they want mainly to communicate with other people and should be offered opportunity to chat with others in real time. An online program for communicating will satisfy them.
- **Criticizers:** These are the customers who always comment, show critical attitude towards all topics.
- **Collectors:** These are customers who visit different media, lead by the idea to collect points or win certain reward, look always for promotion and need to be engaged, stimulated. In this regard, these customers can be offered competitions, lotteries, games, as they have these needs.
- **Spectators:** The percent of these customers is highest. They only read the comments in the online communities, need an opinion of other customers to take certain decision.

- **Passive Customers:** They do not take part in the discussions nor take any position on certain questions.

In order to attract customers to the online community, different rewards could be given. For people who are looking for a place to express their opinion, this might mean giving them opportunity to share their travel experience or expertise on certain problem. The list of motives that customers have is very precious for the marketing and social media managers. It will give an opportunity for self-targeting of customers. In this way, customers agree to become a part of a certain database and to receive information from a certain company and its products. In this way the company creates digital relationships with consumer.

Strategic Step 5: Creating a Sense of Belonging for Customers in the Travel Online Community

As discussed before, the sense of belonging is a very important characteristic of the virtual community. Managers should focus their efforts on offering value for which the customers will feel engaged and being a part of certain travel online community. Creation of the sense of belonging is connected with the social identification of the individual. Being a part of certain online community, the customer agrees that the values and principles of the community are part of his behavior. In this regard, he identifies with the community. It means that he is proud being a part of this community, as he fully identifies with it, as it is with the fans of renowned hotels for example.

In order to create a sense of belonging customers need to have an opportunity to publish their own messages as well as video and audio materials. Any business organization can use the opportunities provided by *Internet* for presenting information in order to create the online experience of the customer. Offering such online experience means the customer will be able to visit the web content again and again.

In this regard, use of video gives opportunity for brand creation and brand recognition as well as for full study of the opportunities of product or service offered. People have always tried being a part of a certain group. Marketing managers need to analyze very carefully the motives of people of using virtual communities. Motivation theorists argue that unobserved needs and desires motivate behavior, the purpose of which is to satisfy needs (Zhong & Luo, 2015). Congruence of user need satisfaction in social media marketing can be achieved by aligning the marketing content, process and goal with users' needs of using social media. To develop their marketing strategy and design appropriate advertising appeals, it is critical that marketers identify and understand the needs and motivation behind social media usage. This will give an opportunity for communication with their targeted audience on a personal, meaningful level through social media (Zhong & Luo, 2015).

The community needs to have an image of a communication channel to which customers can trust. In other words, the virtual travel community is perceived as a valuable information source to which recent and potential customers can rely in their choice of a travel destination (Dencheva, 2014). After the SME business organization knows and understands the motives why customers join the travel community, the next step is to give them an opportunity to share the opinion they have and provide them with different status, based on the motivation they have. For example, people interested in active participation in the travel community can be given a status of an administrator. It will be very motivating if the community manager provides certain reward for the community members. For example, giving full access and administrative rights for certain period for the publishers might be challenging and motivating for them.

How a business model will be organized is crucial for the effectiveness of the travel community. By business model is meant the organization of business processes and way of making a profit as well as attracting customers. After the motivation of customers is defined, it is important to know the opinion of the customers. In this regard, it will be more effective if the company creates a blog or forum to their web-site. In this way the company will have the opportunity to track the behavior of its customers, by showing them interesting events and turning them to loyal. If the virtual travel community focuses not only attracting members but keeping them loyal then it should build an individual approach to a customer by applying various tools of digital communication technologies.

Strategic Step 6: Achieving Personalization for Customers in the Online Community

The development of a personalized content is an opportunity to interact with potential clients on a more personal level. Thus, the following recommendations are aimed to increase the SME personalization issue:

1. Develop a registration option or log in via social networks, so a visitor of the web-site is able to create his/her profile. The form includes personal details and preferences so when the customer logs in, he sees information content according to his preferences. In this case, the customer behavior can be tracked easily by knowing his preferences. Every time he/she logs in the web-site, a personalized content will be sent to him/her according to the personal details and information gathered about him/her. Each web-site is providing this option. However, registration is an opt-in option, which means that it is not obligatory. Some customers prefer to remain anonymous as it is one of the advantages provided by *Internet*.
2. Use *cookies* for user authentication, storage of personal preferences and settings; monitor the user status on the web-site, analyze the history and purchase behavior of the customer. There should be again opt-in option as some customers avoid using cookies.
3. Develop an internal search engine as well as online tour selection section, so a web-site use can easily access and buy a service offered by the respective company. It is a good option for fast and easy navigation in the web-site. As already discussed before, the visual aesthetics is very important tool for the success of the virtual community.
4. Develop an online assistant so that the customer can speak with a travel agent online. Online assistant is very effective especially for customers who prefer to have direct contact with real seller.
5. The web-site can be translated in several languages, based on the targeted customer group.
6. Send personalized e-mails to customer in birthdays or other celebrations. Personalization can be achieved not only by e-mail marketing campaign. If the customer is greeted on some occasion, he/she feels engaged with the community.
7. Send news to customer if he/she has given his consent.
8. Inform about events and tours offered with enclosed tour packages considering clients preferences and previous travel experiences. It is a part of an e-mail marketing campaign.
9. Develop a subscription option on the web-site for the newsletters and send newsletters to the clients about the new trips, destinations, "last minute" propositions.
10. Create a special travel community: "Travel-territory".

The web-site should provide an additional section called "Travel-territory" which is a place where each registered user can have his/her travel territory. This page will allow the tourist and SMEs to obtain supportive information regarding the destinations, make friendships with the other travellers, share experiences and impressions from travelling with the SME travel agency.

The proposed online community should have 2 subsections: *Tourists* and *Meeting Point.*

- **Tourists and SMEs:** This is the page where all the user profiles will be represented. Each user profile consists of Names, About yourself section, Contact information, Friends, Map of destinations, Photo blog and Ratings. One registered user can visit a page, make a friendship and give "likes" to the pictures and blogs of the other registered user.
- **Meeting Point:** A forum, where a user or SME travel agent can create a topic and make discussions, ask questions and follow dialogues. Having a news portal is an important tool to make the customers visit the web-site and check the updated information on daily bases. In addition, there must be options to leave a comment on the post and share the post via social networks. In such a way, the information will be delivered to a wider SME audience and moreover the web-site will gain promotional benefits which can also be managed by SMEs.

Strategic Step 7: Developing a Popular Travel Online Community

The next important step is attracting customers, i.e. developing a popular community which is interesting so that people will be glad to join it. The openness of the online community is a very important characteristic which means easy access to its members. There are various approaches in popularizing it such as search engine optimization. The key moment in using search engines is finding the right word that the customers use. The first goal is that the company can be found easily among other things. It can be achieved by choosing all the words that are related with the product. All business organizations are trying to be on the top of ratings. It means that they can be found easily. After that the customers must find the content of the community useful and easy to use, share it to their relatives, friends and others. Therefore, in the beginning the main goal is traffic generation and focusing efforts on the information presented, on its content.

After it becomes popular, the next aim is to increase the number of likings. If customers like it and recommend it to each other, then the customers will be engaged with it. In order to achieve it, the interest of the customer would be taken into account by updating its content on a regular basis. In this regard, there should be a specialist dealing with it, working on regular posting, updating the information content so that it will be useful to its customers. If the web-site is well organized as navigation and design, and is user-friendly, then it will be recommended to others. In this way the customer base will be increased.

If the hotel creates a blog to its web-site but does not pay attention to the interest and does not present events latest, news, etc, there will be a great risk of distrust among customers. Blogging is one of the ways of promotion, which is considered as an inexpensive marketing activity as compared to traditional advertising. In this way, hotel companies have realized their advantages and have encouraged visitors to write comments on their hotel blogs with rewards (Gulbahara & Yilidirima, 2015).

Strategic Step 8: Measuring the Effectiveness of a Travel Online Community

It logically follows from the first seven strategic steps that virtual travel communities can help increase competitiveness of the SMEs by having direct contact with customers. They can be very useful for SMEs. The virtual travel community can help increase the number of customers by attracting customers not only from the regional country but also foreign customers. In this regard, virtual travel communities can increase the competitiveness of the SMEs by making them more competitive not only on local level but also internationally. On the other hand, if the SMEs focus its marketing efforts on increasing the visits, its effectiveness can be measured by number of likings and clicks. Engagement is proving the interest of customers. It is impossible for a very short time, to increase the likings of the company with less marketing efforts and on low price. Internet allows such opportunity for SMEs. By presenting the company in any social media, it can receive fast feedback and clicking of customer to the main webpage of the company. The power of *Internet* is huge not only in attracting new customers to the company but also in keeping them loyal. In this regard, virtual travel communities provide excellent opportunity for SMEs to focus their attention on loyal customers by the recommendations they made to other future prospects.

In the beginning the main goal is visit and then the next goal is increase of the engagement rate measured by likings, clickings, etc. In order to be engaged in active communication, members of virtual community must be engaged in active communication. In this regard, managers of the community should have different approaches to customers, having in mind the differences in their behavior. In this regard, managers should stimulate the activity of their participants by giving them freedom to express opinion, comment, etc. For example, a guest of the hotel can be invited to share their experience in the book for guests. If there is a forum in the web-site of the hotel or blog, guests can easily publish their photos. It will be quite useful for potential guests of the hotel. In this way, the hotel can support prompt and provide quite feedback with customers that will be useful for both sides, as customers and company.

A lot of SMEs always offer some competition, lottery and other advertising methods to keep customers engaged. Having a dialogue with the company and its customers leads to higher engagement. It is beneficial for both sides. On one side, customers can express very fast their opinion and share experience they have with other potential customers. It is also useful for the company as they can understand the weak points in the customers' service. It is the challenge nowadays to have a long term relationships with customers and to see the customer as a partner. In this regard, interesting topics should be chosen. If the topics discussed in the forum, blog, web-site of the hotel are interesting for the customer, he will be attracted. The name of the company is not so interesting and attractive for the customer but rather its activity, values, etc. Customers are aiming to become a part of certain virtual travel community based on the motives they have. For example, info-seekers are focusing on professional blogs, forums, where they can find help and expert advice.

In order to understand how virtual travel community has optimized the development of SMEs, measuring the effectiveness needs to be conducted. By effectiveness is meant that the goals set by the company have been achieved with the results achieved (see table 1).

As emphasized before, the goals of the SMEs are defined with the right choice of the business model to be applied. If the company aims to use *Internet* as another communication channel to reach its customers, then the marketing managers are using the criteria for cost effectiveness of the campaign. If the company aims at creating brand awareness, then the brand awareness can be measured by number of people, who like and use the social media (Hoffman & Hodor, 2010). Also these people can be fans or make a reservation. Brand engagement can be measured by number of active participants compared

Table 1. Metrics for measuring social media effectiveness

Social Media Application	Brand Awareness	Brand Engagement	Word of Mount
Social network Facebook	•number of members/fans •number of installs of applications •number of impressions •number of bookmarks • number of reviews/ratings and valence +/–	•number of comments •number of active users •number of "likes" on friends' feeds • number of user-generated items (photos, threads, replies) • usage metrics of applications/ widgets •impressions-to-interactions ratio • rate of activity (how often members personalize profiles, bios and links)	• frequency of appearances in timeline of friends •number of posts on wall •number of reposts/shares • number of responses to friend referral invites

Source: Hoffman & Hodor (2010)

with passive. Active users are people who active use the social media. They comment, like or share with friends. These people are engaged in company and can share it with friends. This is good because they made the company more popular for others. Therefore, the effectiveness of social media can be measured by the number of people who use, like, and share it. If many people see the company in social media that means this social media is effective for the company.

If the goal of the SMEs is to use social media for attracting audience, the appropriate measuring criteria can be Facebook Fans. The opportunity provided by social media is to create for free a Facebook page. Once the SMEs create a business page for free on Facebook it is possible to track the number of people that become a fan of this company. It is also possible to track the number of people who visit the web page and the number of new fans. Facebook provides a decent free tool to track all of these stats (Facebook, n.d.). The social media gives excellent statistics who are the customers. Knowing the total number of visits will provide information if the campaign is successful in achieving traffic to the web-site.

FUTURE RESEARCH DIRECTIONS

The effectiveness measuring has been addressed in the last section and this remains to be studied in details in the future works. In order to understand how effective the VTC is, several factors should be further studies. At first, the goals and the choice of business model, followed by the analysis of the competitors. Effectiveness will give proper understanding for the marketing and community managers if the travel online community is meeting its goals and provides ways for optimizing it. In order to understand how effective were the digital marketing campaign, the focus of future research should also be on the entrepreneurial goals. They are quite different, ranging from (personal and business are often intertwined) which more effective digital marketing can help to facilitate. Future research also needs to focus on business sustainability and growth (customer acquisition and retention and combating seasonality), innovative marketing through digital media and autonomy (reducing dependence on marketing intermediaries), personal development, satisfaction in mastering digital marketing and technology and profitability-more effective and efficient marketing.

Future research needs to incorporate how SME companies can benefit from electronic commerce including access to information on tourism services, providing better information on tourism services, and providing convenience for customers.

It is expected that future research will prove that virtual travel communities can be very useful tools in achieving not only customer acquisition but most of all retention. In fact, it is due to the usefulness of the travel communities that people are brought together from different regions, united by their same interests or hobbies. The general idea of the community is sharing knowledge. In this regard, VTC not only helps travelers in making their travel decision but serves also as profitability centers. In other words, future research should explore whether higher level of satisfaction of its members leads to higher profitability of the communities.

CONCLUSION

The travel virtual communities represent an excellent opportunity for increasing the competitiveness of SMEs. The most important factor that needs to be taken into account is how *Internet* will be used by SMEs. Therefore, their choice of business model is very important. The steps of the strategy for applying virtual travel community are various and were elaborated by the author. They depend on the business. The chapter presents steps of strategy that could be taken by SMEs in the hospitality sector. However, the strategy presented can be applied to different companies from lines of business as well. The author presented the main steps to be followed by the entrepreneurs, operating in the *Internet era*. The chapter has further shown that virtual travel communities can optimize the processes in SME business organizations, not only externally but also internally. They provide fast feedback from customers. In this regard, the business organization can react faster and adapt to the requirements and needs of the customers. On the other hand, the community leads to optimization in the company inside. It is performed by close interaction between community managers and managers of other departments so that the company will function more effectively in the new *Internet* marketing paradigm.

Virtual travel communities provide an environment that facilitates users' interaction and search for devices. SMEs from the hospitality industry can change the way in which business is traditionally conducted, with travelers as co-creators and travelers as co-distributors. The travelers are becoming co-creators as they create the content in the travel community by posting photos, videos from the visited destination, by sharing their experience. As a result, a new SME business model has appeared the model of the community. It includes: the traveler, the business organization and the online tourism domain. It is created in most of the cases by the customers, and managed successfully later by the business organization. It is the key of success and future development of SMEs, particularly in the tourism industry.

REFERENCES

Abubakar, A., & Ilkan, M. (2015). Impact of online WOM on destination travel and intention to travel: A medical tourism perspective. *Journal of Destination Marketing & Management*. 10.1016/j.jdmin.2015.12.005

Amaro, S., & Duarte, P. (2015). An integrative model of consumers intentions to purchase travel online. *Tourism Management*, *46*, 64–79. doi:10.1016/j.tourman.2014.06.006

Browning, V., & Sparks, B. (2011). The impact of online reviews on hotel booking intentions and perceptions of trust. *Tourism Management*, *32*(6), 1310–1323. doi:10.1016/j.tourman.2010.12.011

Casaló, L. V., Flavián, C., & Guinalíu, M. (2011). Understanding the intention to follow the advice obtained in an online travel community. *Computers in Human Behavior*, *27*(2), 622–633. doi:10.1016/j.chb.2010.04.013

Dencheva, S. (2014). Electronic business models, adopted by the hospitality industry in Bulgaria. *IIBA conference,Istanbul Conference proceedings*, (pp. 311-317).

Gulbahara, M., & Yilidirima, F. (2015). Marketing Efforts Related to Social Media Channels and Mobile Application Usage in Tourism: Case Study in Istanbul. *Procedia: Social and Behavioral Sciences*, *195*, 453–462. doi:10.1016/j.sbspro.2015.06.489

Hoffman, D., & Fodor, M. (2010). Can you measure the ROI of your social media marketing. *MIT Sloan Management Review*, *1*(52), 323–343.

Jones, R., Alford, P., & Wolfenden, S. (2015). *Entrepreneurial Marketing in the Digital Age: A study of the SME tourism industry.* Retrieved August 7, 2016 from Bournemouth University web site: http://eprints.bournemouth.ac.uk/22932/1/Entrepreneurial%20Marketing%20in%20the%20Digital%20Age%20A%

Kavoura, A., & Stavrianea, A. (2011). Following and belonging to an online travel community in social media, its shared characteristics and gender differences. *Procedia: Social and Behavioral Sciences*, *175*, 515–521. doi:10.1016/j.sbspro.2015.01.1231

Kim, Ch. (2006). *E-tourism: An innovative approach for the small and medium sized tourism enterprises (SMTES) in Korea.* Retrieved August 16, 2016 form OECD web-site: http://www.oecd.org/industry/tourism/34268048.pdf

Kiralova, A., & Pavliceka, A. (2015). Development of social media strategies in tourism destination. *Procedia: Social and Behavioral Sciences*, *175*, 358–366. doi:10.1016/j.sbspro.2015.01.1211

Liu, Zh., & Park, S. (2015). What makes a useful online review? Implication for travel product websites. *Tourism Management*, *47*, 140–151. doi:10.1016/j.tourman.2014.09.020

Manchanda, P., Packard, G., & Pattabhiramaiah, A. (2014). The Economic Impact of Customer Participation in a Firm-Sponsored Online Customer Community. *Marketing Science*, *34*(3), 367–387. doi:10.1287/mksc.2014.0890

Mariani, M., Felice, M., & Mura, M. (2015). Facebook as a destination marketing tool: Evidence from Italian regional Destination Management Organizations. *Tourism Management*, *54*, 321–343. doi:10.1016/j.tourman.2015.12.008

Munar, M., & Jacobsen, J. (2014). Motivations for sharing tourism experiences through social media. *Tourism Management*, *43*, 46–54. doi:10.1016/j.tourman.2014.01.012

Nguyen, K., & Coudounaris, D. (2015). The mechanism of online review management: A qualitative study. *Tourism Management Perspectives*, *16*, 163–175. doi:10.1016/j.tmp.2015.08.002

Öztamur, D., & Karakadılar, İ. (2014). Exploring the Role of Social Media for SMEs: As a New Marketing Strategy Tool for the Firm Performance Perspective. *Procedia: Social and Behavioral Sciences*, *150*, 511–520. doi:10.1016/j.sbspro.2014.09.067

Preece, J. (2001). Sociability and usability: Twenty years of chatting online. *Behavior and Information Technology Journal*, *20*(5), 347–356. doi:10.1080/01449290110084683

Qu, H., & Lee, H. (2011). Travelers social identification and membership behaviors in online travel community. *Tourism Management*, *32*(6), 1262–1270. doi:10.1016/j.tourman.2010.12.002

Sanchez-Franco, M., & Cataluna, R. (2010). Virtual travel communities and customer loyalty: Customer purchase involvement and web site design. *Electronic Commerce Research and Applications*, *9*, 171-182. Retrieved August 26th, 2016 from Research Gate web site: https://www.researchgate.net/publication/220066649_Virtual_travel_communities_and_customer_loyalty_Customer_purchase_involvement_and_web_site_design

Standing, C., Holzweber, M., & Mattson, J. (2016). Exploring emotional expressions in word-of-mouth from online communities. *Information Processing & Management*, *000*, 1–12.

Timmers, P. (1998). Business Models for Electronic Markets. *Electronic Markets*, *8*. Retrieved August 26th, 2016 from University of Utrecht web site: http://www.cs.uu.nl/docs/vakken/ec/Timmers_BMem.pdf

Wang, Y., Shuang, S., & Li, D. (2015). Customer participation in virtual brand communities: The self-construal perspective. *Information & Management*, *52*(5), 577–587. doi:10.1016/j.im.2015.04.003

Zhong, D., & Luo, Q. (2015). Using social network analysis to explain communication characteristics of travel-related electronic word-of-mouth on social networking sites. *Tourism Management*, *46*, 274–282. doi:10.1016/j.tourman.2014.07.007

ADDITIONAL READING

Choi, Ch., Guoxin, L., & Elliot, S. (2013). Understanding service quality in a virtual travel community environment. *Journal of Business Research*, *66*(8), 1153–1160. doi:10.1016/j.jbusres.2012.03.011

Chung, N., & Koo, C. (2015). The use of social media in travel information search. *Telematics and Informatics*, *32*(2), 215–229. doi:10.1016/j.tele.2014.08.005

Gu, B., & Ye, Q. (2013). First Step in Social Media: Measuring the Influence of Online Management Responses on Customer Satisfaction. *Production and Operations Management*, *23*(4), 570–582. doi:10.1111/poms.12043

Hsu, L., Wang, K.-Y., Chih, W.-H., & Lin, K.-Y. (2015). Investigating the ripple effect in virtual communities: An example of Facebook Fan pages. *Computers in Human Behavior*, *51*, 483–494. doi:10.1016/j.chb.2015.04.069

KEY TERMS AND DEFINITIONS

Blogs: Short recordings that are chronologically ordered and contain media materials.

Conversion Rate: The number of customers who turn to buyers.

Electronic Business Model: An architecture for product, service and information flows, including a description of the various sources of revenue, business actors and their roles, their potential benefits for various business actors such as SMEs.

Likings: Criteria for people who express positive attitude towards a certain web page, i.e. they like it.

Personalization: Sending a personalized content to the user according to his/her preferences.

Social Networks: A virtual network aimed to establish relations among its users potentially supporting SME development.

Virtual Travel Communities: An aggregation of people able to exchange ideas, beliefs, experiences online through space and time with potential to support SME development.

ENDNOTES

1. Editors' note: The notion of strategy has been discussed in chapters 14 and 16 from other important points of view.

Chapter 5
Critical Review of SME Regulation Optimization in Serbia:
A Reflection on Harmonization with the EU Acquis

Olgica Milošević
Faculty of Applied Management, Economics, and Finance in Belgrade, Serbia

ABSTRACT

The European Union (EU) and the Republic of Serbia have recognized the importance of SMEs and are developing accordingly the institutional framework of this sector. In the process of EU accession, Serbia has to fulfil political, economic and legal requirements by implementing systemic reforms of its legal system in order to align it with the EU acquis. Some of these requirements have a direct or indirect impact on SME regulations, to which our chapter has been dedicated within a series of our papers on developing the SME regulatory framework. This paper will try to present the importance of innovating legal development of these companies. Furthermore, modern concepts will be introduced together with very important tools of legal forms of SME organization. The intention is to clearly define specific legal goals of SMEs, expressed through targeted legal forms. The modern regulatory framework relating to SMEs should combine the existing solid theoretical base and innovative models from the practice.

INTRODUCTION

The purpose of this chapter is to comparably examine the institutional framework and process that optimizes company law through the adoption of more uniform regulatory models and laws compatible with a policy agreed at a multilateral level of the European Union (EU). The constituent premise is that development of small and medium enterprises (SMEs) can benefit from this process. Given that 99% of all forms of businesses in the EU are SMEs, which provide 67% of jobs and another 85% in newly

DOI: 10.4018/978-1-5225-1949-2.ch005

created businesses, the EU therefore strongly encourages entrepreneurship and SME development (European Commission, 2016a).

Many studies highlight the value of examining legal, regulatory and administrative environment deemed favorable for SMEs. For example OECD (2004) study, featured, *inter alia*, the following:

- Property rights are clearly recognized;
- Contracts are easily enforced;
- A simple, transparent and low-compliance-cost tax system is operational and it is perceived as fair;
- Businesses are able to register with authorities through a simple and inexpensive system, preferably by remote access through the Internet;
- Business licensing requirements are minimized, and when they are enforced, the objective is to safeguard health and safety of consumers and labor rather than only being a source of revenue for local and/or central government;
- Labor regulations are balanced and flexible, protecting the rights of labor and the firm equally;
- SMEs, whether they are exporting or importing, interact with a streamlined customs administration that is efficient, simple and transparent;
- Financial sector regulations (banking, insurance, leasing) recognize SME constraints and provide legal and regulatory instruments that enable commonly available SME assets to be used as collateral;
- Public administrators at local levels appreciate entrepreneurs as contributors to economic growth, treat them fairly and are committed to limiting corruption to a minimum;
- Legislation and regulation is gender insensitive, rule of law and rules of the game apply equally to men and women;
- SMEs can easily set up and join membership organizations;
- Bankruptcy legislation does not impose unduly high penalties on the entrepreneur or the SME.

Clearly, in the field of the administrative burdens such as registration costs, efficiency of commercial courts and required permits it is recommendable to reduce the paperwork and cost involved in setting up SMEs.

Following a brief presentation of the EU approach to SME development and Serbia's progress (European Commission, 2016c), in this field the main context of this chapter is to examine the impact of some unique important principles of company law on SME growth and development. Part of the author's approach is also to critically analyze the most significant facilitating and restrictive factors as challenges or concerns for SMEs in both EU and Serbia. Consequently, the chapter will examine rewards and challenges associated with SME regulatory optimization and harmonization. To enhance this chapter, the author uses the comparative scientific method coupled with logical-axiological research and normative theory of law.

Following the above described content, the chapter discussion will present future research directions and is followed with a conclusion.

Table 1. Development of SME policy and regulatory framework

Institutional Framework	Legislative Simplification and Regulatory Impact Analysis	Public-Private Consultations
• SME definition • Availability of statistic data • Inter-governmental co-ordination in policy elaboration • SME development strategy • SME implementation agency or equivalent • Measures to tackle informal economy	• Review and simplification of current legislation • Legislative guillotine • Use of regulatory impact analysis • The SME test	• Frequency and transparency of public-private consultation • Formal influence of public-private consultations • Representativeness of public-private consultations

Source: Adapted from OECD (2012)

BACKGROUND ON THE EU SME APPROACH AND REVIEW OF PROGRESS IN SERBIA

A typical internationally recognized SME country mapping consists of three sub-dimensions, analyzing these main aspects (OECD, 2012, pp.98-105) portrayed in Table 1.

As an example, let us consider the Republic of Serbia, which began its EU accession process in 2005. It is currently in the negotiation phase, involving 35 chapters. Serbia's task is to fulfill political, economic and legal requirements by implementing systemic reforms of its legal system in order to harmonize it with the EU *acquis* (European Integration Office, 2014). Some of these chapters have a direct / indirect impact on the development of entrepreneurship and SMEs, such as the Company Law, Taxation, Social Policy and Employment, Enterprise and Industrial Policy, Financial and Budgetary Provisions, etc. Although most of these chapters have not been officially "opened", the latest European Commission's Serbia 2015 Report unequivocally shows that Serbia has made a major step forward and is on the right track (European Commission, 2015a). In building SME development policy Serbia is harmonizing regulations by following Small Business Act (SBA) principles. According to (SBA) Fact Sheets, which present an assessment of progress in SBA implementation at national level, the European Commission stressed that administrative and regulatory requirements need to be reduced and simplified in Serbia. Further support is needed for the internationalization of SMEs, focusing on the sectors whose good performance stands out. Access to finance is still generally poor. Bank lending, crucial for SMEs, has become less available. Interest rates have lowered, but remain high. Therefore, alternative sources of financing should be developed. The methodology for the *'SME test'* exists, but needs to be applied systematically (European Commission, 2015b).

As jointly assessed by the Organization for Economic Cooperation and Development (OECD), European Commission and other relevant European institutions, Serbia's key priorities in 2016 are economic competitiveness and export-led grow (OECD, 2016, pp. 311-329). The key mid-term priorities are to establish a progress monitoring system, implement defined SME policy and measure its effectiveness in practice. Building on a well-developed institutional framework for strategic SME policy Serbia has improved its SME environment, especially regarding company registration and e-government, while focusing on the need to analyze results of numerous trainings held on SMEs. Significant progress is seen in promoting SME innovativeness so as to enable entrepreneurs to actively participate in international cooperation programs, which is definitely a big step in the right direction. Access to funds is still a stumbling block for SME development and needs to be strengthened by expanding regulations in key areas such as permits and licenses. Another disadvantage is the fact that environmental policy in the SME area has not been implemented yet and is not considered to be an integral part of the environmental policy.

Obviously, many obstacles to SME development have been eliminated, but enormous work is yet to be done before full alignment of national legislation with the EU *acquis* in this area. In fact, results of focus group discussions and in-depth interviews conducted by the United States Agency for International Development (USAID) pointed out the following problems of Serbian SMEs regarding business environment, duties, taxes and contributions, as shown in the Table 2 (USAID Serbia, 2015).

In summary, fiscal levies need to be optimized and transparent support provided to the private sector by the government, with a focus on effective and horizontal objectives, such as SME support. It is also noted that SMEs mostly "suffer" from poor financial status and liquidity, as well as from regulatory unpredictability. SMEs have a limited access to different financing sources, especially at the beginning of their business operations when there is a lack of reliable data on financial and other features of the business and when company's credit solvency is weak (Beraha, 2016, p. 56).

In order to coordinate activities of various ministries and other organizations to promote entrepreneurship and develop SMEs, Serbian Government formed a Council for SMEs, Entrepreneurship and Competitiveness. This interim working body is composed of representatives of line ministries, businesses and research community. Furthermore, in March 2015 the government adopted the 2015-2020 Strategy to Support Development of SMEs, Entrepreneurship and Competitiveness and the Action Plan for its implementation, fully in line with the EU policy in this area i.e., the SBA for Europe from 2008 (Ćoćkalo, Đorđević, Sajfert & Bogetić, 2011. p. 450). The Strategy's firm baseline reflects information and recommendations from reports and analyses by international institutions and projects, the European Commission and national institutions. Its introduction emphasizes three reasons for its adoption: defining mid-term measures to promote entrepreneurship and competitiveness in Serbia in line with the state of affairs and real needs of the SME sector in Serbia; continuing and building upon the existing policy on developing entrepreneurship and competitiveness; and harmonization with the EU policy in this area. Six strategic goals – *Strategy pillars* – are clearly defined:

1. Boost business environment;
2. Improve access to financing sources;
3. Continuously develop human resources;
4. Build sustainability and competitiveness of SMEs;
5. Enhance access to new markets, and
6. Develop, promote entrepreneurial spirit, stimulate women, youth entrepreneurship and social entrepreneurship.

Table 2. Serbia SME problems with business environment, duties, taxes and contributions

General Business Environment	Taxes, Contributions, Fees
Lack of certainty and predictability	Value-added tax (VAT)
Distrust in state institutions – the judiciary	High labor taxes and contributions
Distrust in state institutions – public authorities	Advance payment of corporate tax
No exchange of information and coordination between authorities	Para-fiscal charges: number and amount
E-Government solutions are helpful	Administrative burdens

Source: Adapted from USAID Business Enabling Project (2015a)

Table 3 illustrates how the Serbian strategic goals are linked with the *Small Business Act for Europe.*

Comprehensive regulatory reform is an ongoing process that has to take place at all government levels, and its success depends on effective implementation of adopted regulations. Obviously, the very process of passing regulations has to be enhanced by increasing transparency and predictability of regulatory activities, and by enabling businesses and citizens to participate more in the adoption of new regulations. For example, the SME definition (see key terms and definitions) used by the authorities is only partially in line with the definition recommended by the European Commission as it does not cover affiliated and partner companies. The SME definition that came into effect on 1 January 2005 is based on three criteria: number of employees, total annual revenue and total assets (balance-sheet assets), and a company needs to meet the criterion relating to the number of employees plus one or both other remaining criteria. Micro businesses also include entrepreneurs operating independently (Stamatović & Vemić, 2009, p. 882). It would, however, be beneficial for SMEs in Serbia if the EU recommendations would be fully implemented in a single legal act, because that would enable them to receive treatment by the authorities and EU programs, for which this definition is binding.

More recently, Serbia's progress in the process of alignment with the EU *acquis* has been demonstrated in several ways. The Executive Summary of Business Survey Serbia 2015 conducted by USAID shows results achieved in implementing the reforms but also a number of important issues remaining (see Table 4).

Table 3. Links between strategy to support development of SMEs, entrepreneurship and competitiveness (2015-2020) and the small business act for Europe

Small Business Act for Europe	Support Development of SMEs, Entrepreneurship, and Competitiveness
Principle 1 - Create an environment in which entrepreneurs and family businesses can thrive and entrepreneurship is rewarded	Pllar 3 - Continuously develop human resources Pillar 6 - Develop and promote entrepreneurial spirit and stimulate women and youth entrepreneurship and social entrepreneurship
Principle 2 - Ensure that honest entrepreneurs who have faced bankruptcy quickly get a second chance	Pillar1 - Boost business environment
Principle 3 - Design rules according to the "Think Small First" principle	Pillar 1 - Boost business environment
Principle 4 - Make public administrations responsive to SMEs' needs	Pillar 1 - Boost business environment
Principle 5 - Adapt public policy tools to SME needs: facilitate SMEs' participation in public procurement and better use State Aid possibilities for SMEs	Pillar 1 - Boost business environment Pillar 2 - Improve access to financing sources Pillar 4 - Build sustainability and competitiveness of SMEs
Principle 6 - Facilitate SMEs' access to finance and develop a legal and business environment supportive to timely payments in commercial transactions	Pillar 2 - Improve access to financing sources
Principle 7 - Help SMEs to benefit more from the opportunities offered by the Single Market	Pillar 5 - Enhance access to new markets
Principle 8 - Promote the upgrading of skills in SMEs and all forms of innovation	Pillar 4 - Build sustainability and competitiveness of SMEs
Principle 9 - Enable SMEs to turn environmental challenges into opportunities	Pillar 4 - Build sustainability and competitiveness of SMEs
Principle 10 - Encourage and support SMEs to benefit from the growth of markets	Pillar 4 - Build sustainability and competitiveness of SMEs Pillar 5 - Enhance access to new markets

Source: Adapted from Serbia Strategy to Support Development of SMEs, Entrepreneurship and Competitiveness (2015)

Table 4. Progress of reforms in Serbia 2011-2015

Significant Improvement	Some Improvement	No Improvement	Worse
• Wage taxes and contributions • Obtaining construction permits • Obtaining business licenses • Inspections surveillance • Enforcement of contracts • VAT payment procedures • Corruption in public administration • Inflation • Exchange rate volatility	• Court procedures & requirements • Para-fiscal charges • Administrative procedures • Court procedures & requirements • Labor law procedures & requirements • Customs and external trade procedures • Payment transactions • Public Procurement • Interest rate volatility • Difficulty in getting a loan • Bank collateral requirements • Interest rates and fees • Gray market employment • Awareness of regulatory changes	• VAT rates • Customs duties • Duration of loans • Duration of loan approval process • Banks' reporting requirement	• Use of external finance • Bank lending • Quality of bank services/ support • Excise taxes • Asset price volatility

Source: Adapted from USAID Business Enabling Project (2015b)

Serbia's progress is also reflected in conclusion of an agreement with the EU confirming Serbia's participation in the COSME Program for the Competitiveness of Enterprises and SMEs until 2020. This is an EU-funded program implemented in the period 2014-2020 with a budget of 2.3 billion Euros to support SMEs in improving access to financing sources, support in internationalization and accessing foreign markets, creating a stimulating business environment for development of competition, and encouraging the spirit of entrepreneurship. The Agreement with Serbia went into effect on 1 January 2016 (COSME, 2016).

In the next, central part, of the chapter the author will consider the impact of some key regulatory principles relevant for SME development in the EU, as well as in Serbia.

AN OVERVIEW OF SERBIAN COMPANY LAW IMPACT ON SME GROWTH AND DEVELOPMENT

This section focuses specifically on the use of key company law principles of relevance for SMEs. The author looks at the theory and practice of company law as well as the main benefits and threats associated with the treatment of SMEs.

The Principle of Completeness of All Forms of Companies

Serbia Company Law (Serbia Company Law, 2015) defines 4 legal forms of companies without leaving the possibility of incorporating a company in a form not regulated by the Law. The Law regulates companies' legal status, including in particular their incorporation, management, status changes, changes of legal form, dissolution and other issues relevant for their status, as well as the legal status of entrepreneurs. Provisions relating to entrepreneurs, who do not have a company status i.e. legal entity status, but are, instead, natural persons, make an impression that they accidently got left out from this Law.

Companies are usually classified into two categories: partnership (general partnership (GP) and limited partnership (LP) and corporation (limited liability company (LLC) and joint stock company (JSC). Their characteristics are presented in Table 5.

Table 5. Ordinary classification of companies in Serbia

Characteristics	Partnership	Corporation
Established in view of the member' status	✓	-
Members are involved in the economic activity	✓	-
Members run business operations directly	✓	-
Minimum capital assets prescribed by law	-	✓
Members are liable for company's obligations with their personal property	✓	-

Source: Own presentation of author

As the above legal forms of companies are defined by the company law in line with the principle of completeness, they should as such have a facilitating effect on legal development of SMEs. One paradox is noticeable, however, that despite the generally accepted term “small and medium-sized enterprises” this law does not explicitly recognize the word “enterprise”, although entrepreneurs are mentioned, which the author does consider inhibiting for entrepreneurship.

The Principle of Essentiality of Intention to Demarcate SMEs as Companies from Institutions

Serbian company law clearly defines the company as a legal entity engaged in an economic activity for the purpose of profit generation, which is also the aim of SME establishment. On the other side, there are also other forms of association into institutions and companies for the purpose of organizing public services. They pursue economic activities that fulfill the needs of citizens and organizations and other legally determined interests in certain areas, and are also regulated by the Law on Public Services (Serbia Law on Public Services, 2015). This way a clear distinction was made between different forms of potential organization based on the purpose they are incorporated for, which shows that optimization is the purpose of the principle of essentiality of the intention to demarcate SMEs as companies from the so-called institutions.

The Principle of Equality of Companies

The principle of equality derives from the constitutional principle of equality and independence of companies. It concerns organizational forms of all companies and enables property and management components of the ownership function, regardless of the ownership basis. This principle is of particular importance from the aspect of economic activity individual companies can engage in and is reflected in easier market access. Introduction and popularization of European business and labor ethic, coupled with legal equality of all companies, increase the chances of SMEs to position themselves better on the market and distinguish them from other companies. This is facilitated by the well-known advantages of SMEs including, among others, better flexibility and dynamics in adjusting to demands of the market and new technologies, and their quick response to changes in demand (Ožegović & Pavlović, 2012, p. 75).

The Principle of Tripartite Interest of the Majority / Minority Owners, Creditors, and Employees

The tripartite principle embraces the modern thinking of comparative law and business practice, according to which in terms of interests a company belongs not only to its owners but to creditors and employees as well who all have certain specific rights set in the company's managing structure. In fact, in modern economy this concept grows from tripartition to multipartition. Therefore, along with the above interests included are also social interests such as the interest of clients and even the very society sociologically speaking, which all together strengthens SME organization.

The Principle of Compatibility of Legal Forms in Developed and Transition Countries

As mentioned earlier, Serbian legislation "knows" only 4 company forms regulated by the Company Law, which also embraces definition of the entrepreneur, as a physical person. This closed system of company forms, the so called *numerus clausus,* may in fact have a restrictive effect on SME development, because it fails to recognize some company forms present in the EU. Therefore, the principle of comparability of legal forms of companies in developed and some transition countries is not favorable enough for SMEs. At present this legislative approach does not have a major impact on SMEs in Serbia because most companies in the EU are also incorporated as limited liability companies. However for a fully-fledged EU member and market economy company legislation needs to be harmonized so as to introduce new forms of companies.

The Principle of Enhanced Liability of Management Board, Controlling Members, and Shareholders

Liability of corporation members is brought down to liability for company's obligations, but only in the amount of invested capital, for they are not liable for company's obligations with their personal assets. An exception is the case when owners use the company as a "shell" for their own unfounded enrichment. This is most often manifested by mixing personal and company's assets, in using company assets for personal purposes and personal benefit, fraudulent abuse of company and its assets and similar actions, followed by untruthful or incomplete presentation of the status of assets in business records. Candidates for the abuse of legal personality include, among others, one-member and no employee companies, companies with close links among members with family ties and highly undercapitalized companies. Abuse of legal personality is a relatively new legal institute and its most common form involves deceiving creditors and abuse institute was invented as a "defensive mechanism". Activation of this institute can be demanded if the charged person:

1. Used the company to achieve otherwise prohibited objectives;
2. Used or disposed with company's assets as their own personal property;
3. Used the company or its assets to cause damage to company creditors;
4. Reduced the company's assets for personal gain or gain of third parties, although they knew or ought to have known the company would be unable to meet its obligations.

In judicial practice this institute has been applied very rarely and very restrictively. It should be regulated in more detail in order to encourage courts to apply it more effectively.

Director's liability is regulated by limitation of authority, prescribed level of due diligence according to which the director is bound to carry out his/her duties in good faith and in a reasonable belief that he/she acts in the best interest of the company. He has an obligation to report personal interest and avoid conflict of interests, keep business secrets confidential and respect disloyal competition.

Status, obligations and responsibilities of the controlling members are defined in detail. A party is the controlling member of the company always when such party owns, independently or with related parties, the majority share in the company's capital assets. The controlling members / shareholders bear unlimited joint and several liabilities for company's obligations even upon deletion of the company from the register. Various limitations have been prescribed for the purpose of protecting company's members and creditors such as, e.g., prohibition of subscription of shares – company's shares cannot be subscribed by its controlled company, nor by a third party acting in its own name and for the account of the controlled company. The controlling company is held liable for damages suffered by the controlled company due to actions upon binding instructions on how to run business, guided by the group's interests. The controlling shareholder having shares that account for minimum 90% of the company's capital assets is bound to buy shares of each of the remaining shareholders at a price set in the same fashion as in payment to disagreeing shareholders. All above protective measures enhance SME development by strengthening the legal institute of liability of the above mentioned groups of parties.

The Principle of Transparency for Company Operations

The transparency principle is highly positioned in the hierarchy of essential principles of the company law. In fact, ensuring transparency is one of the two goals for adoption of the Code of Business Ethics (Serbia Code of Business Conduct, 2006). In addition, transparency of financial information is a legal requirement for participants in the capital market (Todorović, Bunić & Todorović, 2013, p.11). Protecting investors' interests by disclosing relevant and reliable information is one of the segments of harmonization of regulations within the EU.

The transparency principle means that along with the traditional set of financial statements, other reports are also made available to the public in line with the accounting standards. The annual audit should be published by an independent and qualified auditor while external auditors should be responsible to shareholders providing information equally to all participants. This approach builds corporate social responsibility ensuring transparency towards all stakeholders and creating a long-term value through protection of interests of company members/shareholders. Therefore, this principle has a strong impact on building organizational development of SMEs.

The Principle of Modernity, Sophistication, and Liberal Enactment

The Serbian legal system belongs to the system of European continental law, but it also incorporates some institutes of Anglo-Saxon law. The new Company Law is a modern piece of legislation largely aligned with the EU company law. The principle of liberal enactments serves specific interests of all investors who can operationalize their business idea without the claws of imperative norms. Enactment norms are predominant in the law and offer freedom of choice of a solution for each particular situation regardless of the form of company, except partially for the public joint stock company. There is freedom of choice in relation to the solution foreseen by the norms, giving an opportunity to the interested entities to regulate their rights and obligations by themselves differently than as set forth by the law in accordance with the principle of autonomous will. Therefore the Serbian company law should contribute to creation of a legal system complementary to the internal EU market. It facilitates growth and development of SMEs because it optimizes the regulatory framework of SMEs and is in legal correlation with SME Development Strategy and some laws such as the accounting law (Serbia Accounting Law, 2013), and business registration law (Serbia Law on the Procedure of Registration with the Business Registers Agency, 2014). However, in the author's view, the present version of the procurement law (Serbia Public Procurement, 2015) still fails to offer liberal enabling provisions supporting SME development.

ANALYSIS OF FACILITATING AND INHIBITING REGULATORY FACTORS OF SIGNIFICANCE FOR SME DEVELOPMENT IN EU AND SERBIA

In a comparative analysis of company law, especially in the part relating to regulations of some essential issues of corporations, different factors with a substantial impact on growth and development of SMEs in the EU and Serbia can be distinguished. They represent unique opportunities and challenges for SMEs. Therefore, in this section of the chapter the author brings attention to the factors that can either enable or inhibit successful SME development applying the comparative method of scientific research.

Procedure for Establishment, Registration, Change of Legal Form, Mergers, and Acquisitions

This is a factor revealing unambiguous legislative progress in Serbia in legal gravitation towards the EU, with some inhibiting aspects. The Serbian Business Registry Agency (SBRA), which controls all registers at the central level, has increased the number of registers from 15 to 17. Although the procedure of filing the registration application is substantially simplified and the costs of establishing a company significantly reduced electronic company registration is still not in full operation. Optimal time for company establishment is now significantly shorter and it takes just a few steps, meaning that a business can be started up in 24 hours (European Commission, 2015c). The minimum amount of LLC's capital assets was lowered (most common SME form) from 500 EUR to 100 RSD (counter value ~ 1 $). This may makes it easier to start a business, but it does not provide sufficient protection of creditor' rights.

The possibility of changing a legal form and approaching mergers and acquisitions (M&A) represents a great opportunity for growth and development of SMEs through synergy, cost reduction and diversification. Legal aspects of M&A could contribute to enlargement, growth and development of SMEs and with proper assessment of tax risks and burdens can be optimized. Fusion is regulated in great detail by EU and Serbian Company Law, but the intended protection of interests of all categories of affected parties has not been achieved. The protection of rights of stakeholders is high, but there is insufficient protection of employees and stakeholders in ensuring preservation of their past status.

Rights and Obligations of Company Members and Shareholders from the Standpoint of Creditors and the Principle of Fiduciary Duties

The issue of company creditors' protection is always one of the burning issues of economic stability in any country. These issues result in double threat to SMEs in Serbia, both as creditors and debtors. Competitors from the ranks of "big players" can abuse their dominant position to push the SMEs into bankruptcy by not paying invoices, without the true background of their ruin ever becoming known while the bankruptcy proceedings do not provide adequate opportunities for their reorganizations and a new beginning. The new solutions in the Bankruptcy Law (Serbia Bankruptcy Law, 2014) demonstrate readiness of Serbian institutions to create an environment in which creditors' interests would be protected. The Law has enabled legal entities to respond timely and in the most favorable fashion in case of business troubles in the interest of the legal entity and that of the creditors and the company as a whole. It introduces the *Prepared Reorganization Plan* institute for the first time. Amendments to the Company Law also set the imperative legal solidary liability of company's members with "their own assets" proportionately to their share in the company's capital assets. This legal solution is not grounded in the very nature of this company. According to Vasiljević (2012) "it is entirely arbitrary and reflects the understanding of absolute omnipotence of legislation, which does not respond to itself, logic or philosophy of law" (p.12). Some blanket provisions of this Law enable high discretion, for instance, in case of inadmissible payments to a company's member by the company, when the member is liable to return such payments to the company and the company cannot release him/her from that obligation, and with this such obligation can be claimed from a *bona fide* company member only if necessary to settle the claims of company's creditors. Protection of creditors in case of reduction of the company's capital assets is brought down to a written notification of creditors by the company on the decision to reduce the company's assets, who have the option to request protection of their claims, as in the case of status change (Jevremović-Petrović, 2011, p. 248).

The principle of fiduciary duties is a clause of loyalty of members of company bodies and controlling shareholders and members towards the company prescribed by the Code of Business Ethics. It stipulates an obligation to act in the company's best interest (International Finance Corporation, 2008). In author's opinion the above factors should facilitate SME development.

Minimum Founding Capital

The minimum founding capital1 for LLC has an inhibiting impact on SME development because the law stipulates the minimum amount of only 100 RSD (around 1US $), unless a higher amount of capital is foreseen by a special law for companies engaged in specific economic activities. The good side of this solution is that there is no need to assess the value of contributions forming the minimum part of capital assets, in view of the fact that the contributions are in the form of money. This further means that when subscribing a contribution to the company its value cannot be overestimated, which is the fundamental drawback when a non-pecuniary contribution is subscribed to the company. An argument against such solution is that conducting economic activity does not require just money, but also existence of certain assets and rights, so their exclusion from the value of minimum part of the capital assets is not founded, especially in case of a non-pecuniary contribution, which can be acquired only from the founders, but not from third persons. This provision may have a stimulating effect on establishment of new SMEs, but cannot suffice as security to company's creditors, because it is commonly known that the higher the company's capital assets, the higher creditors' security, bearing in mind that a LLC member is not liable for company's obligations with own personal assets. Hence, there is also negative impact on implementation of the bankruptcy law.

Improvement of Corporate Governance and Development of Employee Participation in Management

Improvement of corporate governance, development of employee participation in management has a facilitating SME impact. All companies in Serbia, regardless of their legal form, are subject to a comprehensive package of regulations. In the author's view further development of corporate governance for SME development is recommendable. Collective participation of employees in management consists of three segments:

1. Consultative participation enabling employees to give opinion on certain issues, with final decision made by the employer;
2. Actual participation where special working teams organise work, work independently and make nearly all decisions on their own in conditions of limited oversight;
3. Participation through representatives allows employee participation in the work of managing and supervisory boards and ensures other forms of participation.

Despite the comprehensive regulation of trade union organization and prescriptions from the Labor Law employee participation is not sufficiently developed in Serbia. It works only to some extent in the public sector, while the trade union organization in the private sector is very poorly developed. Given that the General Collective Agreement ceased to be valid back in 2011 (Serbia General Collective Agreement, 2009) and no agreement was reached in a tripartite dialoge on a new one, and that collective agreements are rare in the private sector, protection of workers in Serbia is reduced to the minimum prescribed by the Labor Law. Social dialogue is still an issue because trade unions representatives abandoned the negotiations on amending the Labor Law (Serbia Labor Law, 2014). The law is now aimed at enabling easier employment, flexibility and mobility of Labor force, enlargement of trade unions and increasing convergence with the EU legislation have in fact deteriorated the position of employees. Since all levels of employee participation in Serbia are below the satisfactory level, the author takes the view that institutional measures need to be taken for the development of participation such as in the French model or through co-decision of employees as in Germany and Sweden.

Valuation of Non-Monetary Contributions of the Founders

Valuation of the non-monetary contributions that may be exercised by the founders themselves has a facilitating impact on the development of SMEs in all forms of companies, with the exception of public joint stock companies where the law lists the competent entities to carry out this assessment (certified court expert, auditor or other professional authorized by the competent authority of the Republic of Serbia to perform valuation of certain assets or rights). On the one hand this factor increases cost of establishment, and on the other provides greater protection of shareholders. However, the legislator leaves an option to future company members to entrust valuation of assets and rights to an authorized assessor.

Clear Demarcation of Open and Closed Societies

Clear demarcation of open and closed societies seems to have a restrictive impact on SME development, and a regulatory step backwards was taken in this field. The previous company law clearly distinguished between open joint stock companies, established through public invitation for subscription and payment of shares and the closed where shares were issued only to the founders or to a limited number of other people. The current Serbian law recognizes only public and non-public companies, without a clear demarcation and uses poor legislative technique which makes it difficult to "navigate" in the law.

There are views (Jovanović, 2011, p. 319) that this Law was adopted for one time use during the transitional period of Serbia's EU accession, until a better one is adopted. The company law also has a legal gap with regards to the market price at which a company is bound to repurchase shares from disagreeing shareholders.

Regulation of Ownership and Control Functions, Protection of Minority Shareholders and Members of Corporations

The analysis of the Company Law, Capital Markets Law on and the Law on Takeover of Joint Stock Companies shows that these Serbian laws are now fully aligned with the EU Directives in the area of minority shareholders' rights (Đulić & Kuzman, 2012, p. 96). Therefore, at present, regulation of Ownership and Control functions, Protection of Minority Shareholders and Members of Corporations can have a limited facilitating effect on the development of SMEs, and is a direction to consider in future improvements to the company law. Hence, according to the World Economic Forum's Report Serbia holds the 94th place out of 144 countries covered by this competitiveness Report (World Economic Forum, 2016).

Obiovusly, a good legal framework is not sufficient for shareholders to enjoy effective protection if the regulation is not effectively implemented in practice. Since voluntary enforcement of law is clearly missing, the importance of forced enforcement is multiplied, and is manifested in the form of private enforcement by courts or public enforcement by regulatory and supervisory bodies such as the Securities Commission. A defect of this law is reflected in the fact that private persons are at will to decide whether to initiate enforcement procedure in case of violation of the rules, meaning that protection depends on citizens' private initiatives. This system is clearly not yielding intended results, taking into account that citizens' private initiative largely depends on their perception of the efficiency of courts and the legal system in general, costs, clarity of rules, burden of proof and incentives given to them. A detailed analysis of the practice of public and private enforcement needs to be conducted in the near future so as to identify issues and obstacles and define steps to eliminate them (Đulić & Kuzman, 2012, p. 95) in order to facilitate SME development.

Regulation of essential factors of interest for functioning of SMEs is certainly important, but development of new legal forms is equally important. For the reason, the next section of this chapter the author will focus on further development of new forms of companies in the EU and their possible introduction in Serbia.

FUTURE RESEARCH DIRECTIONS AND INTRODUCTION OF POTENTIAL NEW LEGAL FORMS OF BUSINESS ORGANIZATION FOR SMEs

The author believes that it is very important to continue with research and development of SME regulatory framework in the EU and Serbia. Notably, the European Company, European Economic Interest Grouping (EEIG) and European Cooperative Society already exist in EU member states. The author provides a critical review of these forms of organization and linking of companies that could have a particular importance for medium-sized companies in both EU and Serbia.

The European Company

The European Company (also known by its Latin name *Societas Europaea* or SE) is a type of public limited-liability company regulated under EU law (European Commission, 2016b).

The European Company as a legal institute was introduced in 2004, but has so far been used by a small number of companies for various objective reasons. According to available data of the European Company Database only 2,125 SEs were established until March 2014. The number of European companies is presented in Figure 1.

There are several features differentiating this joint stock company from other company types. First of all, it is a legal entity, a business company and a corporation, with its capital assets divided into shares, shareholders not liable for company's obligations and it has a statute as the articles of association. Founders of SE can only be the existing legal entities, and physical persons cannot appear in the founder's role. SE can be formed only in one of the foreseen ways: merger of two or more existing joint stock companies, formation of the SE as a holding company by a joint stock company, formation of the SE as a daughter company by any companies Conversion of an existing joint stock company established following regulations of a Member State can be converted under certain conditions into SE and formation of SE as a daughter company by the existing SE. Once formed, an SE can also found other SEs.

Figure 1. Geographical distribution of European Companies in EU (per country)
Source: *Own presentation of author*

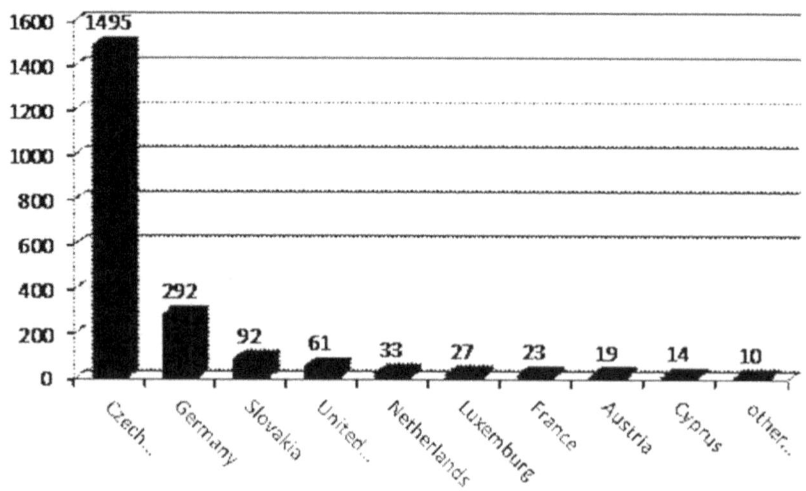

Its main advantage is the possibility of company's restructuring across borders into a form of cross-border merger, which was impossible until by some Member State laws, and there were insuperable difficulties. European companies thus adjust to the economic reality of a single market without meeting strict and complicating conditions of national laws, without costs particularly in relation to liquidation and new establishment in another country (Jeveremović-Petrović, 2007, p. 403). Many economic analysts deem that this institute suits larger companies and is discouraging for SMEs as it can be established by a minimum of two companies from two EU countries with a minimum capital of 120,000 Euros. As noted (Stefanović, 2011, p. 238-244), this automatically eliminates the vast majority of SMEs for which this founding capital is too high.

European Economic Interest Grouping

European Economic Interest Grouping (EEIG) is another form of interest linking of companies for a limited time on a status-legal basis with a view of facilitating and developing performance of the economic activity of members of SME associations and persons as professionals engaged in liberal professions and for establishing a cross-border cooperation. This could benefit the growth and development of SMEs from Serbia (Jevremović-Petrović, 2005, p. 8) and across EU when cooperating with neighboring countries. It is an excellent opportunity for international association not requiring high costs and is not complicated as other possible forms of association such as mergers, acquisitions and joint ventures.

Another advantage of this form of company is creation of links in different EU Member States without losing own legal personality or independence. This means that EEIG represents a legal entity acting on its own behalf and account in business transactions. Inexistence of regulations at its minimum capital assets does not lessen the legal certainty of its creditors, because unlimited joint and several liability of its founders for company's obligations is prescribed. It is important to mention that EEIG is not a simple association, not only for its formal legal personality but also because its founders retain independence in business. They do not transfer their activities and business to the EEIG, nor do they harmonize them with the EEIG's (Stefanović, 2011, p. 261).

EEIG is usually formed in the services sector such as legal practice, public notaries, telecommunications, financial services and real estate transactions. This supra-national legal creation can be founded by minimum two members (physical or legal persons, national or foreign) from at least two EU Member States and as such it facilitates cross border SME development within the EU. It is established in accordance with national legislation of the member state where it is seated upon its entry in the corresponding register based on the agreement on foundation.

The European Cooperative Society

Another supra-national legal creation can be established in the EU titled *The European Cooperative Society* (SCE, for Latin *societas cooperativa Europaea*). As its aim of establishment is to perform certain economic activities in the interest of its owners, not to penetrate the market for profit generation, it has no direct impact on SMEs but indirectly it does. *The European Cooperative Society* may operate throughout the European Economic Area if established by 5 or more persons residing in different member states or by legal entities (i.e. SMEs) established in different member states and its subscribed capital shall not be less than EUR 30,000.

Practice shows that for development of their activities on the territory of other EU Member States SMEs use the generally known form of LLC, based on national law of the country in which they want to perform certain activities i.e. where they have their seat registered. It is a form of trading company most frequent in practice and particularly suitable for SMEs. Interestingly, the EU does not have a special regulation for LLCs because harmonization of legal norms relating to them was never proposed or initiated in the EU bodies. For that reason, this form of companies is subject to national law of the country of their establishment and harmonization is recommended.

Every systemic change of legislation, such as SME regulation optimization, usually has its threats and opportunities. For some of these changes it may be very difficult to define the category they fall in, e.g. with EU membership Serbian SMEs will join the single EU market, which will generate business opportunities for some SMEs while bringing "the end of business" to others for they may not be capable of meeting high technical and other standards regulated by the EU or handling the pressure of increased competition. Advantages of SME regulation optimization are numerous because they are implemented according to plan in line with a national SME Development Strategy. However, harmonization of national legislation with the EU *acquis* also bears certain costs, which could be categorized as follows according to the Standard Cost Model (European Commission, 2015a):

1. **Familiarization Costs:** Those incurred in the course of developing and understanding legislative requirements;
2. **Adoption Costs:** Representing the cost required changes to comply with legislation;
3. **Administrative Costs**: Representing the cost the administrative obligations required by the legislation.

Since there is no unique methodology to assess these costs, it is very difficult to determine them, especially since a difference should be made between costs existing without legislative harmonization (i.e. „business as usual costs") and those incurred due to harmonization (i.e. „administrative burden").

If change is made in Serbia through harmonization with EU law, optimization consequences will be numerous because they affect development of the entire Serbian economy whose prime mover is SMEs. Regardless of negative effects that may appear, the author finds them negligible as compared with the expected benefits.

CONCLUSION

The basic SME characteristics enable them to adjust faster and flexibly to dynamic conditions, changes in consumer needs and demands on the global market. These are also the sources of problems and obstacles restricting their development and not rarely endanger their survival. Experiences of developed countries show that a significant impact of SMEs on growth and employment can be expected only in an organized and stimulating business and legal environment suitable for their foundation, improvement and stability of business. This requires adoption of sound regulations and their efficient enforcement, which will enable fast, professional and objective resolution of problems they SMEs encounter in doing business. Growth and development in the EU and Serbia require development of SMEs whose competitiveness on the global market will be based on knowledge, use of new technologies and innovations. In order to attain this goal in practice, readiness of the SME sector to conquer a market respecting adherence to necessary standards is of extreme importance.

On its way to the fully-fledged EU membership, Serbia has taken serious steps to meet the necessary requirements, some of which directly or indirectly relate to SME development. A comprehensive reform of legislation is currently ongoing for the purpose of its alignment with EU regulations, which goes in favor of SME development in Serbia, since the EU emphasizes this sector by strongly supporting its growth and development.

Analysis of SME regulatory development has shown that relying only on the existing legal forms is not enough to fulfill the needs of this sector and its expansion at growth rates necessary to maintain their share in the market. Finding and developing optimal regulatory forms is the key of further SME success.

The EU and the Serbian Government in particular, need to consider and propose a series of legislative and regulatory measures, instruments and institutions so as to improve the SME regulatory framework. Those regulations that are not operating in accordance with the legal intentions need to be innovated. Therefore, governments, agencies and institutions with the public support should improve the SME sector through regulatory development, tax reliefs, exemption from general economic-legal regulations and granting a relatively favorable status to SMEs. Hence, it can be justifiably assumed that the process of modernization of the SME regulatory framework will continue.

The assumption of continued process of developing legal regulations stems from an estimate that the low level of growth and development of SMEs in countries like Serbia persists. Increased quality of the institutional structure, along with the reform of legislation on financial markets, is a vital precondition for changes in the quality of SME development. Indeed, company law, stock exchange law, securities law, competition law and bankruptcy law analyzed by the author should be better harmonized to serve both SME development and Serbia's EU accession, instead of being obstacles. The fact that 2016 is declared the "Year of the Entrepreneur" shows that Serbia is on the right path. This project has various state programs to build entrepreneurial strengths, assist entrepreneurs starting or developing a started business, and provides financial and non-financial support for long-term development of the entrepreneurial spirit.

SMEs in the EU and Serbia are an important sector of the national economy, which is why the economic policy with other segments of the "policy mix" should support its development. There isn't a unique and universal recipe on how to proceed with legal harmonization for SMEs. One of important recommendations refers to optimizing SME regulatory framework and cooperation with line ministries and development agencies. The Governments' role is to increase institutional and regulatory support to SMEs, offer new instruments and better inform the sector, especially on how SMEs can develop new legal forms of organization.

This chapter examined an area that is extremely important – SME regulatory optimization in the EU and its accession countries such as Serbia. Further, this chapter highlighted the theoretical basis for cross border SME cooperation. Each of these terms were presented from a theoretical basis and examined in the context of new legal SME forms. Ultimately, SMEs must innovate and respond to legal advances as all other business organizations, regardless of their legal status and size, in order to compete.

REFERENCES

Beraha, I. (2016). *Strategija unapređenja poslovnih performansi malih i srednjih preduzeća u lancu snabdevanja automobilske industrije* (Doctoral Dissertation). University "Union-Nikola Tesla", Faculty of Entrepreneurship. Retrieved July 26, 2016, from: http://unionnikolatesla.edu.rs/doktorati/isidoraberaha/disertacija.pdf

Ćoćkalo, D., Đorđević, D., Sajfert, Z., & Bogetić, S. (2011). SMEs in the Republic of Serbia: The developing capacities. *Journal of Applied Engineering Science*, *9*(4), 449–456. doi:10.5937/jaes9-1131

COSME. (2016). *Europe's programme for small and medium-sized enterprises.* Retrieved July 26, 2016, from EU web site: http://ec.europa.eu/growth/smes/cosme/index_en.htm

Đulić, K., & Kuzman, T. (2012). Protection of minority shareholders' rights: legal framework and implementation. *Quarterly Monitor of Economic Trends and Policies in Serbia, 30*, 86-95.

European Commission. (2015a). *Cost of the Cumulative Effects of Compliance with EU Law for SMEs.* Executive summary. Publications Office of the European Union. Retrieved July 26, 2016, from EU web site: ec.europa.eu/DocsRoom/documents/16321/attachments/1/translations/en/renditions/native

European Commission. (2015b). *2015 SBA Fact Sheet-Serbia.* Retrieved July 26, 2016, from EU web site: http://ec.europa.eu/growth/smes/business-friendly-environment/performance-review/

European Commission. (2015c). *Serbia 2015 Report.* Retrieved July 26, 2016, from EU web site: http://ec.europa.eu/enlargement/pdf/key_documents/2015/20151110_report_serbia.pdf

European Commission. (2016a). *Enterprise.* Retrieved July 26, 2016, from EU web site: https://europa.eu/european-union/topics/enterprise_en

European Commission. (2016b). *The European Company – Your business opportunity?* Retrieved July 26, 2016, from EU web site: http://ec.europa.eu/internal_market/company/societas-europaea/index_en.htm

European Commission. (2016c). *The European Commission on Serbia's progress in 2015.* Retrieved July 26, 2016, EU web site: http://ec.europa.eu/enlargement/pdf/key_documents/2015/20151110_report_serbia.pdf

European Integration Office. (2014). *National Program for Adoption Acquis European Union.* Retrieved July 26, 2016, from European Integration Office web site: http://www.seio.gov.rs/upload/documents/nacionalna_dokumenta/npaa/npaa_2014_2018.pdf

International Finance Corporation. (2008). *Corporate Governance – Manual* (updated edition). Retrieved July 26, 2016, from International Finance Corporation web site: http://www.ifc.org/wps/wcm/connect/topics_ext_content/ifc_external_corporate_site/corporate+governance/publications/toolkits+and+manuals/ifc+corporate+governance+manual

Jevremović-Petrović, T. (2005). European business units. In Guide through the EU law (pp. 9-11). Belgrade: Institute of International Politics and Economics.

Jevremović-Petrović, T. (2011). Creditor protection in company law. *Annals of the Faculty of Law, 59*(1), 223–254.

Serbia Accounting Law. (2013). Official Gazette of the Republic of Serbia no. No.62/2013.

Serbia Labor Law. (2014). Official Gazette of the Republic of Serbia nos. 24/2005, 61/2005, 54/2009, 32/2013 & 75/2014.

OECD. (2004). *Networks, Partnerships, Clusters and Intellectual Property Rights: Opportunities and Challenges for Innovative SMEs in a Global Economy*. Retrieved July 26, 2016, from OECD web site: http://www.oecd.org/cfe/smes/31919590.pdf

OECD. (2012). *SME Policy Index: Eastern Partner Countries 2012: Progress in the Implementation of the Small Business Act for Europe*. OECD Publishing. Retrieved July 26, from: http://www.keepeek.com/Digital-Asset-Management/oecd/finance-and-investment/sme-policy-index-eastern-partner-countries-2012_9789264178847-en#page1

OECD. (2016). *SME Policy Index: Western Balkans and Turkey 2016: Assessing the Implementation of the Small Business Act for Europe*. OECD Publishing. Retrieved July 26, 2016 from: http://www.keepeek.com/Digital-Asset-Management/oecd/development/sme-policy-index-western-balkans-and-turkey-2016_9789264254473-en#page313

Ožegović, L., & Pavlović, N. (2012). Management of small and medium size enterprises as a carrier of economic growth. *School of Business*, *1*, 74–84. doi:10.5937/skolbiz1201074O

Serbia Bankruptcy Law. (2014). Official Gazette of RS, No. 104/2009, 99/2011§ other Law, 71/2012 § decision of Constitutional Court and 83/2014.

Serbia Code of Business Conduct. (2006). Official Gazette of the Republic of Serbia no. 1/2006.

Serbia Company Law. (2015). Official Gazette of the Republic of Serbia nos. 36/2011, 99/2011, 83/2014 - § other Law and 5/2015.

Serbia General Collective Agreement. (2008). Official Gazette of RS, No. 50/2008, 104/2008 - Annex I and 8/2009 - Annex II.

Serbia Law on Public Services. (2014). Official Gazette of the Republic of Serbia nos. No.42/91, 71/94, 79/2005 § other law, 81/2005 § correction, 83/2005 - correction i 83/2014 § other law.

Serbia Law on the Procedure of Registration with the Business Registers Agency. (2014). Official Gazette of the Republic of Serbia nos. 99/2011 and 83/2014.

Serbia Public Procurement Law. (2015). Official Gazette of the Republic of Serbia nos.124/2012, 14/2015 and 68/2015.

Serbia Strategy for Supporting the Development of SMEs, Entrepreneurship and Competitiveness. (2014). Official Gazette of the Republic of Serbia, No. 35/2015.

Stamatović, M., & Vemić, M. (2009). Elaboration of the Profound Effects of the Global Economic Crisis on Small and Medium Enterprises in Serbia. In *Proceedings of the 3rd Central European Conference in Regional Science (CERS 2009)*. Retrieved July 26, 2016, from Faculty of Economics, Tehnical University in Košice web site: http://www3.ekf.tuke.sk/cers/cers2009/PDF/02_66_Vemic_Stamatovic.pdf

Stefanović, Z. (2011). *European Union Law.* Belgrade: Faculty of Law Union University and Official Gazette.

Todorović, Z., Bunić, S., & Todorović, I. (2013). Transparency in corporate governance. *Acta Economica, 11*(9), 9–39. doi:10.7251/ACE1319009T

USAID Business Enabling Project. (2015a). *Business enabling project qualitative research-Results of focus group discussions and in-depth interviews*. IPOS. Retrieved July 26, 2016, from site: http://www.policycafe.rs/documents/business/survey/Focus%20Groups%20Report%202015%20final.pdf

USAID Business Enabling Project. (2015b). *Business survey Serbia 2015: Executive Summary*. Retrieved July 26, 2016, from site: http://www.policycafe.rs/Anketa2015/BEP%20Survey%202015%20Executive%20Summary%20-%20english.pdf

World Economic Forum. (2016). *The Global Competitiveness Report 2015-2016*. Retrieved July 26, 2016, from World Economic Forum web site: http://reports.weforum.org/global-competitiveness-report-2015-2016/

ADDITIONAL READING

Anderson, W. (2011). Internationalization opportunities and challenges for small and medium-sized enterprises from developing countries. *Journal of African Business, 12*(2), 198–217. doi:10.1080/15228916.2011.588912

Ayyagari, M., Beck, T., & Demirguc-Kunt, A. (2007). Small and medium enterprises across the globe. *Small Business Economics, 29*(4), 415–434. doi:10.1007/s11187-006-9002-5

Bauer, F., & Matzler, K. (2014). Antecedents of M&A success: The role of strategic complementarity, cultural fit, and degree and speed of integration. *Strategic Management Journal, 35*(2), 269–291. doi:10.1002/smj.2091

Boubakri, N., & Ghouma, H. (2010). Control/ownership structure, creditor rights protection, and the cost of debt financing: International evidence. *Journal of Banking & Finance, 34*(10), 2481–2499. doi:10.1016/j.jbankfin.2010.04.006

Ćeha, M. (2013). Analysis of the Application of the Concept of Corporate Social Responsibility in Local Businesses. *Singidunum Journal of Applied Science, 10*(1), 1–10. doi:10.5937/sjas1301001C

Chowdhury, S. R. (2011). Impact of global crisis on small and medium enterprises. *Global Business Review, 12*(3), 377–399. doi:10.1177/097215091101200303

Crossan, K., & Henschel, T. (2012). Corporate governance: An holistic model for SMEs. *Journal of Management and Financial Sciences, 5*(8), 54-75. Retrieved May 26, 2016, from Warsaw School of Economics site: http://kolegia.sgh.waw.pl/pl/KZiF/publikacje/Journal%20of%20Management%20and%20Financial%20Sciences%20JMFS/Documents/JMFS8.pdf#page=54

Dias Simões, F. (2013). Legal Capital Rules in Europe: Is There Still Room for Creditor Protection? *International Company and Commercial Law Review*, *24*(4), 166–172.

Dukić-Mijatović, M. (2013). The legal position of secured creditors in financial restructuring and bancruptcy, *Law-theory and practice, 30* (1-3), 1-13.

EUR-Lex. Open acess to European Union law. (n.d.). EU law and related documents. Retieved May 26, 2016, from EU web site: http://eur-lex.europa.eu/collection/eu-law.html

Ezhilarasan, M., & Selvalakshmi, M. (2010).Small and Medium Enterprises. *Innovation and Competitiveness of Small and Medium Enterprises*, 36.

Filipović, S. (2003). Limiting factors of development small and medium enterprise in Serbia. *Industrija*, *31*(3-4), 25–36.

Garay, L., & Font, X. (2012). Doing good to do well? Corporate social responsibility reasons, practices and impacts in small and medium accommodation enterprises. *International Journal of Hospitality Management*, *31*(2), 329–337. doi:10.1016/j.ijhm.2011.04.013

Hernández-Cánovas, G., & Koëter-Kant, J. (2011). SME financing in Europe: Cross-country determinants of bank loan maturity. *International Small Business Journal*, *29*(5), 489–507. doi:10.1177/0266242611402569

Jasra, J. M., Hunjra, A. I., Rehman, A. U., Azam, R. I., & Khan, M. A. (2011). Determinants of business success of small and medium enterprises. *International Journal of Business and Social Science*, *2*(20).

Jevremović Petrović, T. (2012). Directors' responsibility to creditors in Company law. *Annals of the Faculty of Law*, *60*(3), 149–169.

Jõeveer, K. (2013). What do we know about the capital structure of small firms? *Small Business Economics*, *41*(2), 479–501. doi:10.1007/s11187-012-9440-1

Jovanović, N. (2012). Why going private is (was) prohibited in Serbia? *Annals of the Faculty of Law*, *60*(3), 113–135.

Jovetić, S., & Ilić, M. (2001). Institutional changes as the basis for small and medium size enterprises development. *Industrija*, *27*(1-4), 21–38.

Komandić, V. (2013). Risk assessment in small and medium-sized enterprises, specifics and differences in approach. *Journal of Applied Engineering Science*, *11*(3), 123–128. doi:10.5937/jaes11-3665

Kontinen, T., & Ojala, A. (2011). International Opportunity Recognition among Small and Medium-Sized Family Firms. *Journal of Small Business Management*, *49*(3), 490–514. doi:10.1111/j.1540-627X.2011.00326.x

Krstić, B., Stanišić, T., & Stojanović, M. (2012). Analysis of state and development factors of small and medium enterprise sector in Serbia, *Facta universitatis - series. Economics and Organization*, *9*(3), 355–368.

Man, T. W., Lau, T., & Chan, K. F. (2002). The competitiveness of small and medium enterprises: A conceptualization with focus on entrepreneurial competencies. *Journal of Business Venturing*, *17*(2), 123–142. doi:10.1016/S0883-9026(00)00058-6

Mulović, A., Hunjet, A., & Kozina, G. (2015). Toward economic reward: Corporate social responsibility comunications of SMEs. *Megatrend Review*, *12*(3), 215–230. doi:10.5937/MegRev1503215M

Obradović, K. (2003). SME business environment in Serbia. *Industrija*, *31*(3-4), 15–24.

Olusegun, A. I. (2012). Is small and medium enterprises (SMEs) an entrepreneurship? *International Journal of Academic Research in Business and Social Sciences*, *2*(1), 487.

Omerzel, G.D., & Antončič, B. (2008). Critical enterpreneur knowledge dimensions for SME performance, *Industrial Management & Data systems, 108*(9), 1182 – 1199. Retrieved May 26, 2016, form:10.1108/02635570810914883

Philip, M. (2011). Factors Affecting Business Success of Small & Medium Enterprises (SMEs). *Amity Global Business Review, 6*(1).

Rubio-Andrada, L., Del Mar Alonso-Almeida, M., & Rodríguez-Antón, J. M. (2011). Motivations and impacts in the firm and stakeholders of quality certification: Evidence from small-and medium-sized service enterprises. *Total Quality Management & Business Excellence*, *22*(8), 833–852. doi:10.1080/14783363.2011.593858

Siems, M. M., Herzog, L., & Rosenhäger, E. (2011). The protection of creditors of a European private company (SPE). *European Business Organization Law Review*, *12*(01), 147–172. doi:10.1017/S1566752911100051

Sousa, M. J. Corporate Social Responsibility in SME. *I Encontro Científico da Unidade deInvestigação e Inovaçãoem Economia e Sociedade (I2ES)–ISLA Santarém*, 11. Retrieved May 26, 2016, form: http://www.eci2es.santarem.unisla.pt/images/Livro%20I%20Encontro%20Cientfico%20I2ES.pdf#page=18

Stošković, M., & Ivanković, J. (2015). Challenges and future of SME sector in Serbia, *Economic horizons,20*(4), 439-454.

Street, C. T., & Cameron, A. F. (2007). External relationchips and the small business: A review of small business alliance and network research. *Journal of Small Business Management*, *45*(2), 239–266. doi:10.1111/j.1540-627X.2007.00211.x

Szydło, M. (2010). Prevention of Forum Shopping in European Insolvency Law. *European Business Organization Law Review*, *11*(2), 253–272. doi:10.1017/S1566752910200046

Wymenga, P., Spanikova, V., Barker, A., Konings, J., & Canton, E. (2012). EU SMEs in 2012: at the crossroads: Annual report on small and medium-sized enterprises in the EU, 2011/12. *Report for the European Commission, Sept.*

KEY TERMS AND DEFINITIONS

Small and Medium-Sized Enterprises (SMEs): Small enterprises are defined as enterprises which employ less than 50 persons and whose annual turnover or annual balance sheet total does not exceed 10 million euros; medium enterprises employ fewer than 250 persons, have turnover below 50 million euros and annual balance below 43 million Euros.

European Union (EU): Unique economic and political union between 28 European countries that operates as a cohesive economic and political block.

EU Acquis: The accumulated of legal regime and body of common rights and obligations binding on all the EU member states.

Legal Harmonization: A process that approximates legal rules within an institutional framework through the adoption of regulatory models and laws more uniform and coherent with a policy agreed at a multilateral level such as the EU in which member states achieve uniformity in laws.

Legal Reform: Process of examining existing laws, and advocating and implementing changes in a legal system, usually with the aim of enhancing justice or efficiency.

Legal Certainty: A principle which holds that the law must provide those subject to it with the ability to regulate their conduct and is internationally recognized as a central requirement for the rule of law.

Micro Businesses: Enterprises and entrepreneurs operating independently which employ less than 10 persons and whose annual turnover or annual balance sheet total does not exceed 2 million Euros.

Legal Optimization: A systematic approach that shifts the focus to legal innovation or what should be different in a process with legal outcomes involving comprehensive structural change necessary for innovation, business outcome, or both.

ENDNOTES

1 Editors' note: The possible connections between company size and founding capital in SME life cycle are also discussed in chapter 10.

Chapter 6
Small and Medium Enterprises in the Slovak Republic:
Status and Competitiveness of SMEs in the Global Markets and Possibilities of Optimization

Peter Malega
Technical University of Košice, Slovakia

ABSTRACT

The main goal of this chapter will be the clarification of SME´s in Slovak republic as the main creator of value in the Slovak republic, which is one of the semi-developed countries in EU. In the Slovak Republic have small and medium enterprises irreplaceable role particularly in job creation and regional development. This chapter has three main sections, which are interlinked. First main section is about the current conditions of SMEs in Slovakia, where are explained in detail business environment in Slovakia and also support for SMEs. Second main section deals with SMEs and international cooperation, where are described business risk, internationalisation of SMEs in EU markets and analysis of SMEs possibilities. Third main section is about entry strategy into foreign markets and competitiveness of SMEs and in this section you can find entry strategies of enterprise on foreign markets and competitiveness of Slovakia and Slovak SMEs.

INTRODUCTION

Considering the fact that the world economy is still recovering from financial crisis, which began in 2008, a number of economies urgently need to create employment opportunities for their citizens. With the creation and growth of SMEs (SMEs), this sector became an important part of most economies, especially taking into account the significant contribution of SMEs to employment and gross domestic product (GDP).

DOI: 10.4018/978-1-5225-1949-2.ch006

Furthermore, entrepreneurship has become the basic cornerstone of effective operation of a market economy. For the Slovak Republic, whose main objective is to reach the level of developed European countries in the shortest possible period of time, business development has extraordinary importance. Business performance became a driver of changes, a measure of economic development, a tool of SME adaptation and a condition to stay within major global economic trends.

The main objectives of this chapter are to evaluate the situation with SMEs in Slovakia and their competitiveness.

This chapter consists of three main sections. The first section focuses on current conditions of SMEs, the business environment and level support to SMEs which is one of the biggest problems in Slovakia. The second part deals with SME growth and entry strategy opportunities in the global market, possibilities of optimization of international cooperation of SMEs. The objective of the final third section is to evaluate an entry strategy into foreign markets for competitiveness of SMEs because this could also improve their position in the EU market.

BACKGROUND WITH LITERATURE REVIEW

While evaluating SME dynamics in EU member states the author observed some encouraging post crisis trends. Despite the fragile economic environment, the latest estimates indicate that SMEs in a growing number of countries are starting to recover and to contribute to economic growth of the respective countries (European Funds and Slovakia, 2014). Although there are still many countries where the situation got worse, the overall average trend is that SMEs record a recovery. While in 2009 SMEs in 22 EU member states recorded a negative growth of real gross value added and employment, the situation in 2011 was more positive, with only three member states finding themselves in such a bad position, while further 13 countries recorded positive real gross value added and employment growth. In 2012 there remained only two member states with a negative growth rate of both indicators (European Funds and Slovakia, 2014) revealing a possible recovery process.

All businesses, SMEs included, enter the market with two goals: meeting the needs of customers and making a profit. Usually there are a lot of potential customers, they are diffused and differ in their purchase requirements. Therefore, instead of business competition across the market with lots of competitors, they have to define the most attractive customer groups which they could efficiently satisfy. The starting point in choosing target groups, to which the company wants to focus on, is to understand the needs, motives and behavior of customers, overall market development as well as future trends (Belás, Demjan, Habánik, Hudáková, & Sipko, 2015).

SMEs provide an irreplaceable contribution to economic growth, employment and competitiveness of each country, but because of their size they are sensitive to changes in the external environment and their productivity is disturbed by existence of business barriers. SMEs are not able to maintain control of markets in which they operate or take higher business risk and they have insufficient capital. It is their valid need to receive active and effective support, especially in specific sectors such as agriculture, science and research, start-up of entrepreneurship and employment of the first employee.

In this regard it can be noted that SME development is one of the priorities of economic development of the Slovak economy. The most important factor of this development is the creation of an appropriate business environment implying simplification of legislation, reducing administrative, social contributions and tax burdens, reinforcement of infrastructure and improvement of access to capital. Therefore, different factors participate in the development of entrepreneurship such as state support, investments related to costs and operation of the businesses (Smit, 2010).

To support SMEs available are EU funds as the most important source for EU regional policy to eliminate regional disparities and for overall improvement of the economy. EU funds are an instrument to equalize the disparity between individual states of the EU including the Slovak Republic. During the first full absorption of EU funds in the 2007 – 2013 programming period, the Slovak economy has undergone some progress. The impact of financial assistance through EU funds in this period is not yet entirely clear, since it was possible to spend these funds until 2015 and the results are not yet available.

EU funds are currently discussed and it is an actual topic among member countries. These funds generally assist weaker regions of the Union. They largely serve regional development and ultimately development of member countries and the EU as a whole.

Related to this aspect is competitiveness measurement through an instrument performed by the World Economic Forum through the Global Competitiveness Index (GCI), (World Economic Forum, 2016). The GCI index is based on 12 pillars of competitiveness and provides a comprehensive review of the competitiveness environment in many countries of the world at different stages of development. It remains the most respected evaluation of national economic environment, competitiveness and ability to achieve sustained levels of prosperity and growth.

The World Economic Forum has been carrying out its annual public opinion research for more than 30 years. The survey will change over time to capture new data relevant for the GCI and other indexes of the Forum. Collected data thus provide a unique source of knowledge and a qualitative view of national economic and business environment allowing comparisons between countries. Based on the creation of sub-indices established is the overall index and identified are the levels of development in terms of competitiveness. In recent years the business environment in Slovakia has been continuously improving, but there still remain areas where progress can be made in the function of competitiveness.

For example, entrepreneurial orientation of Slovak SMEs can be improved implying innovativeness, risk-taking and proactive attitude as the basic characteristics of entrepreneurial orientation. It is assumed that that entrepreneurial orientation is positively related to the profitability of small firms (Soininen, Martikainen, Puumalainen, & Kyläheiko, 2012) also in Slovakia. Authors Eggers, Kraus, Hughes, Laraway and Snycerski (2013) indicated that entrepreneurial orientation is positively related to SME growth, but customer orientation shows a negative association with the growth. Therefore, to drive growth Slovak firms will also need to deploy a strategy centred on entrepreneurial orientation.

Some other findings of the author from which Slovak SMEs can benefit are as follows. Initial employment size, rather than age, was found to have some bearing on the nature of the growth path. Smaller firms grew more often and with more continuity than larger firms (Hillary, 2000). The role of external relationships is one of the key drivers of small business innovation (Boutellier, Gassmann, & Von Zedtwitz, 2013). Innovation as a decisive factor of competitiveness the, is reflected in these forms:

- Restoring and improving a range of products and services and expansion into relevant markets;
- Creation of new methods of production, supply and distribution;
- Indication of changes into management, work organization, working conditions and know-how;
- Skilled labor force.

Research results also indicate that innovation performance is higher in SMEs that are proactive in strengthening their relationships with innovative suppliers, users, and customers (Lasagni, 2012).

Innovative SMEs do perform better in a turbulent environment (Kraus, Rigtering, Hughes, & Hosman, 2012) and it remains to be seen how Slovak SMEs can benefit from this finding. SMEs are smaller, more informationally opaque, riskier, but more dependent on trade credit and bank loans (Dierkes, Erner, Langer, & Norden, 2013) while in Slovakia the author has found problems in access to bank credit abd funding.

Despite innovation potential, small businesses, particularly younger, have little internal cash flow to finance their operations and are also associated with significant asymmetric information (Canales & Nanda, 2012), and in Slovakia the author has identified similar problems (Malega, 2010).The high risk of SMEs is characterized by a high level of debt and a limited capacity to guarantee. As such, the acquisition of commercial loans for such companies is highly problematic for SME *opportunities in Slovakia* (Májková-Sobeková, 2011). In this respect, micro-enterprises are especially predisposed to incorrect selection and moral hazard, and therefore their access to debt financing is limited (Neuberger & Räthke, 2009) which the author also observed in the Slovak mechanical engineering sector (Malega, 2010).

SME MAPPING OF THE SLOVAK REPUBLIC

Entrepreneurship itself is a driver of change, a measure of economic development, an instrument of customization, the instrument of adaptation and ability to keep pace with the major global trends. For the Slovak Republic, whose main objective is to reach the level of developed countries in Europe as soon as possible, entrepreneurship development has outstanding importance. Entrepreneurship has become the basic element of effective operation of its market economy and SME development alike.

SMEs represent 99% of all businesses in the EU. SME definition is important for access to finance and EU support programs targeted specifically at these enterprises. The main factors in determining whether an enterprise is an SME are (see Table 1):

1. Staff headcount and,
2. Either turnover or balance sheet total.

These ceilings apply to the figures for individual firms only. A firm that is part of a larger group may need to include staff headcount/turnover/balance sheet data from that group too. Small, medium and large enterprises can also be distinguished through quality characteristics. According to Kováč et al. (2013) the differences can be seen in defining business goals and orientation, in the way of planning business activities and their funding, in business management and in manager´s approach towards specific situations (see Table 2).

Table 1. Categorization of SMEs in EU

Company Category	Staff Headcount	Turnover	OR	Balance Sheet Total
Medium	< 250	≤ € 50 m		≤ € 43 m
Small	< 50	≤ € 10 m		≤ € 10 m
Micro	< 10	≤ € 2 m		≤ € 2 m

Source: (Kováč, Malega, & Spišáková, 2013)

Table 2. Quality features of enterprises by size

	Small Enterprises	Medium Enterprises	Large Enterprises
Goals	Market-oriented Intuitive improvised	Market-oriented Intuitive Ad hoc strategies	Dominant Systematic
Management	Authoritative Direct	Personal consultation Small teams of specialists	Large teams functional structure
Planning	Minimal	Short term	Long term
Product	Without market research	Irregular market research	Extensive market research
Finance	Own resources Bank sources	Own resources Bank sources	Own resources Bank sources Financial market

Source: Kováč et al. (2013)

SMEs are the backbone of the Slovak economy and have an irreplaceable role particularly in job creation and regional development. Representing more than 99.9% of all Slovak business entities, they are net job creators, employ almost 75% of the active labor power and contribute more than 50% of GDP. SMEs are more prevalent in the Slovak 'non-financial business economy' than in the rest of the EU. SMEs provided more than 70% of jobs and created more than 60% of value added in this sector. The proportion of people employed by SMEs was higher than the EU average by 3 percentage points and the proportion of SME value added exceeded the EU average by 4 percentage points. Basic values for Slovak SMEs are summarized in Table 3 (European Funds and Slovakia, 2014).

Slovakian SME employment forecasts predict a slight fall of about 0.6% per year in 2015 and 2016. In the same period, SME value added is estimated to increase by 2.4% per year. SME performance is likely to outpace the performance of large businesses during the forecast period. The future growth in value added in large enterprises is estimated at 1% per year. Similarly, employment by large businesses is expected to contract at an annual rate of 2.5% in 2015 and 2016.

Quality business environment and conditions for achieving long-term sustainable economic growth are basic prerequisites for SME development and increasing competitiveness of the Slovak Republic on

Table 3. Indicators of development of Slovak SMEs

	Number of Enterprises			Number of Persons Employed			Value Added		
	Slovakia		EU–28	Slovakia		EU–28	Slovakia		EU–28
	Number	Share	Share	Number	Share	Share	Billion €	Share	Share
Micro	377 379	96.4%	92.7%	552 381	39%	29.2%	9	27.1%	21.1%
Small	11 474	2.9%	6.1%	224 291	15.8%	20.4%	6	17.7%	18.2%
Medium	2083	0.6%	1.0%	224 798	15.9%	17.3%	5	16.4%	18.5%
SMEs	**390 936**	**99.9%**	**99.8%**	**1 001 470**	**70.7%**	**66.9%**	**20**	**61.2%**	**57.8%**
Large	469	0.2%	0.2%	415 365	29.3%	33.1%	13	38.8%	42.2%
Total	391 405	100%	100%	1 416 835	100%	100%	33	100%	100%

Source: European Funds and Slovakia (2014)

Table 4. SWOT analysis of the business environment in the Slovak Republic

Strengths	Weaknesses
• Stable macroeconomic and fiscal environment. • Developed and healthy banking sector able to provide credit resources. • Membership in the EU and EMU. • Ideal geographic location for investment and transit of goods and services within Europe. • Continuous growth of the business sector. • Relatively low profit tax burden.	• Higher administrative burden of business due to lower level of digitization and computerization of public administration in relation to the business environment. • Higher tax payroll of employees. • In the past, limited financial resources to implement support programs for SMEs, less-developed tools in the area of venture capital. • Lower innovation performance of Slovak enterprises in international comparison. • Lengthy court activity and related lower enforcement of rights.
Opportunities	**Threats**
• Higher emphasis of SMEs on innovation and the knowledge economy. • Expansion and innovation of existing financial instruments and application of other financial instruments to provide business support. • Government actions for SMEs in order to reduce the impact of the economic crisis. • Systematic education and support of existing and potential SMEs in this area. • Reduction of administrative burden on enterprises. • Systematic and rigorous creation of a regulatory environment for business. • Maximal utilization of the economic potential of the structural funds.	• Impact of the global financial and economic crisis. • Centralization of policy-making and decisions at EU level and related loss of independence in decision-making about certain issues. • Depreciation of the currencies in neighbouring countries to the euro and related lower competitiveness of Slovak export in comparison with V4 countries. • Brain drain (high-quality labor force) to other markets. • Larger lag in innovation and implementation of the knowledge economy. • Delays in the implementation of already defined and adopted recommendations on the EU and also on the Slovak level.

Source: Kováč et al. (2013)

an international scale (Zajko, Chodasová, Jemala & Materák, 2010). In pursuit of business environment quality in Slovakia, the decisive areas are presented by means of SWOT analysis as displayed in Table 4.

Obviously, the business environment in Slovakia has been continuously improving in recent years, but there are still areas for progress. Weaknesses are lower and lengthy law enforcement, dynamically changing legislation without an effective system for evaluating impact on the business environment, existing administrative burdens of enterprises and less developed capital market. Many of these problems are also typical for other EU countries.

Therefore, Slovakia needs to further stabilise its business environment through regulatory impact assessments, appropriate stakeholder consultation mechanisms and the SME test. Public administration needs to be made more transparent and the widely perceived corruption should be addressed. This country should also better respond to SME needs by providing higher quality services. E-government solutions, currently in an early phase, may play an important role in the process. Slovakia needs to continue to boost skills and innovation in SMEs and enable them to achieve their international potential. Following the above described content, the next section of the chapter will proceed with evaluation of support for SMEs in the Slovak Republic.

Snapshot: SME Access to Capital and Enterprise Support in the Slovak Republic

In cooperation with the Regional Consultant and Information Centre, the first contact centres, business and technology incubators, business innovation centres and with the company Fund of Funds, Ltd., the

Slovak Business Agency enables SMEs to gain access to capital necessary for the initiation and development of business activities. Support is also provided by individual ministries (Schwarcz et al., 2011).

For example, the Ministry of Economy manages the Micro-loan program in amounts ranging from 2,500 EUR to 50,000 EUR for the development and survival of SMEs in a competitive environment.

Venture capital funds in Slovakia are managed by the Fund of Funds, Ltd. They are aimed at providing financial support for the creation and development of SMEs, finding suitable partners for co-financing projects and obtaining expert consultation. The venture capital fund is a compound of these organizations: Seed Capital Fund, Regional Seed Capital Fund, INTEG Fund, Fund SISME, Seed Capital Fund, K. S., Slovak Development Fund, Inc., and Slovak Growth Capital Fund, Inc.

Venture capital funds work with unlisted companies to finance initial business activity, innovation and development projects associated with high risk. They relate to SMEs with lack of own resources and experience in business which are not led by professional managers. Investors of venture funds obtain an agreed share in the company for the provision of a certain amount of resources. This investment is not a one-time provision of resources, but a multi-year process of businesses coexistence with the venture capital investor. The investor obtains a minority ownership interest in the company and participates in the adoption of fundamental decisions but normal operation is left to the original owners. If an enterprise fails, the investor loses embedded assets. If the enterprise is successful, the minority share may represent a multiple valuation of investment.

An important part of the current support infrastructure for SMEs are business and technological incubators. This support system in the Slovak Republic is only partially developed. Slovakia, has built several business and technological incubators through the PHARE program. The mission of business incubators is to create suitable start up conditions for entrepreneurs, at least for three years. The mission of technological incubators is to create favourable conditions for the exploitation of research and development, patents, industrial and utility models closely associated with scientific and technical institutions including the method of *Research – based spin – off*. The results are innovative products and services with higher added value. For mentioned activities used is the program "*Support for SMEs through a network of incubators and the implementation of Research – based spin – off*". This program positively contributed to the elimination of regional disparities and in solving high unemployment as observed in an earlier study (Canales & Nanda, 2012).

Slovak Business Agency, Ministry of Economy, Regional Consultant and Information Centre, Business innovation centres and business and technology incubators also manage other programs, which support SMEs in Slovakia. These programs are:

- Support to education and consulting service for selected groups interested in entrepreneurship and start-ups;
- Support of consulting service and education for SMEs;
- Promotion of successful entrepreneurial experience and entrepreneurial education.

Indirect assistance for SMEs with information and consulting service about all matters relates to enterprise operation, development of cooperation in the EU internal market is provided through the program "*Assistance for SMEs in the EU market*." The aim is to increase competitiveness of Slovak SMEs, support innovation activities and penetrate foreign markets. SMEs gain information about functioning of the EU single market and legislation in the field of business.

The Slovak Guarantee and Development Bank, the JEREMIE program and EXIM bank's scheme support exports. A micro loan program is available from the Slovak Business Agency, while the European Investment Bank and the European Bank for Reconstruction and Development provide preferential financing for SMEs through commercial banks.

The Structural Funds and the Cohesion Fund are now also among the main forms of financial assistance for EU Member States. They support economic development and job creation. In the last programming period, i.e. 2007 – 2013, enterprises had the possibility to draw funds under 11 operational programs. These were:

1. Bratislava region;
2. Transportation;
3. Informatisation of Society;
4. Competitiveness and economic growth;
5. Regional operational program;
6. Technical assistance;
7. Education;
8. Research and development;
9. Healthcare;
10. Employment and social inclusion;
11. Environment.

EU funding within each operational program was monitored monthly and evaluated for each period (see Table 5).

Spending and contracting of funds in higher territorial units is presented in the table below which shows specific performance (see Table 6).

Table 5. Review of Spent Funds, 2008 - 2014, in percent

	Programming Period 2007-2013						
Operational Programs	**31.12.2008**	**31.12.2009**	**31.12.2010**	**31.12.2011**	**31.12.2012**	**31.12.2013**	**31.12.2014**
Bratislava region	0.29	5.58	13.76	30.54	40.92	52.34	58.99
Transportation	0.01	6.34	16.85	23.19	31.29	45.87	61.76
Informatisation of Society	0.13	0.40	2.62	9.04	27.31	45.97	65.70
Competitiveness &economic growth	0.05	5.55	17.71	37.73	42.19	48.26	52.48
Regional operational program	0.28	6.33	19.54	38.62	58.00	70.35	76.83
Technical assistance	4.23	14.70	27.96	39.80	45.04	54.87	64.84
Education	0.12	0.85	8.11	13.36	25.18	38.97	60.57
Research and development	0.06	2.77	9.91	20.14	32.23	45.20	61.68
Healthcare	0.28	5.01	19.33	42.49	65.12	81.75	82.64
Employment and social inclusion	1.53	8.55	18.41	34.59	49.57	61.06	74.01
Environment	0.11	1.80	6.31	18.98	31.03	44.36	52.54

Source: European Funds and Slovakia (2014)

Table 6. Operational programs, 2007 – 2013, by higher territorial units in thousands of EUR

Higher Territorial Units	Exhausted Resources (in thousands of EUR)	Contracted Resources (in thousands of EUR)	% Share of Higher Territorial Unit of the Total Volume of Funds
Bratislava region	417,063	1,019,087	4.27
Banská Bystrica region	635,448	1,205,690	6.37
Žilina region	816,588	1,572,067	8.21
Košice region	588,132	980,533	5.84
Prešov region	779,009	1,310,723	7.78
Nitra region	376,579	657,710	3.74
Trenčín region	959,053	1,525,888	9.72
Trnava region	829,001	557,015	3.26

Source: European Funds and Slovakia (2014)

Table 6 shows that the amount of used resources was significantly smaller than the amount of contracted resources.

To implement projects the EU budget reserved 11.36 billion EUR for the Slovak Republic. Except approved assistance from the Structural Funds and the Cohesion Fund, the Slovak Republic had available further 227.3 million EUR for European Territorial Cooperation (ETC) programs.

Also, a sum of Euro 50.25 billion was reserved for the major financing instrument of science and technology in the whole of EU, which is the 7th Framework Program of the European Union for research, technological development and demonstration activities. The benefits of participation for SMEs in this program are as follows:

- SMEs gain access to new research results;
- SMEs gain new contacts that help them to establish business relations;
- SMEs gain new international contacts with research organizations;
- SMEs gain funds for research activities and innovative products, services and processes;
- SMEs reduce commercial risk in projects demanding on innovation.

Gap Analysis - Slovak Competitiveness

As mentioned in the introduction the GCI index is based on 12 pillars of competitiveness and provides a comprehensive evaluation of the competitiveness environment in countries of the world at all stages of development. The pillars are as follows:

1. Institutions;
2. Infrastructure;
3. Macroeconomic environment;
4. Health and Primary Education;
5. Higher education and training;
6. Goods Market Efficiency;

7. Labor market efficiency;
8. Financial market development;
9. Technological readiness;
10. Market size;
11. Business Sophistication;
12. Innovation.

Slovakia worsened considerably its position over the last ten years. In the years 2006-7 the Slovak republic around was 40th, along with Hungary, only a few places behind the Czech Republic. Northern neighbour Poland was at that time behind the Slovak republic on the 45th position. At that time there was a lot of foreign investment and Slovakia was then called the Central European tiger. But then arrived a gradual and after 2009 a rapid decrease in the position almost to the 80th position in 2013. In the last two years, the perception of competitiveness ultimately improved and Slovakia was on the 67th place in 2015.

Compared to last year, one of the biggest problems which entrepreneurs perceive is increased perception of corruption. It is considered as the biggest difficulty among 18% of respondents of the national survey. The perception of corruption is followed by the state bureaucracy, high taxes and labor market restrictions.

It can be stated that gradual decrease of Slovakia's position (except year 2015) is due to the existing business barriers, which are still very strong and are not removed. The image of Slovakia as the reform country is part of history, and since 2004 entrepreneurs do not perceive reforms which significantly improve business conditions. Slovakia's position has now risen for the first time in last six years but is still increasingly lagging behind other V4 countries (i.e. Hungary, Poland and Czech Republic). The main barriers that worsen Slovak competitiveness are in particular the following:

- Inefficient functioning public institutions;
- Poor law enforcement;
- High deficits of public finance and weak efficiency of public expenditures;
- Low quality of the education system;
- Inefficient labor market;
- Bureaucracy;
- Cronyism;
- Corruption.

Enterprises in Slovakia (including SMEs) are in an environment that is characterized by increasing market openness and integration of the economy. This causes importance growth of mutual comparison of individual enterprises, countries and regions (Annoni & Dijkstra, 2013). To ensure successful and long-term progress of SMEs and also their competitiveness (Altomonte, Navaretti, di Mauro, & Ottaviano, 2011) it is important to objectively evaluate the current situation in which Slovak SMEs now operate.

A SWOT analysis of SME competitiveness in Slovakia[1] is displayed Table 7.

In support for above information Slovak SMEs were surveyed by the SBA (Slovak Business Agency). On a sample of 100 SMEs respondents were asked to indicate how high a share exports have in the total turnover. The results are shown in Table 8.

Table 7. SWOT analysis of SME competitiveness in Slovakia

Strengths	Weaknesses
• Labor flexibility. • High motivation to perform among SME owners. • General advantage of SMEs is the knowledge of local markets, respectively ability to flexibly respond to market development.	• Production with high energy intensity. • Lack of orientation in the support for SMEs. • Backwardness in the area of new technologies. • Lack of knowledge and experience in the area of management and marketing. • Prevalence of production with low added value. • Lack of available financial resources in enterprises in the field of intellectual property and business laws.
Opportunities	**Threats**
• Higher utilization of support programs for SMEs. • Wider penetration in the European single market. • Extension of cooperation between enterprises in area of common interest, building of cooperation networks and clusters. • Support of SMEs through the development of research, innovation and technological development, respectively collaboration with research institutions.	• Enhancing technical and technological backwardness of enterprises due to lack of financial resources. • Impact of the financial and economic crisis. • Competition from large enterprises. • Departure of highly skilled workers abroad.

Source: Malega (2012).

Table 8. Share of exports in total turnover among selected Slovak SMEs

Share of Exports in Total Turnover	Number of SMEs
enterprise doesn´t export	66
more than 70%	7
less than 70%	2
less than 50%	4
less than 30%	9
less than 10%	12

Source: Malega (2014)

The most frequent response was that an enterprise doesn´t export, which involved 2/3 of all respondents. This response was significantly represented in the category of micro enterprises with up to 9 employees (80%) and also in all business sectors except for businesses oriented towards industrial production (27%).

To summarize this section, it can be emphasized, that despite significant changes which occurred for SMEs in recent years, there are still several important problems (particularly for small enterprises in Slovakia). The biggest problems of Slovak SMEs are herewith summarized by the author in Table 9.

SLOVAK SME OPPORTUNITIES AND CHALLENGES IN THE GLOBAL MARKET

From the previous sections of the chapter it is already clear that SMEs are a specific form of business. For this reason, when evaluating their potential to penetrate global markets, it is necessary to take this aspect into account. It is not sufficient to have a national business strategy, but a wide-spectrum business environment and to perceive the whole world as a potential target market (Esser, Hillebrand, Messner & Meyer-Stamer, 2013).

Table 9. Main problems of Slovak SMEs

Stage of Development in Firm Life Cycle	Type of Problem
Phase of business start-up	• Lack of input capital • Obtaining contracts • Obtaining a place in the market • Provision of materials, machines, respectively workers • Legislative problems (registration, authorization) • Distrust
Phase of business development	• Financial difficulties • Tax and other burdens • Sales difficulties • Insufficient infrastructure • Relations with local government

Source: Malega (2014)

Business conditions in Slovakia, as well as in the EU, are constantly changing and forming a new business environment. With incorporation of the Slovak Republic in the unified European market Slovak SMEs have an opportunity to expand business activities, but are also faced with potential risks. The positive consequence is the fact that this paves the way for close cooperation with international partners outside Slovakia.

Unquestionably, the positive effects of Slovakia's EU membership include harmonisation of trade barriers, better protection of intellectual property and convergence of standards and certificates in the interest of free movement of goods, services and capital and having support from the EU *structural funds*.

Implementation of business activities in domestic and foreign markets is determined by a number of other important factors including adapting to different economic, cultural and political conditions. SME business management has to carefully plan and then coordinate the steps to foreign markets in such a way that penetration can be successful. Following is a review of specific SME business risks.

Business Risk

The author notes that regardless of whether an enterprise operates only at national level or globally, it faces various economic, business, technological and other risks. Neglect and disregard of risks, multiplies and accumulates them. Risk reduction or elimination means increasing the probability of obtaining an expected profit.

There are many methods and analyses for identification, monitoring, reduction and elimination of risks. SMEs especially don´t have the sufficient resources for the establishment of a risk management department to perform analyses in the field of risk management.

Risks exist and it is not possible to avoid them. It is therefore necessary to underline the importance of rational perception of risk as a part of business and its utilization. In the current environment, characterized by linking national economies, it is much more important to deal with risks on a multinational level, where the effects of risks multiply. It should be noted, that enterprises operating on the global market are undergoing much higher risk, but at the same time have a chance for making higher profit. A review of external and internal risks in international cooperation is provided in Table 10.

Besides risk management[2], decision to penetrate a foreign market also requires the application of principles of international management and marketing, i.e. actions which reflect the multinational conditions.

Table 10. Selected external and internal risks in international cooperation

Type of Risk	Existence	Characteristics	Form of Occurrence	Elimination
Market risk	Exporter	Risk of obtaining and maintaining the optimal market	Risk of sales, purchases and changes in price relations	Marketing research, contact with the market
Non-acceptance of goods risk	Exporter	Non-acceptance of goods risk by importer	Rejection of documents, unacceptance of bill, non-acceptance of goods, cancellation or radical changes in the contract	Confirmed letter of credit, information about importer
Risk of non-payment	Exporter	Displeasure, insolvency of importer, delaying of payment, paying less than actuarial value	Unopened letter of credit, rejection of documents	Documentary letter of credit and collection, bank guarantee, bills, insurance, factoring, forfeiting
Technological risk	Producer	Lack of technologically demanding production parts, the risk of operator and service of new equipment, development of new technologies and products	Incompetence in management, or incompetent work with technological devices, unavailability of high-tech components	Investment in research and development, innovation management, purchase of new hi-tech and IT equipment, education of people
Operational risk	Producer	Seriousness of partner, safety, continuity and healthy working environment	Disobedience of QMS and production control, dissatisfaction of staff and management, distrust of partner	Definition of managerial competencies, collaboration with partner, system of rewards and benefits
Credit risk	Producer	Solvency, respectively liquidity to pay the loan respectively leasing	More loans than is enterprise able to manage	Financial planning, profitability, increase of business performance, increase of sales
Commodity risk	Importer	Risk that the exporter doesn´t deliver the goods that were contractually agreed in the required quantity, quality and time	Production and financial problems, delay in transportation, unreliable business partners	Bank guarantee, certificate of goods quality
Transportation risk	Importer	Risk that the goods will be stolen, destroyed, or damaged during the transport	Independent of human will, arising from human activity, respectively natural character	Inco terms, insurance

Source: Hillary (2000).

International marketing represents a business orientation on foreign markets and involves *optimal* placement of its products in these markets. An SME realizes this target through the deployment of marketing mix elements. The role of international management is to ensure the coordination and management of all activities leading to the successful penetration on foreign markets. The common goal is to ensure the best possible satisfaction of the needs and desires of customers, to reinforce their loyalty with the aim of increasing frequency of their purchases. International marketing also provides SME management the information to choose an optimal business strategy to survive in an international setting.

Table 11. Motives for international cooperation

Motive	Activity	Level
Trade	Export	Commercial
Agreements	Licenses	Contractual
Capital interest	Establishment of consortium	Participatory
Integration	Direct investment of production	Integrating
Autonomy	Establishment of other enterprises at all stages of business activities	Autonomous

Source: Hillary (2000)

For a better understanding of international business management and process of penetration on foreign markets as well as its different phases it is important to understand the reasons that motivate SMEs. Motives of international cooperation are provided by the author in Table 11.

Internationalisation Initiatives for SMEs

International strategy is based on the detection of the interaction opportunities for SMEs in world markets. Hence, their development is gradually derived from the operating range on the broader market (penetration of new markets), from the growth of business agility with extension not only of business activities but also production activities in areas with more favourable conditions (Nummela, 2011).

If an SME is confident that domestic customers will be a sufficient target then it is looking for a local strategy. This enterprise will be interested in penetration of EU to the Slovak environment and impact of EU regulations on SMEs in Slovakia.

Similarly, if an enterprise is active in an industry with quite an international structure, it delivers universal-use and easily transportable product on the market, or already knows that in the future local markets will not be sufficient. Therefore, it works on its international Euro-strategy. Protection of local positions for this enterprise is not the solution and also it would not be effective or sustainable on a long term basis. An enterprise has to find resources to penetrate new markets. It must include EU to its strategic analyses and plans in the true sense, with the rules of the internal market, the transfer of business in other member states, delivering services abroad and similar.

Consequently, global enterprises are companies which not only deliver to the whole world, but are consciously deployed worldwide, dislocate their participation to take global advantages in places rich with raw materials, labor force, and intellectual activities. These companies effectively exploit benefits provided by states in the field of taxation, investment protection and less demanding business environment.

An enterprise will obviously start to analyse the specific and at first glance complicated EU environment before going global. This approach is summarized by the author in Table 12.

Table 12 shows that the EU internal market is not itself a mechanism, but rather one of the main tools to achieve a number of objectives, such as:

- Balanced and sustainable growth that respects the environment;
- High level of employment and social protection;
- Higher level and quality of life;
- Economic and social consistence.

Table 12. Components of the EU environment

Politics and Law	Economics	Society
Stability of political regime	Economic level – GDP (gross domestic product) per capita	Size of the consumer market
Government stability	Economic cycle	Demographic trends
Legal culture and regulations	Openness to external influences, degree of integration	Dominant values and traditions
Administrative culture and administration	Currency stability	National-cultural homogeneity and multiculturalism
International relations, responsibilities, alliances	Interest rate	Education and qualification
Economic and fiscal policy	Exchange rate	Work ethics and discipline
Foreign trade policy	Price level and inflation	Strength of civil society and their organizations
Support of competitiveness, export and business infrastructure	Unemployment	Employment structure
Competition protection	Investments	Social/Income structure of society
Commercial law	Consumption	Labor market and social dialogue
Labor law	Access to credits and other resources	Consumer patterns and trends
Environmental protection	Availability and price of raw materials	Dominant and generation lifestyle
Consumer protection	Availability and price of energy	Consumer costs
Protection of industrial and intellectual property		Availability of suppliers
Technical legislation		Communication infrastructure

Source: Hillary (2000)

An EU internal market also helps meet above goals by initiation of economic performance, participating in the EU as a whole. It is implemented by these actions:

- Assistance to more efficient allocation of production factors;
- Enablement to produce on a larger scale and more efficiently;
- Support for greater competition which increases economic performance;
- Support for incentives to invest.

Further Analysis of Slovak SME Opportunities in International Cooperation through the EU market

Despite advantages of international cooperation, this also brings significant risks that have been accentuated in previous sections of this chapter. For this reason, large enterprises more rapidly and flexibly engage in the international activities (Rugman, Oh, & Lim, 2012). The largest representation of Slovak enterprises engaging in international activities has been among medium enterprises (73%) followed by small enterprises (58%) and finally micro-enterprises (43%). This relationship is clearly visible mainly in foreign trade. From the total number of SMEs 24% of micro-enterprises export, 38% of small enterprises and 53% of medium enterprises. 28% of micro-enterprises import, 39% of small enterprises and 55% of medium enterprises. SMEs that work longer are more intense in both imports and exports. Enterprises that operate at least four years export more than 15%, while enterprises that operate at least 25 years or more export nearly 30%.

A substantial number of Slovak SMEs perform international activities within the EU single market and only 13% of SMEs are active in markets outside the EU. Statistically, 1 of 4 SMEs exports in the single market and only 1 of 8 in the markets outside the EU. Imports from countries outside the EU account for 14% of Slovak SME imports.

In analysis of internationalized SMEs there appears obvious correlation between enterprise size and activities related to cooperation with foreign enterprises (foreign direct investment – FDI, technology cooperation, subcontractor of foreign enterprise or enterprise, which has foreign subcontractor itself). The largest share of SMEs in other forms of internationalization was recorded among medium enterprises, in all four activities, small enterprises were recorded less frequently, and even least active in these activities were micro-enterprises, which can be seen in the author's presentation in Table 13.

The importance of SME internationalization for an economy can be monitored and evaluated through different indicators. It is crucial if the SME internationalization brings new jobs and to determine what impact SMEs have on employment. SMEs that are involved in international activities generally create more jobs. When groups of SMEs involved in several forms of internationalization are compared to those that are not involved in these activities, it is possible to perceive differences in employment growth:

- SMEs which are both importers and exporters, achieve 10% growth in employment, those that are not at the same time importers and exporters, achieve 3% growth in employment;
- SMEs that are active in all modes (import, export, FDI, technical cooperation and subcontracts) achieve 6% increase in employment compared with 2% growths among other SMEs;
- SMEs that enter into relations with foreign countries through foreign direct investment (FDI) reached employment growth of 16%, others reached only 4%.

The above information leads to the conclusion that greater SME internationalization is a necessity. Level of programs and utilization of development tools to support SME internationalization is very limited, because less than 10% of internationally active SMEs use this support. Many SMEs don´t even know the opportunities offered by public support for their participation in international activities. The existence of this support is realized by 22% of Slovak SMEs and only 10% of enterprises that are not internationally active.

Table 13. EU SMEs participation in international cooperation in %

	Investment Abroad	Technological Cooperation with Enterprises Abroad	Subcontractor of Foreign Partner	Enterprise with a Foreign Subcontractor
Micro	2	7	7	7
Small	6	12	11	12
Medium	16	22	17	16
SMEs total	**24**	**41**	**45**	**45**

Source: Malega (2012)

44% of SMEs in the EU are engaged in all forms of internationalization, while in Slovakia it is 58%. The relative share of exports and imports is as follows:

- Compared to the EU average, which is 25%, the share of Slovak SMEs in exports is above this average with a value of approximately 28%.
- The share of SMEs in Slovakian imports is over 40%, while the EU average is 29%.

It is evident that the Slovak SMEs are involved in internationalization mainly in the area of EU member states while in the area of non-EU markets activities of Slovak SMEs are very low.

DESIGN OF FOREIGN MARKET ENTRY STRATEGY FOR SME COMPETITIVENESS

It proceeds from the author's analysis that before an enterprise starts to participate in external economic relations in the form of entry into foreign markets, it must answer three basic questions:

1. Are conditions (external and internal) created to enable successful realization of activities in the international market?
2. What kind of foreign market is best for the enterprise?
3. What form of action on the international market is optimal for the enterprise?

Finding answers to these questions is about establishment on the international market and the associated decision-making process. This can be divided into three separate stages (Figure 1).

It should be noted that between doing business in the domestic and foreign markets, there are certain differences as summarized by the author in Table 14.

Figure 2 shows the modes, in the author's view, as to how an enterprise can proceed in the decision making process on entry into foreign markets.

Figure 1. Decision-making process with entry into foreign markets
Source: *Author*

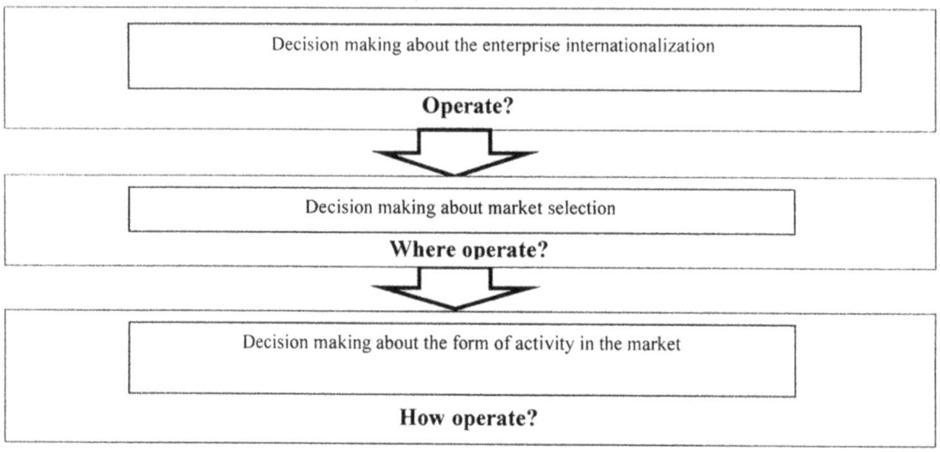

Table 14. Differences in doing business between domestic and foreign markets

Business in the Domestic Market	Business in the Foreign Market
One language, culture, nationality	Many languages, cultures and nationalities
Homogenous market	Heterogeneous market
Easier access to information	Complicated access to information
Political factors are rather unimportant	Political factors are more decisive
Independence from government interference	Dependence on government intervention
One currency	Different currencies
Stabilized and known business rules	Various, often unstable and unclear business rules
Management accustomed to the division of responsibilities and financial control	Management with great autonomy and low financial control
More stable and familiar environment	Unknown and unstable environment

Source: Malega (2010)

Figure 2. Opportunities in decision-making with entry into foreign markets
Source: *Author*

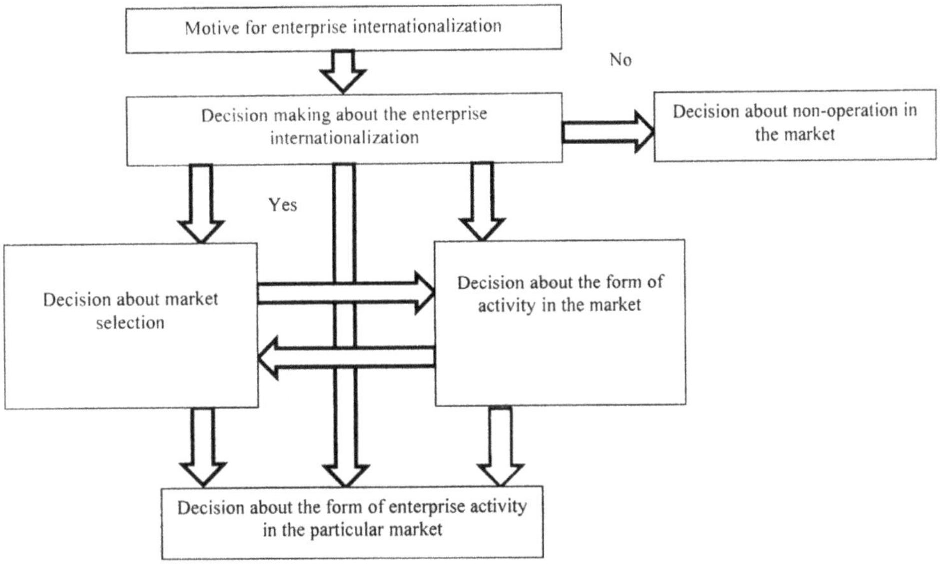

FUTURE RESEARCH DIRECTIONS

If a computer simulation is programmed, it would also clearly show that SMEs in Slovakia have very large potential still not sufficiently used in the EU and broader. However, removing high administrative burdens for business, improvement of digitization and computerization of public administration in relation to the business environment is indeed required if change is to be made through competitiveness and optimization of Slovak SME´s in the global markets.

This can be the key to success for the Slovak Republic, because this weakness decreases the rating of the country as a whole and of course for Slovak SMEs it is also a very important barrier. In case of reorganization of legal environment in this segment, it will open other markets for Slovak SMEs and help them to increase their competitiveness.

If new surveys or interviews are conducted in SMEs, they will probably show that the Slovak Republic is strongly oriented on the machine industry, in particular automobile production, as it already established three automobile factories (Volkswagen, KIA, Peugeot-Citroen) followed shortly with another (Land Rover) becoming the first in the world by the number of cars produced per capita (in the year 2015). This is a very big problem for SMEs, which consequently therefore have a very difficult position in the Slovak Republic. The starting point should be better state support which is already operating and should become the reality until the end of the 2017.

Finally, new research will probably show that despite a number of positive signals, which have already been implemented for SMEs, the biggest problem for competitiveness is the lack of finance, which is a problem in every sector of the economy and not only in the Slovak Republic. Because of this reason, Slovak SMEs are increasingly motivated for obtaining support from the EU, which for them is an invaluable source of additional finance which otherwise can´t be obtained.

CONCLUSION

Entrepreneurship has become the basic element of efficient operation of the market economy. It is a driver of change, measure of economic development, instrument of customization and the ability to keep pace with the major global trends. In each country SMEs have an irreplaceable contribution for economic growth, employment and competitiveness. But because of their size they are sensitive to changes in the external environment and their productivity depends on existence of business barriers. Concerning the fact that SMEs are not able to maintain control of the markets in which they operate, they take a greater business risk and have insufficient capital. Therefore, active and effective support is necessary. For a lot of them the chance to operate in the international market is a big opportunity. This form of operation has a lot of advantages, but also disadvantages and on these advantages SMEs must react and manage them.

Creating conditions for a quality entrepreneurial environment to achieve long-term sustainable economic growth is a prerequisite for business development and increasing competitiveness of the Slovak Republic on the international scale. For this country, whose main objective is to reach the level of developed countries in Europe as soon as possible, business development for finding new customers has irreplaceable importance. There are a lot of potential customers who are scattered and differ in their purchasing requirements. Therefore, instead of competing across the market with lots of competitors, enterprises have to define the most attractive customer groups in the market for them that they could efficiently satisfy. The starting point of the target groups for which the enterprise wants to focus on are the customer needs, motives and behavior of customers and also the overall market situation, developments and future trends in the market.

This chapter summarized the approaches for a methodology to improve the business environment and competitiveness in the Slovak Republic in order for this country to compete in the domestic and international market.

Slovak SMEs have a lot of troubles in their activity, such as barriers and strong orientation on automotive industry (large enterprises) which is a big problem because state support is oriented primarily on large enterprises. Despite these facts, Slovak SMEs are competitive and they play a significant role in the Slovak business environment.

REFERENCES

Altomonte, C., Navaretti, G. B., di Mauro, F., & Ottaviano, G. (2011). Assessing competitiveness: how firm-level data can help. *Bruegel Policy Contribution, 16*.

Annoni, P., & Dijkstra, L. (2013). *EU regional competitiveness index*. Luxembourg: Europe.

Belás, J., Demjan, V., Habánik, J., Hudáková, M., & Sipko, J. (2015). The business environment of small and medium-sized enterprises in selected regions of the Czech Republic and Slovakia. *E+ M Ekonomie a Management*, (1), 95.

Bialková Šimová, D. (2014). *Efficiency of amount incurred to support SMEs from EU funds*. Bratislava, Slovakia: TU SjF Košice.

Boutellier, R., Gassmann, O., & Von Zedtwitz, M. (2013). *Managing global innovation: uncovering the secrets of future competitiveness*. Springer Science & Business Media.

Canales, R., & Nanda, R. (2012). A darker side to decentralized banks: Market power and credit rationing in SME lending. *Journal of Financial Economics, 105*(2), 353–366. doi:10.1016/j.jfineco.2012.03.006

Dierkes, M., Erner, C., Langer, T., & Norden, L. (2013). Business credit information sharing and default risk of private firms. *Journal of Banking & Finance, 37*(8), 2867–2878. doi:10.1016/j.jbankfin.2013.03.018

Eggers, F., Kraus, S., Hughes, M., Laraway, S., & Snycerski, S. (2013). Implications of customer and entrepreneurial orientations for SME growth. *Management Decision, 51*(3), 524–546. doi:10.1108/00251741311309643

Esser, K., Hillebrand, W., Messner, D., & Meyer-Stamer, J. (2013). *Systemic competitiveness: new governance patterns for industrial development*. Routledge.

European Funds and Slovakia. (2014). *The experience of the past for future development I.* NSRR of SR for programing period 2007-2013.

Hillary, R. (2000). *Small and Medium Sized Enterprises and the Environment: Business Imperatives.* Greenleaf Publishing. Retrieved May 12th, 2016 from web site: http://reports.weforum.org/global-competitiveness-report-2015-2016/

Kováč, M., Malega, P., & Spišáková, E. (2013). *SMEs*. Košice, Slovakia: TU, SjF.

Kraus, S., Rigtering, J. C., Hughes, M., & Hosman, V. (2012). Entrepreneurial orientation and the business performance of SMEs: A quantitative study from the Netherlands. *Review of Managerial Science, 6*(2), 161–182. doi:10.1007/s11846-011-0062-9

Lasagni, A. (2012). How can external relationships enhance innovation in SMEs? New evidence for Europe. *Journal of Small Business Management, 50*(2), 310–339. doi:10.1111/j.1540-627X.2012.00355.x

Májková-Sobeková, M. (2011). Analyse of barriers and factors of SMEs funding in Slovak republic. *Economic Magazine*, (10), 1033-1049.

Malega, P. (2010). External business environment searching as the important tool for decision-making. *Managerial Forum, 8*(2), 79-81.

Malega, P. (2012). Competitiveness of Slovak enterprises and possibilities of the comparison. *Managerial Forum, 12*(1), 17-22.

Malega, P. (2014). Entry strategy into foreign markets of SMEs. *Management of Companies.*, *4*(2), 89–93.

Neuberger, D., & Räthke, S. (2009). Microenterprises and multiple bank relationships: The case of professionals. *Small Business Economics*, *32*(2), 207–229. doi:10.1007/s11187-007-9076-8

Nummela, N. (2011). *International Growth of SMEs*. New York: Routledge.

Rugman, A. M., Oh, C. H., & Lim, D. S. (2012). The regional and global competitiveness of multinational firms. *Journal of the Academy of Marketing Science*, *40*(2), 218–235. doi:10.1007/s11747-011-0270-5

Schwarcz, P. (2011). *European Union programs and their funding*. Nitra, Slovakia: Slovak University of Agriculture in Nitra.

Smit, A. J. (2010). The competitive advantage of nations: Is Porter's Diamond Framework a new theory that explains the international competitiveness of countries. *Southern African Business Review*, *14*(1), 105–130.

Soininen, J., Martikainen, M., Puumalainen, K., & Kyläheiko, K. (2012). Entrepreneurial orientation: Growth and profitability of Finnish small-and medium-sized enterprises. *International Journal of Production Economics*, *140*(2), 614–621. doi:10.1016/j.ijpe.2011.05.029

World Economic Forum. (2016). Retrieved August 17th, 2016 from WEF web site: http://reports.weforum.org/global-competitiveness-report-2015-2016/

Zajko, M., Chodasová, Z., Jemala, Ľ., & Materák, M. (2010). *Management of SMEs*. Bratislava, Slovakia: STU in Bratislava.

ADDITIONAL READING

Balkyte, A., & Tvaronavičiene, M. (2010). Perception of competitiveness in the context of sustainable development: Facets of sustainable competitiveness. *Journal of Business Economics and Management*, *11*(2), 341–365. doi:10.3846/jbem.2010.17

Blanke, J., & Chiesa, T. (Eds.). (2013). The Travel & Tourism Competitiveness report 2013. Retrieved August 16th, 2016 from World Economic Forum web site: http://www3.weforum.org/docs/WEF_TT_Competitiveness_Report_2013.pdf

Bristow, G. (2010). *Critical reflections on regional competitiveness: Theory, policy, practice*. Routledge.

Brockhoff, K. (2012). *Industrial research for future competitiveness*. Springer Science & Business Media.

Camagni, R., & Capello, R. (2010). Macroeconomic and territorial policies for regional competitiveness: An EU perspective. *Regional Science Policy & Practice*, *2*(1), 1–19. doi:10.1111/j.1757-7802.2010.01016.x

Croes, R. (2010). Measuring and explaining competitiveness in the context of small island destinations. *Journal of Travel Research*.

Delgado, M., Ketels, C., Porter, M. E., & Stern, S. (2012). *The determinants of national competitiveness (No. w18249)*. National Bureau of Economic Research. doi:10.3386/w18249

Dijkstra, S., & Giaccaria, P. (2013). *Local development and competitiveness* (Vol. 59). Springer Science & Business Media.

Dunning, J. H. (2013). *Multinationals, Technology & Competitiveness (RLE International Business)* (Vol. 13). Routledge.

González-Pernía, J. L., Peña-Legazkue, I., & Vendrell-Herrero, F. (2012). Innovation, entrepreneurial activity and competitiveness at a sub-national level. *Small Business Economics*, *39*(3), 561–574. doi:10.1007/s11187-011-9330-y

Grandori, A. (Ed.). (2012). *Interfirm networks: Organization and industrial competitiveness*. Routledge.

Hitt, M. A., Ireland, R. D., & Hoskisson, R. E. (2012). *Strategic management cases: competitiveness and globalization*. Cengage Learning.

Huggins, R., & Williams, N. (2011). Entrepreneurship and regional competitiveness: The role and progression of policy. *Entrepreneurship & Regional Development*, *23*(9-10), 907–932. doi:10.1080/08985626.2011.577818

Kráľovič, J., & Vlachynský, K. (2002). *Financial management*. Bratislava, Slovakia: IURA EDITION, Ltd.

Majková-Sobeková, M. (2008). *Funding opportunities for SMEs in Slovakia*. Brno, Czech Republic: Tribun EU, s.r.o.

Malega, P., & Janeková, J. (2015). *Economics of SMEs*. Košice, Slovakia: TU, SjF.

Martin, R., Kitson, M., & Tyler, P. (Eds.). (2012). *Regional competitiveness*. Routledge.

Mihok, J., Janeková, J., Vidová, J. (2005). *Business Economics*. Košice, Slovakia: East Slovakia printer, Inc.

Rugman, A. M., Oh, C. H., & Lim, D. S. (2012). The regional and global competitiveness of multinational firms. *Journal of the Academy of Marketing Science*, *40*(2), 218–235. doi:10.1007/s11747-011-0270-5

Sahlberg, P., & Oldroyd, D. (2010). Pedagogy for economic competitiveness and sustainable development. *European Journal of Education*, *45*(2), 280–299. doi:10.1111/j.1465-3435.2010.01429.x

Schwab, K., & Sala-i-Martin, X. (Eds.). (2011, December). The global competitiveness report 2011-2012. Geneva: World Economic Forum.

Şener, S., & Sarıdoğan, E. (2011). The effects of science-technology-innovation on competitiveness and economic growth. *Procedia: Social and Behavioral Sciences*, *24*, 815–828. doi:10.1016/j.sbspro.2011.09.127

Timmer, M. P., Los, B., Stehrer, R., & De Vries, G. J. (2013). Fragmentation, incomes and jobs: An analysis of European competitiveness. *Economic Policy*, *28*(76), 613–661. doi:10.1111/1468-0327.12018

KEY TERMS AND DEFINITIONS

Business Risk: Chance that a company will achieve lower than anticipated profit or experience a loss rather than make a profit.

Competitiveness: Ability of a firm or a nation to offer products and services that meet the quality standards of the local and world markets at prices that are competitive and provide adequate returns on the resources employed or consumed in producing them.

Entry Strategy: Planning and implementation of a strategy of delivering goods or services to a new target market.

EU Market: An association of countries trading with each other without any internal borders or other regulatory obstacles to the free movement of goods and services, without restrictions or tariffs, that came into effect on January 1, 1993.

International Cooperation: The interaction of persons or groups of persons representing various nations in the pursuit of a common goal or interest.

Internationalization: The designing of a product in such a way that it will meet the needs of users in many countries or can be easily adapted to do so.

Regional Operational Programs: An implementation tool to achieve the aims and actions of the EU Cohesion Policy particularly meant for reducing disparities in development of a territory and to contribute to the priorities of smart, sustainable and inclusive growth.

ENDNOTES

1. Editors' note: It is perhaps useful here to make a kind of comparison with what has happened in other countries. A fuller discussion of Polish, Italian, Finish and Spanish SME competitiveness is made in chapters 1, 2, 3, 7 and 14.
2. Editors' note: For a fuller treatment of risk management through insurance see a similar discussion in chapter 9.

Section 2

Optimal Pricing, Financing and Insurance Management Strategies for SMEs

Chapter 7
Initial Price Strategies of Polish Micro and Small Enterprises:
An Application of Game Theory for Industrial Organization of the SME Sector

Mariusz Maciejczak
Warsaw University of Life Sciences (SGGW), Poland

Adrian Słodki
Warsaw University of Life Sciences (SGGW), Poland

ABSTRACT

The sector of micro, small and medium size enterprises is important for any economy. It is important also for Poland. Analyzing the industrial organization of this sector it was confirmed that the owners and managers of such companies are applying strategies, which are rational from their point of view, but not from the perspective of real market conditions. It is argued therefore that the game theory is for them a solution in enhancing competences and performance of their organizations. Based on randomized sample of Polish micro and small companies the chapter aimed to find out if the managers apply the game theory rationales when choosing price strategy when enter the market. It was confirmed that they do not and that they don't play Nash equilibrium in the strategic interaction when it comes to the price level. There was applied maxmin strategy, which maximises the worst - case scenario from the game. Thus there is a real chance that if entrepreneurs would analyze the situation with respect of game theory, their strategies would be more accurate and provide better outcomes.

INTRODUCTION

Industrial organization is concerned with the functioning of markets and industries, in particular the way firms compete with each other. However, in the mainstream economics the study of how markets operate is the object of microeconomics. But, as stated i.e. by Tirole (1988, pp. 7-12), the main reason for considering industrial organization as a separate subject, respectively in a heterodox way of economic

DOI: 10.4018/978-1-5225-1949-2.ch007

thought, is its emphasis on the study of firm strategies that are characteristic of market interaction, i.e. price competition, product positioning or research and development.

Entrepreneurs and managers, especially small, incl. micro, and medium size enterprises (SMEs) can be guided by many reasons in choosing a strategy from different strategic options. They are applying strategies, which are very often rational from their point of view, but not from the perspective of real market conditions. Therefore, in the wake of knowledge-based economy, to elicit the tacit knowledge and face it with rational guidelines the use of directions based on rational and optimal choices, i.e. game theory, seems to be a useful solution for them in enhancing competences and organizational performances. The use of game theory guidelines in such companies is however at infancy.

Regardless if it is conscious or not, the game that every company must play, and is playing, is the game of price strategy, i.e. while entering the market. Entrepreneur can be guided by different reasons in choosing this strategy. If an owner or a manager of SMEs tries to create an image of a cheap company, the price level will be set very low. If the company is trying to seem expensive and sophisticated and offers high quality products or services, the price level will be set rather high. If the entrepreneur won't think about the image but just focus on price margin, the price level can be similar to the competitors. Entrepreneurs from the SME sector can have different visions about their company. But they usually have the same goals (see Audretsch & Beckmann, 2007, pp. 45-47): to generate profit and try to eliminate competitors. In achieving these goals, game theory seems to be a reasonable tool to use in establishing an initial price strategy.

As far as Polish micro and small enterprises are concerned, it was shown by the previous studies (i.e. Smolarek, 2012, pp. 53-68) that they have very limited preferences that allow to examine their initial price strategies. Usually in a typical Polish micro or small company the manager is also the owner. This is very important for the study, because such a person is able to answer questions about company's strategy and conditions in the first months of activity. Also, micro and small-sized companies are usually younger than medium or large firms. The younger the company, the more accurate will be the answers about its initial price strategy. So far there has not been research carried out about the initial price strategies of Polish micro and small enterprises. And this work is trying to fill this gap.

The primary goal of this chapter is to present how rationally, from the perspective of game theory, Polish micro and small size enterprises behave once they develop their initial prices on the market. Additionally it aims both: to present in which areas of their activities SMEs could strategically use the game theory achievements as well as to present the importance of micro, small and medium size enterprises for the Polish economy.

DATA AND METHODS

Apart from an in-depth critical literature review the author has applied a primary research on Polish micro and small companies, which were analysed by the game theory assumptions. For the purpose of the study there were 153 Polish entrepreneurs questioned through in depth interviews executed in 2013 and 2014. The sample was selected randomly. All researched companies operated locally in the Mazovia province of Poland. The selection of the companies was limited by two main restrictions: number of employees (not more than 50 workers in the company) and time of presence on the market (at least one year of activity).

There was used a structured questionnaire. In the survey, on the question of employment in the company the respondent had four options to choose: 0-5 – this is the group of self-employed and micro companies that hire less than 6 employees; 6-10 – this is the group of micro enterprises that employ up to 10 employees; 11-50 – this is the group of small enterprises, the largest companies in the analysed sample. In the sample, more than half (53%), whereas companies of self-employed entrepreneurs, and micro enterprises that employ less than 6 workers. Next was the group of "larger" micro enterprises (36%), that employ 6 and more employees. The smallest response to the questionnaire was from the small companies – 11%.

Second restriction was time experience on the market. Under this consideration only entrepreneurs that run their business for more than one year were interviewed. The question asked in the survey was to point out the price strategy from the start of the business activity and to evaluate it; to specify if it was successful for the company, and if the set goals were achieved. The entrepreneur that freshly started the business can precisely respond about his or her price strategy, but he or she cannot properly evaluate it because of the lack of experience. To respond to the question about the age of the company, the respondent had four options to choose from. First, the company exists on the market less than a year. Those companies were excluded from further analysis due to the reasons presented before. Other three options were from one to five years, from five to ten years and more than ten. Division by age of the company is quite regular. The largest group consists of companies that are present on the market from one to five years (38%). Next are the entrepreneurs that run their business more than five years but less than ten (34%), and the smallest group are companies that are present on the market for more than ten years (28%)

The purpose of asking entrepreneurs about their education was to prove that they haven't heard about game theory. The question was not phrased directly "do you know anything about game theory?" for two reasons. Firstly, to avoid confusion, the questionnaire was tested on a selected, small group of entrepreneurs. The question, phrased directly, raised lots of questions. Second reason, to diminish risk of untrue responses; it is a common fact, that people when asked about their knowledge in some area tend to try look smarter. The question phrased indirectly solves those two problems. The answer gives just the probability that entrepreneurs know about game theory and shows the structure of education in the sample. Even if person would lie in the question, it won't be that harmful as it would be with question phrased directly. In the question, respondent had four options to choose from; elementary education, high-school education, National Vocational Qualification and university education. The biggest group in the sample were entrepreneurs with high- school education (40%), the smallest group, entrepreneurs with elementary education (12%). The shares of company owners who have higher education or National Vocational Qualification were similar, respectively 23% and 25%. On the grounds of those findings it is possible to assume, that 23% of the analysed group had a possibility to learn about game theory, because this topic is touched only in universities and not in an elementary or high school.

From the methodological point of view there was applied a game theory approach in the analysis of the collected data. It was assumed that game theory looks at behaviour of self-interested participants in strategic interactions and also how those interactions should be structured in order to lead to the best outcomes.

To find out if Polish micro and small-sized company managers used game theory in taking a price strategy when they entered the market there was created a model that simulates the situation. However due to the lack of respondents who chose strategy of higher prices, the model was simplified and a player could take two strategies. There was applied the strategy that maximises player's worst case scenario payoff, assuming that others wish to hurt the player while the other agent's strategy was to diminish other players' payoffs. For solving the model there was used a maxmin strategy, which maximises the worst-case scenario from the game. This strategy has particular sense in zero-sum games.

From the game theory point of view the key ingredients of used game are players, actions and outcomes. Those three elements create a model that shows possible behaviour of participants and can be solved by using methods that game theory offers (Jackson, 2010, pp. 20-45). A player in the game is the participant that makes the decision; it can be a person, a company, a government or even an animal. It all depends on the situation that is being studied. A player is the self- interested participant of the situation or self-interested agent. Self-interested implies that every player has own preferences, which doesn't necessarily mean that self-interested agents don't care about other participants. Every player has a utility function, which is a mathematical measure that reflects how much an agent likes or dislikes the given situation. It quantifies the degree of preference across alternatives and explains the impact of uncertainty. Every player takes action that maximises its expected utility (Dixit & Nalebuff, 1993, pp. 58-61). Accordingly, an action is the choice of the player that leads to a certain outcome. Finally an outcome of payoff is the utility that a player receives after the game. A payoff is what motivates the player. It depends on the preferences of the participant, regardless of whether a player cares about profit or other players.

Basic ingredients of the Nash equilibrium are some predictions of what other players are going to do and choosing an optimal strategy in response to other players' behaviour which leads to the best outcome (best response). If one player knows how other players will behave, it is possible to tell what the best response is. If $a_{-i} = <a_1, \ldots, a_{i-1}, a_{i+1}, \ldots, a_n>$ is the action sequence of all other players; without player a_i; the entire action vector is $a = (a_{i-1}, a_i)$ (Roth, 2015, pp. 8-15).

$$a_i^* \in BR(a_{-i}) \text{ if } \forall a_i \in A_i, u_i(a_i^*, a_{-i}) \geq u_i(a_i, a_{-i}).$$

Players best response (a_i^*) must be a part of the set of best responses (BR) when every other player plays a_{-i}, if and only if, for any a_i which is any action that player is willing to take from all the actions possible (A_i), the player's utility ($u_i(a_i^*, a_{-i})$) for playing that strategy, is equal or greater than any other action that player can take (($u_i(a_i, a_{-i})$)). The a_i^* is player's best response. However, it is not always possible to know what other players will do. The notion of best response is used to define the Nash equilibrium which is a set of actions one for each of the agents such as each is the best response to the others.

$a = <a_1, \ldots, a_n>$ is a ("pure strategy") Nash equilibrium if $\forall i, a_i \in BR(a_{-i})$.

Set of actions or action profile a is a Nash equilibrium if and only if for every agent i, the action a_i is the best response for the remaining actions that other players can take (Leyton-Brown & Shoham, 2008, pp. 45-63). Assuming the above, the strategy is a choice of action. There are two different strategies s_i and s_i' that player i can take. S_{-i} is the set of all possible strategy profiles that other players can take.

s_i strictly dominates s_i' if $\forall s_{-i} \in S_{-i}, u_i(s_i, s_{-i}) > u_i(s_i', s_{-i})$

s_i strictly dominates s_i' if for every other strategy profile of other agents, the utility that player i receives when he or she plays s_i is greater than the utility when he or she plays $s_i^{'}$. The player is always happier when he or she play s_i than $s_i^{'}$ because it brings more units of utility.

s_i very weakly dominates $s_i^{'}$ if $\forall s_{-i} \in S_{-i}, u_i\left(s_i, s_{-i}\right) \geq u_i\left(s_i^{'}, s_{-i}\right)$

s_i very weakly dominates $s_i^{'}$ if for every other strategy profile of other agents,

the utility that player i receives when he or she plays s_i is equal or greater than the utility when he or she plays $s_i^{'}$. The player is always at least as happy playing s_i as he or she is playing $s_i^{'}$. Strategy s_i very weakly dominates $s_i^{'}$ because the utility that strategy brings the player can be greater than or equal as the strategy $s_i^{'}$ (Dixit & Nalebuff, 1993, p. 67-88).

The dominant strategy is the strategy that is better than all the other possible strategies. It is the best strategy to choose; it always brings the largest amount of utility no matter what other agents do. Strategy profile consisting of dominant strategies for every agent must be the Nash equilibrium. If all players play a dominant strategy, the outcome is the Nash equilibrium, because every player has the best outcome for themselves and no one wants to change the strategy.

As Dixit, Nalebuff (1993, p. 72-79) indicate, maxmin and minimax strategies correspond with each other. Each player's maxmin value is equal to the minimax value; for both players, the set of maxmin strategies coincides with set of minimax strategies. They produce the same result as a method of finding the Nash equilibrium in general 2x2 games. And the maxmin value of the game for the player is the minimum payoff guaranteed by a maxmin strategy.

APPLICATION OF GAME THEORY FOR INDUSTRIAL ORGANIZATION IN THE SME SECTOR

It is accepted and confirmed i.e. by Vos (2005, pp. 989–999) that there exists a managerial gap in many micro, small and medium-sized enterprises to formulate their strategies. Additionally not much research has been carried out on SMEs companies, which indicates, how these companies formulate, implement, manage and evaluate their strategic options and strategies. As of yet, little attention has been paid on rationality of decision making in case of their existing strategies too. But it is shown by Villa & Antonelli (2009, pp. 61-115) that managers of SMEs have poor skills in reflecting upon their companies strategically. These limited skills make companies in a sense out of control.

Nonetheless, SME managers have a number of strategic orientations they can adopt in their attempts to maximize the chances of obtaining assumed goals. These include an entrepreneurial or innovative orientation, a marketing or a market orientation, just to mention a few quoted by different authors (i.e. Auerswald & Branscomb 2003, p. 61-65; Feldman 2003, pp. 91-96)[1]. In deciding which of these alternatives to pursue, firms must evaluate their internal resources and the environment, to find an optimal strategic fit. In making this choice, a theoretic game perspective could be used as a good decision support tool (Kleindl, 1999, pp. 265-274).

Game theory is a complex subject that spans games of static (one-time) and dynamic (repeated) nature under perfect or imperfect information. There is however a limited number of researches on its applications, which offer solid background for decision making of entrepreneurs from micro, small and medium companies. The literature review allowed the author to distinguish few areas where entrepreneurs and decision makers, especially from micro and small companies, could get support from scientific and applied works based on the game theory. These are however issues related to management of financial situation, coordination of cooperation and of course competition.

One of these areas is industrial organization of the supply chain. Cachon and Netessine (2004, pp. 13-59) as well as Hennet and Yesemin (2008, pp. 399-405) showed that in practice, decisions on the structure of the supply chain cannot be considered optimal since they are taken locally by legally separated organizations and they are subject to time constraints of minimal duration of contracts and financial agreements. SMEs managers must address the issues of raising capital, hiring competent qualified personnel, investing on industrial and administrative premises and building a reliable business plan. A key factor influencing these strategic decisions is the choice of products and markets, with two possible trends: diversification or specialization. Many different models have been proposed to represent the local autonomy of production units within supply chains. As an organizational structure, a supply chain is a system composed of different units having their own resources, methods and goals (Monostori et al., 2015, pp. 4-15). In this context, negotiation is necessary to achieve collaboration of partners with different and often conflicting objectives. Such characteristics are of major concern in game theory, under the classical decomposition into cooperative (or coalitional) and non-cooperative games. The model referred as “non-cooperative” consists of players with different preference relations or utility functions. Their actions follow a strategy which takes into account information (generally imperfect) on the other players’ actions and preferences. Of particular importance in non-cooperative games theory are the problems of existence, uniqueness and characterization of an equilibrium set of actions, which is locally optimal for all players. Under different rules, a game may lead to different equilibrium profiles: the Nash equilibrium if the game is symmetric, the Stackelberg equilibrium in the case of two players with a leader and a follower (Hennet, 2008, pp. 86-96).

As noted in (Cachon & Netessine, 2004, pp. 45-59), most supply chain models based on game theory use a non-cooperative approach. In this setting, a key result for decentralized supply chains is sub-optimality of Nash equilibria. Sucky (2007), stated that competition reduces however efficiency (pp. 3638-3651). A large number of studies have proposed contracts between suppliers and retailers to coordinate the system and reduce the optimality gap through cost and risk sharing (Cachon & Netessine, 2004, pp. 50-56).

Another application area is management of the company’s financial strategies. As indicated by Wang & Xu (2010, pp. 434-438) financing has become the biggest obstacle and the “bottleneck” in constraining the development of SMEs. They need to look for external sources of capital. In taking the loan game theory offers support in the scope of negotiation strategies but also in establishing collaterals. It was confirmed that at present, most of the bank loans for SMEs are based on fixed assets mortgage. But for many SMEs, their fixed assets meeting the demands for mortgage are very limited. What they do have, are current assets such as inventory, bill of rights (e.g. warehouse receipt), which account for the majority of their total assets. This could be a kind of trade financing based on commercial credit.

The literature devoted to game theory application to SMEs decision also shows that changing the game is hard. There are many potential traps. The game theory might be to help managers recognize and avoid these traps. It is advised i.e. that it is far more rewarding to be a game maker than a game taker, as well as to look for both win-win and win-lose strategies – is more profitable. Game theory solutions point out that to understand the effect of any particular strategy, managers need to go beyond their own perspective, be more allocentric rather than egocentric (Brandenburger & Nalebuff, 1995, pp. 57-74).

THE SME SECTOR IN POLAND

SMEs are a highly diverse population of enterprises, and are present in every nook and cranny of the economy, with activities ranging from the production of artisan food to the production of high tech space exploration equipment, from retail services to the provision of highly specialised professional services, from focusing primarily on serving domestic customers to focusing mainly on the export markets. According to recent studies of the European Commission in the European Union (28 Member States) micro, small and medium sized enterprises accounted for 99,8% of all enterprises in the non-financial business sector. In comparison, for every km^2 of land surface the EU has an average of 5 SMEs. Moreover, in 2014 SMEs employed almost 90 million people - 67% of total employment, and generated 58% of the sector's value added. Almost all SMEs (93%) are micro SMEs employing less than 10 people. About three quarters of SMEs are active in the five key sectors: wholesale and retail trade, manufacturing, construction, business services and accommodation and food services. It is estimated that SMEs in EU generated 71,4% of the increase of employment in 2014 in the non-financial business sector, accordingly value added grew by 3,3% and employment by 1,2%, while in 2013 value added grew by 1,6% and employment declined by 0,5% (Muller et al., 2015, pp. 4-12).

The structure of Polish enterprises has remained practically unchanged in recent years (see table 1). From 2008 the micro-enterprises accounting for around 95%, small enterprises for around 4%, and medium enterprises for around 1%. Large enterprises accounted for 0.2% of all enterprises. In 2016 there were registered in Poland 1450 thousand micro enterprises, 55.3 thousand small and 13.7 thousand medium firms. The SME sector therefore accounts for 99.8% of all companies, which is the same as the EU average. In the period 2008-2016 there were observed no significant changes in the structure of the enterprises in Poland.

Also, the structure of employment is Polish companies remained without significant changes (see Table 2). In the period 2008-2012 SMEs employed about 69% of all employees, while in the following years this number decreased to 68%. In this number the micro companies accounted for the largest share, i.e. about 53% in the period 2008-2016. Accordingly the small companies employed about 20%, and medium sized companies about 27% of workforce.

Considering the value added that SMEs bring to national economy in Poland the input accounts for about 50% in the years 2008-2016 and is not changing significantly in the year the year comparison (see Table 3). The micro companies brought in 2016 about 29.5 billion euro to the national accounts, i.e.

Table 1. Number of enterprises in Poland in the years 2008-2015

Size Class	2008	2009	2010	2011	2012	2013	2014	2015
0 – 9	1 464 089	1 358 017	1 392 002	1 431 525	1 426 044	1 402 323	1 446 554	1 447 731
10 – 49	51 403	47 985	49 758	53 021	55 195	55 402	55 985	55 654
50 – 249	15 567	15 559	15 447	15 266	15 049	14 471	14 489	14 123
250 +	3 134	3 078	3 083	3 009	3 014	3 056	3 114	3 047
Total	1 534 193	1 424 639	1 460 290	1 502 821	1 499 302	1 475 253	1 520 142	1 520 556
All SMEs	1 531 059	1 421 561	1 457 207	1 499 812	1 496 288	1 472 196	1 517 028	1 517 508

Source: Muller et al., 2015

Table 2. Employment in Polish enterprises in the years 2008-2015

Size Class	2008	2009	2010	2011	2012	2013	2014	2015
0 – 9	3 214 347	3 048 935	2 998 170	3 058 650	3 003 819	2 892 603	3 001 152	3 011 868
10 – 49	1 122 407	1 086 725	1 090 171	1 122 123	1 154 137	1 140 857	1 163 831	1 168 850
50 – 249	1 629 887	1 621 901	1 607 878	1 577 418	1 550 950	1 506 826	1 524 841	1 507 972
250 +	2 701 274	2 617 432	2 644 285	2 593 557	2 583 679	2 627 925	2 704 834	2 690 154
Total	8 667 915	8 374 993	8 340 504	8 351 748	8 292 585	8 168 210	8 394 658	8 378 843
All SMEs	5 966 641	5 757 561	5 696 219	5 758 191	5 708 906	5 540 286	5 689 824	5 688 690

Source: Muller et al., 2015

Table 3. Value added of Polish enterprises in the years 2008-2015 (m euro)

Size Class	2008	2009	2010	2011	2012	2013	2014	2015
0 – 9	31 289	24 207	26 293	29 142	25 551	26 215	27 439	27 934
10 – 49	25 442	19 272	21 850	23 888	23 630	24 294	25 441	25 808
50 – 249	40 349	32 858	35 712	37 088	36 580	37 911	39 608	39 634
250 +	87 336	72 490	82 064	86 920	85 599	88 378	90 813	91 044
Total	184 416	148 826	165 920	177 038	171 360	176 798	183 302	184 419
All SMEs	97 079	76 336	83 856	90 119	85 761	88 420	92 489	93 375

Source: Muller et al., 2015

30% of all SMEs. Accordingly small companies generated 27.2 billion. euro of value added, i.e. 27.8% and medium size companies 41.2 billion euro, i.e. 42%.

As indicated by the OECD studies the SME sector in Poland is dominated by significantly smaller size companies, with a vast majority of microenterprises (Potter & Protto, 2010, pp. 12-17). According to national statistics, in 2014 new enterprises accounted for approximately 8,7% of entities in the national economy. Natural persons conducting economic activity constituted the largest group among the newly established entities (more than 80% of their total number). In terms of industry sectors, the most intensive growth in the number of newly registered companies occurred in the following: Other service activities (up by 34%), transportation and storage (up by 7,4%) and construction (up by 5%). As regards active companies, i.e. those companies which actually pursue economic activities, according to the most recent data available (i.e. the data for year 2014), their number has decreased slightly, reaching the level of 1.,77 million (Ministry of Economy of Poland, 2015, pp.3-26).

Potter & Proto (2010) reported the small share of technology based and/or SMEs engaged in high value added activity in Poland. Accordingly it was concluded that this sector is characterized by a small share of technology oriented and high value added firms in the economic structure. This has been also confirmed by the latest studies on innovation performance, which show that Polish SMEs perform poorly on most innovation indicators, reflecting underdeveloped market-oriented innovation systems at the national and regional levels, as well as failure to recognise the potential importance of innovation to competitiveness in the mind-sets of SME managers. According to the latest Innovation Union Scoreboard, Poland is among the countries which rank as moderate innovators (Hollanders et. al, 2015, pp.7-25).

The PARP surveys (2015, pp.5-35) confirmed that almost half of entrepreneurs indicate that they are operating under the conditions of fierce competition. The highest percentage of companies experiencing strong competition (65%) was recorded in the trade sector. The strategy that was most often indicated as the most efficient for coping with competition was range extension, which means placing new products/services on the market or innovative activity. Very few entrepreneurs (3%) thought that the most efficient way was to cut costs - which means to reduce employment, lower wages and similar. If such an attitude becomes a permanent part of the operation policy of Polish companies, one cannot expect a real breakthrough in the sources of competitiveness that will define Polish enterprises in the coming years.

STRATEGIES OF POLISH MICRO AND SMALL ENTERPRISES

The structure and the people – owners and/or managers of SMEs influence the most both short and long term functioning of companies. When Poland is concerned, the research of Smolarek (2015, p. 65) showed that about 60% of SMEs have a strategy. Strategy is defined by identification of opportunities and threats from the environment. The biggest opportunities are identified in demand increase and use of EU funds for development. Accordingly, when threats are identified competition from domestic and international markets decrease the demand. The main strategic activities were connected with reduction of costs, modification of present and introduction of new products (Smolarek 2015, pp. 53-68).

However, other studies proved that the strategic approach and actions are not an issue of concern. This comes mostly from the SME focus on their current financial situation, which is especially in regard to liquidity issues and profitability (Orłowski et al., 2010, pp. 17-19). Kaczmarek (2009, pp. 260-271) shows that strategies used by SMEs from Poland use passive and conservative strategies due to the risks that are assessed by entrepreneurs as high. As argued by Strychalska-Rudzewicz & Wysocki (2013, pp. 67-78) the type of used strategies, if identified as strategic actions, depend on the size of company, the scope of activities as well as type of activity (Rosłon 2011). It was confirmed that production companies are keen to use more precise strategies then service companies (Strychalska-Rudziewicz & Wysocki, 2013, pp. 67-78). Additionally, Kordel (2015, pp. 98-103) identified that the more mature SME is the more strategically it is oriented. Other studies indicate that 36% SMEhave developed a strategy dedicated mostly to the marketing issues (price, distribution, sale, etc.), (Rubicom, 2011, pp. 5-16).

ELABORATION OF THE MODEL

To find out, if micro and small-sized company managers used game theory in taking a price strategy when they entered the market, the author created a model that simulates the situation. The extensive form of the model is presented in Figure 1, where:

C_N – new company;
C_O – old company;
$a_1, a_2, \ldots, a_9$ – payoffs;
j – action of player $C_{O;}$
k – action of player $C_{N;}$
$R0$ – player is indifferent, no change in price level;

Figure 1. The extensive form of the model
Source: *compiled by author*

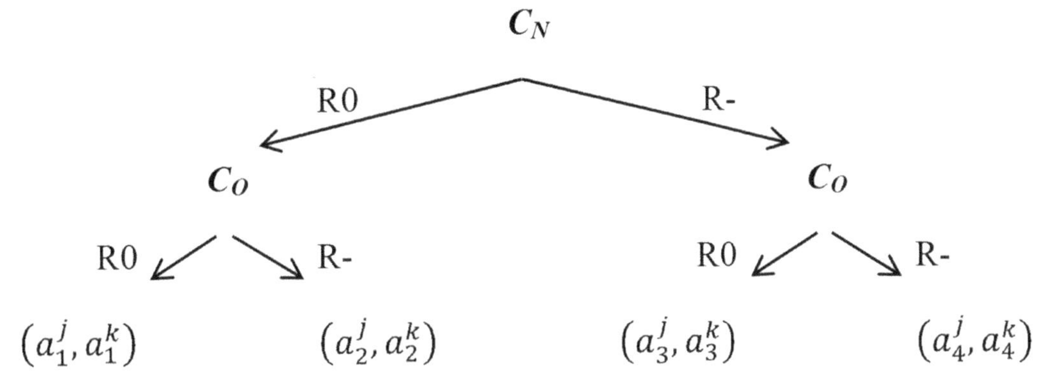

R- - reduction of the price level.

The payoffs are averages based on the survey result and are calculated as follows:

$$C_o = \frac{W_0}{W_G},\ C_N = 1 - C_0;$$

where:

C_o – payoff of the old company (already on the market);
C_N – payoff of the new company;
W_0 – number of won games by old company (C_O) using certain strategy in reply to certain strategy;
W_G – total number of games (with certain strategies).

It is very important, that the model refers to a specific situation, when a company enters the market, and the key interest is the new company and its strategy which makes it a finite game.

The whole point of creating the model is comparison of the probabilities of taking each action by players. The probabilities were calculated by using a maxmin strategy method, which takes the following form:

$$\max_{s_N} \min_{s_O} (s_N(R0)s_O(R0)a_1^k + s_N(R0)s_O(R-)a_2^k + s_N(R-)s_O(R0)a_3^k + s_N(R-)s_O(R-)a_4^k)$$

where:

s_N *(R0)* – probability that player C_N will enter the market with the same price level as competitors;
s_O*(R0)* – probability that player C_O won't change the price level in response to player C_N strategy;
s_N *(R-)* – probability that player C_N will enter the market with lower level of price than competitors;
s_O*(R-)* – probability that player C_O will lower the price level in response to player C_N strategy.

Table 4. Values of payoffs in the model: The normal form

		C_N	
		RO	*R-*
C_O	*RO*	41.51%; 58.49%	57.69%; 42.31%
	R-	46.81%; 53.19%	44.44%; 55.56%

Source: Compiled by authors on the grounds of own research

The payoffs were calculated on the grounds of data from the survey, and for each strategy were as presented in table 4.

The payoffs represent average effectiveness of the strategy for the player, e.g. when new company (C_N) decides that their initial price level will be the same as the competitors and, competitor (C_O) reply to their strategy by lowering the price, the strategy is successful for the new company 53.19% of a time and for the old company 46.81% of a time.

The purpose of maxmin strategy is to find out how players should randomise in a Nash equilibrium; by using the payoffs from the survey it is possible to calculate randomisation that will show a Nash equilibrium in the game.

$$\max_{s_N}\min_{s_O}(s_N\left(R0\right)s_O\left(R0\right)0.5849+s_N\left(R0\right)s_O\left(R-\right)0.5319+\left(R-\right)s_O\left(R0\right)0.4231+s_N\left(R-\right)s_O\left(R-\right)0.5556)$$

$s_N(R\text{-}) = 1 - s_N(RO)$

$s_0(R\text{-}) = 1 - s_0(RO)$

$$\min_{s_O}(s_N\left(R0\right)s_O\left(R0\right)0.5849+s_N\left(R0\right)\left(1-s_O\left(R0\right)\right)0.5319+$$
$$\left(1-s_N\left(R0\right)\right)s_O\left(R0\right)0.4231+\left(1-s_N\left(R0\right)\right)\left(1-s_O\left(R0\right)\right)0.5556)$$

$$\min_{s_O}\left(0.1855\cdot s_N\left(R0\right)s_O\left(R0\right)-0.1325\cdot s_O\left(R0\right)-0.0237\cdot s_N\left(R0\right)+0.5556\right)$$

Player C_N plays with probability $s_N(RO)$ and wants to find out what will be the reply of C_O. The function is expressed by $S_0(RO)$.

$f(s_0(RO)) = (0.1855 \cdot s_N(RO) - 0.1325) \cdot s_0(RO)$ - 0,0237 $\cdot s_N(RO) + 0.5556$

First derivative: $f'(s_0(RO)) =$ 0,1855 $\cdot s_N(RO) - 0.1325$

$f'(s_0(RO)) = 0$

$0.1855 \cdot s_N(RO) - 0.1325 = 0$

$s_N(RO) =$ 0,7143 which means that $s_N(R\text{-}) = 0.2857$

The new company entering the market should randomise 71.43% to 28.57% between entering the market with a price at the same level as competitors and offering a lower price. 71.43% of a time the price level should be the same as that of the competitors and 28.57% the price level should be lower than the competitor to play a Nash equilibrium.

Respectively, to calculate the probability of player C_O $S_0(R0)$, the function is expressed by variable $s_N(R0)$.

$$f(s_N(R0)) = (0.1855 \cdot s_o(R0) - 0{,}0237) \cdot s_N(R0) - 0.1325 \cdot s_o(R0) + 0.5556$$

First derivative: $f'(s_N(R0)) = 0.1855 \cdot s_o(R0) - 0.0237$

$$f'(s_N(R0)) = 0$$

$$0.1855 \cdot s_o(R0) - 0.0237 = 0$$

$s_o(R0) = 0.1278$ which means that $s_o(R\text{-}) = 0.8722$

A company present on the market defending its position should lower the price 87.22% of a time, and leave the price level unchanged 12.78% of a time (see Table 5).

RESULTS AND DISCUSSION

On the grounds of conducted research and calculations, the answer to the question if the managers of analysed micro and small companies from Poland apply the game theory in choosing price strategy when they enter the market would have to be "no". According to the calculations based in the research they don't play Nash equilibrium in the strategic interaction when it comes to the price level.

An entrepreneur who enters the market randomises between choosing the strategy of lower prices and the same prices as competitors respectively 48.37% and 51.63%, which is close to the convenient randomisation 50/50. There are several possible explanations to this statistics.

One, the easiest to deduce, is that entrepreneurs randomise 50/50 when they enter the market. They offer a price and wait for the results. If the results are not satisfactory they change (or not) a price strategy more accurate to the realities of the market. This is a very inconsiderate strategy because by entering the market an entrepreneur should attract in some way the customers (OECD, 2008). Price strategy is usually an elementary factor for customers and that is why it should be carefully considered and not chosen by a flip of a coin.

Table 5. Collation of results from survey and the model

	Reality		**Model**	
	R0	*R-*	*R0*	*R-*
C_N	51,63%	48,37%	71,43%	28,57%
C_O	65,36%	34,64%	12,78%	87,22%

Source: Compiled by authors on the grounds of own research

The other reason can be that entrepreneurs have too big expectations. Large part of the examined sample of entrepreneurs assessed their strategy as non-satisfactory. Wrongly set goals can be the reason why they specifically fixed a price. Entrepreneurs are confident, that their company will be a huge success and they conquer easily the market eliminating a large part of the competition. This reason seems to be credible, because all interviewed companies existed on the market for more than a year, which means that the strategy they obtained at the beginning was not necessarily a wrong one. It just didn't bring expected outcomes quickly enough or the expected results were out of reach. Questions in the questionnaire were designed to allow entrepreneurs to assess their strategy subjectively. The questions were asked in that way, because if entrepreneurs were asked about specific numbers they would not know the answer, or they would not be willing to share such confidential data.

To play the Nash equilibrium, entrepreneurs should randomise 71.43% to the strategy of the same prices and 28.57% to the strategy of lower prices. New entrepreneurs seem to try and play a Nash equilibrium, because a slightly larger group choses the strategy of the same prices as the competitors. On the grounds of the data from the author's model, new entrepreneurs should more confidently enter the market and offer a present price. Thanks to that they would achieve their goals quicker and the tough beginning period would be more manageable.

The author's model focuses on t new entrepreneurs, but due to the nature of the model, the group of old entrepreneurs and their defence strategy had to be examined as well. According to the research, group of entrepreneurs that are facing more new competition often choose the strategy of not changing the price (65.36%) rather than lowering the price level as a defence strategy (34.64%). The same as with new entrepreneurs there are several explanations to such statistics.

One is that old companies ignore new competition on the market. The way of thinking can be that entrepreneurs are confident with their position on the market and believe that their company is "rock-solid" and no one can take away from it loyal customers. Unfortunately for most companies, large share of companies are not that loyal and small and micro companies tend not to be "rock- solid". Again, an irrational attitude to the price strategy can be pernicious.

Another explanation, the same as with new companies, can be a wrong assessment of the strategy's outcome. Since the strategy is assessed subjectively, the entrepreneurs evaluate the strategy badly. But the problem could not lay in the strategy itself, because facing a new competitor always seems harmful. Once again, all the questioned companies were running their business while being examined; hence their strategy was more or less correct. The reason for the dissatisfaction of the entrepreneurs could not be necessarily the strategy, because they defended themselves.

In comparison with the model outcomes, the strategy of old entrepreneurs is completely different. To play the Nash equilibrium they should choose the strategy of lowering the price 87.22% of a time. Based on the results from the model, managers should be less confident of their strong position on the market and devote more time into choosing the proper response to the competition.

In the nature of the game presented in the chapter is that the situation for both players is difficult and the goal is to receive as much utility as possible. For solving the model the author used a maxmin strategy, which maximises the worst - case scenario from the game. There is a real chance, if entrepreneurs would analyse the situation with respect to game theory that their strategies would be more accurate and provide better outcomes.

FUTURE RESEARCH DIRECTIONS

The presented research was conducted on the randomized sample of micro and small companies from the central part of Poland. It discovered the ex post evidence of the applied strategies of initial price settings. It would be important to confirm what happen when the managers knows in advance the guidelines of game theory, assumes ex ante the prices according to them, and apply them in a real market setting. As it is underlined, successful strategy cannot depend just on one firm's position in industry, capabilities, activities, or what it has (Brandenburger & Nalebuff, 1995, pp. 57-74). It depends on how others react to your firm's moves, and how others think it will react to theirs. By fully understanding the dynamic with others, the firm can recognize win-win strategies that make you better off in the long term, and signal tactics that avoid lose-lose outcomes. Moreover, if the managers properly understand the game, they can take actions to change the rules or players of the game in their favour. One way a firm can change the game and capture more value is by changing the value other players can bring to it (Brandenburger & Nalebuff, 1995, pp. 57-74).

If SME change is made through the *Initial Price Strategies and an Application of Game Theory*, consequences might flow into two directions. Firstly, in order to make decisions about strategies the owners or managers would need to execute more precise analysis of the market environment, including in depth review of the competitors or substitutors. This will certainly result in the increase of knowledge of the market, which seems to be a sine qua non condition in setting up strategic options. Secondly, having acquired more knowledge about the market, being more conscious precise decisions could made by the owners or managers. These include optimization, from the point of view of game theory, of decisions, in certain market conditions. And although using game theory's equilibria guidelines for decisions' determination would be a significant value added, it needs to be pointed out that even higher value added, and more practical value could be achieved from better adjustment of decisions made to specific market conditions.

The problem of application of game theory guidelines for SME strategies development is in an initial stage of scientific recognition. There is an urgent need to fill in the knowledge gaps about this issue (Snyman et al., 2013, pp. 1-20). This includes not only the questions about how the knowledge on the market is gathered and used, how information is processed, what strategic options are considered, but also, on other hand, what is the level of game theory knowledge among SME managers and owners, what is their opinion about applicability of this theory, just to mention a few topics. Therefore, new surveys or interviews are needed to be conducted in SMEs to confirm the Initial Price Strategies and an Application of Game Theory. They should be performed on less randomized samples and be more diversified according to specific characteristics of sectors or industries. These researches would provide better insight into how initial prices are adjusted to highly demanding and uncompromising markets.

This approach however will show quite static situations, and one needs to remember that the market games are subject to dynamic changes and turbulences. In order to make the picture more dynamic an application of the model applied in this chapter in a computation system will allow more exact evaluations as well as development of simulations. These solutions might also be important from the point of view of the so-called born-global companies. As shown by Onkelinx & Sleuwaegen (2010, pp. 1-45) SMEs from their inception internationalize their activities, very often becoming global from their beginning. Thus, especially for such companies initial price strategies are important and need to be considered from a dynamic perspective.

All in all, apart from scientific values and new knowledge generated, the future researches would result in providing real evidence that game theory principles are useful even form micro and small entrepreneurs. This knowledge could be used by SME owners and managers to better set up their initial price strategies as well as assess benefits and costs of such actions. And through that their businesses would better adjust to the market, especially with initial process, and they would gather more knowledge and be more conscious about decisions made, which simply will reduce risk of doing the business. The risk reduction can also be beneficial to other players on the market, which integrated into value chains and can benefit from application of game theory principles as shown by recent evidence (Yueliang & Nan, 2015, pp. 31-36).

CONCLUSION

Micro, small and medium sized enterprises represent the entrepreneurial approach of the society. Entrepreneurs, owners and managers of these companies very often start the business without knowledge about how the market is functioning, what rules to follow, how cooperation and competition affects their operations. They learn from experience, which in many cases is painful. Therefore, it is important that once starting the company they should obtain some advice how to operate in order to minimize the chance for failure or, in other words, increase chances for success.

One of key elements in the set of knowledge for SMEs entrepreneurs is how to set up the initial price once entering the market. This price is positioning the company on the market, builds its perception in the eyes of consumers and competitors and is a pre-condition for its development. Based on the conducted research it might be argued that researched micro and small companies in Poland apply a very conservative strategy of the same prices as the competitors. However as shown in the results of the calculations based on the game theory principles new entrepreneurs should more confidently enter the market and offer a present price. Thanks to that they would achieve their goals quicker and the tough beginning period would be more manageable.

It is therefore also argued that the basic knowledge from the game theory might be helpful for SMEs to ensure their more secure entrance on the market and longer lasting thereinafter. Conducted analysis also confirmed that benefits from this kind of knowledge application would be profitable to the whole economy of Poland. It is due to the fact that the SME sector plays a crucial role in the economic development of the country. And this conclusion could also be more generally related to the economic situation of each economy. The more knowledge that generates rational behaviour among owners and managers, including better alignment of business to market conditions of coordination, cooperation and competition, the better industrial organization and development.

REFERENCES

Audretsch, A. B., & Beckmann, A. M. (2007). From Small Business to Entrepreneurship Policy. In D. B. Audretsch & I. Grilo (Eds.), *Handbook of Research on Entrepreneurship Policy* (pp. 45–47). London: Edward Elgar Pub. doi:10.4337/9781847206794.00007

Auerswald, P. E., & Branscomb, L. M. (2003). Start-ups and Spin-offs: Collective Entrepreneurship Between Invention and Innovation. In D. M. Hart (Ed.), *The Emergence of Entrepreneurship Policy: Governance, Start-Ups, and Growth in the U.S. Knowledge Economy* (pp. 61–65). Cambridge, UK: Cambridge University Press. doi:10.1017/CBO9780511610134.004

Brandenburger, A. M., & Nalebuff, B. J. (1995, July-August). The Right Game: Use Game Theory to Shape Strategy. *Harvard Business Review*, 57–74.

Cachon, G. P., & Netessine, S. (2004). Game theory in supply chain analysis. In D. Smichi-Levi, S. D. Wu, & Z. Shen (Eds.), *Handbook of Quantitative Supply Chain Analysis: Modeling in the eBusiness Era* (pp. 13–59). Boston: Kluwer. doi:10.1007/978-1-4020-7953-5_2

Dixit, A. K., & Nalebuff, B. J. (1993). *Thinking Strategically: The Competitive Edge in Business, Politics, and Everyday Life* (pp. 59–79). New York, NY: W. W. Norton & Company.

Feldman, M. P. (2003). Entrepreneurship and American Research Universities: Evolution in Technology Transfer. In D. M. Hart (Ed.), *The Emergence of Entrepreneurship Policy: Governance, Start-Ups, and Growth in the U.S. Knowledge Economy* (pp. 91–96). Cambridge, UK: Cambridge University Press. doi:10.1017/CBO9780511610134.005

Hennet, J.-C. (2008). SMEs in supply chain networks: cooperation and competition. In A. Villa (Ed.), *Road Map to the Development of European SME Networks Towards Collaborative Innovation* (pp. 86–96). Berlin: Springer-Verlag.

Hennet, J.-C., & Yesemin, A. (2008). Supply chain coordination: A game-theory approach. *Engineering Applications of Artificial Intelligence*, *21*(3), 399–405. doi:10.1016/j.engappai.2007.10.003

Hollanders, H., Es-Sadki, N., & Kanerva, M. (2015). *Innovation Union Scoreboard 2015* (pp. 7–25). Brussels: European Commission, Directorate-General for Internal Market, Industry, Entrepreneurship and SMEs.

Jackson, M. O. (2010). *Social and Economic Networks*. Princeton, NJ: Princeton University Press.

Kaczmarek, A. (2009). Strategie działania małych i średnich przedsiębiorstw. *Zeszyty Naukowe Małopolskiej Wyższej Szkoły Ekonomicznej w Tarnowie, 2*(13), 260-271.

Kleindl, B. (1999). A game theoretic perspective on market-oriented versus innovative strategic choice. *Journal of Strategic Marketing*, *7*(4), 265–274. doi:10.1080/096525499346378

Korderl, Z. (2015). Strategie sektora MSP w Polsce. *Logistyka*, *3*, 98–103.

Leyton-Brown, K., & Shoham, Y. (2008). *Essentials of Game Theory: A Concise, Multidisciplinary Introduction*. New York, NY: Morgan Claypool Publishers.

Ministry of Economy of Poland. (2015, September). *Entrepreneurship in Poland*. Retrieved July 26th, 2016, from Ministry of Economy of Poland website: https://www.mr.gov.pl/media/15459/Entrepreneurship_in_Poland_2015.pdf

Monostori, L., Valckenaers, P., Dolgui, A., Panetto, H., Brdys, M., & Csanád Csáj, B. (2015). Cooperative Control in Production and Logistics. Report prepared by the IFAC Coordinating Committee on Manufacturing and Logistics Systems.

Muller, P., Caliandro, C., Peycheva, V., Gagliardi, D., Marzocchi, C., Ramlogan, R., & Cox, D. (2015, November). *SME Performance Review 2014/2015 (pp. 4-12)*. Retrieved July 26th 2016, from European Commission, Directorate-General for Internal Market, Industry, Entrepreneurship and SMEs website: http://ec.europa.eu/growth/smes/business-friendly-environment/performance-review/index_en.htm

Onkelinx J. & Sleuwaegen L. (2010, October). Internationalization strategy and performance of small and medium sized enterprises. *NBB Working Paper, 197*, 1-45.

Orłowski, W., Pasternak, R., Flaht, K., & Szubert, D. (2010). *Procesy inwestycyjne i strategie przedsiębiorstw w czasach kryzysu*. Warsaw: PARP.

PARP. (2015, April). *Report on the condition of small and medium-sized enterprise sector in Poland in 2012-2013*. Retrieved July 26th, 2016, from Polish Agency for Enterprise Development website https://badania.parp.gov.pl/images/ReportonSMESectorinPoland_2014.pdf

Potter, J., & Proto, A. (2010). OECD Review of SME and Entrepreneurship Issues and Policies at National and Local Levels in Poland). Paris: OECD.

Rosłon, J. (2011). Mikro i małe przedsiębiorstwa w Polsce – wybrane aspekty funkcjonowania i rozwoju. *Zeszyty Naukowe Uniwersytetu Szczecińskiego, 81*, 139–147.

Roth, A. (2015). *Who Gets What and Why: The New Economics of Matchmaking and Market Design*. Denver, CO: Houghton Mifflin Harcourt.

Rubikom Strategy Consultants. (2011). *Kondycja Marketingu w przedsiębiorstwach sektora MSP*. Warsaw: Rubikom Strategy Consultants.

Smolarek, M. (2012). Strategie konkurencyjne małych przedsiębiorstw. *Zeszyty Naukowe Wyższej Szkoły Humanitas, 1*, 53–68.

Snyman, H. A., Kennon, D., Schutte, C. S. L., & Von Leipzig, K. (2013). Formulating a strategic framework to promote SME development. In *SAIIE25 Proceedings*.

Strychalska-Rudzewicz, A. & Wysocki, J. (2013). Strategie małych i średnich przedsiębiorstw w działalności innowacyjnej. *Organizacja i kierowanie, 2,* 67-78.

Sucky, E. (2007). A model for dynamic strategic vendor selection. *Computers & Operations Research, 34*(12), 3638–3651. doi:10.1016/j.cor.2006.01.006

Tirole, J. (1988). *The Theory of Industrial Organization*. Cambridge, MA: MIT Press.

Villa, A., & Antonelli, D. (2009). *A Road Map to the Development of European SME Networks*. Berlin: Springer. doi:10.1007/978-1-84800-342-2

Vos, J.-P. (2004). Developing strategic self-descriptions of SMEs. *Technovation, 25*(9), 989–999. doi:10.1016/j.technovation.2004.02.014

Wang, L., & Xu, Y. (2010). Game Theory Analysis on SME Financing Under FTW. *Orient Academic Forum,* (pp. 434-438).

Yueliang, S., & Nan, L. (2015). Simulation of Game Model for Supply Chain Finance Credit Risk Based on Multi-Agent. *Open Journal of Social Sciences, 3*(01), 31–36. doi:10.4236/jss.2015.31004

ADDITIONAL READING

Avinash, D., & Nalebuff, B. (1991). *Thinking Strategically*. New York: Norton & Co.

Binmore, K. (1992). *Fun and Games: A Text on Game Theory*. Lexington: D.C. Heath & Co.

Binmore, K. (2007). *Playing For Real: A Text On Game Theory*. New York: Oxford University Press. doi:10.1093/acprof:oso/9780195300574.001.0001

Cavusgil, S. T., & Knight, G. (2015). The born global firm: An entrepreneurial and capabilities perspective on early and rapid internationalization. *Journal of International Business Studies*, *46*(1), 3–16. doi:10.1057/jibs.2014.62

Cobb, B. R., Basuchoudhary, A., & Hartman, G. (2013). Mixed Strategy Nash Equilibria in Signaling Games. *Theoretical Economics Letters*, *3*, 52–64. doi:10.4236/tel.2013.31009

Dutta, P. K. (1999). *Strategies and Games. Theory and Practice*. Cambridge, MA: MIT Press.

Osborne, J., & Rubinstein, A. (1994). *A Course in Game Theory*. Cambridge, MA: MIT Press.

KEY TERMS AND DEFINITIONS

Game Theory: The study of mathematical models of conflict and cooperation between intelligent rational decision-makers. It is a major method used in mathematical economics and business for modelling competing behaviours of interacting agents.

Industrial Organization: An economic theory that is concerned with the workings of markets and industries, in particular the way firms compete with each other.

Nash Equilibrium: A concept of game theory where the optimal outcome of a game is one where no player has an incentive to deviate from his chosen strategy after considering an opponent's choice.

Small and Medium Size Enterprises: Businesses whose staff headcount and either turnover or balance sheet total fall below certain limits. Those limits differ by country.

Stackelberg Equilibrium: An equilibrium in which one company acts as a leader and one another company as a follower, the leader firm moves first and the follower firm follows it later; it differs with Nash equilibrium and has applications in supply chain and marketing channels.

Strategy: A high level plan to achieve one or more goals under conditions of uncertainty.

Supply Chain Model: A system of organizations, people, activities, information, and resources involved in moving a product or service from supplier to customer.

ENDNOTES

1 Editors' note: This corresponds to the clear presentation of market orientation importance presented in chapter 1.

Chapter 8
A Further Look at Working Capital Optimization in Medium-Sized Firms:
Concepts and Evidence

Milan B. Vemić
Higher School of Academic Studies "DOSITEJ", Serbia

ABSTRACT

SMEs battling with financial volatility require financial and business ambidexterity. To them working capital optimization is one partial solution to improve performance. Examined concepts will not develop a spectacular model but rather quantitatively - qualitatively upgrade existing theories. The treatise will demonstrate that in stochastic working capital decisions important are varieties of alternatives. A structural model of working capital optimization is developed with practical examples for financial decision making under uncertainty. Emphasis will be placed on structuring management problems with the analysis of preferences using probability tools. Additional light is shed on coupling static and dynamic indicators, systemically approaching inventory and cash management, sales and purchasing, costs and profitability. Working capital optimization can contribute to management of financial risk and have an overall impact in medium sized enterprises. Conclusions and recommendations will aim to contribute to the overall body of knowledge on optimized financial decision making.

INTRODUCTION

The Small and Medium-sized Enterprise (SME) finance theory is the result of long and fruitful efforts of economists to account correctly for an extremely wide range of research results, which previously existing theory of the firm could not sufficiently explain. It is not generally acknowledged, however, that new approaches in SME and entrepreneurship financing represent a radical change, not only in the content of scientific knowledge, but also in the fundamental theoretical framework on the basis of which such knowledge can be applied in business. The true extent of this development of theoretical

DOI: 10.4018/978-1-5225-1949-2.ch008

framework has possibly been clouded by the contrast between the relatively descriptive and smoothly understood definitions in which the neo-classical and even modern theories of the firm have regularly been expressed, with the quite abstract and mathematical inference in which this chapter formulates working capital optimization in medium enterprises. So sharp is this divergence that a significant number of economists were led to assume that working capital optimization entails relinquishment of classical financial management tools, and that instead, there should prevail rigorous mathematical optimization which can, in some enigmatic way, forecast precisely SME financial performance. However, further development of working capital optimization provides significant advantages. (See Vemic, 2015, for an exploratory case study.) These are fundamentally different attributes from those appearing in the theories of the firm. To produce such a conception of the working capital optimization theory in medium enterprises at a relatively practical level, in correspondence with static and dynamic ratios of financial analysis, is the main objective of this chapter.

The exact essence of the new working capital optimization concepts will be developed through the whole chapter, particularly in the section on the development of the structural model. First, the origin of the working capital optimization theory in medium enterprises is approached through static and dynamic indicators modified with author's Toolbox for *evolution* of working capital management optimal functionality of structure. Second, the challenges are addressed through key factors of static and dynamic working capital optimization. The third main section involves the development of the structural model of working capital optimization in medium enterprises with a mathematical notation and with practical applications.

In correspondence with the prevailing plan outlined above significant emphasis is placed on showing how working capital optimization concepts can be applied practically without abandoning the classic static and dynamic financial analysis. It has been the author's objective throughout this chapter to portray the main concepts of the working capital optimization theory in medium enterprises using both mathematical and non-mathematical inference in expectation that the reader will thereby perform more contemplation and undertake further exploration in this appealing area of scientific research.

BACKGROUND AND ORIGIN OF THE WORKING CAPITAL OPTIMIZATION THEORY IN MEDIUM ENTERPRISES

"Optimization" originates in the Latin word *optimus*, meaning "best". What is an optimum and how to approach it? The author assumes that each individual optimization criterion is a new angle of seeking an optimal solution in the process of transformation of business objectives into limiting conditions. This will be proved through the Bayes' theorem later on. Acknowledging optimization potential, it will also be assumed that it is sought by the management for short-term, mid-term or long-term strategies (financing and other) according to basic criteria harmonized in the SME involving a set of limiting conditions. Highlighting the working capital allocation problem through the Markowitz portfolio and diversification theory (Markowitz, 1952) leads the author towards the ultimate goal which is to ensure an optimal plan and maximal needed production with given resources (Kantorovich, 1985). Second, considered is an extended model of total company assets, thus including fixed asset utilization as well. An optimum can be the least costs of financing working capital or a solution where the overall cost of business finance is lowest. This financially optimal solution is approached through a holistic business field and balance sheet of the whole enterprise. In this model off-balance sheet assets and liabilities are excluded although these items are significant amounts in larger enterprises and financial institutions that have assets under management. Logically, it follows for medium sized firms that an optimal portfolio represents the best combination of fixed and variable assets based on the real correlation coefficient (Froehlich, Hawranek, Lettmayr, & Pichler 1994, pp.137-138), between the coefficient of total turnover of working capital and the ratio of the total turnover of fixed assets. Mensuration of working capital focuses on Days Sales Outstanding (DSO), Days Payable Outstanding (DPO), Days Inventory Outstanding (DIO), Days Working Capital (DWC) and Cash-to-Cash Cycle (C2C) while fixed asset utilization involves optimization of capacity utilization, equipment reliability, percent of breakdowns and flexibility in terms of range of products that can be produced (Kaplan & Norton, 2004, p.72). The author's proposed holistic yields correlation coefficient is the covariance divided by the product of standard deviation of the returns of fixed and variable assets as two subfactors, quantitatively signifying efficiency of sources of finance:

$$p = \frac{Cov(r_{fa,}r_{va})}{\sigma_{fa}\sigma_{va}} \tag{1}$$

Assuming a factor approach to asset allocation correlating yields will produce an optimal solution in choosing between factors having differing degrees of limitation. General equation (2) according to Kantorovich (1985) is (p.215)

$$\sum_{i=1}^{n+m} c_i a_i^s \leq 0, (s = 1,...,r) \tag{2}$$

Kantorovich's (1985) following equation for an optimal solution (p.215), is now used to focus on the working capital problem:

$$\sum_{i=1}^{n+m} c_i a_i^s = 0, (s = 1,...,r) \tag{3}$$

where $a_1...a_{n+m}$ = types of working capital, $c_1...c_{n+m}$ = objectively conditioned cost of capital, while s = total number of possible sources of working capital

As shown in equation (3) cost of capital optimization coupled with a factor approach to variable asset allocation (Markowitz, 1952) through maximal diversification of all disposable sources of working capital will maximize the main assumed SME objective which is profit maximization. This inductively infers the claim that optimization and diversification enhance SME.

Before further elaboration of the model of working capital management and its subsystems in the strict sense it is necessary to cross-examine the pre-defined mutually interdependent working capital criteria. Clearly, the manifestation of these criteria is not identical in all medium-sized enterprises and there are significant differences in the structure of current assets of industrial and service companies. Therefore, the design of the final model of working capital for medium-sized enterprises shall require further research of financial structure of specific companies in order to determine the relationship between these criteria and the limits of tolerance of them. This argument acknowledges the proved assumption

Table 1. Working capital criteria impacting financial policy and sources of finance

	Ex-ante and ex post-management functions and interventions	Key working capital factors	Impact on performance
1.	Static & Dynamic Financial Analysis Ratios	Cash/net working capital/inventory	Implicative is strong influence of working capital factors on practical performance and firm profitability (ROA & ROE).
2.	Financial planning and budgeting	Inventory turnover; percent stock outs	
3.	Management accounting and operations management	Current assets; current liabilities; Buyers/receivables/cost of goods sold	
4.	Analysis and audit of financial reports	Suppliers/debt/cash conversion cycle	
5.	Financial reporting and control of transactions focusing on GAAP and risk	Relations with large firms, banks and financial transactions are critical	

Source: Own presentation of author

that corporate financial structure and dividend policy is irrelevant for the value of the firm (Modigliani & Miller, 1958). Having working definitions of the best results obtainable under specific conditions still requires determination of diverse criteria specific for working capital and policies in the external environment. The various interdependent working capital management factors in medium enterprises are herewith summarized in Table 1.

These are the main quantitative - qualitative criteria forming a unique parallelogram of financial forces, directly affecting working capital optimization. With their diversity estimating total working capital can be quite complex. It is particularly difficult to estimate the value of unfinished production, semi-finished and finished products as components of earnings cycle and C2C. The significance of working capital indicators will now be discussed and evaluated.

Significance of Static Indicators of Criteria for Working Capital Optimization

Relevant transition economy studies highlight the prominent significance of working capital for competitive position of SMEs (Sobczyk, 2006, pp.194-208), and form a correspondence principle with Porter's (1998) three generic strategies for achieving above average performance in an industry: cost leadership, differentiation and focus (p.11). In author's view relevant are also the quality of company's internal structure, its external structure and ultimately the competencies of individual employees highlighted as lagging and leading indicators for entrepreneurial endeavors under conditions of uncertainty (McGrath & MacMillan, 2000). Following the above holistic yields correlation coefficient the lagging and leading static indicators for comparing own and competition current business performance are now developed. A summarized approach of optimizing effects of company's structure through improvement of relevant coefficients is given in Table 2.

Table 2. Coefficients of working capital optimization for lean trail development

Liquidity indicators: Current Ratio; Quick ratio; Net working capital (fund)
Indicators of *activity* (efficacy): DSO, DPO, DIO, DWC and C2C
Indicators of *financial structure*: Internal-External Financing; Interest burden (EBIT – interest expense)
Indicators of *profitability* (profitability): Gross profit, net profit, EBIT, ROE, ROI
Indicators of *market value*: Dividend rate, dividend ratio, earnings per share (EPS), price earnings ratio (P/E); stock pricing book value.

Source: Own presentation of author.

A lean enterprise will consider all coefficients (see Table 2) and try to develop a lean trail with these gravitational rankings:

$$\text{Lagging X-}\Delta\text{X} \rightarrow \text{Stagnant X} \rightarrow \text{LeadingX+}\Delta \rightarrow \text{Lean} \rightarrow \text{Optimal}.$$

Obviously an optimal range of financing working capital derives not only from one type of working capital but rather from the multitude of them and various types have direct influence on each other and on other production factors as well. One of the indirect measures proposed in order to optimize this chain of funding is to determine the degree of convergence between what a medium company plans to achieve on the one hand and what it actually achieves. The following table provides a categorization of possible ranking of performances (yield and production possibility curves), that can be used for determining the degree of convergencedepending on the degree of knowledge of the working capital algorithm.

The ordinal ranks (see Table 3) are 1, 2, 3, 4, 5 (1 = poor; 8 = excellent). Indirectly evaluated performance 1 indicates a much worse-than-expected performance compared to ranking of 8 which is a far better result than expected compared to the competition. It is clear that in lean enterprises these results can be displayed as standard deviation of results against the planned yield of certain types of working capital. Therefore, diversification is effective whenever the yields on certain types of working capital are not perfectly positively correlated. Research has shown (Bodie, Kane, & Marcus, 2007) that this concept is succinctly supported (pp.173-174). Although these convergence ideas are only implicit in the present form of the working capital optimization theory, the author wishes to suggest here in a speculative way that the successful extension of the above indirect method to the domain of computational analysis of scoring models for working capital selection may perhaps be produced as a concrete system of measurement.

Dynamic Performance Evaluation and Evolution of Interdependent Subsystems

In advanced financial analysis (Carleton & Lerner, 1966), dynamic performance evaluation consists of providing weekly plans of working capital requirements. Business goals should then set these levels at the end of the month (also annualized), including their optimal level based on optimized securing of working capital at minimal costs. To achieve that, Kantorovich (1985) discovered a method of gradual

Table 3. Indirect method of determining the degree of convergence between planned and achieved

Financing Target	Possible Managerial Impact	Ordinal Rank
$R_{max}=\Psi(F1+F2+...Fn\text{-}1+Fn)$	Catastrophic	1
	Extremely unfavorable	2
	Very unfavorable	3
	Unfavorable	4
	Negligible	5
	Favorable	6
	Very favorable	7
	Extremely favorable	8

Source: Own presentation of author.

correction (improvement) of plan. He stated, "should a plan prove not be optimal it is then determined to replace it with a plan which yields bigger production for each product" (Kantorovich, 1985, p.39). This will be elaborated in detail through practical applications of the Bayes' probability criterion.

In accounting practice for dynamic performance evaluation, financial statements for analysis typically include a balance sheet, income statement (the statement of comprehensive income or profit and loss account), statements of cash flows and changes in equity, notes to the financial statements and statistical annexes. These reports, usually annual, should be in accordance with International Accounting Standards satisfying users such as suppliers, customers, unions, financial analysts, statisticians, economists, tax and legislative authorities. Their preparation, subject to an auditor's report, adheres to certain fundamental accounting postulates not specifically cited as their acceptance is presumed. Consequently, some authors (White, Sondhi, & Fried, 2007) emphasized the need for greater uniformity of such financial statements across national borders impacting medium enterprises as well. However, for a dynamic evaluation of performance a company can prepare financial statements in a variety of ways that best suit the needs of internal and external stakeholders.

One practical example of a potential dynamic subsystem for evaluation interdependent with optimal working capital management is the Kanban system, built in 1953 by the Toyota Motor Company (Ohmae, 1982). Higher quantities than actually necessary are not ordered and data from where to order, timing of manufacturing process and how to produce is optimized. This prevents excessive transport and production of unnecessary goods, directly optimizing required working capital levels. A dynamic optimal working capital management approach is systematized in Table 4.

Table 4. Toolbox for evolution of working capital management optimality of structure

Subsystem	Current performance	Optimal subsystem development	Future performance	Performance of lean competition
Financial analysis	Ad hoc/lagging	Improved financial analysis	Regular	Regular financial analysis and planning accompanied with an Internal Control System and internal audit with optimized receivables in the CRM, JIT, TQM or Kanban optimal functionality of structure and warehouse monitoring.
Financial planning	Ad hoc or lagging	Improved financial planning	Quarterly	
Buyers	Ad hoc	Improved relations	CRM based	
Receivables and days payable	Uncollected receivables	Reduced receivables / Optimized days payable	Shift to optimal maturity structure	
Relations with banks/ transactions	Ad hoc or lagging	Improved relations with banks and transactions	Customer Relationship Management (CRM)	
Inventory/ inventory turnover	Ad hoc or lagging	Reduced inventory/ Increased turnover	Optimal inventory/ functional structure	
Analysis and checking reports	Ad hoc check ups	Efficient analysis and checking of financial reports	Lean, lead indicator/ optimal structure	
Reporting and controlling	Ad hoc reporting	More efficient reporting and controlling	Lean, lead indicator/ optimal structure	
Internal control and internal audit	Nonexistent	More efficient audit and introduced internal control	Functionality of structure is optimal	
Accounting role	Non managerial	More efficient accounting	Management accounting	
Profitability	Ad hoc or lagging	Increased profitability	Transitioning to cost effectiveness and maximal profit	

Source: Own presentation of author.

Table 4 shows how working capital management based on dynamic assessment of performance interacts with CRM. This subsystem of knowledge management comprises these elements: market information management, sales channel management and monitoring of demand, creating and managing marketing campaigns and financial monitoring of sales campaigns, analysis of data on customers and banks, interaction with current and potential clients and banks.

The Definition of Working Capital Optimization Principles for Medium-sized Enterprises

The capacity of medium-sized enterprises to plan and evaluate the level of necessary working capital exceeds the capacity of small businesses and potentially provides their qualitative gravitation towards large enterprises. Therefore, specific impact is far more dynamic with medium-sized enterprises than with the small. In fact, evidence from East Germany from 90's also confirms that the smaller the firm the less it is able to attract public money for the financing of its investments and operations (Brezinski, 1998, p.138). Consequently, it is necessary to consider various funds management efficiency, impacting the C2C, either via the index of performance or by determining the optimal level of working capital. The first indicator represents the relative ratio between the positive and negative variations in available resources in relation to the pre-planned level. Available and planned working capital actually means current assets that already exist in their generic forms of cash, securities, receivables and inventory, as well as those that are planned which is determined based on a desirable and acceptable coefficient of their turnover (Farkas, 2005).

The actual funding volume depends primarily on the size and activities of the company, credit availability, as well as attitudes towards risk and profit. If a company strives to achieve a turnover of six times a year for a specific source of working capital, for example cash, then the planned cash level at maturity of financial liabilities should equal the amount of two months' cash requirements. All other sources of working capital in medium-sized enterprises can be measured similarly.

The author's second approach proceeds from the need to achieve desired quality level for working capital which logically requires efficient real time cost management. The working capital funding quality depends on the level of effectiveness of staff to obtain capital, the system of work and efficiency of receivables collection. Empirical evidence from IBM (Rogers, Lukens, Lin, & Jon, 2007) proves that good governance is being developed in an integrated finance organization increasing effectiveness in driving synthesis of information enterprise-wide on the basis of globally mandated standards, *Standard Chart of Accounts*, common data definitions and standard common processes enterprise-wide (p.20). In this model reducing operating costs naturally increases available working capital, affects cost price and profitability through improved C2C which can be optimized in several differing ways. Similarly, according to Cooper (1995), "Unlike kaizen costing, which focuses on reducing a product's cost through increased efficiency in the production process, target costing focuses on reducing the cost of a product through changes in its design" (p.135-136). This approach involves calculation of the target selling price obtained by adding expected cost and expected margin.

Analogously, it follows that the author's approach to modeling the quality of working capital is twofold:

1. Proceeding from the function of profit as the difference between revenues and expenditures, *therefore a structured target costing approach*, and
2. Proceeding from the balance between the marginal cost of capital and the internal rate of return.

The author interprets these two relations mathematically by starting with a profitability interval from X1 to X2 (for $R \rhd C$). The optimal level of working capital taken as X0 is defined as the maximum difference between revenues and expenditures. The condition is met for the maximum value of X for which the first derivative (Tranter, 1975, p.125) equates the growth of revenue and expenditure in medium enterprises, with the additional condition that the second derivative of the revenue function is lower than the cost function. Hence, a good quantitative measure of the optimal level of working capital equates marginal cost of capital and internal rate of return:

$$\max\left(\mathrm{R}-\mathrm{C}\right)\ \text{for}:\ \left(\mathrm{R}-\mathrm{C}\right)'=0\rightarrow \mathrm{R}'=\mathrm{C}';\ \left(\mathrm{R}-\mathrm{C}\right)''\ <\ 0\rightarrow \mathrm{R}''<\mathrm{C}''\rightarrow IRR=MCC \tag{4}$$

The formula explicitly implies that these requirements are met when the company yield curve is concave and cost curve is convex which in practice exists in profitable companies. Similarly revenues are equal to the cost of production in the point where the function of total income meets total expenses. Maximum performance requirement is satisfied at the limiting condition point where the cost of inventories and other working capital burdens are least and therefore an optimal level of working capital is determined by the balance of the marginal cost of capital and the internal rate of return.

Since it is more important to consider changes in additional costs incurred than to focus on average costs, it is advisable to make use of the marginal cost of capital and the internal rate of return as management instruments. Based on the defined experiential model that emerges from research results it is argued that measurement values and relevant provision of information is of central importance. Hence, costs and revenues should be optimized for various levels of production volume or services provided, compared with companies in the same sector and predicted on the basis of reliable market research. For example with linear regression models (Farkas, 2005). Differences in working capital metrics relative to benchmarks of similar companies, as well as own company historical benchmarks, quickly reveal improvement opportunities. There are many mathematical and statistical methods for calculating deviation from the optimal use of working capital. Once optimized, it is recommended to keep it at that level for as long as all the other circumstances of the organization remain unchanged.

An important optimization prerequisite is certainly stimulating regular financing because optimal continuity of total funding incorporates optimization of working capital financing. In contrast losses are incurred due to inadequate cash and stock management. These are indeed phenomena that the author observed in recent research (PWC, 2015). Also, despite improving subsystems of working capital enterprises also encountered strong distrust of banks because of suboptimal collection of receivables in the previous period or uncertainty about the use of inventories was too high (Ernst & Young, 2013). As a result it is not coincidental evidence from Austrian SMEs that banks' documentation requirements became much (30%) or slightly (28%) tougher during this period (World Bank Centre for Financial Reporting Reform and the Austrian Institute for SME Research, 2013) which is an opportunity for medium-sized enterprises compared to the small ones.

Deductively proposed overarching principles of optimal working capital management are as follows:

- Inclusion of the subsystem of working capital in an integrated system of financial management, with relevant financial, non-financial norms and standards, and accounting policies, as well as the permissible standard of deviation from estimated optimum working capital;

- Providing longer data time series on turnover index for various forms of working capital;
- Organized provision of working capital following financial market research;
- Integrated management of DPO, DSO, DIO, DWC, C2C and time required to issue and deliver invoices so to avoid unnecessary costs burdening the working capital;
- Analysis of the creditworthiness of customers and suppliers through an internal control system focusing on conditions of sale and purchase with most successful and valuable partners. For example, if the sector standard is that current liabilities are settled within 30 days, while a medium company approves a loan up to 15 days it may consider extending a deadline;
- Dynamic C2C, particularly with inventories and accounts receivable, because the shorter the cycle and the more working capital an enterprise generates the less it borrows;
- Systematic monitoring and evaluation of working capital pressure on company credit capacity;
- Ensuring dedicated personnel in financial management to achieve anticipated levels of funding through the knowledge of the planned financial level and effects;
- Having financial management personnel acquainted with alternative possibilities of financing scarce working capital;
- Training of finance officers in financial evaluation and improvement of working capital utilization and relevant remuneration of their performance.

KEY OPPORTUNITIES AND CHALLENGES IN WORKING CAPITAL OPTIMIZATION

In this section the author highlights his multiangular approach towards characteristical mutual feedbacks and methods of maximizing marginal utilities between pairs of opposites within a complex parallelogram of forces statically and dynamically impacting optimal working capital management of medium enterprises.

Optimizing the Management of Manufacturing, Sales and Purchasing

Optimizing manufacturing and subsequently obviously sales implies reaching higher profits per sale and higher sales per customer. Companies that constantly increase the size of these two optimization factors explicitly contribute to an overall optimization of the business. This optimization leans on the right model of excellence, such as that of the European Foundation for Quality Management (EFQM), and is intended for all organizations that want to be more competitive regardless of industry, size, structure or maturity of the business. Consequently, an optimized medium enterprise ensures a strategic marketing policy is optimally linked with working capital management on the basis of these key elements:

- Optimal size and number of sales staff performing regular comparison with competition, ensuring customer stability;
- Optimal allocated marketing and promotion costs inclusive with after sales and guarantee services, retention of old and acquisition of new customers whereby in this duality, according to Kotler (1988), "profit optimization theory leads to finding an optimal level of total marketing costs" (p.307). The author shall address this in the section on cost optimization.

- Optimal client management system with contracting, support, research and client satisfaction surveys followed by process based resolving of complaints and monitoring the number of returned products;
- Optimal size of warehouses and inventory levels for an optimum capacity.

Accordingly, the usual manufacturing optimization rule is to equate marginal revenue and marginal cost (Nicholson, 1997).

Optimizing DSO, DPO, DIO, DWC, and C2C

Collection of receivables potentially exacerbates any SME indebtedness and hinders working capital quality. The smaller medium-sized enterprises, with a weak bargaining position, are certainly faced with long waiting times for collection of receivables and with relatively short deadlines for the settlement of their obligations, though stark differences exist. In contrast to Scandinavian countries where payments are executed much faster building a good company image towards suppliers (Blom, 2004, p.69.), this isn't the case in Southern Europe (Danske Bank & Ernst & Young, 2009, p.13). The 2009 SME survey implemented by the Serbian Republic Agency for SME Development on a sample of 427 medium-sized enterprises is in agreement with this finding. It reveals a mismatch between DSO and DPO: 62% of respondents met their obligations to suppliers in less than 30 days (31% in less 15 days), while only 33% managed to collect receivables in the same period of time and 14% in less than 15 days, (Ministry of Economy and Regional Development, 2009). A deductive conclusion is that most Serbian SMEs have not introduced policies and business procedures that optimize DSO, DPO, DIO and DWC which implicitly negatively impacts C2C.

Practically, this optimization incorporates optimal management of customers and receivables as two central categories of current assets and income. Elements of the subsystem are tightening credit standards of medium enterprises and struggle for a better bargaining position versus large companies in order to improve receivables and inventories financing. This requires cluster analysis for networking between relevant medium enterprises in a specific industry or as part of a chain of new value creation in homogenous customer segments (Kaplan & Norton, 2004, p. 109). Evidence from Europe also shows that harmonization of the turnover coefficient for customers with the turnover coefficient for suppliers can reduce the time gap between current liabilities and receivables, decreasing the C2C (PWC, 2013, p.9).

Highlights of Inventory Management Optimization

The author has in mind an optimization of a subsystem of the internal control system based on a medium enterprise policy of fluctuations in permissible inventory levels, disregarding for the moment the chosen method of valuation. The tools of this approach guarantee keeping inventory within predetermined boundaries and its dynamic adaptation to internal and external management factors. Practically the inventory data for tracking levels, orders, sales and deliveries circulates optimally throughout the enterprise, optimizing management of finance, production and sales.

As such inventory management becomes a central component of operational and strategic management conditioning manufacturing and sales that generate cash flow. A consistent program of optimizing inventories explicitly considers time and cost factors of DPO, DSO, DIO (Ernst & Young, 2015), and therefore implicitly of DWC and C2C. The following parameters are typically optimized for all five components:

- Time required to prepare an order, required by suppliers to deliver and transport time;
- Optimum size of the order;
- Qualitative and quantitative control of incoming goods;
- The costs of acquiring, keeping and shortages of supplies;
- The size and cost of the optimal production batch;
- When to order, knowing uncertainties of inventory and order execution. For example, an optimization manager should be aware that LIFO method produces less profit than the FIFO method (most frequent) in periods of rusubg prices and more profit in periods of declining prices.

In dynamic organizational and production systems, needs are not always known and constant. It is therefore a very complex problem to define optimal inventory without operational research involving several mathematical models of which the following three are very important. First, lowest inventory costs with constant satisfied demand are now given by the equation

$$\mathrm{Co} = \sqrt{2QTc_1c_2} \tag{5}$$

Similarly, lowest inventory costs with constant and unsatisfied demand are now given by the equation

$$\mathrm{Co} = \sqrt{2QTc_1c_2}\,\sqrt{\frac{c_3}{c_2 + c_3}} \tag{6}$$

Finally, lowest expected inventory costs with stochastic (aleatory) demand are now given by the equation

$$\mathrm{C}_0 = \mathrm{C}_2\sum_{q=0}^{S_0}(\mathrm{S}_0 - \mathrm{q})\mathrm{p}(\mathrm{q}) + \mathrm{C}_3\sum_{q=S_0+1}^{\infty}(\mathrm{q} - \mathrm{S}_0)\mathrm{p}(\mathrm{q}). \tag{7}$$

where parameters represent: Co = lowest costs; C1 = costs of one delivery; C2 = unit warehousing costs; T = time period (360 days); C3 = costs of unsatisfied demand; Q = quantity of demanded goods; So = optimal series; P(q)= probability of demand;

Optimizing inventory policies and levels implies that the warehouse is filled in a timely fashion prior to an increased and accelerated production cycle ensuring that the company is able to produce at lowest cost which is optimized by establishing an economic order quantity (EOQ) for each product (Chaston, 1990, pp. 126-127) in the C2C. An optimal inventory management program should therefore increase the inventory turnover coefficient because:

- Reduced are costs of storage and interest costs on goods in stock;
- Reduced is the risk associated with any price increases;
- Reduced are losses related to malfunction and waste;

- Improved is liquidity when funds invested in stock are released;
- Increased are profitability and all other qualitative factors of business activity.

SMEs often hold higher inventory levels compared to large firms. This argument is supported by evidence from Hungary where larger firms have many sources of funds and may need less working capital in relation to the total assets or sales (Farkas, 2005, p.3). The next common SME constraint is that contracted delivery of goods comes into question due to possible difficulties and delays in production (related to inventories), or that urgent orders of large customers cannot be met due to unoptimized working capital. Here, the origin of correlation between finance, inventory, marketing and productivity in SMEs is abductively inferred. However, companies strive to increase productivity regardless of size but at the cost of suboptimal inventory. In fact, according to Mortensen, & Bloch (2005) "enterprises considered large in most developing countries usually operate at suboptimal production scales, with higher unit costs and far from optimal efficiency" (p.137).

Highlights of Optimizing Cost Management and Profitability

Any effort to optimize working capital requires satisfactory cost hygiene. This implies a clear, accurate and comprehensive understanding of the structure and movement of costs that enables an accurate analysis and appropriate management intervention.

Assuming a profit maximizing behavior medium-sized enterprises should optimize short-term cash and working capital demands to meet long-term strategic objectives. This ensures flexibility and develops ability to survive global economic crises for growth in the aftermath. Therefore optimal cost management offers an opportunity to reconcile short-term and long-term goals. This model compels SMEs to identify and manage actual fixed and variable costs, to carry out benchmarking of costs depending on their size, market position, business model and strategic imperatives. How rare and difficult this management task can be, is best confirmed by the fact that only 20.00% of the surveyed 32 *medium-sized construction companies* in Slovenia (compared to 31% of large and 40% of small businesses) used a method of benchmarking by activities of the business process (Friedl, 2008).

A successful cost optimization program will ultimately produce clear indicators allowing evolutionary changes and dynamic growth. This position of the author coincides with the results of research on the target group 32 construction medium enterprises in Slovenia from 2008 (Friedl, 2008). From the sixteen then identified methods of evolutionary change the majority of medium enterprises then selected cost optimization as a tool. 76.70% of the respondents then confirmed to be using a method of optimizing costs by activities of business processes.

The above approach requires an initial step in determining which accounting records are maintained for the purpose of developing modern cost accounting. This implies streamlining records and possibly excluding some records and transactions that have no direct impact on increasing management effectiveness and efficiency in order to do the right things and do them right (Müller-Stewens & Lechner, 2005). Consequently the economics of financing of a competitive firm is improved either through the reduction of fixed and variable costs or through an increase in revenues for the same type of expenditure. A more precise definition of effective cost management is now approached followed by elements of process optimization. With achieved effectiveness (E1) measure achieved is gross margin with the employment of all factors of production using the following *partial indicator of optimization* of effective cost management expressed as a percentage

$$E1 = \frac{\mathrm{Re}\,venue - Cost\ \ of\ \ goods\ \ sold}{\mathrm{Re}\,venue} \bullet 100\% \rightarrow \max(E) \tag{8}$$

As a prior condition of working capital optimization, *partial* effective cost management is typically fostered in one of the four direct ways:

1. By increasing output with unchanged costs;
2. By achieving higher sales prices for goods and services;
3. By reducing labor costs;
4. By reducing the purchase price of non-labor factors of production.

More comprehensive explicit and implicit cost optimization involves these questions:

- Are costs in accordance with mission, vision, aims and objectives of the medium enterprise?
- Are costs sustainable and in accordance with an acceptable degree of risk?
- Are costs in accordance with firm size, market position and the desired direction of development?
- Do costs contribute to the achievement of strategic objectives?
- Are there interventions to adjust, reduce or increase the individual costs?
- Are there opportunity costs concerned with the real sacrifice in undertaking a certain investment, such as the loss of interest income or depreciation of machinery used for a capital project?
- Is the cost of expenditure equal to marginal consumption of products which according to some authors (Skoko, 1986) represents the requirement for optimal costs ensuringe maximum profit in a competitive firm?

Realized effectiveness also reveals how business operations are affected by factors on which the company has no influence (for example prices of inputs which directly affects costs), and which therefore directly influence working capital. For companies that produce a variety of products, quantities cannot be automatically added and are multiplied with prices and then summed up. In the numerator quantity of used products is multiplied with prices and added. Calculation of cost management effectiveness under influence of purchase and sales prices is therefore used *as a partial indicator of optimization* because according to Barcus & Wilkinson (1995), "Satisficing is often applied in situations for which solutions cannot be feasibly developed by the optimizing approach, even though other available solution may be likely to provide higher payoffs" (p. 12-15). Partial indicators lead to the following more general definition:

Definition 1: For a medium enterprise "*E*", with efficiency "*E1*", its business process is optimal if it maximizes the difference between input and output prices; that is:

$$E1 = \frac{\sum P_{sales} \bullet Q_{product}}{\sum P_{purchase} \bullet Q_{raw-materials}} = \rightarrow \max(E) \tag{9}$$

Equation (9) shows that so long as the input and output prices are included economic structure of this enterprise can be optimized. Optimization of working capital portrayed in this way actually means optimizing cost-effectiveness. According to Kantorovich (1985) this is achieved at minimum cost per unit of objectively conditioned evaluation of production factors (pp.60-63), which in this case is a standard cost. Optimum represents the greatest difference between the objectively reduced components of relatively fixed costs and the organizationally conditioned increase of fixed costs per unit product. The goal of cost effectiveness optimization is definitely to optimize costs in such a way to achieve the highest possible value of the total revenue with the least cost of elements of production or that achieved cost effectiveness gravitates towards objective costs: $E1 \leq E0$.

The ultimate result of cost effectiveness optimization derives from its causal link with profitability optimization as a broader measure of business performance. These two symmetric criteria optimize a medium enterprise into a lean organizational form. Similarly, Von Neumann and Morgenstern (1944) originally discovered that in zero-sum two person games players "are in relationship of symmetry or duality to each other" (p.104). Therefore, cost effectiveness optimization is explicitly a condition of optimal profitability the goal of which is to achieve a greater gain with the least committed funds or least average total costs. Analogously, gravitation towards objective cost effectiveness enables gravitation towards objective profitability: $P1 \leq P0$, under an important condition that the cost effectiveness principle is not a principle of an absolute optimum. Obviouisly, a duality exists in working capital optimization theory so that optimization may be approached with either of the two addressed criteria forming the *primal problem or the dual problem*.

Highlights of Optimizing Management of Net Working Capital and Cash

For this aspect of optimization the definition of net working capital as the difference of current assets and current liabilities is used. Clearly, it follows that the company must individually optimize the total current assets and current liabilities before the assessment of the maximum and minimum needed net working capital.

Under this aspect of optimization the author assumes own resources or equity that represents a guarantee, because it protects creditors against loss of borrowed capital and provides security for receivables. Middle sized companies should therefore focus on monitoring optimal cash balances starting with influence of organizational readiness on cash optimization given by the equation

$$Q_{opt} \div Q_{actual} = N_{opt} \div N_{actual} \rightarrow \Delta Q \rightarrow Q_{opt} - Q_{actual} \rightarrow \Delta Q = Q_{actual}\left(\frac{N_{opt}}{N_{actual}} - 1\right) \tag{10}$$

where Q is cash level, N is organizational readiness measured through attitudes toward change.

Striving for cash optimization implies that the actual volume of cash gravitates towards optimal level, and that the difference between actual and optimal level approaches a tangent when it satisfies the property that $\Delta Q_{opt} - Q_{actual} \rightarrow 0$. Practical cash optimization can be achieved by identifying, measuring and optimizing costs in short-term business strategy of medium enterprises which is given by the equation

$$Q_{actual} \rightarrow Q_{opt}, \Delta Q \rightarrow 0, \Delta \mathrm{T} \rightarrow 0 \tag{11}$$

Actual cash volume will certainly depend on the size, goals and activities of the company, the availability of funding sources and attitudes towards management, organizational readiness, profit and risk.

Highlights of Optimizing Management of Existing and New Products

This subsystem of optimization relates to effective planning and management of necessary working capital for both existing and new products within the alternative entrepreneurial production possibilities frontier. Thus, it reveals both actual and potential investment efficiency. And while with existing products the subsystem ensures the flow of current expenditure of working capital to finance the ongoing process of production, with the new products a company should reckon with investments in research and development and these expenditures can sometimes consume a large portion of circulating working capital. In pharmaceutical and chemical industry in particular, these investments represent a very demanding outlay, often unattainable for medium-sized enterprises which leaves the bulk of research and development of these sectors to large companies. Managing the assortment of existing and new products and optimal allocation of resources in the technical sense also requires detailed planning of necessary working capital particularly if prices of raw materials and sale vary. In the first instance the production possibility frontier denotes all combinations of products that can be produced with disposable production factors. Marginal analysis or the ratio of change in one product to a marginal change in another allows us to determine the equilibrium relative price for goods "X"and "Y". Therefore, for efficiency in the output mix, according to Nicholson (1997) "the rate of product transformation between two goods is given by the ratio of good's marginal costs" (p.283)

$$RPT = \frac{MC_x}{MC_y} \text{, therefore profit maximization}$$
$$\text{will result from efficient choice of output at equlibrium relative price } \frac{P_x^*}{P_y^*} \tag{12}$$

Following the above combination of products which maximize gain financial optimization involves mensuration of total revenue, total costs and finally total gain on technically optimized combinations of products. Hence, according to Lange (1958) total gain achieved for an optimal program by an enterprise can be mathematically expressed as follows, and for an optimal program gain is maximal (p.253):

$$L = \sum_{i=1}^{n} p_i x_i - \sum_{i=1}^{n}\sum_{j=1}^{m} a_{ij} q_j x_i = \sum_{j=1}^{n}\left[p_i - \sum_{j=1}^{m} a_{ij} q_j\right] x_i \rightarrow \sum_{j=1}^{n}\left[p_i - \sum_{j=1}^{m} a_{ij} q_j\right] x_i = \max \tag{13}$$

DEVELOPMENT OF THE STRUCTURAL MODEL OF WORKING CAPITAL OPTIMIZATION IN MEDIUM ENTERPRISES WITH A MATHEMATICAL NOTATION

The author assumes one of the possible models of quantitative economic analysis which systematically and quantitatively establishes the financial linkage of elements in the complex parallelogram of forces that influence optimal working capital management of a medium size company. The main problem is the selection of essential variables for the desired description and their linkages in the form of mathematical relationships. In the general business case, the relations linking a variable Y are called the dependent (or measured, observed, described) variable connected with a large number of independent (or controlled, explanatory) variables Xi. Generalization and abstraction will be used as scientific methods in this section in order to observe deeper common factors of certain variables. The statistical basis of the structural model will be a table of mutual influence of elements (Menger, 1996) further developed into a unique systematical input-output table of financial structure of working capital for a specific period of time. In order to explore the linkages elements of the model are divided into a number of separate but interconnected organizational units. Thus the structural model conceived shows the transformation of individual categories of working capital into total working capital. It leads to a dynamic movement of the chain entropy (Kukoleča, 1980) where on the one hand buyers and receivables provide propulsion for cash, stocks and manufacturing and on the other hand the total financial result of a medium enterprise is produced.

In principle, the total spending of working capital involves own consumption, for example for energy and salaries and whatever is directed towards the market. If the total consumed working capital is marked with the symbol X1, its payment and collection of other enterprises are marked with the symbol Xij, where the index "i" marks an enterprise that makes payments and index "j" an enterprise receiving working capital, and if its conditionally speaking internal sales are marked with the symbol Yi, a linear equation of generated working capital in medium enterprises is expressed as

$$Xi = xi1 + xi2 + ...Xin + Yi \quad (14)$$

This is the formalization of the problem of working capital linear programming. And furthermore, if certain types of working capital are marked with numbers from 1 to n, then the total spending of the current assets of the company can be effectively presented as a system of full balance equations in a developed form that in principle can always be solved to give the equilibrium of enterprise working capital

$$\begin{array}{l} Xi = x11 + x12 + ...x1n + Y1 \\ X2 = x21 + x22 + ...x2n + Y2 \\ \ldots\ldots\ldots\ldots\ldots\ldots\ldots\ldots \rightarrow Xi = \sum_{j=1}^{n} Xij + Yi(i = 1,2,...n) \\ Xi = xi1 + xi2 + ...xin + Yi \\ \ldots\ldots\ldots\ldots\ldots\ldots\ldots\ldots \\ Xn = xn1 + xn2 + ...xnn + Yn \end{array} \quad (15)$$

Theoretically this formula would be completely correct if, instead of spending the entire working capital probability of its employment was used. However, in practice it is extremely rare that these probabilities are available to medium-sized enterprises, and particularly rare are hypotheses about likely changes over time in conditions of uncertainty. Therefore, an equation is received which presents the total volume of working capital distributed on internal consumption and external payments. Starting from the balance equation approach a logical conclusion is reached that financing total working capital can be calculated as the sum of all types of working capital disaggregated and mathematically presented in the following way:

$$\sum_{i} Xi = \sum_{i} \sum_{j} xij + \sum_{i} Yi \quad (16)$$

All the basic forms of working capital previously defined theoretically are included as a complex process outcome. Ultimately, in the envisaged organizational structure, heads of departments such as the head of finance and marketing should know the structural model, while financial officers need to study it well and learn how to competently determine, track and evaluate the position and role of all individual types of working capital in medium enterprises.

Development and Notation of the Stochastic Model of Working Capital Management Optimization

Obviously, depending on available data, the use of statistics and probability theory becomes inevitable in order to reach a precise projection of working capital in medium-sized enterprises. On the other hand, the author's quantitative approach certainly does not exclude the need to use qualitative assessment of certain operating assets and liabilities that would be based on experience and tacit knowledge of financial managers. It has previously been explained that what is mathematically equivalent for risks is certainly not always psychologically acceptable (Taleb, Goldstein, & Spitznagel, 2009). And while mathematical inference (Stokes, 1900) reduces the number of abitrary elements in a given financial context there is further evidence that the problem of working capital optimization is not just of quantitative nature. In fact, qualitative analysis of certain types of working capital and of their mutual linkages with costs significantly facilitate the optimization of financial management and ambidextrous decision-making in particular (O'Reilly, & Tushman, 2004). This argument was and subsequently and independently also supported by empirical evidence of an IBM study which shows that 76% of dextrous CEOs replace fixed costs with variable costs (Berman, 2010, p.57).

Following is brief case example on the degree of ambidextrous optimization of working capital through the use of random (aleatory) events. Therefore this treatise goes further in order to structure the discussed process by using stochastic models and random variables (values) whose distribution in the statistical set of N elements corresponds to the parallelogram of financial forces in medium enterprises.

If the set of all possible outcomes or results of optimal working capital management is denoted by the symbol Ω and its elements are denoted by the symbol ω then ω is one possible result. Assume that the set can be finite or infinite. A random financial event A is defined as a subset of Ω optimal working capital management if and only if it implements an outcome ω which belongs to a subset of A.

The set of all events that correspond to exploratory working capital financing will be called *the field of complex financial events* and is denoted by the symbol ϕ. The field ϕ is not just any set of events whose elements are subsets A, B, C of set Ω. It is natural to require the condition: $\Omega \in \Phi, and, \varnothing \in \Phi$. As an element of the set Ω actually takes place the author will call it a certain event or a financially secure event, while an empty set ϕ denotes an impossible financial event.

Assume a very frequent medium enterprise situation with receivables collection, although it is clear that taking a loan or the issuance of stock can assume the same practical implications. If in the working capital model receivables collection represents an event A and implies settlement of payables in event B then an expression is used: $A \subset B$. This means that whenever collection is implemented a payment to suppliers is settled. From the standpoint of ω subsets of the set Ω of optimal working capital management the expression $A \subset B$ denotes that A is a subset of B (implying logical subordination).

Consider now the operation of Union and crossing of two financial events A and B which define this event: $C = A \cup B$. Collection is denoted by symbol A and taking a working capital loan is denoted by

B. Thus, it is clear in this case that the event C that represents the binding or immobilization of company resources and can is achieved only if at least one of the two events A and B, or both take place.

Similar to subsets A and B, iterations can be made for the remaining key criteria of working capital optimization. However, the most precise outcome of any criteria is provided by the law of probability. Therefore, the use of random variables requires a distribution that must be used which complicates the practical applicability of probability theory based on random return variables.

On the other hand, in practical financial life the total distribution is usually unknown, i.e. the law of probability. In such cases, it turns out that specific numerical characteristics are sufficient to solve some practical problems of applied theory of probability. Of course, numerical characteristics are determined by certain rules so that instead of the entire probability distribution, i.e. the law of probability of random variable, it is necessary to observe some financial parameters in order to sufficiently describe the observed variable. Similar to investing funds which requires determining the cost of capital and performing measurement of expected financial effects, in the use of random variables one must first measure certain characteristics through a defined method. Certain key financial parameters of the random variable shall be determined through the method of mathematical expectation.

First a formal definition of mathematical expectation is required on the n-dimensional random variable X, which later will be used to determine the mathematical expectation of an unknown demand of working capital. Consider a discrete set of financial statistical events with a final set of possible values and their functions of the observed characteristics. Weighted arithmetic mean can be expressed and defined with the following ratios:

$$\overline{X} = \frac{f1x1 + f2x2 + ...fnxn}{N} \quad \text{or} \quad \overline{X} = \frac{\sum_{i=1}^{n} X_i f_i}{\sum_{i=1}^{n} f_i} \quad \text{or} \quad \overline{X} = \frac{1}{N}\sum_{i=1}^{n} fixi \tag{17}$$

Respectively according to the law of probability, in decision-making under uncertainty it is necessary to determine the likelihood of aleatory events and in this case for a set of values [xi, pi] the management decision on working capital is of aleatory probability. This number is a random variable, since in another series of *"f"* variables and independent experiments, the value of x1 will not be registered exactly *"f1"* times but rather in *"p1%"* cases, x2 will be registered in *"p2%"* of cases and according to the Bernoulli law of large numbers.

Relative frequency $\frac{fk}{f}$ of the event *{x=xk}* is grouped as probability.Arithmetic mean shall then represent the average value of appearance of X; that is:

$$\overline{X} = \frac{1}{N}\sum_{i=1}^{n} xi \cdot pi \tag{18}$$

However in this case it is not a random variable already but rather an internal product of the aleatory variable X (*working capital*) and the corresponding probabilities, which, if uniformly and continuously converges as a random variable, has a sum that is called the mathematical expectation and is denoted by the expression E (X). In the form relevant for presentation in this treatise the nominal definition of mathematical expectation in the form of a synthetic syllogism is:

Definition 2: For a random discrete type of variable X with the distribution of aleatory variables of working capital {x1, x2, ... xn}, which behave according to the law of probability p (x) → {p1, p2, ... pn} their sum is called mathematical expectation of working capital, expectation of random of variable X, and is denoted by E (X), and is also called the mean value of X; that is:

$$E(X) = p1x1 + P2x2 + ...pnxn \rightarrow \sum_{i=1}^{n} pixi \tag{19}$$

Consequently, mathematical expectation represents a certain mean value (Nelson, D. 2003, p.153), in this case of working capital, around which all the other values of the random variable are grouped.

Problem 1: If the random variable X of average working capital is defined with a simple matrix of payment

$$X = \begin{pmatrix} 1 & 2 & 3 & 4 \\ \frac{1}{9} & \frac{3}{9} & \frac{2}{9} & \frac{3}{9} \end{pmatrix},$$

And its mathematical expectation (mean value of working capital) is calculated as follows:

$$E(X) = \sum_{i=1}^{n} pixi \frac{1}{9} \cdot 1 + \frac{3}{9} \cdot 2 + \frac{2}{9} \cdot 3 + \frac{3}{9} \cdot 4 = \frac{25}{9}$$

However, when it comes to a continuous schedule the final sum of working capital is replaced by definite integrals which converge, so that then the expected value of the variable aleatory is constant, whose probability law "*f(x)*" of expected value is given by the equation

$$E(X) = \int_{-\infty}^{+\infty} xf(x)dx \tag{20}$$

The first distinct phase in the cumulative function of working capital financing financing extends from the beginning of the flow of time (t0) to the point where maximized funding is achieved (t1), otherwise still unattained in most researched medium-sized enterprises. At this stage, the overall financing is progressively growing and at the end of the first phase reaches an inflection point, changing from progressive into digressive growth. The derived function of finance is also growing and the end of the first phase reaches its maximum. As pointed out earlier, however, (Vemic, 2015), it is is abductively inferred that the dynamic high-growth medium-sized enterprises are gravitating towards this point of inflexion which could, for example, be achieved through mergers.

The second financing phase starts at a point where the finance function is at a maximum and extends to the next inflection point where marginal financing equals zero allowing financial innovation to come into force1. The overall financing increases digressively and the end of this phase reaches its maximum.

It is further abductively inferred that dynamic enterprises are at the end of the second phase. Marginal financing declines during and at the end of that phase, in point (t2) it is equal to zero. The third phase extends from point (t3) and its characteristic is that the total financing is declining while marginal financing becomes negative. Thus, the total financing of working capital *as a process of growth toward large enterprise status,* in time, can be expressed in nominal terms at constant prices through a definite integral as follows:

$$E(X) = \int_0^n y(t)dt \tag{21}$$

The total funding, up to the point where decreasing funding starts (in some cases sub-optimal), may be nominally expressed in constant prices through the following definite integral:

$$E(X) = \int_0^1 y(t)dt \tag{22}$$

The results of author's research clearly confirm that the financing of future development of medium-sized enterprises should be completely different from suboptimal because the expected skills, networks and entrepreneurial infrastructure are different. This inductive conclusion was in fact confirmed in a parallel study of small, medium and large enterprises in Singapore (KPMG, 2009).

However, in contrast to Singapore, the financial infrastructure in transition economies is is under-developed and shallow which is due to the general lack of social, economic and legal conditions for a functioning financial market. Clearly, much larger total investments are required to transition from either missing or digressive financing into growing sources of finance for overall development of medium-sized enterprises. This represents a major theoretical and practical challenge for otherwise a normal feature of capital that channels itself into areas with a developed economic and social infrastructure in the pursuit of profit maximization (Brigham & Houston, 2001).

The question of impact of quality of certain categories of working capital is especially significant for analysis of overall immobilization of sources of finance in medium enterprises. It will therefore be demonstrated that it is possible to make optimal working capital decisions by using mathematical expectations in relation to the results of certain actions. Although the criteria may differ in practice let us consider the working capital decision through the following profit maximization example.

Problem 2: Let us suppose that a medium enterprise plans to produce one type of product. Every product sold yields a profit of *"a"* monetary units, and each unsold product brings a loss of *"b"* monetary units. The necessary working capital is not known, for this enterprise it represents a random variable *"X"* whose distribution is not known. Let us also suppose that X uniformly and continuously converges as a random variable with a density of $\varphi(x)$. The values of variable X of a discrete type are large numbers e.g. {2000, 2001, 2002, ...}.

Before the funding starts it is necessary to decide on the volume of required working capital. Among several possible criteria assume the following: it is necessary to provide as much working capital as possible in order for mathematical expectation to be maximal. Profit is a random variable, denoted by *"Z"*, because it is a function of the demand for working capital "*X*". Z also depends on production costs that are not a random variable. Therefore: *Z=fy(X)*, whereby for clarity reasons the variable "*y*" is expressed as a parameter. The objective is to find such $Y=Y_0$ so that mathematical expectation of working capital $E(Z)=E[fy_0(X)]$ is maximal.

Next, assume the following elements under the terms of the problem:

$$Z = \begin{Bmatrix} ay \text{ if } X \geq y\text{, i.e. demand is greater than supply} \\ aX\text{-}b(y\text{-}X) \text{ if } x \lhd y. \end{Bmatrix}$$

Therefore, the function *fy*(•), which is at stake, is as follows

$$fu(x) = \begin{Bmatrix} ay & x \leq y \\ (a+b)x\text{-}by, & x \lhd y. \end{Bmatrix}$$

Obviously X≥0, so the function fy(x) will be placed for X≥0.

$$E(Z) = \int_0^\infty fy(x)\varphi(x)dx = \int_0^y \left[(a+b)x - by\right]\varphi(x)dx + \int_y^{+\infty} ay\varphi(x)dx.$$

E (Z) is a function of y. Let's explicitly express it as E (Z) = g (y). Unlike a satisfactory decision which proceeds from a set of criteria which describe the minimally satisfactory alternative, an optimal decision for $y=y_0$, which is the result of applying formalized procedures in finding choices within a financial system of internal controls, is such that $g'(y_0) = 0$

Let us conretize the problem with following conditions: a=300, b=100, X variable: **U** (2,4) distribution. With a countable set of Rx=[2,4] based on upper criteria, consider that the units of demand for working capital are given in thousands of monetary units. The actual demand for working capital is a random variable with a set of values {2000, 2001,…, 4000}. Therefore, the mathematical expectation of working capital is expressed as follows:

$$\varphi(x) = \begin{Bmatrix} \frac{1}{2}, 2 \leq x \leq 4 \\ 0\text{,NA} \end{Bmatrix} \rightarrow E(Z) = \frac{1}{2}\int_2^4 f_y(x)dx$$

The function from this example has the following conditions required for calculating mathematical expectation of gain:

$$f_y(x) = \begin{Bmatrix} 300y & \text{for } y \lhd 2 \text{ and all } x \rhd 0 \\ 400x\text{-}100y & \text{for } 2 \leq y \leq 4 \text{ and } x \lhd y \\ 300y & \text{for } 2 \leq y \leq 4 \text{ and } x \rhd y \\ 400x\text{-}100y & \text{for } y \lhd 4 \text{ and all } x \rhd 0. \end{Bmatrix},$$

therefore

$$g(y) = E(Z) = E\left[f_y(x)\right] = \left\{\begin{array}{l} y \lhd 2 \qquad : \frac{1}{2}\int_2^4 300 \\ 2 \le y \le 4 : \frac{1}{2}\int_2^y (400x\text{-}100y)dx \\ \frac{1}{2}\int_y^4 300\ y\ dx = \text{-}100y^2 + 700y - 400 \end{array}\right\}$$

$$\text{and } \left\{ y \lhd 4 \ : \ \frac{1}{2}\int_2^4 (400x\text{-}100y)\ dx = \text{-}100y\text{+}1200 \right\}$$

It follows that g(y)= has a maximum for some y:

2<y<4 where $g(y) = -100y^2 + 700y - 400 \rightarrow (-100y^2 + 700y - 400)' \rightarrow y_0 = 3.5$ and $g(y_0) = 825$.

Therefore, in deciding for working capital at this level of financing which is actually 3,500 the mathematical expectation of gain for this working capital level is maximal and amounts to 825.

Overview of Some Specific Mathematical Models of Working Capital Optimization Under Stochastic Conditions

The main feature of decision-making in conditions of stochasticity is choosing a working capital financing strategy with lesser or greater uncertainty. Considering the number of possible alternatives to maximize their results, medium-sized enterprises decide in conditions of information asymmetry about the environment and with risk. There are many business situations when a company is forced to position itself in this manner: when deciding on working capital under the influence of external factors, with adoption of new sources of working capital finance and when financial managers are faced with an imperfect set of business information. Stochasticity is exacerbated by the fact that most medium size firms are not listed which would allow measurement of alpha for non-specific market risk or beta coefficient for stock volatility (Downes & Goodman, 2003).

Other examples are obtaining working capital under inflationary conditions where the factor of stochastic behavior of interest rates and exchange rates becomes fundamental, then under conditions of the global economic crisis. Bodie et al. (2007) point out that "this particularly happens when terms of trade rapidly change with continuous rotation of recession and recovery" (p.366) causing, in the view of the author, demand volatility which significantly reduces the chances of optimizing the costs of funding sources. Here the author is particularly interested in working capital decisions without going in detail of various forms although there may indeed be distinct differences between inventory and receivables, or with fixed assets.

In all cases studied so far multiple actions and reactions can be assumed in the form of entrance into the business system and outcomes in the form of financial interventions. Depending on these results potential gain or loss for medium-sized enterprises is established and can conveniently be displayed as a matrix or a vector of financial complexity. The general case of presenting any potential financial problem can be displayed as a single input vector: P = (p1, p2, p3, ... pi, ..., pm). Consequently potential financial solutions or effects can be displayed as an output vector: C = (c1, c2, c3, ... ci, ..., cm). This approach clearly develops into a correlation model of problems and goals or aggregate inputs and outputs (effects) of financial information, earlier categorized as factual, prognostic, normative, exploratory, hypothetical, logical, explicative and instrumental (Sierpińska & Niedbała, 2003, pp.27-28).

Clearly, the above way of elaborating stochastic working capital decision making reveals the significance of having a variety of available alternatives. The simplest example is the choice of decisions on whether to approve or not approve working capital completely or partially approve, or to take a negative decision. A negative decision may mean investing in alternative areas such as fixed assets, bonds or term deposits. These methods are a function of the overall working capital optimization and financial risk management in medium enterprises.

Now, using Laplace's, Wald's, Hurwitz's, Savage's and Bayes' criterion assume and apply the following four inter-connnected steps of financial analysis and working capital optimization:

1. **Financial Problem Structuring:** The enumeration of possible alternatives, and determining the payment status;
2. **Analysis of Financial Uncertainty:** Assigning probabilities to all financial states;
3. **Intuition of Utility or Financial Preference:** Assigning preference for risky consequences in applied theory of games (Bernoulli, 1954; Von Neumann, & Morgenstern, 1944);
4. **Selection of optimal financial intervention** by the criterion of expected monetary value or utility.

Problem 3: As an entrepreneur, a financial manager examines the development alternatives of a medium-sized enterprise. If he identifies such an opportunity, he initiates and leads the development and financing change. He determines where and how the company will use scarce financial resources. As a negotiator, the financial manager represents an organization in negotiations with banks and other business partners of the joint venture and drives eliminating any financial problems relations. When obtaining data needed for decision-making he identifies interdependence and then selects the best option. For the next input output matrix, reflecting the complexity of the management process presented in Table 5, let us find the best financial alternative with the application of Laplace's, Wald's, Hurwitz's and Savage's criteria.

Table 5. Input-output matrix for testing best working capital alternatives

A/T	Outcomes		
Alternative decisions	T1	T2	T3
A1	140	50	25
A2	110	70	15

Source: Own presentation of author.

A. **Working Capital Optimization Through an Applied Laplace Criterion**: According to this criterion every outcome has an equal chance to take place in the future (Charreton, & Bourdaire, 1985, p.24). All probabilities are the same and for his strategy the financial manager selects a sequence which fulfills following Laplace's criterion

$$\max\left[\frac{1}{n}*ai1+\frac{1}{n}*ai2...\frac{1}{n}+ain\right] \text{ n- number of events} \quad (23)$$

In the given example the probability that individual financial circumstances will take place is calculated as follows:

$$P(Ti)=\frac{1}{n}=\frac{1}{3}=0.33 \rightarrow \mathrm{A}_1=0.33\cdot 140+0.33\cdot 50+0.33\cdot 25=46.2+16.5+8.25=70.95$$
$$\mathrm{A}_2=0.33\cdot 110+0.33\cdot 70+0.33\cdot 15=36.3+23.1+4.95=64.35$$
L:max(70.95;64.35)=70.95. A_1 is financially more attractive.

It is obvious that Laplace's criterion suggests the selection of the first alternative working capital decision because it offers the maximum expected value of effects.

B. **Working Capital Optimization Through an Applied Wald Criterion:** Under this criterion of statistical estimation (Wald, 1950) of pessimism or the maximin criterion (Charreton, & Bourdaire, 1985, pp.22-24), which is closely related to, and draws on, the theory of the zero-sum two-person game, any intervention of the financial manager has a certain minimum level of security. In fact this criterion defines risk aversion. For every action *"ai"* the weakest outcome *"wij*"* is identified and selected is such an action *"ak"* for which this value is the greatest. The task of financial managers lies in the responsibility to adopt a decision that maximizes the minimal effects of the condition of Wald's criterion

$$\max(xi)\min(Tj)(\sum_i wij\ xi) \quad (24)$$

From the above example the minimum outcome for both alternatives is determined as follows:

A1;min w1j=25
A2;min w2j=15
max(xi) min (wij)=max(25,15)=25

After appropriate successful comparison of two available obtained is a result by Wald's criterion indicating that A1 variant should again be chosen for working capital.

C. **Working Capital Optimization Through an Applied Hurwitz Criterion:** Represents a kind of compromise balance between the previous two criteria because it takes into account the optimism of the financial manager that is expressed by the number alpha (α), such that 0 <α <1 (Charreton & Bourdaire, 1985, p.24). The parameter α (index of pessimism-optimism) in fact characterizes the financial manager's attitude toward risk. If you take that alpha closer to 1, the financial manager obviously cares about the reaction of the business and financial environment, and if alpha is closer to 0 then he does not care about the reaction. It clearly stems from the given definition that with optimism the manager solves the problem of working capital information asymmetry and that in pessimism he creates it. For a working capital strategy he chooses a sequence for which the following condition from Hurwitz's criterion holds

$$\max\left[a * Ai + (1-\alpha) * ai\right] \tag{25}$$

where the parameters are: Ai – maximal element or most favorable consequnece of the sequence; ai – minimal element or most adverse consequence of the sequence. From the same example the better working capital alternative using Hurwitz's criterion, is determined as follows, with an assumption that: α=1/3

$$A_1 := \frac{1}{3}\cdot 140 + \frac{2}{3}\cdot 25 = 46.67 + 16.67 = 63.33$$

$$\rightarrow A_2 := \frac{1}{3}\cdot 110 + \frac{2}{3}\cdot 15 = 36.67 + 10 = 46.67 \rightarrow \max \text{hi} = \frac{1}{3}\cdot 110 + \frac{2}{3}\cdot 15 = 36.67 + 10 = 46.67$$

It is clear that the maximum outcome (hi) is achieved with the first financial decision A1. Obviously the main criticism of this criterion is that it only takes into consideration the maximum and minimum values while in reality it is difficult to determine the optimization coefficient alpha (α) for working capital.

D. **Working Capital Optimization Through an Applied Savage Criterion of Minimal Regret or Missed Gain:** The Savage criterion (Savage, 1954), as applied here, is based on the awareness that decisions of financial managers can not affect the state of the business environment, while the consequences for him arise from his selection of intervention. Therefore, it is reasonable to compare the consequences of different actions that he could take in relation to a particular financial state. A term introduced to formalize such an analysis is "regret" corresponding to a subjective expected utility that occurs when found out what he as a manager eventually missed with his wrong choice. The notions of utility and expected utility-maximization were originally introduced by Bernoulli (1738). According to Bernoulli (1954),"To do this the determination of the *value* of an item must not be based on its *price,* but rather on the *utility* it yields" (p.3). In order to apply the Savage criterion regrets resulting from a decision must first be calculated which is achieved with Savage's coefficient

$$r_{ij} = \max_k \omega_{kj} - \omega_{ij} (r_{ij} \rhd 0) \tag{26}$$

This coefficient represents the difference between the best possible outcomes of a financial state "*j*" and the consequences if an action "*i*" is selected. Matrix reduction is performed in order to find the leading element of each column and subtract it from all the other elements and itself from itself. This produces a matrix of alternative losses or regrets. It becomes a new risk matrix. The leading element of each row in the new matrix is found next, written aside and found is the minimum of these maximum elements (Minimax). At that moment, that number represents the optimal strategy for a financial manager. Based on data from the given table formed is a simple regret (decision) matrix, reflecting the complexity of the management process with 6 elements arranged in 2 rows and 3 columns, as follows:

$$\mathrm{R}=\begin{pmatrix} 0 & 20 & 0 \\ 30 & 0 & 10 \end{pmatrix}, \rightarrow \mathrm{A}_1 \max_{r1j} = 20, \mathrm{A}_2 \max_{r2j} = 30, \rightarrow \min \max = \mathrm{r}_{ij} = 20, \ \mathrm{A}_1 \text{ is selected}$$

Therefore, using a minimax choice based on regret, the best course would be to select the first row with the minimal possible loss, ensuring a regret of no worse than 20.

E. **Working Capital Optimization Through an Applied Bayes' Probability Criterion:** Unlike previous the Bayes' criterion introduces subjective probabilities based on experience of financial managers who choose an alternative working capital with minimal determined or anticipated risk. An experience acquired after the introduction of a source of finance, can indeed assist in assessing the initial probability of events related to the repeated use of a financial product. These initial probabilities are the result of previous experience of financial managers and can change over time. Obviously, if financial market research is carried out it can modify the previous know-how. Hence, the initial probability becomes the new probability, then denoted as final. The transformation of the initial into final probability is executed through the sequential use of the Bayes' formula for optimal decision-making (Şipoş, 2009, p.308).

Let A* represent an algebraic structure of financial events with a defined probability function *P: A^*→R*, and let events, $B_i (i = 1,2,...n) \rightarrow P(B_i) \rhd 0, (i = 1,...n)$ comprise a complete complete set of disjunct events. Assume also that,

$$A \in A^* \rightarrow (A \cap B_1) \cup (A \cap B_2) \cup ... \cup (A \cap B_n) = A$$

With sets of disjunct financial events assume that

$$\mathrm{P(A1A_2)} = \mathrm{P}(\mathrm{A}_1) + \mathrm{P}(\mathrm{A}_2) \rightarrow P(A \cap B_1) = P(A) .$$

Using formula of conditional probability total probability is now given by the conditions

$$P(A \cap B) = P(B)^* P(\frac{A}{B}) = P(A)^* P(\frac{B}{A}), \text{ if } \mathrm{P(A)} \rhd 0 \rightarrow \text{total probability: } \sum_{i=1}^{N} P(B_i)^* P(\frac{A}{B_i}) = P(A). \quad (27)$$

The total probability formula enables calculation of a financial event *"A"* when the probabilities and financial hypotheses for *"B"* and conditional probabilities for event *"A"* in relation to event *"B" are known*. Bayes' theorem (Charreton, & Bourdaire, 1985, pp.52-59) for finding probabilities in more complicated financial management examples may now be derived in the following way:

if $P(A) \rhd 0$, and $P(A)^{*}P(\frac{Bi}{A}) = P(B_i)^{*}P(\frac{A}{B_i}), \rightarrow P(\frac{B_i}{A}) = \frac{P(B_i)^{*}P(\frac{A}{B_i})}{P(A)}$, with i=1,...n.

We have the Bayes' formula which is now given by the equation

$$P = \frac{B_i}{A} = \frac{P(B_i)^{*} P(\frac{A}{B_i})}{\sum_{i=1}^{n} P(B_i)^{*} P * (\frac{A}{B_i})}; \text{ with i=1...n.} \tag{28}$$

This equation is called Bayes' formula and it will be proved with logical consistency that it can be effectively applied in solving and optimizing various practical working capital problems involving conditional probability of a hypothesis from set *"B"*, when it is known that a financial event A has taken place together with one of the hypotheses from set *"B"*: $B_i (i = 1, 2, ...n) \rightarrow P(B_i) \rhd 0, (i = 1, ...n)$

Problem 4: Sources of finance of permanent working capital in a medium enterprise are funded by three shareholders, in the following proportion of 1:3:5. The relevant probability that the cost of capital is acceptable for management is in the following proportion 0.6; 0.7; 0.8. Fortunately a decision is made to accept the source of finance at an acceptable cost of capital. Using Bayes' criterion let us calculate the probability that the offered acceptable cost originates from the third source of funding?

The total number of possible options related to the financing permanent working capital is:

1+3+5=9, and the following train of events is given:

B_1 = sources of finance originate from first shareholder $P(B_1) = \frac{1}{9} = 0.111$

B_2 = sources of finance originate from second shareholder $P(B_2) = \frac{3}{9} = 0.333$

B_3 = sources of finance originate from third shareholder $P(B_3) = \frac{5}{9} = 0.56$

The following a priori working capital probabilities are given:

$\frac{A}{B_1}$ = source of finance acceptable and from first shareholder $P\frac{A}{B_1} = 0.6$

$\frac{A}{B_2}$ = source of finance acceptable and from from second shareholder $P\frac{A}{B_2} = 0.7$

$\frac{A}{B_3}$ = source of finance acceptable and from third shareholder $P\frac{A}{B_3} = 0.8$

Adjusted probability that cost-effective working capital offered is exactly from the third of the sources of finance, is as follows:

$$P(\frac{B_3}{A}) = \frac{P(B_3) * P(\frac{A}{B_3})}{P(B_1)P(\frac{A}{B_1}) + P(B_2)P(\frac{A}{B_2}) + P(B_3)P(\frac{A}{B_3})} = \frac{0.56 * 0.8}{0.111 * 0.6 + 0.333 * 0.7 + 0.56 * 0.8} = \frac{0.448}{0.748} = 0.599$$

Obviously, the obtained result is *a posteriori probability* in Bayesian inference for this source of finance of working capital. The Bayes' theorem and formula enable prior estimates of working capital probability to be modified when new financial information becomes accessible in which case a posteriori probability becomes a priori, called a *conjugate prior* (Nelson, 2003, p.26). This may be the greatest optimization advantage of this subjectivism non-exempt tool in medium enterprises.

FURTHER RESEARCH DIRECTIONS

The useful, new, or counterintuitive about proposed working capital ideas and models relate to great expectations from optimal organization of finance and accounting jobs, standard skills set of finance officers and development of internal control systems for financial analysis leading to optimum combination of working capital in order to improve the overall financial management and performance of medium-sized enterprises. This conclusion inductively coincides with other consecutive recent studies (CFO, 2013, 2014; Danske Bank and Ernst & Young Global Limited, 2009, 2013; Ernst & Young, 2015; KPMG 2005, 2009, 2012; PWC, 2013, 2014, 2015) and proves that a rich research agenda lies ahead of us.

The next area of additional inquiry relates to some very specific limitations of strict accounting and audit of financial ratios, sometimes complex for SMEs, resulting in lower objective quality of financial statements impacting optimization. As already discovered (Klapper, Sarria-Allende, & Sulla, 2002), firms in Latvia and Russia are missing turnover variables and firms in Slovenia and Croatia only report total liabilities and do not include a breakdown of their debt structures such as maturity (p.7). Here, with "creative accounting" researched SMEs tend to pay lower taxes, to "hide" information on their inefficient work, to achieve the best possible price for negotiations on the sale of the company or to attract new investors (shareholders and creditors) through deliberate deception. Clearly the truthfulness and clarity of statements that comply with accounting regulations is a precondition of reduction of fraud and any abuse no matter how complex and costly the requirements are. SMEs can be overcome this issue by the availability of expertise to assist the company in accounting. Readers are thus led to think that future research should develop a close relationship between accounting policies, internal control systems and depreciation methods in medium sized enterprises providing new instruments for their growth and development.

An inquiry into options for preserving integrity in the value of business assets in times of market turbulences, exchange rate devaluation or inflation is also recommended. These are important indirect factors of sustainable enterprise development.

In conditions of the global economic crisis required is an institutionalization of accounting solutions for maintaining the value integrity of business assets particularly in high growth SMEs. It derives logically that the institutionalization of accounting solutions is a condition of orderly internal control through which the management control function preserves the integrity of business assets incorporating optimized working capital.

Finally, what is also required is a greater understanding of the implications of proposed models. Future empirical research should be expanded more deeply, sectorally and to the comparative study of medium enterprises in contrast to both small and large firms. That would reveal new scientific knowledge and facts on what actually lacks in medium-sized enterprises in terms of working capital optimization management.

CONCLUSION

Let us now summarize the ideas of this treatise. This analysis suggests some of the ingredients of successful working capital optimization. First, we have established the relevance of DSO, DPO, DIO, DWC and C2C under a reasonably general set of conditions that are debatable. Second, there exists a certain system of estimating working capital costs correlated with an optimal plan which satisfies the profitability principle of minimum costs such that the total cost of financing working capital is minimal. If an enterprise takes full advantage of optimizing DSO, DPO, DIO, DWC and C2C and achieves minimal costs of financing within a profitability principle, not underestimating any savings, then its working capital plan is optimal. We can conclude that all these elements must be included to obtain consistent and optimal financing. Medium enterprises battling to optimize working capital need to keep in mind the limited potential of financial ratios that show static moments of a business and abstracting a moment in time. It is therefore critical for them to select the best standard for comparison within comparable sectors and the most appropriate moment of time in which to measure working capital categories keeping in mind the dynamics of organizational and economic systems.

In order not to break down because the markets are inefficient and to move away from doing what most models do which is using the past to predict the future the author's proposed solutions emphasize focusing on real time data and holistic linkages between all key identified systemic factors of medium enterprise financing and development. This treatise has also demonstrated that working capital combinations are a critical component of the financing policy based on the phenomenon of continuous qualitative change or alternative substituting of working capital forms.

Studied literature and proposed conceptual framework confirm the significance of working capital optimization. Cited and analyzed references combined with the research from this chapter distinctly, analytically and synthetically validate that there is currently no process of development of specific theories, methods and practices of innovative working capital optimization in medium-sized enterprises as independent firms. Additional work is therefore needed to develop practical working capital criteria of optimality for these firms which will ultimately make a choice whether or not to use them by applying the procedures of decision making inherent in medium-sized enterprises.

REFERENCES

Barcus, S. W. W., & Wilkinson, J. W. (1995). *Handbook of management consulting services*. McGraw-Hill.

Berman, S. (2010). *Capitalizing on complexity*. IBM Global Business Services.

Bernoulli, D. (1954). Exposition of a new theory on the measurement of risk. *Econometrica*, *22*(1), 23–36. doi:10.2307/1909829

Blom, R. J. (2004). *Handboek Midden- en Kleinbedrijf*. Graydon Nederland BV.

Bodie, Z., Kane, A., & Marcus, A. (2007). *Essentials of investments*. New York, NY: McGraw-Hill Higher Education.

Brezinski, H. (1998). The investment behavior of SMEs in East Germany in the 90s, some insights from a survey. In L. Csaba (Ed.), *The Hungarian SME sector development in comparative perspective*. CIPE.

Brigham, E., & Houston, J. (2001). *Fundamentals of financial management*. San Diego, CA: Harcourt. Inc.

Carleton, W. T., & Lerner, M. (1966). *A theory of financial analysis*. San Deigo, CA: Harcourt, Brace & World, Inc.

CFO. (2013). CFO/REL Working Capital Scoreboard. *CFO Magazine*. Retrieved July 2, 2014, from CFO web site: http://www3.cfo.com/working-capital/2013-06

CFO. (2014). CFO/REL Working Capital Scoreboard. *CFO Magazine*. Retrieved October 2, 2014, from CFO web site: http://ww2.cfo.com/cash-flow/2014/07/running-in-place-working-capital/attachment/14jul_workingcap_charts/

Charreton, R., & Bourdaire, J. M. (1985). *La décision économique*. Presses universitaires de France.

Chaston, I. (1990). *Managing for Marketing Excellence*. New York, NY: McGraw-Hill.

Cooper, R. (1995). *When lean enterprises collide: competing through confrontation*. Harvard Business Press.

Danske Bank & Ernst & Young. (2009). *Working Capital Management: A Survey of Nordic Companies*. Copenhagen, Denmark: Danske Bank A/S & Ernst & Young.

Downes, J., & Goodman, J. E. (2003). *Finance and Investment Handbook*. Barron's Educational Series.

Ernst & Young. (2013). *Working capital management report*. Retrieved July 2, 2014, from Ernst & Young web site: http://www.ey.com/Publication/vwLUAssets/Working_capital_management_2013:_all_tied_up/$FILE/All_Tied_Up_2013_DE0430.pdf

Ernst & Young. (2015). *Working capital management report*. Retrieved May 27 2016, from Ernst & Young web site: http://www.ey.com/Publication/vwLUAssets/EY-all-tied-up-working-capital-management-2015/$FILE/EY-all-tied-up-working-capital-management-2015.pdf

Farkas, A. (2005). Benchmark of Working Capital Performance. In *Proceedings-3rd International Conference on Management, Enterprise and Benchmarking (MEB 2005)* (pp. 65-78). Óbuda University, Keleti Faculty of Business and Management.

Friedl, P. (2008). *Model izbire metod evolucijskega načina spreminjanja organizacij: primer gradbene panoge* (Unpublished doctoral dissertation). Univerza na primorskem, Fakulteta za management, Koper, Slovenia.

Froehlich, E. A., Hawranek, P. M., Lettmayr, C. F., & Pichler, J. H. (1994). *Manual for small industrial businesses: Project design and appraisal*. UNIDO General Studies Series.

Kantorovich, L. V. (1959). *Ekonomicheskii raschet nailuchshego ispolzovania resursov* [The Best Use of Economic Resources]. Acad. of SC., USSR, Centar za kulturnu djelatnost Zagreb.

Kaplan, R. S., & Norton, D. P. (2004). *Strategy maps: Converting intangible assets into tangible outcomes*. Harvard Business Press.

Klapper, L. F., Sarria-Allende, V., & Sulla, V. (2002). *Small and medium-size enterprise financing in Eastern Europe* (Working Paper No.2933). Retrieved June 15, 2013, from World Bank website: http://info.worldbank.org/etools/docs/library/28751/SME%20financing%20Eastern%20Europe.pdf

Kotler, P. (1988). *Marketing Management: Analysis, Planning, and Control*. Zagreb: Informator.

KPMG. (2005). *Working Capital Management Survey: How European companies manage their working capital*. Zurich: KPMG Europe LLP & Mitglied des KPMG – Netzwerks unabhängiger Mitgliedsfirmen, die KPMG International Cooperative.

KPMG. (2009). *Singapore Cash Management Survey*. Singapore: KPMG Advisory Services Pte. Ltd.

KPMG. (2012). *Annual Working Capital Survey*. Retrieved July 2, 2014, from KPMG web site: http://www.kpmg.com/DE/de/Documents/Annual-working-capital-survey-2012-KPMG.pdf

Kukoleča, S. (1980). *Osnovi teorije organizacionih sistema. In Privredno finansijski vodič*. Oeconomica.

Lange, O. (1958). *Wstęp do ekonometrii*. Panstwowe Wydawnictwo Naukowe.

Markowitz, H. (1952). Portfolio selection. *The Journal of Finance*, *7*(1), 77–91.

McGrath, R. G., & MacMillan, I. C. (2000). *The entrepreneurial mindset: Strategies for continuously creating opportunity in an age of uncertainty* (Vol. 284). Harvard Business Press.

Menger, C. (1996). *Investigations into the Method of the Social Sciences*. Ludwig von Mises Institute.

Ministry of Economy and Regional Development, Republic Development Bureau, Republic Agency for Development of SME and Entrepreneurship. (2009). *Report on small and medium-sized enterprises and entrepreneurship 2008*. Belgrade: Ministry of Economy and Regional Development Republic Development Bureau Republic Agency for Development of SME and Entrepreneurship.

Modigliani, F., & Miller, M. H. (1958). The cost of capital, corporation finance, and the theory of investment. *The American Economic Review*, *48*(3), 261–297.

Mortensen, P. S., & Bloch, C. W. (2005). *Oslo Manual-Guidelines for collecting and interpreting innovation data*. Organisation for Economic Cooporation and Development, OECD.

Müller-Stewens, G., & Lechner, C. (2005). *Strategisches Management: Wie strategische Initiativen zum Wandel führen, 3., aktualisierte Auflage*. Stuttgart: Schäffer-Poeschel.

Nelson, D. (2003). *The Penguin dictionary of mathematics*. London, UK: Penguin UK.

Nicholson, W. (1997). *Intermediate microeconomics and its application*. The Dryden Press.

O'Reilly, C. A., & Tushman, M. L. (2004). The ambidextrous organization. *Harvard Business Review*, *82*(4), 74–83. PMID:15077368

Ohmae, K. (1982). *The mind of the strategist: The art of Japanese business*. New York, NY: McGraw-Hill.

Porter, M. E. (1998). *Competitive advantage: creating and sustaining superior performance*. The Free Press. doi:10.1007/978-1-349-14865-3

PWC. (2013). *European Working Capital Annual Review*. Retrieved July 2, 2015, from PWC web site: http://www.pwc.be/en/publications/2013/capital_market_survey_2013.pdf

PWC. (2014). *Cash for Growth. Annual Global Working Capital Survey*. Retrieved May 27, 2016, from PWC web site: http://www.pwc.com/gx/en/financial-services/publications/assets/working-capital-2014.pdf

PWC. (2015). *Annual Global Working Capital Survey*. Retrieved May 27, 2016, from PWC web site: http://www.pwc.com/gx/en/business-recovery-restructuring-services/working-capital-management/working-capital-survey/2015/assets/global-working-capital-survey-2015-report.pdf

Rogers, S., Lukens, S., Lin, S., & Jon, E. (2007). *Balancing Risk and Performance with an Integrated Finance Organization: The Global CFO Study 2008*. IBM Corporation in cooperation with The Wharton School and Economist Intelligence Unit.

Savage, L. J. (1954). *The foundations of statistics*. John Wiley & Sons.

Sierpińska, M., & Niedbała, B. (2003). *Controlling operacyjny w przedsiębiorstwie. Centra odpowiedzialności w teorii i praktyce*. Warszawa: PWN.

Şipoş, C. (2009). Optimization models for managerial decisions. *Megatrend Review, 6*(2).

Skoko, H. (1986). *Optimalne kombinacije faktora čovjek-stroj, istraživanje na modelu tekstilne industrije u Hrvatskoj*. Informator.

Sobczyk, G. (2006). *Strategie konkurencji małych i średnich przedsiębiorstw*. Wydawnictwo Uniwersytetu Marii Curie-Skłodowskiej.

Stokes, G. (1900). The Theory of Mathematical Inference. *The American Mathematical Monthly, 7*(1), 1-8. Retrieved July 7, 2016, from http://www.jstor.org/stable/2971489 doi:1

Taleb, N. N., Goldstein, D. G., & Spitznagel, M. W. (2009). The six mistakes executives make in risk management. *Harvard Business Review*, *87*(10), 78–81. PMID:19968059

Tranter, C. J. (1975). *Advanced level pure mathematics*. Hodder and Stoughton Educational.

Vemic, M. B. (2015). Strategic Data-Based Optimization of Working Capital Management in Medium-Sized Firms: A Real-World Case Study. In Strategic Data-Based Wisdom in the Big Data Era (pp. 15-50). Hershey, PA: IGI Global. doi:10.4018/978-1-4666-8122-4.ch002

Von Neumann, J., & Morgenstern, O. (1944). *Theory of games and economic behavior*. Princeton University Press.

Wald, A. (1950). *Statistical decision functions*. John Wiley & Sons.

White, G. I., Sondhi, A. C., & Fried, D. (2003). *The analysis and use of financial statements* (Vol. 1). John Wiley & Sons.

World Bank Centre for Financial Reporting Reform (CFRR) and the Austrian Institute for SME Research. (2013). A Survey of SME Accounting and Reporting Practices in Austria. Vienna: World Bank Centre for Financial Reporting Reform (CFRR) and the Austrian Institute for SME Research.

ADDITIONAL READING

Acs, Z., Parsons, W., & Tracy, S. (2008). *High-Impact Firms: Gazelles Revisited.* Office of Advocacy, the United States Small Business Administration.Small Business Research Summary No. 328. Retrieved March 14, 2015, from web site: http://www.hrp.org/Site/docs/ResourceLibrary/High-Impact_Firms_Gazelles_Revisited.pdf

Bailey, R. E. (2008). Corporate Finance: The Modigliani-Miller Theorems. *Economics of Financial Markets.* Retrieved July 13, 2016, from http://down.cenet.org.cn/upfile/36/200523225142185.pdf

Benninga, S., & Sarig, O. (1996). *Corporate finance: a valuation approach*. McGraw-Hill/Irwin.

Cole, A. R. (2008). What do we know about the capital structure of privately held firms? Evidence from the Surveys of Small Business Finance. Retrieved July 13, 2016, from https://mpra.ub.uni-muenchen.de/24669/1/MPRA_paper_24669.pdf

Fiaschi, A. (2009). *Managerial discretion and optimal financing policies with cash flow uncertainty.* (Working Paper No.3) University "La Sapienza", Rome (Italy). Retrieved July 13, 2016, from La Sapienza Doctoral School of Economics web site: https://web.uniroma1.it/phdschooleconomics/sites/default/files/allegati/Managerial%20discretion%20and%20optimal%20financing%20policies%20with%20cash%20flow%20uncertainty.pdf

Klapper, L. (2005). The Role of Factoring for Financing Small and Medium Enterprises. Retrieved July 13, 2016, from World Bank website: http://siteresources.worldbank.org/INTEXPCOMNET/Resources/Klapper,_The_Role_of_Factoring_for_Financing_Small_and_Medium_Enterprises.pdf

Livingstone, J. L., & Grossman, T. (2001). *The portable MBA in finance and accounting*. John Wiley & Sons.

Prentice-Hall, inc. (1952). *Encyclopedic dictionary of business*. Prentice-Hall.

Spencer, L. T. (2011). *Accelerating Job Creation in America: The Promise of High-impact Companies.* Office of Advocacy, the United States Small Business Administration. Small Business Research Summary No. 381. Retrieved March 14, 2015, from SBA web site: https://www.sba.gov/sites/default/files/HighImpactReport.pdf

Times, F. London Business School, University of Chicago Graduate School of Business, & ea. (1998). Mastering finance: the complete finance companion. Pearson.

Van Horne, J. C., & Wachowicz, J. M. (2008). *Fundamentals of financial management*. Pearson Education.

KEY TERMS AND DEFINITIONS

Average Cost: The arithmetic mean of the cost of all the units of materials or products received.

First In-First Out (FIFO): Method of inventory valuation where goods first acquired are the goods first sold.

Generally Accepted Accounting Principles (GAAP): Rules and guidelines observed by accountants when preparing comparable financial statements (balance sheets, profit and loss statements, financial notes, and disclosures).

Gross Working Capital: Resources invested in firm's cash, accounts receivable, inventory and other current assets.

Highest In-First Out (HIFO): method of inventory valuation. First consumed goods shall be calculated and valuated at highest prices at which they were received.

Just in Time Method of Inventory Valuation: *Lean enterprises* do not need to hold stock and with supportive supplier networks tangentially optimize orders "just in time" with total quality management.

Last In First Out (LIFO): Method of inventory valuation. Goods most recently acquired and units on hand are those first acquired.

Lean Enterprise: A new organizational entity conceived in Japan having know-how to increase management effectiveness and efficiency through lean production methods tangentially approaching optimal enterprise solutions.

Marginal Analysis: A decision-making tool assisting management to maximize profits by optimizing a complex financial system through marginal cost-benefit manipulation of its working capital variables.

Medium-Sized Enterprises: In the EU firms with 50-249 employees, turnover not exceeding EUR 50 million and balance sheets not exceeding EUR 43 million, while in the United States SMEs are firms with fewer than 500 employees.

Merger: A combination of two or more enterprises one of which keeps its identity in contrast to acquisition which is purchasing of all or part of a business.

Net Working Capital: Current assets minus current liabilities.

Optimum Capacity: Level of output at lowest cost.

ENDNOTES

1 Editors' note: This propensity is essentially discussed in chapter 10 on financial innovation in medium enterprises while Figure 1 of chapter 16 graphically portrays the role of innovation in sustainable enterprise development.

Chapter 9
Insurance as an Optimization Tool for Risk Management in Small and Medium-Sized Enterprises

Katarina Ivančević
Law School of Union University Belgrade, Serbia

ABSTRACT

The SME risk management optimization model includes appropriate methods for identification, analysis and classification of risk assessment and developing a plan to overcome the risks and control the applied reaction to risk. Insurance is a classic way of a risk management process and is applicable to specific situations. The subject of this study is exposure and risk management in small and medium-sized enterprises (hereafter: SMEs) which are eligible for transfer of insurance coverage. This chapter discusses the process of risk management in SMEs by transfer to insurance and indicates which risks and in what way can be transferred to the insurer. In this chapter we analyze the supply of domestic insurers in Serbia, and insurance requirements, variants that offer coverage to overcome the particular risks to which SMEs are exposed. In the elaboration and research of mentioned theses the chapter will use methodological procedures of analysis and synthesis, the inductive-deductive method, and if necessary, the comparative method of scientific research.

INTRODUCTION

The purpose of this chapter is to analyze the risks which SMEs transfer to insurance companies, and the value of risk optimization management in these businesses. The underlying premise is is that SMEs in Serbia fail to manage risks in an optimal way. The chapter starts from the premise that SMEs have difficulty in understanding insurance products, and that they need both support and assistance in this regard. In this sense, the chapter deals with main characteristics of insurance, with the aim to show how SMEs can optimize risk management and what the role of insurance companies could be. In an effort

DOI: 10.4018/978-1-5225-1949-2.ch009

to identify issues that occasionally arise and point to possible solutions the author will compare the experience of countries with an advanced insurance market for SMEs, Great Britain and Australia in particular, with that of Serbia. Experience shows that in advanced economies insurance market is very competitive. Next, the chapter will examine offers of insurance products in the Serbian market and the use of broker insurance services. To enhance this chapter, the author will consider issues of importance to SMEs using as an example insurance services in Serbia and possible differences in relation to countries in which there exists a developed insurance market. Insurance sector examples will be presented to support the approach, but the focus will be on opportunities for SMEs. Following the above main content, the chapter discussion will include future research directions. Finally, the chapter ends with a conclusion and key terms and definitions.

BACKGROUND

Small and middle sized enterprises (SMEs) are considered the key drivers for growth and development of national economies. They contribute to growth of employment, particularly in the sectors of transport, civil engineering, communications and tourism. (Jeremić, Živić, 2011, p. 78) For example, in United Kingdom SMEs play a vital role in the health of the economy and according to the Financial Conduct Authority (2015) "99.3% of private sector businesses are SMEs." Similarly, in the Serbian economy 99.8% of business entities are SMEs (Chamber of Commerce and Industry of Serbia, 2016).

SMEs are exposed to numerous business risks. Most organizations must take risks in order to develop, so not all risks are inherently a negative phenomenon (Udovičić & Kadlec, 2013). SMEs regularly face certain threats and numerous risks which endanger their operations. This can have unfavourable economic consequences. Exposure to risks, especially negative ones, must be taken into account also in good times, when profits are increasing (Simons, 1999). SME business success depends on decisions concerning protection of particular risks and measures that should be adopted in order to control and reduce risks. In that respect, quality management of each company includes the organization of monitoring and management of risk.

Realizing SME significance the Boston Consulting Group carried out an extensive survey of 2,500 small businesses in six largest SME insurance markets: U.S., UK, Italy, France, Germany, and Japan. The research explored the insurance decision-making processes of small business owners and found that SMEs are more open to direct interaction and outreach from insurers than many experts have expected. SMEs expect targeted outreach initiatives, improved value proposition for brokers and agents and a "*hybrid direct*" model that provides semi customized offerings, advice, and information (Hoying, Agarwal, Rapsch, & Yasushi, 2014).

At a time when businesses, particularly SMEs, face the consequences of the economic crisis and climate change increases the frequency of extreme events, it is important to discuss issues related to security as a model for overcoming the risks to which SMEs are exposed. Insurance provides a certain amount of security in terms of decrease of potential loss that may occur through realization of risks covered by the insurance. The main objective of risk optimization in SMEs, especially after the occurrence of adverse events, is to establish at least a minimum of business activities in the shortest possible period of time (Njegomir, 2011, p. 41). The additional benefit is that an insurer provides support to SMEs, stimulates better and safer SME operations.

In the whole of European Union there is emphasis on the problems of financing for SMEs, but significantly less attention is paid to insurance for SMEs. The EU Commission has noted that insurance requirements are often difficult to comply with and that in some cases SMEs and professionals in particular have found it hard to obtain insurance coverage for cross-border activities (Jeanmart, 2015).

According to available insurance indicators, Serbia is significantly below the average of EU member states. For example, with the value of the realized premium Serbia is on the 86th place in the world (National Bank of Serbia, 2016, p. 9). However, this data does not include the premiums collected from businesses and there is in fact no data on SME insurance to evaluate. The development of the insurance market reveals the level of awareness and need to conclude insurance contracts. In countries where insurance culture is more developed insurance of the business sector is increasingly provided. For example, in Australia only 26% of all SMEs do not have any form of general insurance. And while in Serbia texts are rare that deal with risk management and insurance for SMEs Australia provides detailed analysis of data with basic insurance information of relevance to SMEs (Insurance Council of Australia, 2008).

SIGNIFICANCE OF SME RISK MANAGEMENT THROUGH INSURANCE CONTRACTS

Conclusion of an insurance contract requires understanding rules on which the insurance and main features of this service are based. The lack of required knowledge may complicate understanding of the insurer's offer and lead to a conclusion of a contract which may not provide adequate risk coverage. An optimal common aim is to envisage the SME insurance also from the perspective of advantages provided by insurance and not only from the perspective of insurance expenses which is what SMEs usually do.

SMEs evaluate the recurrence and intensity of damage and maximum amount of loss that may occur for each separate risk. Ranking of risks and damages that may occur enables a risk manager to choose the most adequate methods or their combination in the risk management process. SMEs may be protected from some risks by insurance. According to Dickinson (2001), "most risks that a company faces cannot be insured or hedged and must be retained or financed internally" (p. 363). Therefore, insurance is one of the methods of partial financial control applied to provide funds for coverage of loss (Andrijašević & Petranović, 1999, p. 14).

In addition, not all risks are insurable. Financial risks which produce impacts measurable in monetary terms are suitable for insurance as a method of loss distribution (Vaughan &Vaughan, 2000, p.17). It is common that only pure risks are accepted for insurance. Insurance neither provides protection of individuals from speculative risks which they have voluntarily accepted (Njegomir, 2011, p.24). The insurer undertakes only a part of the potential loss that he is able to bear in the course of the calculation period. Thus, some risks can only be covered by specific and not all kinds of insurance. Consequently, risk of war and strike risks are covered in transportation insurance.

When SME risk manager determines the risks which will be covered, he should decide which risks will be covered by insurance and which will be retained. "Some insurable risks could be prevented or their impact reduced through efficient loss-prevention or control systems while some could be retained and financed within the company" (Dickinson, 2001, p. 360). Therefore, insurance represents only one component of the risks management process. And not all risks suitable for insurance should be insured or they should not be fully insured (Ivančević, 2006, p. 415) Priority is given to those risks which SMEs are legally bound to insure pursuant to the legal provisions or contractual obligations it had undertaken.

This includes risks characterized by high intensity and low frequency because their realization could cause huge losses to SME.

In most cases the total risk of loss is not transferred to an insurer. Consequently, insurance indemnity will not completely cover the loss. The most important consequence for the insured SME is that, provided it is covered by the insurance, the amount of loss shall be reduced and the great loss shall be distributed into acceptable parts (Sneyd, 2003, p. 105). Laymen may believe that payment of premium is a waste of money if the insured event is not realized within the contracted period and if no loss takes place. However, even if the insured SME does not sustain any losses, it has received something in return for the premium paid: the insurers promise that, in case of insured event occurrence, it will be indemnified. The advantage of insurance is that the insured may recover the money invested in the insurance, by the indemnity payment in the current year in excess of the total amount paid for premium during previous years.

Besides providing a certain level of security, insurance can also improve the quality of SME operations. When deciding to insure a company against certain risks, insurers chose to insure successful companies. During the validity of the insurance contract period, they encourage clients to avoid activities exposing them to excessive risks, thus encouraging actions in the optimal interest of stakeholders (Kočović, Šulejić, 2002, p. 46). For example, insurance terms and conditions emphasize compliance with legal liabilities of the insured, such as the obligation of the regular bookkeeping in line with the provisions of the law. In some countries, insurers try to assist SMEs to improve risk management practices and knowledge of the benefits of insurance. For this purpose several publications have been prepared in the UK. AXA Insurance has already published "Guidance on preparing continuity plans for small businesses" involving business continuity planning for small businesses. Also, Association of British Insurers (ABI) issued a guidance ''Insurance for Small Businesses a guide to protecting your business''. However, in many cases there is still lack of risk awareness SMEs and according to Crichton (2009) "they are not meeting the very basic risk management standards to meet their insurance cover requirements." Risk management costs require the respective budgeting to cover all necessary activities. Generally, SMEs are not able to hire *full*-time *risk managers, as it is set forth in the concept of risk management.* Therefore, insurers are recommended to engage more and support SMEs in the process of risk management. In fact, experience of developed countries shows that when the insurer provides this type of support SMEs can reduce their loss ratio to 10% while increasing customer retention by more than 15 percent (Hoying et al., 2014).

Opportunities in Undertaking Optimal Preventive Measures in SMEs

Cooperation with the insurer and risks transfer to an insurance company may be useful for SMEs in many ways. Disbursal of compensation also contributes to the improvement of SME business environment and performance, and prevents the occurrence of certain risks covered by insurance (Mrkšić, Petrović, & Ivančević, 2007, p.7). Insurers also encourage SMEs to undertake preventive measures aimed at prevention of occurrence of certain risks. Prevention funds may be used for the purchase, replacement or rectification of equipment and protective devices, for employee training in proper use of equipment and for protective measures (Ilkić, 2015, p. 49).

Fire prevention measures are a particularly important example. Fire risk is a typical risk where coverage is provided through an insurance contract.

Training of insured personnel is important because unprofessional and inadequate behaviour represents a risk which can be reduced to a minimum through preventive training. These courses are also interesting for the insured, since in this way security level is raised, with an impact on the amount of premium which provides significant budget savings. Confederation of Fire Protection Association (CFPA) Europe has therefore composed an integral system of training of the insured employees to familiarize them with a fire risk assessment methodology (Vidaković, 2010). Through application of this methodology in industrial, commercial, administrative and residential facilities, adequate risk assessment and prevention level of future risks is provided. The installation of high quality equipment is encouraged (electrical equipment, heating and similar) in accordance with applicable standards, as well as prevention and equipping facilities with fire protection devices. It is required that all stocks and especially materials with explosive or flammable properties are stored in line with law. Hence, the insurer's liability shall be excluded if the insured has not taken these protective measures.

Emerging Fundamental Risks in SME Insurance

Fundamental risks affect material goods, lives and health of many people. They involve floods, windstorms, earthquakes, wood fires and similar. Regularly, these risks affect large segments of the SME sector. Their realization results in catastrophic losses for some businesses. Risks stemming from weather phenomena and climate change nowadays require a special approach. It is considered that climate change represents the greatest threat to modern economy and particularly to agriculture (Petrović & Grujović, 2016). The number of these events grows globally each year. In fact, there were about 14% more natural catastrophe events in 2014 than in the previous year, which was the record for one year (Swiss Re, Sigma, 2014). Similarly, financial losses caused by floods in May 2014, in Serbia alone, exceeded 1.53 billion Euros. The loss recovery was mainly financed from budgetary funds, donations and foreign credit lines. The insurance business participated in indemnity with only 2.4%, i.e. with about 37 million Euros (Business & Finance, 2015). There are no data on losses in Serbia suffered from SMEs, but it can be assumed that they were greatly affected by losses from floods and that, like the majority of economy and population, have not been insured against this risk. In other countries the situation is similar. For example, according to Figueres (2016) "70% of natural disasters that affect people's lives and livelihoods are still not covered by insurance, most often because the right combination of financing and policy are absent or uncoordinated" (p. 42). These figures speak for themselves in terms of SME insurance opportunities.

Given the number and diversity of SME activities in all countries the author assumes that they are significantly affected by the impact of climate change alone. However, their problems are not given enough attention by the media or adequate support from state and local authorities. Even in the UK only a small number of SMEs received flood assistance from the local administration. Damages are mostly covered from their own resources or from their insurer. According to Crichton (2009) "most of the help for small businesses comes from their insurance companies", however many SMEs in the UK were unable to renew their insurance. Therefore, it is estimated that extreme cases of floods could lead to SME bankruptcy (Crichton, 2009).

Losses resulting from the natural catastrophes are vast. Classic insurance is not able to take over and cover all these risks. Participation of the state in respective risk management is demonstrated through the creation of different funds for insurance against these risks, cooperation between state and the insurer, introduction of obligatory insurance against certain risks which are considered catastrophic and subsidies or assistance in case of such risks. Development and implementation of policies that encourage demand for insurance products against catastrophic risks and stimulate good practices of managing catastrophic risks are important. Accordingly, compulsory insurance against storms was introduced in

France, Spain and Belgium which is contracted together with fire insurance (Saković, 2015). Turkey introduced compulsory insurance from earthquakes (Ivančević, 2006, p. 478)

There is no secure way to prevent natural catastrophes, but their devastating effects can be mitigated in part. Preventive and protective measures against loss occurrence are commonly applied. Accordingly, embankments, barriers, concrete walls are constructed and monitoring of meteorological data is organized. Some other activities are also taken. However, according to Crichton (2009) "SME are as yet not well prepared for the impacts of climate change and indeed many do not take the threat seriously, despite the fact that the smaller the business the more vulnerable it is to the impacts of climate related disasters while larger companies are more likely to describe their strategy regarding climate change and are probably more likely to have the resources to do so" (p.2). Only a partial loss can be covered by insurance and insurers have the right not to pay 100% of the claims costs. Which means a typical SME customer may not have the funds to meet the deficit. It is not possible to completely envisage all the consequences occurred by realization of these risks. Therefore, insurers limit coverage. If these risks were fully overtaken, premiums would be extremely high and therefore unacceptable for the majority of insured SMEs.

CHALLENGES OF SME PROPERTY INSURANCE

SME Property insurance requires rationality. Namely, the insurance of all property against all risks to would be too expensive and unnecessary. Decision on risks which will be transferred to insurance should be based on evaluation of the frequency and intensity of particular property risk exposure. Insurance of property assessed as low intensity risk with low likelihood of risk occurrence is not required. Such risk can be completely eliminated or significantly reduced through SME preventive and protective measures. The high frequency risks, including those of low intensity, as well as low frequency risks having great intensity should be insured. In the risk assessment processes the general statistical data should be combined with own experience. The place of insurance occupies an important place in the decision-making process. Some risks can be realized in some places to a lesser or larger degree, while elsewhere they cannot be realized. Duration of the insurance contract (short-term, long-term or multi-year) shall be matched to the risk type and effects.

Special insurance conditions and premium tariffs outline the contracting modalities of property insurance coverage. The insured sum represents the upper limit of insurer's liability and it must correspond to the actual value of the subject matter of insurance. Type of insurance coverage and the sum insured should be selected in accordance with the chosen risks and subject matter of insurance. Insurer's liability can be proportional and non-proportional. Under proportional coverage those parts of property exposed to high intensity risk (fire risk, e.g.) are insured because they can cause total loss of the insured object. In case of proportional insurance indemnity it is important that the insured object value is equal to the amount of the sum insured.

In order to avoid under-insurance some form of non-proportional insurance may be contracted as the insured is not always interested to insure property to the highest actual value. Under *The first loss insurance clause* each loss up to the agreed sum insured is contracted to cover the maximum possible loss that can be predicted for the insured SME facility. When applying this contracting modality, it is important to properly assess the amount of the maximum possible loss. It is favorable for the insured since the insurance premium is significantly lower than for other types of coverage. In case of low intensity risks with lower impact and limited loss (e.g. water leakage from installations) it is assumed that it would be optimal to contract SME insurance coverage through the first loss insurance. Robbery and burglary insurance is also contracted in that way. Special form of the first loss insurance is the *agreed value insurance* applied for high intensity risks which may result in total loss of the insured item. It is applied for the insurance against fire risk and some machinery breakdown. Premium calculation is based on asset book value as of 31 December of previous year. Therefore, it is required to maintain orderly bookkeeping which is an added value in SME development. When insurance based on *tax value* is contracted the insured and the insurer jointly determine the value of the insured item and respective indemnity accordingly. This is a way to insure items having a subjective value for insured such as fine art paintings, jewelry and collections.

Insured Event Prevention and Rescue of Insured Object

The insured is obliged to take all necessary measures in order to prevent risk occurrence covered by the insurance. Should the insured event take place, the insured shall be obliged to take applicable measures (reporting to the competent authority, etc.) or specially stipulated measures, in line with possibilities and situation occurred in order to limit consequential damage. Should the insured, without justified reason, fail to fulfill the above mentioned obligations, insurer's liability shall be reduced by an amount equal to the increase of damage due to such omission (Serbia, Law on Contracts and Torts, Article. 926).

The advantage for the insured, that SMEs should take into account, is the legal liability of the insurer to reimburse the expenses, losses and other damages caused by the insured's attempt to eliminate the immediate risk of insured event occurrence, i.e. reasonable attempt to limit its actual consequences. Reasonable expenses are those expenses which are lower than the loss that could have been expected if the expenses had not been sustained. This liability exists also if all the attempts of the insured have been unsuccessful. In addition, in case that the indemnity and the reimbursement of these expenses are higher than the sum insured, the insurer is obliged to reimburse them in full amount. In that way the client is encouraged to take necessary measures to prevent the insured event occurrence and to reduce its adverse effects.

SME Participation in Loss and Exclusion from Insurance

Through exclusions from insurance coverage and franchise contracting SMEs are encouraged to handle their property with care, to use it properly and improve working environment. The insured is entitled to retain the portion of risk in the self retention insurance. The respective decision is based on statistical methods (Andrijašević & Petranović, 1999, p.20). Insurance against particular risks foresees obligatory participation of the insured in the possible loss (franchise). This is contracted through percentage franchise and determination of minimal participation in the loss for certain risks (robbery, burglary, machinery breakdown and some other risks in domestic transport). In this way, the insured reduces the costs of the premium.

Insurance excludes certain risks, such as vulnerable facilities of the low-rise construction category (kiosks, cabins, containers) and wooden structures. Legal entities and entrepreneurs whose movable property (inventory, equipment and supplies) is located in such facilities are sometimes refused insurance contract in line with the specially designed conditions for SMEs.

Legal Protection Insurance

Bearing in mind the complexity of legal system, SMEs need to engage professional assistance for any proceeding before the court or other dispute resolution entities in order to protect their rights. They will bear certain expenses accordingly. Legal protection insurance allows the insured SMEs to transfer the risk against occurrence of legal costs to the insurer, based on the premium paid. A party who loses the dispute partly or completely is obliged to reimburse the litigation costs to the opposing party, pursuant to court decision, in total or proportionally to the success achieved. (Serbia Law on Civil Procedure, Article. 153) Concluded legal protection enables the insured SME to be covered for expenses incurred during the potential litigation. This insurance type may have a positive effect for insurer as well. If the insured is successful in litigation, the insurer has a possibility to reimburse the expenses from the party which has lost litigation, i.e. from its insurer if that party has concluded a legal protection insurance contract.

Under this type of insurance, the insured has great influence on insured event occurrence, differently than under other types of property insurance. Therefore, insurers stipulate the participation of the insured in loss (franchise) and right of insurer to terminate the contract if two or three insured events occur during the same year (Ivanjko, 2005, p. 40).

This insurance type is developing and the new forms of coverage are offered. According to Serbian law, the legal expenses insurance is designed as a type of non-life insurance. This insurance covers legal expenses, lawyer's expenses and other proceedings costs (Serbia Insurance Law, 2014, Article 9). In Serbia there are no specialized insurers who deal only with legal expenses insurance. It is common that the legal expenses insurance is offered together with some other insurance type, most frequently with liability insurance.

Legal innovation is important since in clauses regulating dispute conduction, insurers should become authorized to require certain behavior from insured, not only in terms of litigation, but also their out-of-court behavior (Šulejić, 2005, p. 396).

However, Serbia has not demonstrated adequate interest in legal protection insurance. This is one of the reasons why the offer of domestic insures is poor. This type of insurance should be encouraged by the state. In return, the state would reduce expenses sustained for free legal assistance. Lawyers are interested in development of this insurance type, since it provides secure costs settlement. Due to the costs of the court and attorney, SMEs often renounce litigation. However, this insurance type may be important for SMEs, since it could enable better recovery of outstanding debts and better protection in legal proceedings initiated against them. SMEs should request from insurers to provide legal insurance coverage wider than coverage provided by classic insurance types. Or engage a lawyer in the pre-dispute proceedings in terms of preventive legal advice in order to avoid controversial situation occurrence, i.e. possible disputes, and obtain advice which could contribute to successful out-of-court dispute resolution. This form of legal protection could be realized in cooperation with lawyer's chamber, in the tradition of comparative legal experiences such as in Germany (Belanić, 2014, p. 32).

This kind of insurance is the most developed in Germany, with a large number of insurance types, high coverage amount and relatively low premiums (Belanić, 2014, p. 31).

Insurance of Persons

SME employers who contract collective insurance against the consequences of casualty for their employees are concerned for their employees and demonstrate corporate social responsibility. This contract can be subject to collective negotiations between employees and employers. Therefore, by paying the insurance premium SME employer provides some protection. In that case, the indemnity received by the employee is included in the total indemnity amount which employer is obliged to pay to the employee if liable for the injury which occurred. SME employer is entitled to insure all employees, regardless of the kind of work contract. Scope of insurance coverage is determined according to the requirements and needs.

An employer is liable to insure employees for occupational diseases and work-related diseases, in order to ensure compensations. (Serbia Law on safety and health at work, Article 53, 2015). Legislator did not set forth the minimum sum insured, enabling employer to conclude contract in line with his financial possibilities. It is necessary to bear in mind that most insurance companies in Serbia excluded infectious diseases, occupational diseases and other diseases from insurance terms and conditions and did not consider them as causalities. Therefore they are not covered by causality insurance. Some insurers have special conditions for occupational diseases. Some of them offer this insurance through stipulation of special clause with collective insurance of employees against the consequences of casualty.

THE SME INSURANCE CONTRACT

Upon deciding to use insurance as a risk prevention method, SME managers need to take measures in order to obtain optimal coverage with the lowest premium. There are numerous insurance products suitable for various needs and situations on the market. Adoption of appropriate decisions on the selection of insurance products requires having a certain level of financial literacy, knowledge and understanding of the basic characteristics of insurance. It is necessary to know the way the insurance system functions, as well as the way in which insurance treats risk transferred to it according to the insurance contract. Insurance service is sophisticated and insurance products are specific and complex. The lack of SME knowledge and clear understanding of the insurer's offer, conditions and the scope of the covered risks can turn into a difficulty and cause rejection of insurance.

On some occasions, the insured may decide to change the insurer, for example if another insurer offers a lower premium. Prior to making this decision, the insured should check if the scope of coverage provided by the other insurer at a lower price is completely the same as the existing one. A lower price does not imply the same conditions, even though the product may have the same or similar name. It seems that for the majority SMEs having the right cover is equally important as the price of the insurance policy (Financial Conduct Authority 2015, p.14). In more developed markets SMEs are looking for a lower price, but will also change insurers if they can obtain better coverage or satisfy a specific need, not just receive a better price (Hoying et al., 2014).

On some occasions, the savings provided by changing the insurer can really be just short-term. In fact, if the insured SME's technical result in a certain period of time becomes negative due to frequent claims within the coverage period, the insurer shall increase the premium by applying the bonus-malus system (BMS). In case of compulsory automobile insurance, the insured cannot benefit from the change of the insurer. All insurers offer the same premium and they all apply BMS in a centralized and organized manner, regardless of the chosen insurer. Interestingly, research has shown how the performance of SMEs does not significantly affect the volume and change in insurance coverage (Insurance Council of Australia, 2008).

Pre-Contract Information is Fundamental

As a professional, an insurer has knowledge and information which is important for the conclusion of contracts, and which, without a preliminary presentation and clarification to the counterparty, is either unavailable or difficult to understand. Pre-contract and contract information is aimed at informing the insurance users of all the facts they need to know of the product, contractual conditions and most of all the parties' rights and obligations (Ivančević, 2015, p. 236).

The right decision can be made only by the customer, or the service user, who has been properly informed (Weatherill, 2001, p. 180). A potential policyholder should assess the information received and based on it decide whether to conclude the proposed contract or not and under what conditions (Ebers, 2004). Additional customer information and clarifications also influence the final decision. Information must not make a false impression of the insurance products. Information should be unambiguous, comprehensible, clearly stated, timely and extensive enough. However, it should not be too extensive, in order not to confuse the recipient.

The Serbian law has established the minimum content of information that an insurer is required to make available to a policyholder prior to the conclusion of a contract and stipulates the manner of making this information available (Serbia Insurance Law, Articles 82-84, 2015). Unlike decisions in comparative law that limit the obligation of precontractual duty to inform a policyholder only in the case of consumer status, the Serbian legislator has decided to impose an obligation for an insurer to report in advance to all potential policyholders (Mrkšić, Petrović, & Ivančević, 2014, p. 184).

In comparative law, the obligation of mutual provision of information of an insurer and a policyholder in the process of the conclusion of an insurance contract is also considered a duty (Borselli, 2012). A sanction for omission to provide information opens possibility of insurance contract cancellation, given that there was a defect of will during its conclusion. If a contract is concluded under fraud, a policyholder is entitled to request damage compensation (Serbia Law on Contracts and Torts, Articles 61, 65, 111, 268). Serbia Insurance Law Act also provides sanctions for failure to provide information. In the course of supervision, National Bank of Serbia is authorised to order an insurance company to eliminate illegalities and irregularities in any operations. A fine is envisaged as a sanction in the case that the prescribed information is not provided or not provided in a prescribed manner. An insurance company can even be left without a work permit if it gives false information about its operations or information that may mislead users of insurance service (Serbia Insurance Law, 2014, Articles 260 & 214).

Most often, insurers offer coverage of certain risks in several ways. Doing so allows policyholders to choose a variant of protection that suits them in substantive and financial terms. However, most small-business owners consider commercial-insurance offerings extremely complex and difficult to understand (Hoying et al., 2014). SMEs' managers have difficulties in choosing an optimal insurance product since they usually lack specific knowledge. For them it is not easy to understand insurance conditions that define the scope of risk coverage, the scope of insured risks and the insurance exclusions. When closing an insurance contract, it is necessary to study insurance conditions of several insurance companies in order to find coverage that provides the SME client with optimum protection. Many SMEs accept inadequate insurance coverage which does not provide optimal compensation of loss. These observations were made in the UK (Financial Conduct Authority, 2015). It was also noted that managers of SMEs who have suffered damage from floods, did not understand the scope of coverage obtained by concluding an insurance contract, or that the coverage is limited (Crichton, 2009). Insurance in case of flood, torrent and high water was an additional risk within the basic insurance covering the damages in case of fire. Those who suffered damage and had only the basic insurance package could not get coverage for the damage caused by floods.

Significance of Hiring Insurance Brokers for Optimal Coverage

When procuring insurance services SMEs can benefit from hiring a licensed insurance broker or a company specialized in insurance broking. SMEs have more complex insurance needs than retail clients but are not always more sophisticated buyers of insurance. Hiring a broker increases the chances of SME managers to obtain optimal insurance coverage according to their needs and financial resources. In United Kingdom many SMEs rely on the services of brokers to help buy insurance. However, in Serbia intermediaries in the collection of the total life insurance premium in 2015 accounted for only 9% and their share in the distribution channels for all types of insurance was only 7% (National Bank of Serbia, p. 14).

Indeed, the Serbian legislator has defined specific obligations for insurance brokers (Serbia Insurance Law, 2014, Article 94). This obligation is also present in comparative law and regulated in accordance with EU directives. For non-compliance with these obligations, a sanction is envisaged in the form of a fine (Serbia Insurance Law, 2014, Article 261). A broker reports to an ordering party or a third person for any damage caused by his omission to perform the obligation, or for its incorrect performance (Ivančević, 2014).

Brokers provide highly professional services specifically expressed when providing information and advice to the SME clients. The duty to advise a policyholder is also established for a broker. The broker is obliged to furnish the insured with all necessary advice and guidance in terms of circumstances valuable for closing and implementing the insurance contract. The duty of information means providing information that describes a product or a service (Ivančević, 2010) which relates to presentation of certain facts. Advisory service is preceded by an analysis of facts from collected information and provided by a person who advises. Recommendations should help the advised person to make a decision that is in his best interest.

As an expert in this field, a broker is obliged to analyze the dangers that the insured could face, define the specific insurance needs based on data provided by the insured, propose adequate insurance coverage, specify his advice and proposal for signing the contract with a specific insurer and mediate in the insurance procuring process in order to provide optimal coverage. Many SMEs in UK don't have contracted optimal insurance coverage especially because they did not understand. Obviously lack of adequate advice is a problem and consequently many SMEs do not use brokers to arrange their insurance (Crichton, 2009). Indirectly, it can be concluded that the situation in Serbia is similar.

Possible contribution of a broker does not end with the conclusion of a contract because during the term of the insurance contract there will appear other significant tasks (Ivančević, 2014, p.158). The last risk management phase is controlling and possibly revising the acts, procedures, measurements and plans for eliminating the risks. An SME risk manager can receive significant support from a broker in terms of risks transferred to the insurance company. After the contract becomes effective, the broker is obliged to review the content of the policy and contract, audit the performance of the contract and make a proposal for contract changes and annexes in the insured SME's optimal interest. The broker must provide necessary assistance to the insured while the contract is effective, as well as in case of the event covered by the insurance. The broker must make sure the insured has taken all the measures that enable it to exercise and protect contractual rights. During the contract validity all SME managers and employees must be aware of contractual obligations in order to prevent losing insurance coverage due to contract breaches.

The risk manager must be aware that insurance decreases the loss over a certain period of time. This way, static risk becomes reduced risk in terms of transferring the risk to the insurance company. For the insurer, the insured static risk becomes dynamic risk since two alternatives are possible: gain or loss (Sneyd, 2003, p.103) When the insured risk occurs, the insurer might not pay the necessary amount in a short period of time. This happens for various reasons, such as untimely insurer's reaction, the lack of resources, or an oversight by the insured SME, which represents a breach of the insurance contract and causes loss of the insurance coverage. By hiring a broker these new risks for SMEs can be reduced (Financial Conduct Authority, 2016). The broker is obliged by law to protect the insured's optimal interest during the closure of the insurance contract. He should clarify reasons why a certain insurance company was proposed, take care of insurance company's professional and other references, introduce the insured to all legal and economic connections with the specific insurer and provide professional support during contract realization in order to avoid any oversights by the insured.

Sometimes involvement of the broker does not guarantee that SMEs will be provided adequate coverage. A review of insurance brokers in the UK found that their service when arranging insurance for SMEs is not always consistent with customers' expectations (Financial Conduct Authority, 2015). Hence, the Association of British Insurers has published a brief guide whose purpose is to encourage SMEs to verify that the broker works in their best interests (ABI, 2016).

A Review of SME Packages in Serbian Insurance Companies

Most insurers work with SMEs according to general and special conditions for particular types of insurance products. Some insurance companies in Serbia are developing their marketing strategy on the basis of introducing new SME insurance products while others are focused on offering classic insurance support for SMEs. Therefore, some Serbian insurers have come to understand the essence of SMEs, their specific activities and are in fact offering special insurance packages. The author shall present the basic characteristics of the insurance products tailored to SMEs needs.

Dunav Insurance Company manages "*Business Bonus Packages*", which include different solutions for insuring certain activities of SMEs (Dunav, 2016). This insurer designed packages for retail, production and artisan industry, catering services (restaurants, cafes, and hotels), service industry such as beauty and spa services, as well as for the office, firms and agencies. Insurance contracts are closed for the period of up to one year. When it comes to price of insurance, the premium is lower compared to standard coverage offered within these packages. In case the insured SME decides to take two or three types of coverage in addition to basic package, 5% discount is granted. For all four types of coverage, 10% discount is granted. The insurer shall assess the damage within 24 hours and quickly process the request and pay for damages. The coverage that was completely or partially expended by the damages can be arranged again, within the remaining period and under favourable conditions.

Risk coverage can be arranged by combining risks from several packages. The basic insurance package includes compulsory coverage and a number of optional types of coverage in order to provide a comprehensive insurance protection. The basic package provides property protection against the consequences of: fire and other dangers, burglary theft and robbery, glass breakage, machines breakdown and other dangers as well as t employees insurance in case of an accident. The optional types of coverage are chosen according to specific needs of an SME industry sector: to insure machines, equipment and installations in case of a failure; business profit and income costs in case of a work interruption due to fire or to have comprehensive general liability insurance. Bonus types of coverage *Dunav Insurance Company* are also foreseen for each package. Each package offers the following bonus two types of coverage:

1. Increasing the insured sum for stocks in the selected period by 20% of the insurance sum for the stocks on each individual account;
2. Costs covering the door lock change due to key loss – 200 Euros per event and per year. Bonus coverage for employees' belongings due to burglary theft and robbery is offered with most packages. The coverage is 50 Euros per employee and per event and 200 Euros per year.

Property insurance is arranged as the insurance sum for each of the risks in the specific percentage of the basic-total insurance sum and with the individual risks with the exhausted insurance sum. The insured also participates in covering damages through franchise deductible. Minimal participation is foreseen for certain risks (burglary theft and robbery, machine breakdown and other dangers, as well as the optional coverage for the work interruption due to fire and other dangers).

Accident insurance covers the death risk in case of an accident and the total permanent disability in three combinations of the insured sums.

Grawe Insurance Group offers SMEs their1 ''Entrepreneur'' package, (Grawe, 2016). Forty different kinds of SMEs can be insured through the package, such as shops, pensions, restaurants and cafes, gyms, kindergartens, schools, retirement homes. It includes property insurance ''at first risk'', without an expert's assessment. Insurance within the package can be agreed on for the period of one to ten years, with the possibility or renewal. Premium can be paid on instalment plan, whereas an instalment cannot be lower than 30 Euros. The package provides insurance of: office space in objects and condominiums (up to 250 m^2 or up to 500 m^2 for certain SMEs, used for entrepreneurial purposes), equipment, machines and furniture, goods, stocks and merchandise.

According to client's choice, insurance provides protection against the following risks: fire and natural disasters, flooding, burglary theft and glass breakage. Insurance sum varies from 10,000 to 75,000 Euros, whereas equal insurance sum is valid for each of the insured risks.

Wiener Stadtishe insurance company offers the *„Insured business"package* intended for micro businesses and SMEs, Wiener (2016). It provides SMEs with the possibility to insure their property through a compulsory package for equipment, stocks, third parties, employees' belongings, and money and cash assets. Within the optional package, the insured can insure its own and the third party's construction sites, as well as investments into others' construction sites.

Basic package provides protection against the following risks: fire, thunder strike, explosion, windstorm, hail, civil unrest, snow pressure, outflow of water from installations. The coverage can be expanded for the case of work interruption due to any of the risks from the basic coverage. The coverage can be additionally arranged for material damages caused by the following risks: flood and high water, burglary theft and robbery, cash on hand in the cash register of the premises exclusively against the risk of robbery, as well as the machines, appliances and equipment breakdown.

The coverage may also be expanded in terms of optional package covering for damage caused by glass breakage at the construction site and waterworks and sewage installations breakdown at the construction site. As additional coverage, insurance of portable equipment can also be arranged, as well as general liability insurance for damages to the third parties, with the possibility to expand it to the employer's liability.

AXA insurance offers firms, micro firms and SMEs specially designed insurance package called *''Complete BUSINESS''* (AXA, 2016). The insurer provides damage assessment within 48 hours from the claim and payment of 50% of the damage in advance, so that the insured could eliminate the consequences of damage and continue with operation.

This package covers the insurance of office space, employees, equipment, stocks, glass and advertisements, SME's guests and service users, money, computer equipment, as well as SME's liability towards third parties and employees. Three insurance packages are available: Standard, Superior and VIP.

VIP package provides insurance against the basic risks: fire, hail, thunder strike, explosion and other; outflow of water from installations, flood, high water, snowmelt, glass breakage, general liability, burglary theft and robbery, machine and installation breakdown, computer equipment insurance. Superior package does not cover for the machine and installation breakdown or for computer equipment insurance. Besides these two risks, standard package does not cover burglary theft and robbery.

Optional coverage can be arranged for all the packages, including the protection against the consequences of landslide, earthquake, employee insurance against work-related injury whereas treatment costs are covered, employer's liability insurance, and protecting insured's users against work-related injury.

Besides basic and additional coverage, bonus coverage is offered as well: insurance coverage for fixed costs caused by work interruption due to fire, insurance coverage for renting a new space due to work interruption caused by fire or explosion, insurance coverage for the production of marketing materials necessary for business promotion after business continuation, coverage for the construction and erection works, coverage for the damage on summer and winter gardens of the cafes and restaurants and key person insurance.

Therefore, offers of Serbian Insurers targeting SMEs are various. Dunav insurance has included in the basic package the employee insurance against accidents, but for most insurers it is contracted on demand. With AXA Insurance employees insurance may be stipulated as collateral, as well as accident insurance for customers insured.

Most insurers offer SMEs property insurance to the sum insured but Grawe Insurance offers first risk. Insurance from liability to third parties in Dunav can be arranged as an optional coverage, while in Wiener Stadtishe this can be additional coverage and liability insurance to employees. Bonus coverage are offered in packages of Dunav Insurance and AXA Insurance.

Variety offer Serbian Insurers shows that SMEs have a choice but also a difficult task to choose among various options. In order to conclude a contract it is essential to obtain assistance from insurers or brokers.

Review of coverage of certain risks through packages is shown in Table 1[1].

FUTURE RESEARCH DIRECTIONS

Overcoming risks through insurance in the normal course of SME business is left to assessment and evaluation of their managers. Insurers are recommended to engage more and support SMEs in the process of risk management. On its side the state can encourage and streamline SMEs to insure against financial loss by providing subsidies for insurance or other financial incentives such as introduction of tax relief for SMEs if they transfer certain risks on insurance companies.

There is an imperative need for the state to identify and take specific measures, using suitable methods, in order to protect from damages that are the result of fundamental risks. The SME sector is considered to be further threatened by the fact that as a rule the state does not allocate funds for the rehabilitation of

Table 1. Examples of SME risk coverage in packages

Type of Risk	Dunav	Grawe	Wiener Stadtishe	AXA
Fire	Basic	Basic	Basic	Standard
Thunder strike	Basic	Basic	Basic	Standard
Explosion	Basic	no	Basic	Standard
Windstorm	Basic	Basic	Basic	no
Flood and high water	Basic	no	Additionally	Standard
Hail	Basic	Basic	Basic	Standard
Torrent	Basic	no	Additionally	
Snow pressure	no	no	Basic	no
Snowmelt	no	no	no	Standard
Earthquakes	no	no	no	Collateral
Landslides	no	no	no	Collateral
Outflow of water from installations	Basic	Basic	Basic	Standard
Glass breakage	Basic	Basic	no	Standard
Burglary theft and robbery	Basic	Basic	no	Superior
Civil unrest	no	no	Additionally	Standard
Robbery during the manipulation of money at the box office	no	no	Additionally	no
Machine and installation breakdown	Basic	no	No	VIP
Computer equipment insurance	no	no	No	VIP

Source: Own presentation of the author.

the consequences on incurred losses so that they are actually left to themselves and their own resources. For this purpose some of the solutions that have given good results in other states, or some of the suggestions of experts can be applied.

It would be important to reinforce awareness of SMEs about offers, opportunities and benefits of insurance. Insurers will need to simplify product offerings and become proficient at providing customized counsel. The state and the educational system have special responsibility to develop and improve models of financial education among citizens. Socially responsible behavior of insurance companies can be expressed through the support of the state, educational institutions and non-governmental organizations in the process of financial education modeled on activities carried out in other countries. Clearly, the state also has an interest in developing proper training of SMEs. In this way it contributes to the realization of the social function of insurance that is manifested through financial security, raising living standards and creating a favorable impact on the overall social welfare.

CONCLUSION

In developed countries the share of SMEs in total premium is much more significant than in Serbia and this affects the number of SMEs and the culture of security. In their business operations SMEs in Serbia underutilize risk insurance. Most often the payment of insurance premiums is considered an unnecessary and not a necessary business expense. Insurance can provide protection from many unforeseen events that may cause losses in business. The chapter has showed that SMEs in the process of risk management have similar weaknesses in Serbia and in the countries with developed markets. SMEs need assistance in risk management as well as appropriate advice and technical support from insurance brokers. In fact, SMEs are in need of the same kind of support that is normally provided to consumers.

Properly choosing risks that will be transferred to insurance, selecting appropriate ways of providing insurance coverage is in the function of SME development. Insurance therefore contributes to the stability and security of SMEs. Through preventive and repressive measures insurers encourage conscientious and responsible business of their policyholders. Consequently, insurance becomes high-quality support for clients, particularly by providing necessary suggestions, advice and in assessing individual risks in SMEs.

SME businesses operate in a tough and changing environment. Their operations are still held to the same standards of corporate governance, employee welfare and liability as in larger companies. They often have to manage risks with fewer resources in less time besides concerns of the day to day business. SMEs therefore require a suitable packaged insurance policy which will protect their assets, liabilities and employees from the risks associated with operating a business. Insurers should be able to adapt and improve their products aimed at the needs of SMEs.

When evaluating the insurer SMEs need to study and understand the insurance terms, comparing the amount of premium for a particular volume of required insurance coverage for individual insurers. Conclusion of appropriate insurance contracts that will meet their actual needs requires expertise and quality assessment of many factors. Hiring a professional person, a broker, is a good option for SME risk optimization, because, as a rule, these entities do not have their own experts, or are exposed to complex risks in their business operations.

It is realistic to conclude that optimization of risk management for SMEs contributes to the stability of their business, their development and overall social security, encourages their investment in innovation and consequently allows greater degree of their overall market exposure.

REFERENCES

ABI. (2016). Retrieved August 21st, 2016 from Association of British Insurers web site: https://www.abi.org.uk/

Andrijašević, S., & Petranović, V. (1999). *Ekonomika osiguranja*. Zagreb: Alfa d.d.

AXA. (2016). Retrieved March 25, 2016, Axa Insurance web site: from https://www.axa.rs/

Belanić, L. (2014). Povijesni razvoj osiguranja pravne zaštite. *Evropska revija za pravo osiguranja*, *2*, 26-34.

Borselli, A. (2012). Cognosceat emptor: On the Insurer's Duty to Inform the Prospective Policyholder in Europe. *Revija za pravo osiguranja*, *2*, 45-54.

Chamber of Commerce and Industry of Serbia. (2016). Retrieved August 21, 2016, from http://www.pks.rs/PrivredaSrbije.aspx?id=20

Crichton, D. (2009). *Coping with Climate Change: Risks and Opportunities for Insurers. Chapter 8: Small and medium enterprises*. Retrieved March 25, 2016, from http://www.cii.co.uk/media/4043823/ch8_smes.pdf

Dickinson, G. (2001). Enterprise risk management: Its origins and conceptual foundation. *The Geneva Papers on Risk and Insurance. Issues and Practice*, *26*(3), 360–366. doi:10.1111/1468-0440.00121

Dunav. (2016). Retrieved August 15, 2016, from *Dunav Insurance Company* web site: http://www.dunav.com/

Ebers, M. (2004). Information and advising requirements in the financial services sector: Principles and Peculiarities in EC Law. *Electronic Journal of Comparative Law*, *8*(2), 1-16. Retrieved March 15, 2016, from http://www.ejcl.org

Figueres, C. (2016). *Sustaining momentum on climate action*. Insurance Europa. Retrieved May 15, 2016, from http://www.insuranceeurope.eu/annual-report-2015-2016

Financial Conduct Authority. (2015). *Handling of insurance claims for Small and Medium-sized Enterprises (SMEs), Thematic Review TR15/6*. Retrieved May 19, 2016. from https://www.fca.org.uk/static/documents/thematic-reviews/tr15-06.pdf

Financial Conduct Authority. (2016). *SMEs and insurance brokers*. Retrieved March 10, 2016, from: https://www.the-fca.org.uk/consumers/smes-insurance-brokers

Grawe. (2016). Retrieved August 15, 2016, from Graw Insurance Group web site: http://www.grawe.rs/rs/index.htm

Hoying, T., Agarwal, R., Rapsch, S., & Yasushi, S. (2014). *Mining the Untapped Gold in SME Commercial Insurance*. Retrieved March 10, 2016, from https://www.bcgperspectives.com/content/articles/insurance_growth_mining_untapped_gold_sme_commercial_insurance/

Ilkić, Z. (2015). Osiguranje mašina od loma, upravljanje rizicima i preventiva u Srbiji. *Evropska revija za pravo osiguranja*, *14*(2), 45-51.

Insurance Council of Australia. (2008). *Non-Insurance in the Small to Medium sized Enterprise sector, December 2008*. Retrieved March 10, 2016, from http://www.insurancecouncil.com.au/assets/report/non-insurance%20in%20the%20small%20to%20medium%20sized%20enterprise%20sector.pdf

Ivančević, K. (2006). Upravljanje rizikom u osiguranju. In Krizni menadžment: Prevencija krize, (hrestomatija), (pp. 413-438). Beograd: Univerzitet u Beogradu, Fakultet bezbednosti.

Ivančević, K. (2010). *Pravna zaštita potrošača korisnika usluge osiguranja i bankarske usluge* (Unpublished doctoral dissertacion). Univerzitet u Beogradu, Pravni fakultet.

Ivančević, K. (2014). Posredovanje u osiguranju. In Uslužni poslovi (pp. 143-166). Pravni fakultet Univerziteta u Kragujevcu.

Ivančević, K. (2015). Finansijsko obrazovanje potrošača i delatnost osiguranja. In Reforme i novi izazovi u pravu osiguranja (pp. 230-252). Udruženje za pravo osiguranja Srbije, Palić.

Ivanjko, Š. (2005). Osiguranje pravne zaštite. In Aktuelnosti građanskog i trgovačkog zakonodavstva i pravne prakse (pp. 39-49). Neum: BiH, Pravni fakultet u Mostaru.

Jeanmart, N. (2016). *Cross-border insurance, Insurance Europa, Annual Report 2015-2016*. Retrieved June 10, 2016, from http://www.insuranceeurope.eu/annual-report-2015-2016

Jeremić, Lj., & Živić, M. (2011). Uloga malih i srednjih preduzeća u ekonomskom razvoju i zaposlenosti u Evropskoj uniji. *Ekonomski vidici, 91(1)*, 77–85.

Kočović, J., & Šulejić, P. (2002). *Osiguranje*. Beograd, Srbija: Ekonomski fakultet Beograd.

Serbia Insurance Law. (2014). Official Gazette of Serbia, no. 139/2014.

Mrkšić, D., Petrović, Z., & Ivančević, K. (2007). *Neživotna osiguranja*. Beograd: Intermex.

Mrkšić, D., Petrović, Z., & Ivančević, K. (2014). *Pravo osiguranja*. Beograd, Srbija: Pravni fakultet Univerziteta Union.

National Bank of Serbia. (2016). *Insurance Supervision Department Report for 2015*. Retrieved March 10, 2016, from National Bank of Serbia web site: http://nbs.rs/internet/english/60/index.html

Njegomir, V. (2011). *Osiguranje*. Novi Sad, Srbija: Ortomedics book.

Petrović, G., & Grujović, M. (2016). *Ekonomske štete od elementarnih nepogoda u Srbiji i šumadijskom okrugu*. Retrieved March 10, 2016, from scindeks-clanci.ceon.rs/.../1452-44571502099P.pdf

Saković, L. (2015). Osiguranje od elementarnih nepogoda: Ruku na oči, prognoze pod tepih. *Biznis & finansije, 118/119*. Retrieved March 10, 2016, from: http://bif.rs/2015/09/osiguranje-od-elementarnih-nepogoda-ruku-na-oci-prognoze-pod-tepih/

Serbia Law on Civil Procedure. (2013). Official Gazette of Serbia, no. 49/2013; decisions of Constitutional Court, § 74/2013; § 55/2014.

Serbia Law on Contracts and Torts. (2003). Official Gazette of the SFRY nos. 29/78, 39/85, 45/89, decision of Constitutional Court 57/89, Official Gazette of the FRY 31/93, Official Gazette of Serbia and Montenegro no. 1/2003.

Serbia Law on Safety and Health at Work. (2015). Official Gazette of Serbia, no. 101/2005, 91/2015.

Simons, R. (1998). How risky is your company? *Harvard Business Review*, *77*(3), 85–94. PMID:10387580

Sneyd, M. R. (2003). *Insurance. Professional Reading Skills*. Pearson Education Limited, Longman.

Šulejić, P. (2005). *Pravo osiguranja*. Beograd: Dosije.

Swiss Re. Sigma. (2014). *Natural catastrophes and man-made disasters in 2013: large losses from floods and hail; Haiyan hits the Philippines*. Retrieved March 10, 2016, from http://www.swissre.com/sigma/?year=2014

Udovičić, A., & Kadlec, Ž. (2013). Analiza rizika upravljanja preduzećem. *Praktični menadžment, 4*(1), 50-60.

Vaughan, E., & Vaughan, T. M. (2000).*Osnove osiguranja, upravljanje rizicima*. Zagreb: Mate d.o.o.

Vidaković, M. (2010). Smanjenje obima štete kroz obuku osiguranika. In Osiguranje u CEFTA regionu, (pp. 273-287). Zlatibor: Privredna komora Srbije.

Weatherill, S. (2001). Jusifying the Limits to Party Autonomy in the Internal Market - EC Legislation in the Field of Consumer Protection. In S. Grundmann, W. Kerber, & S. Weatherill (Eds.), *Party autonomy and the role of information in the internal market* (pp. 173–196). Berlin: Walter de Gruyter. doi:10.1515/9783110873030.173

Wiener. (2016). Retrieved March 15, 2016, from Wiener Stadtishe Insurance Stadtishe web site: http://en.wiener.co.rs/

ADDITIONAL READING

Banks, E. (2004). *Alternative risk transfer: Integrated Risk Management through Insurance, Reinsurance and the Capital Markets*. England: John Wiley & Sons Ltd. doi:10.1002/9781118673270

Crocker, K. J. (2003). Risk and Risk Management. In D. J. Meyer (Ed.), *The Economics of Risk, Michigan USA*. Kalamazoo, MI: W.E. Upjohn Institute for Employment Research. doi:10.17848/9781417505937.Ch1

Hoyt, R. E., & Liebenberg, A. F. (2011). The value of Enterprise Risk Management. *The Journal of Risk and Insurance*, *78*(4), 795–822. doi:10.1111/j.1539-6975.2011.01413.x

Jovanović, M. (2012). Osiguranje od svih rizika na tržištu osiguranja u Srbiji, *Revija za pravo osiguranja*, XI, 4, 69-74

Klobučar, D. (2007). *Risk menagment i osiguranje*. Zagreb: Tectus.

Mundy, C. (2001). Enterprise risk: Who knows my business better than me? *Balance Sheet*, *9*(2), 12–15. doi:10.1108/09657960110695295

Pavić, D. (2009). *Ugovorno pravo osiguranja – komentar zakonskih odredbi*. Zagreb: Tectus.

Petranović, V. (1984). *Osiguranje i reosiguranje*. Zagreb, PA: Informator.

Stulz, R. M. (2009). Six ways companies mismanage risk. *Harvard Business Review*, *87*(3), 86–94. PMID:19630258

Terungwa, A. (2012). Risk Management and Insurance of Small and Medium Scale Enterprises (SMEs) in Nigeria. *International Journal of Finance and Accounting*, *1*(1), 8–17. doi:10.5923/j.ijfa.20120101.02

KEY TERMS AND DEFINITIONS

Insurance Beneficiary: One who will receive the benefit in the event of an insured risk.

Insurance Premium: Amount a policyholder or contractor is obliged to pay as compensation for the provision of insurance protection and represents a measure of risk.

Insurance Policy: The basic written document that accompanies an act of insurance and evidence of the conclusion of insurance containing all essential ingredients of a contract.

Insured Subject Matter: The subject clearly stated in the insurance policy specifying what is ensured. The existence of the object of insurance with certain characteristics is an essential precondition for the conclusion and duration of insurance.

Insureable Risk: The risk covered by insurance if an insured event takes place provided possible loss is plainly explained, it is accidental, it is part of a similar group of risks and the insurer must be able to compute the probability of a loss.

Over-Insurance: Insurance sum is greater than the actual value of property insured.

Property Insurance: Indemnity for damage caused by realization of the insured risk to the extent suffered.

Risks Prevention: Measures aimed at eliminating the possibility of the event of loss such as installation of alarm systems which reduces the risk of burglary and which simultaneously reduces insurance costs.

Sum Insured: The amount paid to the insured if the insured event occurs representing the upper limit of the basic obligations of insurers.

The First Risk, *First Loss Insurance*: Property insurance policy that grants only partial insurance where the policyholder agrees to accept an amount less than the full value of damaged, destroyed or stolen items/property. In return, the insurer agrees to not penalize the policyholder for under-insuring their goods or property.

Underinsurance: The sum insured is less than the value of insured property (in case of damages paid to the extent of the sum insured).

ENDNOTES

1. Editors' note: This is just an example of the combination of possible insurance products described in earlier sections of this chapter.

Chapter 10
Financial Innovation in Medium-Sized Enterprises Optimizes Their Gravitation Towards Capital Markets:
Financial Future in Perspective

Milan B. Vemić
Higher School of Academic Studies "DOSITEJ", Serbia

ABSTRACT

A major problem for medium enterprises is ensuring that their financial management exercises prudence in the attraction of alternative sources of financing in transition into middle tier, larger enterprise status and towards stock exchange. Difficulties in attracting capital, the high cost of bank credit, absence of robust financial and business information reporting capacities may represent obstacles. One partial solution is capacity development for access to stock exchanges and having these institutions with a strong allocation capacity to efficiently finance these businesses. Though underdeveloped in transition economies, they are potentially important institutions in successful financial performance of businesses. In a review of conditions for handling risk and uncertainty, developing capacity for access to stock exchanges through working capital combinations, this treatise demonstrates that medium enterprises could benefit from this approach. Regulatory reforms to facilitate the use of stock exchange financing and well-designed financial management models are recommended.

INTRODUCTION

Modern Small and Medium-sized Enterprise (SME) finance theory is unique in two respects. First, it incorporates sources of finance that differ completely from stage to stage of development in the firm life cycle which is both a constraint and an opportunity. Second, the role of exchanges and innovative sources of financing is much less pronounced than that required in corresponding problems of medium

DOI: 10.4018/978-1-5225-1949-2.ch010

size enterprise financing. As a result, there has been a predisposition to suboptimal level of financing and financial market development. In this chapter, special emphasis is placed on developing the guiding financial principles that are useful not only to apply our ideas to a new SME problem, but also when we wish to discern the general properties of financial markets without performing complicated mathematical calculations. The development of special techniques of mathematical combinations in finance to obtain quantitative outcomes in complex financial problems is, however, also necessary. It is, therefore, assumed that the reader is reasonably familiar with basic concepts of advanced mathematics.

Following the elaboration of sources of finance as a factor of growth of medium enterprises, significant amount of attention is given to the steps by which exchanges could get involved in this process. Here, the determinants and risks are discussed that might lead not only to replacement but rather correspondence of different alternative sources of finance. Therefore, the practical and theoretical SME finance developments are presented in such a way as to point out this interconnection and to demonstrate that each new source of finance is either based on growth or innovation or else follows logically from the previous stage of development in the firm life cycle. In this manner, SME finance theory can be made to seem less like an odd and perhaps inconsistent recipe. An integral part of the author's concept for conceiving the theory on a basis that is not too abstract for practitioners includes an extension and completion of certain ideas on existing sources of finance and financial innovation. Moreover, following the discussion of potential financial innovation indices, broader financial innovation in the development of medium-sized enterprises is portrayed as an opportunity for their growth. Also discussed are some of the mathematical implications of SME finance theory showing that these might lead to a fascinating improvement in the proposed specific approach to medium enterprises as compared with that suggested by the general SME approach. The reader will notice that a proposed codification and typology of these modifications derives from this attempt to solve the problems of financing. The most significant benefit of this diversification is that it makes the reader think more specifically about the subject of medium size enterprise financing, thus furthering his comprehension of these businesses. Indeed, it is also hoped that the reader will thereby find help in the supplementary texts referred to throughout the different parts of this chapter.

BACKGROUND ON ACCESS TO FINANCE AS A GROWTH OPPORTUNITY FOR MEDIUM-SIZED ENTERPRISES

The author recognizes that various countries define SMEs differently and that cross comparison can be difficult. Medium-sized enterprises are defined as enterprises that employ fewer than 250 persons and either have an annual turnover that does not exceed EUR 50 million, or an annual balance sheet not exceeding EUR 43 million, which follows the European definition (European Commission, 2015a, p.11). The precise definition however, does not impact the overall findings of this treatise.

Compared to small firm financing medium-sized enterprises should be less risky due to lower levels of perceived risk and uncertainty. This is followed by their greater investment readiness. Although the standard measure of risk is the value of variance in relation to expected return (Eales, 1995, p.15) the basis for above is author's resource based view of a stronger financial base and market share of medium-sized enterprises.

Likewise, medium firms hesitate less in accepting qualified non-financial and more advanced management consultancy services that reduce the impact of risk. These independent and objective high-quality advisory services help them identify and analyze management problems including business planning, marketing, distribution, competition, financial issues, human resources, team building, process improvement, identification of any internal and external organization strengths, opportunities, weaknesses and threats (Barcus & Wilkinson, Eds. 1995). These services are therefore unique multiplier agents (Meredith, 1987, pp.125-139). In fact, evidence from Serbia reveals the potential of medium enterprises because 46% of them use consulting, compared to 32% of small businesses, 15% of micro-enterprises and 13% of sole proprietors which are another and smaller category than SMEs, (Ministry of Economy and Regional Development, 2009, p.47).

Commercial banks also use standard techniques in accordance with the stages of development in the SME life cycle requiring different types of products and services where again medium enterprises are better positioned than small. Preferably guarantee funds should share the risk of financial results (Firoozmand, Haxel, Jung, & Suominen, 2015, p.18) with these clients by focusing on monitoring and evaluation in each of the stages of their growth and development. And while entrepreneurship inherently involves risks and uncertainties, there is a substantial difference between the two categories. Risk is something that can be measured while uncertainty cannot. In author's view the risk of financial management also involves opportunities or chances of controlling exposure to a monetary loss or reward. This was appropriately defined by Hisrich who stated "Entrepreneurship is the process of creating something new assuming the risks and rewards" (Hisrich & Peters, 2002, p.10), developing into optimal financial management of SMEs. It remains discussible whether and to what extent medium enterprises are better *factor positioned* in relation to small firms, or large firm intrapreneurship, as that may also be industry specific.

When solving growth problems and bottlenecks that inevitably arise, medium in contrast to small enterprises are more focused on research and innovation instead of just survival. A research survey of conditions, problems and needs of Serbian SMEs for 2008 shows that proportionally the highest number of innovative companies are medium-sized enterprises 39.11%, and lowest is among sole proprietors 12.77%, (Ministry of Economy and Regional Development, 2009, p.97). Similarly, among Polish enterprises which participated in innovation activities co-operation in the years 2011-2013, 31.2% of industrial medium, 23.3% of service medium, 18.2% of small industrial and 20.2% of small service firms participated, (Dzida et al., 2014, p.90). Per analogiam medium-sized enterprises are able to express a higher degree of creativity and innovation which gravitates management processes towards optimization, middle market mezzanine, formal stock exchanges and large enterprises.

This suggests the structure of firm's financial growth cycle as summarized in Table 1 correlating specific stages of business development with optimal sources of finance followed by most prominent funding institutions identified.

Consequently, the 2012 survey of SME financial reporting practices in Austria confirms that 85% of firms in the expansion stage report investment requirements (World Bank Centre for Financial Reporting Reform and the Austrian Institute for SME Research, 2013, p.17). This suggests correlation between firm level financial innovation, enterprise growth and development of exchanges*. The author goes on to propose a more limited definition of financial innovation which logically derives from the Table 1.

Definition 1: Financial innovation in a medium sized enterprise is the renewal, improvement or enlargement of a broad range of sources of finance with the development and establishment of new methods of financing as the firm tangentially leaves the start-up stage and enters the third stage of development requiring risk capital for product introduction and expansion.

Table 1. Sources of finance of SMEs by stage of development in firm life cycle

Life Cycle	Types of Sources of Finance	Select Start Up/Expansion Sources	Online lenders
"Seed" stage (0-2 years)	• Informal equity • Loans if available • Founders, friends, family • Crowdfunding • Online lending	• US: SBA; small company funds • EU: Horizon 2020 instrument • Ohio Pre-Seed/Seed Plus Fund Capitalization Program (PFCP) • KFW ERP Capital for start-ups	OnDeck Biz2Credit Lending Club Fundera Boefly Accion Dealstruck Kabbage Pay Pall Lighter Capital A/R Financing Funding Circle Fundation Daric Raiseworks IOUCentral Prosper Swift Capital
Start-up (early) stage (3-4 years)	• Founders, Start -up funds • Suppliers, buyers • Microfinance institutions • Business angels • Savings Credit Cards • Factoring • Commercial/ Industrial Loans, guarantee schemes • Leasing for equipment • NGOs, donors • Informal equity • Venture capital	• SBIC: debt, equity, mezzanine debt with equity features • US: Ex-Im Bank's Supply-Chain Finance Guarantee Fund • Innovation Investment Fund (IIF) Australia • European Investment Fund's Credit Enhancement Activity • Long-term French Investment Program of CDC Enterprises • KfW Entrepreneur Capital • KfW's Promise programme • Spanish FTPYME Securitization Scheme • Rabobank Innovation Capital • Italy District Bonds • Standardized Middle Tier SME Mezzanine of NRW.BANK • Middle Tier SME French Innovation Dev. Contract • Middle Tier SME FIDEME Mezzanine Fund in France • Low tier SME German Silent Partnerships by MBG • Low tier SME Austria AWS Equity Guarantee • Low tier SME Swedish Innovation Loan from ALMI	
Expansion stage attracting low and middle tier SME *and involving high growth dynamic medium sized enterprises* (5-24 years)	• Banks/ Loan funds • Formal equity • Leasing, Factoring • Venture funds • Mezzanine finance • Development & Guarantee Facilities • Funds of private sector non-profit entities • Mid and long term credits • Loan subsidies in less developed, rural areas • Open-end funds • Research Commercialization grants		
Business succession and exit (25 years or more)	• Late mezzanine finance • Debt investments • Private equity & Securitization • Suppliers, Stock market • Guarantees for mezzanine finance, LBO, MBO,MBI	• The Business Development Bank of Canada (BDC) • RBC Royal Bank • UBS Group AG • Credit Suisse Group • Finance Wales Mezzanine Investments	

Source: Revised list of Firoozmand et al. (2015) and European Commission (2007).

Implied is that a specific financial innovation must be new to a medium enterprise, but it does not need to be new to the market. It does not matter if the innovation was originally developed by surveyed or other enterprises. It poses an analytical problem: whether and with what capacity can medium-sized enterprises take advantage of innovative financial instruments, diversify access to finance and gravitate towards stock exchanges in order to grow. This is the problem that this treatise sets out to model and answer coupled with a broader definition of financial innovation later in a specific context.

SOME DETERMINANTS AND MECHANISMS OF STOCK EXCHANGES AND CAPITAL MARKETS OF SIGNIFICANCE TO MEDIUM-SIZED ENTERPRISES

And while available research evidence shows that stock exchanges, in both transition and developed economies, do not participate with complex programs of financing medium-sized enterprises as such, inhibiting their growth, there is significant advocacy for support to high growth firms (Business Development Bank of Canada, 2015), to ambitious entrepreneurship (Stam, Hartog, Stel, & Thurik, 2009) and German "upper" sized firms (Venohr, Fear, & Witt, 2015). SME policy also fails to see the potential of these sources of finance. Consequently no specific measures exist that would enable the introduction of stock exchange financing for *ordinary* medium-sized enterprises as such. Practical foundations for the activation of complex programs to facilitate this form of funding and initial measures will be summarized as follows:

- SME financial performance reports should be expanded with specific findings on mid-sized enterprises and circulated to these businesses;
- Access of medium-sized enterprises to stock exchanges should be facilitated by reducing the threshold of capital required; i.e., according to Article 17 of the Rules on Listing of Belgrade Stock Exchange minimum amount of capital of the issuer for listing A (Prime Market) is EUR 20 million, or four times more than the upper limit for medium-sized enterprises under the Accounting Law and automatically excludes them from this combination;
- Elect an SME council representative in the management body of stock exchanges with a status of observer or advisor to begin with;
- Involve stock exchanges and SME policy representatives in the cross comparison study and functional benchmarking of stock exchange financing models and try to harmonize solutions;
- Banks and accounting associations should consider intensifying initiatives to better justify to SMEs the advantages and strengths of securitization which would support the growth of this market;
- Encourage through economic policy measures to include more medium-sized stock companies in quotation of stock exchanges;
- Increase in SME transparency of business reporting so management would not conceal operating results can facilitate the development of innovative finance products such as mezzanine finance;
- Introduce promotional programs which would train high growth medium enterprises in innovative financial instruments such as mezzanine finance;
- Create special guarantee funds for fast growing medium-sized enterprises in order to attract mezzanine finance like for example through The Private Equity Growth Capital Council (PEGCC) or Network of European Financial Institutions for SMEs (NEFI);
- Through training and awareness reduce the fear of company management from losing control through trading shares with voting rights.

Now assume that these initial measures were in place. The working capital needs of medium enterprises still exceed their own and capacity of stock exchanges to service precautionary measures (to cover unanticipated expenditures) and speculative intentions (to finance uncommonly attractive opportunities) which can be financed with the near cash of marketable securities such as bonds and commercial papers. Such capacity is rare in transition economies, and as observed for example by research evidence Eastern

European firms rarely use long-term financing (Klapper, Sarria-Allende, & Sulla, 2002). Therefore, financing efficiency of a stock exchange is limited and potentially relevant. Assume that markets adapt quickly to new publicly available information and stock prices reflect knowledge and expectations of investors (Downes, & Goodman, 2003). However, proceeding deductively from the results of author's research (Vemic, 2015) that the availability of business information (and capital) as an increasingly significant resource is less accessible for medium-sized firms than the large (and more so than in small firms), that there are no quality conditions for on-line networking and a pool of knowledge readily available, that transaction costs are greater, then, per analogiam, the following two logical deductive conclusions can be stated:

1. In terms of a business information set available to large enterprises, banks and stockbrokers the stock exchange market is generally efficient. As such it operates in accordance with economic principles, potentially makes profit and fulfills its basic function of allocating available capital at market conditions without being governed solely by enterprise size.
2. In terms of a business information set available to medium-sized enterprises in transition economies both the stock exchange and the capital market are not efficient. Although it is also potentially profitable the capital market does not allocate enough free sources of finance for these companies especially if business information is lacking.

In terms of current efficiency of the stock exchange and capital market in transition economies and eligibility of SMEs to access stock exchange finance it is necessary to strengthen the position of medium-sized enterprises towards banks and large enterprises as an intermediate step of their financial strengthening. Particularly because their funding needs are expected to be growing. That would certainly increase the overall efficiency of stock exchanges and their capitalization particularly if the micro sized companies point of view is taken into consideration and platforms are developed to facilitate their financing as well.

Further Elaboration of Risk and Uncertainty in Stock Exchanges for Medium-Sized Enterprises

In the context of medium sized enterprises significant are different degrees of uncertainty. These are correlated with types of required financial innovation in Table 2.

Table 2. Degrees of uncertainty and different types of financial innovation

Degree of Uncertainty	Different Types of Financial Innovation
Actual uncertainty	Fundamental Financial Research & Fundamental Investments
Very high degree of uncertainty	Radical Internal Financial Innovation & External Process Change
High degree of uncertainty	Significant Financial Innovation & Radical Own Process Change
Medium degree of uncertainty	New Generation of Financial Products and Services
Small degree of uncertainty	Financial Product Imitation & Process Internal Modification
Negligible degree of uncertainty	New Financial Management Models & Financial Differentiation

Source: Own presentation of author.

Although different variations are possible, a very close relationship between degree of uncertainty and financial innovation will be established by the reader in the next section of the treatise on innovative financial diffusion. However, it seems instructive to first apply available definitions of general risk classification in the banking business in order to elaborate the key risks in capital markets for medium-sized enterprises:

1. Credit risk implies a business cannot repay a loan payment at maturity of annuity at full book value. Criteria by which a firm is appraised in deciding whether or not credit will be approved are known as the 5 Cs of credit: Character (the readiness of the borrower to repay loan); Capacity (the ability of the borrower to earn enough to repay the obligation); Capital (net worth of the borrower sufficient to support operations); Collateral (assets pledged to secure the loan which can be liquidated if default takes place); Conditions (current and future expected business conditions of the firm and the industry), (Brigham & Houston, 2001, p.728). This risk inherently depends on yield power and property structure of the company. Standard deviation is the most common measure of volatility or risk which measures the absolute variation or dispersion around the expected or average interest revenue. Closely linked to the credit risk is the risk of investment in the context of the capital market which is herewith defined as a medium enterprise achieving sub-optimal revenues preferably channeled to investment in shares. In contrast to small businesses this feature is a smaller risk while the ability to invest in shares is higher. This is normally reflected in risk premiums for financing which are lower for medium-sized enterprises than for small enterprises.
2. Interest rate risk (market or price risk) refers to the adverse financial consequences due to the revaluation of interest rate for the loans or dividend on equity. However, unlike small businesses the overall financial potential and in the narrow sense working capital of medium-sized enterprises reveals their potentially greater capacity to cope with the potentially adverse financial consequences in relation to small firms. The opposite risk coefficient can be observed when comparing medium-size firms with large firms which have access to multiple sources of finance at more favorable conditions. That is, in the gap analysis the gap is negative if changes in lending rates (interest payments) affect medium firm liabilities proportionally more than that of a large firm from the same sector of the economy. Similar gap analysis can be performed between comparable medium-sized firms or in relation to small firms.
3. Exchange rate risk refers to internationalized medium enterprises that are able to take loans denominated in hard currency, which is increasingly the case in transition economies. In addition loans in currency might be taken also by domestic companies that are not trading with foreign partners, simply to reduce the costs of loans. Operationally exchange rate risk exposure is a measurement of elasticity, as follows

$$\text{Elasticity}=\frac{\%\text{ change in cash-flows}}{\%\text{ change in exchange rate}}=\frac{dV}{dS} \tag{1}$$

where V is the value of firm expressed in terms of a stream of present and future cash flows and S is the exchange rate expressed as home currency/foreign currency. For example, in 2008 47% of all Serbian medium-sized enterprises were involved in export and 61% in imports (Ministry of Economy and Regional Development, 2009), but with relative stability of the currency in recent years this elasticity did not cause drastic adverse consequences. Nevertheless the potential elasticity of risk exposure was largest for pure exporting and importing firms. In contrast to a smaller measure of elasticity for *large* multinational firms that match their foreign currency revenues and costs (Bodnar & Marston, 2002), revealing their competitive financial advantage over SMEs.

For Danske Bank & Ernst & Young (2009)*liquidity risk* is of key importance (p.2), and following the still lingering global economic crisis which started in 2007 (Greenspan, 2008) this factor became one of the most important problems of medium-sized enterprises. It is reflected in suboptimal financial management resulting from mismatch between DSO and DPO. These businesses are therefore insufficiently able to settle debt at maturity which directly jeopardizes their optimization and investment capacity in the capital market and puts them at risk in competition with large firms.

4. Internal or operational risk causes adverse financial consequences that occur as a result of management (goal, process, result) errors and inefficiencies. Based on the theoretical model of optimal working capital management[1], elaborated in the author's first chapter, this risk is attributed to the absence of internal control systems and suboptimal organization of the finance department.
5. A management hazard is the inability distinguishing risk from negligence in medium enterprises which is again related to the internal control system and its capacities.

The chapter on working capital optimization has already established that various types of working capital (cash, marketable securities, accounts receivable and inventories) can be displayed as a parallelogram of financial forces in medium enterprises, and laid down were the foundations of optimization with a notation. The same parallelogram of space can be used in the form of this equation synthesizing the above capital markets risks

$$R = \psi(F1 + F2 + F3 + F4 + F5 + ...F_{n-1} + F_n) \quad (2)$$

Ultimately financial decision making is being implemented in the following phases:

- Determining possible directions of activity;
- Gathering relevant business information;
- Selecting best activities to attain goals, defining alternatives, finding answers to the "what if";
- Implementing activities;
- Undertaking corrective or alternative activities;
- Activities involving evaluation.

A company should execute above transactions in a timely fashion ensuring acquisition and placement of working capital, dividend policy at a profit, increasing company value (shareholder wealth), generating cash flow and maintaining solvency. To survive in a globalized and competitive economy a company must also take into account the interests of other stakeholders such as customers, employees, suppliers, financial public and the state. Otherwise, potential conflict of interest between shareholders and management may create an agency problem. In most cases there is no conflict of interest between shareholders and management, however, in the following four cases, conflicts are very likely and might affect transition of medium enterprise into a large one through issuance of stock (Hart, 1995, pp.147-151), or through a management buy-out (MBO) which is an option very interesting for high growth SMEs:

1. When management itself determines excessive salaries that threaten working capital and dividends;
2. All undistributed corporation profit is used to finance development;
3. Company management avoids engaging in risky activities that can bring about higher profits and losses, that is, when the company has achieved significant market power or monopoly position and can seek other goals besides profit maximization;
4. In an MBO the same players are active on both sides of the deal. In one role managers are insiders, with very deep knowledge of the company's current state of business and its future perspectives. In the other role they lead the team considering an acquisition at an optimum price.

Besides risk, uncertain operations and business results also cause insufficient presence of medium-sized enterprises in capital markets. In contrast to the view that a company, its capital and assets only serve to generate income the capital market also allows valuation and estimation of its cash flow performance. Medium enterprises should produce enough valid historical data to assess their cash flows more accurately in relation to larger companies which have long operated in the capital market. Such financial and market status should identify the transaction costs, the level risk of their business and degree of uncertainty of stock issuance for their development.

In the absence of a developed capital market, as is still the case in transition economies, an alternative valuation option is "goodwill" as a part of the company value that exceeds the net asset value, or the value estimated through a working capital model. Goodwill occurs as a result of income beyond the normal level based on the trust of customers and the empirically proven reputation. Its appearance in medium-sized companies should be a natural course of development in relation to small businesses because it reflects a capacity for investment. However, with sub-optimally capitalized medium firms that have insufficient market history and other types of assets such as fixed assets without an active secondary market, their goodwill is very difficult to assess without the secondary financial market which allows constant buying and selling of already issued securities acquired by investors. Hence, this market is also called the transactional ("second-hand") market, through which these firms should gradually grow and transition into large enterprises. If they operate exclusively in the same region without participating in the wider market this is difficult. But for internationalized medium-sized enterprises, an assessment of goodwill is a great opportunity to optimize all working capital components. This is executed on the basis of international accounting standard no. 38 which requires that an intangible asset may be recorded in the balance sheet if the cost of the asset can be measured reliably and if any future economic flows attributable to the asset will be turned into company inflows (Lev, 2001, p.151). These are the three alternative methods of risk and uncertainty assessment that medium-sized enterprises can facilitate in the above processes of working capital optimization (particularly interest-bearing marketable securities) for gravitation towards stock exchanges:

1. An estimate based on known risk and uncertainty for comparable or similar mid size business;
2. An estimate based on known risk and uncertainty for similar business in a comparable large enterprise;
3. An estimate based on the degree of financial innovation in medium enterprises identified as free, uninhibited, imaginative and non-standard thinking about the different alternatives of marketable securities in working capital management and internal control processes.

Obviously, financial innovation involves simple or complex changes in financing businesses, represents a new way of solving financing, leads to improving the practice of planning and operations or adoption of new financial instruments to improve business efficiency.

Various phases in the adoption of financial innovations can be classified into the following 5 groups:

1. Awareness of existence of financial instruments and the collection of relevant data;
2. Testing, evaluation of financial instruments, and relevant training;
3. Definitive adoption of the financial instrument;
4. Practical implementation of the financial instrument including risk and uncertainty assessment.
5. Evaluation of successfulness of the financial instrument measured appropriately.

The proposed risk and uncertainty assessment for medium-sized enterprises would ideally be presented for management in the form of a practical risk control matrix containing total integral knowledge, respecting the complexity of the management process and these internal and external factors:

- The state of the local and national market;
- Legal and regulatory changes that can reduce the level of business risk;
- Level of risk in the region in which to invest;
- Competitive readiness of a medium enterprise and its financial structure;
- Additional investments for talented personnel, research and development, new entrepreneurship;
- The degree of convergence towards a large company size;
- The degree of convergence of small businesses towards an observed medium-sized enterprise;
- Desired future state of a medium enterprise;
- Critical success and failure factors of a medium enterprise;
- Most competitive structure of critical business processes and their desired future state;
- Necessary resources and inputs for each critical process;
- Supplier capacity for necessary inputs and alternatives.

With uncertainty being the main characteristic of modern business, risk management requires a set of activities common to each of its manifestations. These are: risk assessment, grouping mentioned risks, risk control, risk transfer, financial portfolio modeling, i.e. optimization, and especially the process of establishing stock exchange based financing and insurance from risks. At the same time this approach establishes a relationship of correlation of risk to potential gain and then to equity, which eventually ensures liquidity for settlement of current liabilities and solvency for ability of enterprises to bear more global business risk.

Integration of Medium-Sized Enterprises in Gravitation towards Stock Exchanges

An underused form of financial innovation in the transitional context is certainly integration of medium-sized enterprises, possibly through MBOs, to enhance their overall competitiveness and growth. This approach also clearly provides a common solution to allocating resources to the securities markets and commodity exchanges. Such a strategy would reduce their transaction costs and enable earlier security issuance in the firm life cycle. Mergers are not only a risk but an opportunity to leverage own goodwill.

The International Financial Reporting standard (IFRS) – 3, on *business combinations* regulates this as *pooling* of separate companies or businesses into one reporting entity offering a chance to reduce risk and replace inefficient management. In fact all businesses face this problem. Creativity in the modeling of business risk would also allow for risk dispersion across multiple owners reducing the probability of failure. In solving business or financial problems and creating opportunities for the development it should be noted that there is no entrepreneurship without innovation and creativity. That is, risk is inherent to business but companies and financial institutions can manage it. Introduction of an optimal working capital management model is in the function of reducing risk and attracting capital for the integration of medium-sized enterprises also involving stock or equity financing. Not underestimating that the real criterion for making the finance available are the creative impulses for innovation under complexity, the track record of the management team and the financier's gut feeling (Stacey, 1996, p.271).

POTENTIAL CARRIERS OF DIFFUSION OF INNOVATIVE FINANCIAL INSTRUMENTS AMONG MEDIUM-SIZED ENTERPRISES

Of course, more evidence is needed before it can be inferred that financial innovation leads medium enterprises to stock exchanges. It also needs to be explored what gravitates firms of this size to additional and new sources of finance, whether it's product, process, marketing or organizational innovation (Mortensen & Bloch, 2005, p.17) or according to O'Reilly and Tushman (2004) who mapped innovation as incremental, architectural or discontinuous (p.77). Making optimal combinations of these types of innovation in finance could indeed be quite challenging in building capacity for equity financing and critical for fixed assets which should be financed on a long-term basis.

Evidence is found that financial innovation reveals alternatives and reduces risk (IBM, 2006, pp.17-18). Hence, in this section, integrating Rogers' (1962) earlier landmark classification (p.150), and Kanter's (2006) discovered innovation traps and remedies (pp.80-83), provided is a brief overview of the main actors of financial innovation diffusion. Table 3 illustrates symbolically and how the author sees the traps and remedies in entrepreneurship of medium enterprises in their gravitation to stock exchanges.

Table 3. poses an analytical challenge: what other entities assume the role of a financial innovation actor and are the gains worth the costs? Banks innovating traditional bank services, government regulation and progress in new technology were intentionally left out of Table 3. They have historically represented important drivers of financial innovation and will probably remain leaders in the aftermath of the global financial crisis. However, the non-bank financial institutions, some of which are medium-sized, as well as other medium- sized enterprises demonstrating financial strength ratios could be systemically important in the adoption and diffusion of financial innovation. Unfortunately researchers can still not observe very significant innovators, the early adopters and the early majority among medium enterprises, particularly in transition economies. Evidence (Vemic, 2015) suggests that finance is performed in the old conventional way. If innovation means financing new financial infrastructure, new financial processes, instruments and practices, followed by better financial results as a result of gravitation of medium firms towards exchanges, then the information available to the author suggests that there are few examples of this kind. The practice convincingly shows that development funds, leasing companies, guarantee funds and new credit products of banks do not significantly improve the financial infrastructure. Previous exploratory study of the author explicitly shows (Vemic, 2015), that there are not enough of these institutions or significant diffusion of financial innovation. The reality of the entrepreneurial environment compels us

Table 3. Potential pattern recognition for adoption and diffusion of financial innovation in medium enterprises mapped with Rogers', Kanter's and own categories

<table>
<tr><th>MAIN FINANCIAL INNOVATION ACTORS</th><th>DEFINITION AND ROLE</th><th>INNOVATION TRAPS AND REMEDIES</th></tr>
<tr><td>Innovators</td><td>High growth intensively market oriented dynamic medium enterprises (Gazelles).</td><td rowspan="7">Strategy Lessons: Widen the search and broaden the scope for hurdles too high and scopes too narrow.
Process Lessons: Add flexibility to planning and control systems when controls are too tight.
Structure Lessons: Facilitate close connections between innovators and mainstream businesses when connections are too lose and separations too sharp.
Skills Lessons: Select for leadership and interpersonal skills, surround innovators with a supportive culture of collaboration if leadership is weak and communication too poor.</td></tr>
<tr><td>Early adopters</td><td>Pioneers among medium enterprises in an innovative approach to financing entrepreneurship.</td></tr>
<tr><td>Early majority</td><td>First formed majority of medium enterprises with innovative finance thinking.</td></tr>
<tr><td>Later majority</td><td>Mature majority of medium enterprises with innovative finance thinking.</td></tr>
<tr><td>Laggards</td><td>Decelerators of the development of entrepreneurship who often think about finance the old way.</td></tr>
<tr><td>Opponents</td><td>Businesses and entities which object to innovative financing of entrepreneurship and prepare ground for the underminers.</td></tr>
<tr><td>Underminers</td><td>Businesses and entities which undermine & derail innovative financing of entrepreneurship by defending the old ways of business financing.</td></tr>
</table>

Source: Revised list of Rogers (1962), Kanter (2006), with author's input and presentation.

to characterize potential adopters of financial innovation as laggards who decelerate the development of entrepreneurship through gravitation towards exchanges. If there is faster diffusion of innovation and the development of other enabling factors of entrepreneurship development this might lead to having a more pronounced early majority and might reduce the costs for financial innovation actors. This treatise should therefore be viewed as a first attempt at developing a general theory of financial innovation for medium enterprises.

Development of Indicators for Measurement and Diffusion of Financial Innovation in Medium-Sized Enterprises

The first step in achieving better financial results in development of medium-sized enterprises is to measure the current level of knowledge, qualifications and skills. This would provide baseline information and such cohorts could be used for further measurements of progress. Author suggests these four indices of financial innovations that do not exhaust the list of criteria:

1. Index of recent innovations under which the author means the ratio of new financial instruments in relation to their number in the previous year;
2. Index of old innovations, where "credit" is an "old" well known innovation and where the index / ratio of diffusion can easily be displayed as a ratio of credit in relation to the volume of taken credit in the previous year;
3. Index of radical game-changing non-incremental financial innovation in relation to achieved step changes in products, services or models;
4. Index of current radical financial innovation in relation to it in a prior period of time.

These four indices collectively make up a composite diffusion index within which different financial products will have different relative speeds of diffusion due to the differences in concentrations.

The second most important step is how further to measure the diffusion of financial innovation in medium enterprises. With its technological development index the United Nations pioneered an attempt to develop a set of complex indicators of the level and dynamics of technological innovation in modern development.

Applying relevant approaches to measuring human development index and the index of specific technological development of the United Nations Development Program (Fukuda-Parr & Shiva-Kumar, 2003, p.229), further development of an index of diffusion of diffusion of financial innovation is attempted.

Measurement of innovation performance modifying the Fukuda – Parr i Shiva - Kumar approach

$$\text{Indicator of index}=\frac{\text{Real value - observed minimal value of new financial instruments}}{\text{Observed maximal value - observed minimal value}} \tag{3}$$

Innovative financing of medium enterprises can be further measured with several other indicators depending on which financial instrument is used. The European Commission, for example, measures Venture capital investment as percentage of GDP (European Commission, 2015b, p.9). Nevertheless, in the author's opinion, calculating venture capital as a percentage of total sources of SME finance is also significant. Recent financial innovation can also include leasing, factoring, investment in new technology or share of early stage financing.

Measurement of financial innovation difusion through leasing

$$\text{Indicator of index}=\frac{\text{Value of equipment purchased on leasing}}{\text{Total value of equipment pruchased}} \tag{4}$$

Measurement of financial innovation difusion through factoring

$$\text{Indicator of index}=\frac{\text{Value of invoices sold through factoring to a third party}}{\text{Total value of invoiced sales in a particular period}} \tag{5}$$

Proposed indicators give a measure of firms' propensity to introduce new or significantly improved financial products or services. It includes those firms that introduced products that were already available on the market. Similar indicators can be developed for completely new and unavailable products.

Broader Financial Innovation in the Development of Medium-Sized Enterprises

Further development of above indicator of indices shall refer to improvements in financial structure or practices and gives a measure of the significance of non-technological innovation, or "intangible" innovation. A formal definition of broader financial innovation is herewith provided by the author.

Definition 2: Broader financial innovation in a medium enterprise includes new or significantly improved forms of work organization changing internal efficiency, financial management and the way of accessing financial markets and resulting in robust financial and business information reporting capacities, sound track record, stable cash-flows and an experienced management team.

These are some of the main creative changes and activities that may involve broader financial innovation for capacity building, gravitation towards middle-tier, upper-tier, large enterprise status and equity-based financing:

- Implementation of new or significantly improved financing strategies for categories such as capital raising or receivables;
- Implementation of sophisticated financial management techniques such as knowledge management systems;
- Implementation of fundamental changes in management structure, e.g., introduction of cross-functional financial teams, outsourcing of major business functions such as logistics and transport;
- Implementation of improvements in financing models or strategies, e.g., diversifying the ratio of fixed assets to variable assets to access new sources of finance, attracting new support services from banks to expand existing financing arrangements.

The author abductively assumes that large enterprises demonstrate a higher ratio of broader financial innovation activities, particularly in implementing financing strategies and financial organization models, in which non-dynamic SMEs report less performance. Also, further research on the potential in medium enterprises will raise their awareness on the importance of improving their competitive financial position in the market by introducing new financial products and services.

DEVELOPMENT OF MATHEMATICAL PRINCIPLES OF COMBINING WORKING CAPITAL FOR CAPACITY AND GRAVITATION TOWARDS STOCK EXCHANGES

All previous elaboration logically leads to a conclusion that the combination of financial resources represents the foundation of the theory of risk management. Before discussing the combination of working capital, the author previously lays out the mathematical principles of combinations that are applicable to all factors of production, including capital. Depending on the method of arrangement of elements in tuples and different arrangements of elements in grouped tuples the combinatorics of working capital can be based on: permutations, combinations (Tranter, 1975, pp.45-48) and variations which are permuted combinations. To permute means to arrange different categories of working capital among themselves so that all categories remain within a tuple (budget) while the tuples are not repeated when rearranging. Each resulting tuple is called a permutation. Therefore considered are net working capital, cash, stocks and customers (accounts receivable) as the basic forms of working capital, which could also be used for stock exchange based financing, and mark them with 1,2,3 and 4, then the permutations are the following:

$$P=(1,2,3,4)=\begin{pmatrix} 1234 & 2134 & 3124 & 4123 \\ 1243 & 2143 & 3142 & 4132 \\ 1324 & 2314 & 3214 & 4213 \\ 1342 & 2341 & 3241 & 4231 \\ 1423 & 2413 & 3412 & 4312 \\ 1432 & 2431 & 3421 & 4321 \end{pmatrix}$$

The general formula for the number permutations composed of n elements, which are all different, is obtained by multiplying the number of *(n-1)* mutually different permutations with n elements is

$$P_n = P_{n-1} * n = 1*2*3*4...*(n-1)*(n-1) = n! \tag{6}$$

In building working capital permutations, form the above example, comprised of 4 different categories each category may appear as many times on each of the four places in the upper tuple as there are permutations comprised of 3 categories. Therefore, the number of permutations involving four categories is as follows:

$$P_4 = P_3 * 4 = 1*2*3*4 = 4! \tag{6a}$$

The basic derived formula of permutations enables multiple appearance of one element or category according to the following equation

$$P_{\frac{n}{2}} = \frac{n!}{a!} \tag{7}$$

This form can be applied with multiple occurrence of one of the various forms of working capital. If from a group of "n" categories of current assets, one, two, three, four and so on categories are set aside and consolidated in tuples, and if consolidation is performed from left to right from higher to lower rank, then it has to do with combining working capital. Such tuples can be combinations of the first, second, third rank and so on. When each category of working capital occurs only once financial planning is performed as a management function using combinations with repetition that can be unlimited and limited. The equation

$$C^r = (a,b,c) \tag{8}$$

can imply that, when all categories are different, combinations of r rank are being constructed from working capital types *a,b,c*. However, with the previous example of working capital classification assume the combination of working capital by using the following general formula of constructing a third rank combination of working capital without repeating

$$C_n^3 \frac{n(n-1)(n-2)}{1*2*3} = \frac{n(n-1)(n-2)}{3!} \tag{9}$$

Hence, it is possible to develop combinations of the first, second and third rank without repetition. Therefore the solution is as follows:

$$C^1 = (1,2,3,4) = \left|1,2,3,4\right| C^2 = (1,2,3,4) = \begin{vmatrix} 12 & 23 & 34 \\ 13 & 24 & \\ 14 & & \end{vmatrix} \rightarrow C^3 = (1,2,3,4) \begin{vmatrix} 123 & 234 \\ 124 & \\ 134 & \end{vmatrix} \tag{9a}$$

Similarly, combinations of working capital with repetition can be built by connecting each category of working capital with itself or with higher categories. In the opinion of the author this further explains the analyzed optimization phenomenon of finding the best diversified solution and reduces the number of arbitrary elements in the given financing approach. The general formula for constructing working capital combinations with repetition is therefore

$$C_r^r = C_{n+r-1}^r \tag{10}$$

With permutation of combinations of working capital of the same rank combinations are built as variations which can be without repetition, with limited or unlimited repetition of various categories of working capital.

Further Iterations of Qualitative and Quantitative Changes in Medium-Sized Enterprises Facilitating Gravitation towards Stock Exchanges

Qualitative changes, alternative, partially or completely substituted forms of working capital or fixed assets ultimately lead to optimal financial management and potential gravitation towards stock exchanges. For example, when converting cash to inventory, or retained profit for purchasing equity the company essentially replaces one quality of working capital with others. Likewise, it seems wisest to assume this for fixed assets as well.

If receivables increase it only means that instead of cash on bank account of funds are owed to the company. Similarly, short-term sources of finance can be partly replaced with long-term sources which in turn improve the financial structure. In other words, a financing strategy as a way to create sustained value for shareholders (Kaplan & Norton, 2004, p.29), can be replaced with a more favorable strategy. There is also the possibility of conversion of the variable costs into fixed. It logically follows that the forms of transformation for the purpose of growth are numerous.

Today, in the era of the battle for orderly collection of receivables from large companies, it is precisely financial combinations that will significantly contribute to these efforts of medium sizes enterprises. For example, they may include a specific management intervention to improve their coefficient of optimum liquidity, or managing all the various risks when a category of working capital being is converted into securities or cash when necessary. Proceeding from sources of finance as a parallelogram of financial forces, let us assume a symbolic expression of unique complex financial combinations as follows:

First combination $K1 = \psi(F1 + F2 + F3 + F4 + F5 + ...F_{n-1} + F_n)$ (11)

Second combination $K2 = \psi(F1 + \_\ \_\ \_ + F3 + F4 + F5 + ...F_{n-1} + F_n)$ (11a)

Third combination $K3 = \psi(\_\ \_\ \_ + F2 + F3 + F4 + F5 + ...F_{n-1} + F_n)$ (11b)

Concretizing this general pattern means that different forms of sources of finance can be used *in concreto*. Indeed it is possible to agree that such replacements among forms of working capital or fixed assets are characteristically dynamic among high growth SMEs, and that in each business entity there are opportunities to make changes in working capital forms stemming from rationalization and need to carry out alternative and qualitative improvements.

Causes of substitution can be very different, and particularly important are commercial and financial risks of running a business. For example when existing technology and equipment are replaced and modernized which has long-term significance for financing a business. New technology is usually more expensive and requires more investment so it is not easy to choose a suitable financial moment to finance the replacement that often represents a radical innovation in comparison to conventional production methods. Finally, consider uncollected receivables from a critical customer (commercial and financial aspects) with simultaneous replacement of obsolete equipment (production and financial aspects) which causes fundamental adjustment or optimum combination of working capital that ensures the business success. Obviously in the new economy optimal and innovative combinations should be taken with a grain of salt where the movement of working capital takes the form of a tendency and where the management tries to achieve as much leadership as possible in dynamically changing business conditions.

Suggested substitutions should be viewed in their mutual conditionality and coherence revealing that growth of receivables from major customers can inhibit the capacity of medium and especially small-sized enterprises to be competitive, to flexibly modernize equipment and grow. Finally, by reducing the market value of a company, large uncollectible receivables inhibit the capacity for mergers, consolidations and acquisitions so critically needed for sustainable growth. Finally, qualitative changes in training finance department staff ability to monitor, modify and optimize working capital potentially have central influence on intangible value thus building capacity for growth (Mohnen & Nasev, 2009) inherently leading medium enterprises to stock exchange financing and large enterprise status.

CODIFICATION OF INNOVATIVE WORKING CAPITAL COMBINATIONS FOR CAPACITY BUILDING AND GRAVITATION TOWARDS STOCK EXCHANGES

Probability of solving financial problems depends on the ability of entrepreneurs to simultaneously find innovative financial products, to effectively use existing financial resources and adapt to changes in the environment. The path of medium enterprises is often painful but essential for survival. This process consists of several individual barriers to financial innovation and of diversified financial arrangements contributing to the development of tacit financial knowledge and skills in the innovation processes (Lev, 2001, p.156) which are directly in the function of optimization and growth of assets required by formal exchanges. Therefore, the capacity to innovate leads to productivity and growth ultimately steering the firm towards formal stock exchanges. The thorny path to creative and innovative combining of working capital for growth and gravitation towards stock exchanges includes the approaches presented in Table 4.

Table 4. Innovative working capital combinations for gravitation towards stock exchanges

Creative Combining Approach	Development and Possible Conduct of Working Capital Combination Activities in Medium Enterprises in Concreto
Adapt or Abandon	Adapt medium-sized enterprises financial management to new conditions or abandon a type working capital financing if necessary to avoid existing obstacles.
Change	Abandon one form and introduce a new form of financing working capital.
Combine	Combine working capital means that any new funding mix is made from a known structure of financing.
Convert	Change working capital into another form, from one use, function, or purpose to another: *convert bank sources into nonbank sources.*
Design	Design working capital means to create a portfolio for existing business or initiate a new venture where its use is more rational.
Develop	Achieve applied improvements in conditions and manner of financing some form of working capital and / or processes in the finance department.
Divide	Make independent a finance department section, thus optimizing working capital.
Diversify	Commit working capital for a concentric or horizontal corporate strategy to access a new market or create a new product without abandoning existing products or a portfolio management tool where an investor holds an assortment of different investments to counterbalance the poor performance of other ventures.
Eliminiate	Eliminate certain financial products because they injure or threaten the working capital integrity in the company.
Empower	Make a part of finance department independent and / or empower finance officers.
Expand	Take into account operational constraints expanding the scope of working capital commitment in order to achieve medium enterprise business policy objectives.
Improve	Improve the functionality of the existing categories of working capital.
Link	Link related working capital refers to linking processes, such as collection of receivables and executing payments, logically followed by relevant linking of employees in the finance department.
Modify	Modify the structure of financing working capital or restructure financial department organizationally.
Multiply	Achieve synergistic effects by combining different sources of funding and placing working capital of medium enterprises.
Reduce	Reduce the volume of financing of a certain category of working capital.
Reorganize	Rearrange sections in the financial department changing organizational solutions so working capital is more efficiently and productively committed for development.
Reshape	Break down walls, barriers by changing and innovating conditions of financing working capital.
Rotate	Rotate working capital to replace existing forms financing solutions with innovative funding commitments.
Scale up	Apply across the board specific processes / practices in working capital management in medium-sized enterprises
Set apart	Set apart viable from those tasks that are financially less viable. Logically precedes working capital optimization.
Shift	Transfer part or all of one category of working capital into another form.
Shrink	Reduce the scope of operation of certain categories of working capital.
Term	Temporary cession of a right to use part of medium enterprise working capital.
Upgrade	Improve how financial department responds to working capital needs.
Variate	Permute previously built working capital combinations.

Source: Own presentation of author.

In Kornai (1971) it is established that the process of decision making in organizations includes applicable, researched, acceptable and appropriate sets of alternatives. Hence, it derives that the multitude of offered combinations of activities help in overcoming barriers and that they can cause multiple changes in the use of working capital. The central issue is the ability to generate synergies between different types of combinations which certainly involves creativity and innovation in the formation of funds for development and prosperity of medium firms. Resulting effects of specific changes in the combination of working capital are classified analytically in Table 5.

All mentioned changes in creative combining, the order of which was not determined randomly, characteristically have a consequence of affecting all other changes in the system of financing working capital. Logically, structural and positional changes in the development of medium-sized enterprises are the necessary preconditions to achieve quantitative and qualitative changes. This conclusion can be demonstratively proved through the simple number of SMEs and the establishment of wider integrations of enterprises into holding companies. It is clear that without structural changes it is not possible to increase their number and contribute to forming more dynamic and larger companies, which can provide desired quantitative and qualitative changes. Structural changes are also an essential precondition of positional changes and for increasing competitiveness of internal and external structure. In such a scenario, it is possible to talk about a complex developmental change within which mentioned innovation in the process of working capital optimization is the key component enabling the gravitation of medium-sized enterprises towards stock exchanges.

In fact properly modeled financial departments in medium-sized enterprises should have the task of developing their own structure in order to activate the enabling components of growth. That responsibility involves optimizing inventories and linking DSO with DPO as key components of working capital combination (Danske Bank & Ernst & Young, 2009) in the C2C. Some like Brigham & Houston (2001) would also indicate considering a moderate, relatively aggressive or conservative approach to company's current assets (pp.746-748) whose highlights are as follows:

- **Moderate Approach:** Optimize maturity of sources of finance with the maturity of the financing;
- **Aggressive Approach:** Optimize by using short term sources of finance to finance permanent working capital;
- **Conservative Approach:** Optimize by using long-term financing and equity capital to finance permanent and temporary working capital needs.

Table 5. Typology of changes in working capital combinations for medium enterprises

General Typology of Internal and External Development Changes	Relevance of Specific Typology of Combining Working Capital for Stock Exchange Gravitation
Structural changes	Changes in organizational structure of financial department and intensity of connections with other enterprises
Positional changes	Financial competitiveness of observed medium enterprise in relation to other firms
Quantitative changes	Change in volume of working capital use including changes in enterprise financial structure leading to mergers and acquisitions
Qualitative changes	Substitution of used working capital types Partial transition towards stock/equity based financing Integration of medium enterprises Improvement of educational and qualification structure in financial department

Source: Own presentation of author.

Clearly, a moderate approach to financing allows the use of short-term loans to finance short-term working capital and a conservative policy includes financing of not only fixed but also long-term working capital with debt based on a variety of marketable securities. For simplicity, previously defined foundations of combining working capital are here reduced to two basic variants of formation and financing of company assets, i.e. to the moderate and conservative financing policy. It seems neither method is an effective way for optimal gravitation to stock exchanges and therefore an alternative approach is required – and this treatise should be viewed as an attempt that optimizes different sources of finance. Several studies discussed hereinafter point the need for that direction.

This poses an analytical problem: compared to moderate financing with debt is it obvious that conservative policy gravitates towards the zero debt level that allows marketable securities? It turns out that in the EU – as shown in the European Commission Survey on the access to finance of enterprises - debt dominates over equity financing among SMEs because 61% preferred bank loans in 2015, while another 17% preferred loans from other sources such as trade credit while only 9% preferred equity

investments and additional 9% preferred other types of equity financing. Among the different SME types, gazelles not surprisingly seemed to prefer equity investment the most (13%) and according to the same survey gazelles were also more innovative (Doove, Kwaak, & Span, 2015). In the US recent evidence shows that venture capital, angel investment, crowdfunding and online funding are prominent supplements of classic SME lending (Firoozmand et al., 2015). Taking a comparative new look at a target group of 62 medium enterprises from Serbia, a non-EU country, these firms occasionally resorted to moderate policies of borrowing working capital, compromise - conservative policies with own internal-financing dominated, but unfortunately no diversification of the portfolio towards shares and bonds was observed (Vemic, 2015, p.20).

From the standpoint of potential of any medium-sized enterprise, internal and own funding is only a path to realize short-term survival. Occasionally, efficient and rational borrowing can provide medium-term survival, while compromise-conservative policy of diversification of the portfolio towards stocks and bonds, coupled with crowdfunding and online lending platforms can provide a more viable strategic long-term option. This especially applies for financing fixed assets and permanent current assets which should both be financed long-term. In these two cases, the risk of short-term financing is dual: first, short-term interest rates oscillate more significantly than long-term interest rates and second, yielding to short-term debt year after year compels SMEs to sharper volatility in their financing costs as a consequence and makes them more vulnerable in comparison to large firms that have easier access to sources of finance.

FURTHER RESEARCH DIRECTIONS

This treatise demonstrates that researchers and practitioners can look more deeply into the area of alternative enterprise financing that supplements credit markets. Ample research agenda lies ahead: to investigate further how the findings can be implemented in various countries, to investigate efficiency of proposed alternative enterprise financial management models and to evaluate various possible government economic policies that facilitate market access for medium enterprises.

The human resources aspect of creating a culture of entrepreneurship and financial innovation needs to be given a high priority in future research.

Author's analysis of the financial growth cycle and integration of various types of finance reveals that future research may include exploring how sources of finance for medium sized enterprises may change over their life cycle, in reaction to market and government policy changes, during times of volatility in private or public markets, and as management systems continue to improve.

It seems that raising awareness about possible models of stock exchange based financing is the best way to improve the strategy and investment capacity of both medium enterprises and stock exchanges. Ministries of Finance, associations of banks and other financial organizations should organize a few well targeted visits to leading stock exchanges in transition for the purpose of acquainting medium sized enterprises with this potential. It is essential that preparation of capacity building visits includes providing detailed information to host countries about the visitors, interesting stock exchange products and contacts at the highest level with the public sector, entrepreneurs and companies. This would help medium enterprises from transition economies to develop a deeper understanding of critical success factors in these markets.

There is some evidence from working capital surveys consulted (Danske Bank & Ernst & Young, 2009) that financial management in medium enterprises is suboptimal and that use of knowledge could be improved. This requires further research, concentrating on those variables which can potentially describe in more detail the diffusion, adoption and especially measurement of financial innovation which would increase the capacities of funds in medium enterprises.

Correlation between firm level financial innovation and its growth needs to be further studied.

The next key issues is who are the early adopters and holders of innovation in the financing of medium enterprises, what are their attitudes and how can other companies apply positive experiences in this field. From the study of the existing literature, as well as from researching practical high growth entrepreneurship (Business Development Bank of Canada, 2015) an impression is made that some medium enterprises have previously adopted some innovations in their development. It is important to identify the leaders of financial innovation thought for medium firms. These development hypotheses and suggested researches should be published in a broader analysis that potentially represents a new open area of relevance for medium enterprises.

There finally remains creation of information frameworks on medium enterprises and stock exchanges in the countries of various regions. This could include market materials, solutions for communication infrastructure, promotion and marketing activities, government regulations and possible rural development subsidies. Consequently, this would enable different market players to collect and analyze activities of the main stock exchange events and opportunities for medium enterprises.

CONCLUSION

It could be said that the basis of development of medium enterprises is further access to the financial markets. Developed capital markets facilitate the connection of subjects which have excesses of financial resources to the medium enterprises which lack financial resources, i.e. they connect two important macroeconomic and financial categories – savings and investments. While, through money markets, enterprises obtain resources to maintain current liquidity, through capital markets they acquire resources to finance business development.

Stock exchanges facilitate smooth and uninterrupted development of a national economy with a multitude of entities including medium enterprises. All the events in real processes of the economy are displayed on them. They represent one of the main instruments of a market economy based on enterprise development. There are various financial institutions doing business of interest for medium enterprises including banks, savings and credit associations, insurance companies, pension funds, investment funds and of course stock exchanges. All these institutions exist in the developed economies and their intensive growth and development followed the globalization of the world economy and capital markets. As such stock exchanges represent very significant intermediary financial organizations or institutional investors which collect disposable resources from their investors and then place them into short term and long term investments and different financial forms. Therefore, a supplement and an alternative for bank credits is *equity*, a financing method that could be more suitable than a typical bank credit, especially for the high growth dynamic medium enterprises. In transition economies the inability to use equity has posed a major obstacle to the development of effective markets and enterprises alike. Although governments admit the existence of this problem, they have not developed economic policy instruments to address the root causes of insufficient financing of medium enterprises. The development of this sort of financing will certainly accelerate the reform and development of their financial sector and would facilitate development of medium enterprises as such. A question obviously arises at this juncture: if stock exchanges are an important step towards enterprises financing and growth why aren't medium enterprises developing capacity to access the stock exchanges? In developed economies the role of not only stock exchanges but capital markets as well is very extensive. It needs to be recognized and addressed that in developing economies a major obstacle derives from inappropriate regulatory systems. Therefore a well functioning financial market is clearly the potential engine of enterprise development and dynamic transformation of medium enterprises.

This treatise has presented information derived from working capital and access to finance surveys in order to build determinants of medium enterprise transitioning towards stock exchanges including the development of working capital combinations using various mathematical models.

By defining some key diffusion indexes of the financial system in development, entrepreneurship and financing of medium enterprises this treatise has confirmed the fact that it is possible to measure the index of their development closely focusing on some conditions for financing and gravitation toward the formal stock exchanges.

The above analysis suggests that besides the role of stock exchanges medium enterprises also have a role to play in developing their capacity for various sources of finance. Both players can add value in both enterprise and stock exchange development, particularly in developing and transition economies. Thus an infusion of equity into medium enterprises is necessary to get the process started. The author does not wish to imply here that anyone has ever developed a concrete and successful example of such a medium sized enterprise theory, but only states that such a theory of their finance is, as far as the author knows, conceivable. If the empirical data is any guide, stock exchanges and medium enterprises will follow. Further elaboration of adequate interventions in both stock exchanges and medium enterprises will then become applicable indeed complementing each other in providing a complete description of the financial system as a whole.

REFERENCES

Barcus, S. W. W., & Wilkinson, J. W. (1995). *Handbook of management consulting services*. McGraw-Hill.

Bodnar, G. M., & Marston, R. C. (2002). Exchange rate exposure: A simple model. *International Finance Review*, *3*, 107–115. doi:10.1016/S1569-3767(02)03010-8

Brigham, E., & Houston, J. (2001). *Fundamentals of financial management*. Harcourt. Inc.

Business Development Bank of Canada. (2015). *High-Impact Firms: Accelerating Canadian Competitiveness.* Retrieved May 17, 2016 from BDC web site https://www.bdc.ca/EN/Documents/analysis_research/high-impact-firms-accelerating-canadian-competitiveness.pdf

Danske Bank & Ernst & Young. (2009). *Working Capital Management: A Survey of Nordic Companies*. Copenhagen, Denmark: Danske Bank A/S & Ernst & Young.

Doove, S., Kwaak, T., & Span, T. (2015). *Survey on the access to finance of enterprises (SAFE).* European Commission. Retrieved May 27, 2016, from EC web site: http://europa.eu/DocsRoom/documents/14321/attachments/1/translations/en/.../native

Downes, J., & Goodman, J. E. (2003). *Finance and Investment Handbook*. Barron's Educational Series.

Dzida, L., Orzechowska, U., Piotrowska, J., Pokorska, A., Rozkrut, D., & Wegner, M. (2014). *Działalnosc innowacyjna przedsiebiorstw w latach 2011-2013*. Główny Urzad Statystyczny, Urzad Statystyczny w Szczecinie. Retrieved July 7, 2016 from Central Statistical Office of Poland web site: http://stat.gov.pl/obszary-tematyczne/nauka-i-technika-spoleczenstwo-informacyjne/nauka-i-technika/dzialalnosc-innowacyjna-przedsiebiorstw-w-latach-2011-2013,2,8.html

Eales, A. B. (1995). *Financial risk management*. McGraw-Hill.

European Commission. (2007). *Mezzanine Finance: Final Report, 2007*. Retrieved May 29th 2016 from the University of Mannheim European Documentation Center web site: http://edz.bib.uni-mannheim.de/daten/edz-h/gdb/07/roundtable_mezzanine_2007.pdf

European Commission. (2015a). *User guide to the SME Definition.* Retrieved May 27, 2016, from EC web site: http://ec.europa.eu/growth/smes/business-friendly-environment/sme-definition/index_en.htm

European Commission. (2015b). *Innovation Union Scoreboard.* Retrieved May 27, 2016, from EC web site: http://ec.europa.eu/growth/industry/innovation/facts-figures/scoreboards/files/ius-annex-h_en.pdf

Firoozmand, S., Haxel, P., Jung, E., & Suominen, K. (2015). *State of SME Finance in the United States in 2015*. TradeUp Capital Fund and Nextrade Group LLC. Retrieved May 23, 2016 from TradeUp web site: http://www.tradeupfund.com/uploads/2/6/0/4/26048023/state_of_sme_finance_in_the_united_states_2015.pdf

Fukuda-Parr, S., & Kumar, A. S. (2003). *Readings in Human Development: Concepts, measures and policies for a development paradigm*. New Delhi: Oxford University Press.

Greenspan, A. (2008). *The age of turbulence: Adventures in a new world*. London, UK: Penguin Books.

Hart, O. (1995). *Firms, contracts, and financial structure*. Oxford, UK: Clarendon Press. doi:10.1093/0198288816.001.0001

Hisrich, R. D., & Peters, M. P. (2002). *Entrepreneurship*. New York, NY: McGraw-Hill/Irwin.

IBM Global CEO Study Team. (2006). *Expanding the Innovation Horizon: The Global CEO Study 2006*. IBM Global Services.

Kanter, R. M. (2006). Innovation: The classic traps. *Harvard Business Review*, *84*(11), 72–83. PMID:17131564

Klapper, L. F., Sarria-Allende, V., & Sulla, V. (2002). *Small and medium-size enterprise financing in Eastern Europe* (Working Paper No.2933). Retrieved June 15, 2013, from World Bank website: http://info.worldbank.org/etools/docs/library/28751/SME%20financing%20Eastern%20Europe.pdf

Kornai, J. (1971). *Anti-equilibrium: On economic systems theory and the tasks of research*. Amsterdam: North-Holland.

Lev, B. (2001). *Intangibles: Management, measurement, and reporting*. Washington, DC: Brookings Institution Press.

Meredith, G. G. (1987). Management services for small enterprises. In Small Enterprise Development: Policies and programmes (pp. 125-139). Geneva: International Labor Organization.

Ministry of Economy and Regional Development, Republic Development Bureau, Republic Agency for Development of SME and Entrepreneurship. (2009). *Report on small and medium-sized enterprises and entrepreneurship 2008*. Belgrade: Ministry of Economy and Regional Development Republic Development Bureau Republic Agency for Development of SME and Entrepreneurship.

Mohnen, A., & Nasev, J. (2009). *Growth of Small and Medium Sized firms in Germany. In Betriebswirtschaftliche Forschung und Praxis [Business Research and Practice]* (Vol. 5). Academic Press.

Mortensen, P. S., & Bloch, C. W. (2005). *Oslo Manual-Guidelines for collecting and interpreting innovation data*. Organisation for Economic Cooporation and Development, OECD.

O'Reilly, C. A., & Tushman, M. L. (2004). The ambidextrous organization. *Harvard Business Review*, *82*(4), 74–83. PMID:15077368

Rogers, E. (1962). *Diffusion of Innovations*. Free Press of Glencoe.

Stacey, R. D. (1996). *Complexity and creativity in organizations*. Oakland, CA: Berrett-Koehler Publishers.

Stam, E., & Hartog, C., Stel van, A., & Thurik, R. (2009). *Ambitious entrepreneurship, highgrowth firms and macroeconomic growth.* EIM Research Reports. Retrieved July 13, 2016, from Research Gate web site: https://www.researchgate.net/publication/46455754_Ambitious_Entrepreneurship_High-Growth_Firms_and_Macroeconomic_Growth

Taleb, N. N., Goldstein, D. G., & Spitznagel, M. W. (2009). The six mistakes executives make in risk management. *Harvard Business Review*, *87*(10), 78–81. PMID:19968059

Tranter, C. J. (1975). *Advanced level pure mathematics*. Hodder and Stoughton Educational.

Vemic, M. B. (2015). Strategic Data-Based Optimization of Working Capital Management in Medium-Sized Firms: A Real-World Case Study. In Strategic Data-Based Wisdom in the Big Data Era (pp. 15-50). Hershey, PA: IGI Global. doi:10.4018/978-1-4666-8122-4.ch002

Venohr, B., Fear, J., & Witt, A. (2015). *Best of German Mittelstand-The World Market Leaders*. Academic Press.

World Bank Centre for Financial Reporting Reform (CFRR) and the Austrian Institute for SME Research. (2013). A Survey of SME Accounting and Reporting Practices in Austria. Vienna: World Bank Centre for Financial Reporting Reform (CFRR) and the Austrian Institute for SME Research.

ADDITIONAL READING

Christensen, C. M. (2003). *The innovator's dilemma: the revolutionary book that will change the way you do business*. New York, NY: Harper Business Essentials.

European Commission. (2016). *Commission Staff Working Document. Capital Markets Union: First Status Report.* Retrieved July 13, 2016, from EC web site: http://ec.europa.eu/finance/capital-markets-union/docs/cmu-first-status-report_en.pdf

Fredman, A. J., & Wiles, R. (1998). *How mutual funds work*. Upper Saddle River, NJ: Prentice Hall Press.

Friedlob, G. T., & Plewa, F. J. Jr. (1996). *Understanding return on investment*. New York, NY: John Wiley & Sons.

Gompers, P. A., & Lerner, J. (2002). *The venture capital cycle*. MIT press.

Harwood, A., & Konidaris, T. (2015). *SME Exchanges in Emerging Market Economies A Stocktaking of Development Practices.*(Policy Research Working Paper 7160). Retrieved March 12, 2013, from World Bank Finance and Markets Global Practice Group web site: http://www-wds.worldbank.org/external/default/WDSContentServer/WDSP/IB/2015/01/12/000158349_20150112155857/Rendered/PDF/WPS7160.pdf

International Finance Corporation. (2012).*Why Banks in Emerging Markets Are Increasingly Providing Non-financial Services to Small and Medium Enterprises.* Retrieved May 25, 2015, from IFC web site: http://www.ifc.org/wps/wcm/connect/topics_ext_content/ifc_external_corporate_site/ifc+sustainability/learning+and+adapting/knowledge+products/publications/publications_report_nonfinancial-services

Timmons, J. A., & Spinelli, S. (2003). *New venture creation: Entrepreneurship for the 21st century*. New York, NY: McGraw-Hill.

Toland, M., Saffer, A. P., & Heatly, B. (1995). *Stimulating the Growth and Development of Small and Medium Size Enterprises Through Financial Sector Policy Reform in Poland*. Development Alternatives, Inc. Retrieved March 12, 2013, from U.S. Agency for International Development web site: http://pdf.usaid.gov/pdf_docs/PNABW060.pdf

Williamson, J. P. (Ed.). (1988). *The Investment banking handbook* (Vol. 11). New York, NY: John Wiley & Sons.

World Federation of Exchanges. (2015). *WFE Report on SME Exchanges.* Retrieved March 23, 2016, from WFE web site: https://www.smefinanceforum.org/sites/default/files/post/files/WFE%20Report%20on%20SME%20Exchanges.pdf

KEY TERMS AND DEFINITIONS

Capital Market: A broader structured market that includes the stock market and other modalities for trading different financial products such as bonds, derivative contracts such as options, various loans and other debt instruments, and commodity futures.

Change: Systematic and durable improvement of organizational quality.

Credit: Contractual agreement in which a borrower receives something of value now and agrees to repay the lender at some date in the future, generally with interest.

Equity: A stock or any other security representing an ownership interest presented in the balance sheet as an amount of funds contributed by the owners (the stockholders) plus the retained earnings (or possibly losses).

Expansion Capital: Capital to fund the development of an established business.

Factoring: Financial intermediation aimed at purchasing receivables from a company where a factor is basically a source finance that agrees to pay the value of the invoice reduced at a discount for commission and fees.

Innovative Financing: Various sources of finance available to high-tech start-ups in their early stages of growth - includes seed capital funds, informal investors, banks and venture capital funds.

Leasing: A contract granting use of real estate, equipment, or other fixed assets for a specified time in exchange for payment, usually in the form of rent.

Market Inefficiency: A market state of imperfection in which prevailing prices do not mirror all publicly available information about a security being such that some players obtain knowledge of conditions before others, or when some players do not accurately analyze accessible relevant information.

Mezzanine Finance: Hybrid forms of finance interdependent with debt and equity such as subordinated loans, participating loans, profit participation rights, convertible bonds and bonds with warrants particularly helpful in the expansion phase of upper-tier high growth medium-sized enterprises.

Risk: The chance return on investment will be different than expected with a possibility of losing some or all of the original monetary value.

Start-Up Capital: The capital to fund a start-up used for renting offices, hiring personnel and initiating sales based on business plan with market research products or services developed, and management team in place.

Stock Exchange: A structured market operating on the basis of a set of rules in which bonds, notes and shares of publicly held business entities are issued and exchanged either through exchanges or over-the-counter markets.

Uncertainty: A state of business and financial knowledge where the order or essence of events is unknown, the consequences, degree and weight is uncertain, and dependable probabilities on outcomes of events cannot be designated.

Venture Capital: High-risk, high-return investment for financing rapid growth of new technology and potentially high growth SMEs.

ENDNOTES

* This conclusion, the validity of which is fairly evident, will be drawn again later from more detailed discussions.

1 Editors' note: See chapter 8 for a more detailed treatment of working capital optimization in medium enterprises.

Section 3

Optimal Strategic Management for SMEs

Chapter 11
Overcoming the Barriers of Strategic Planning, Implementation, and Monitoring in Turbulent Business Environment:
A Qualitative Study on Finnish SMEs

Binod Timilsina
University of Vaasa, Finland

ABSTRACT

Strategic management has been highly emphasized in the literature as an enabler of competitive advantage. However, the managers and strategic leader often face challenges in systematic strategic planning, implementation and monitoring. Therefore, the objective of this research was to explore the current practices of strategic planning, implementation and monitoring, and pinpoint its barriers. The study has also given emphasis to identify the potential solutions in overcoming the barriers of strategic planning, implementation and monitoring. Likewise, the study has investigated the competencies of a good strategic planner. The study was conducted among the managers of Finnish SMEs. Methodologically the study adopts qualitative study following thematic and inductive approach. After a rigorous analysis of collected data and basing on resource based view and industry organization theory, a framework has been presented for systematic strategic planning, implementation and monitoring. Finally, the research limitation and future research possibilities have been discussed.

DOI: 10.4018/978-1-5225-1949-2.ch011

INTRODUCTION

Successful strategic planning and implementation has long been identified as one of the key aspect for firm's growth and survival. This is for a few reasons: first, strategies are the plan that drives organizational resources towards the desired goals (Azevedo, Almeida, van Sinderen, & Pires, 2015); second, strategy is a connecting bridge between the firm's operating environment and the firm (Grant, 1991; Ralston, Blackhurst, Cantor & Crum, 2015); third, the strategic plan provides organizational stability by assisting leaders and managers in managing the change (Bryson, 1988); and fourth, strategic plan helps a firm to neutralize or overcome the competitor's move. However, because of the constantly changing business environment and rapidly changing customer needs, the plan made today may become useless the day after without the notice of management. Therefore, to meet the challenges of rapidly changing business environment, frequently changing customer needs, and for better reaction to the competitors' move, a firm should have more sophisticated, flexible, and innovative framework for effective strategic planning, implementation and monitoring.

It is assumed in business practice, that firms having better capabilities in utilizing resources should have an advantage over their competitors which results in higher earnings. But gaining competitive advantage through resource deployment is almost impossible without a proper plan, policies, and commitment from the management. It has been argued in literature, that strategic planning helps to reposition the competitive landscape (Drucker, 1954); it is critical to competitive advantage and firm performance (Mintzberg, Ahlstrand, & Lampel, 2009; Sirmon, Hitt, & Ireland, 2007). Consistent with these views Scholes and Johnson (2005) argue that strategic planning and implementation allows better interaction between a firm and its operating environment, and provides an opportunity to reconfigure the firm's resource structure to overcome the challenges posed by the changing business environment. Despite this expectation, organizations often face challenges to keep up with the speed of change in the business environment. This might be due to:

1. Weak implementation of strategies (Waterman, Peters & Phillips, 1980);
2. Lack of a better strategy (Mankin & Steele, 2005).

Therefore, in turbulent times "the balance between operative measures and strategic direction setting becomes critical" (Naujoks, 2010, p. 104). In their study Higgs and Rowland (2005) found that almost more than 70% of changes in strategies were unsuccessful. Similarly, Franken, Edwards and Lambert (2009) claimed for 34% of failure rate in strategy implementation. On average companies achieve only 63% of the financial performance (strategic objectives) included in their strategic plan (Mankin & Steele, 2005). Likewise, the successfully-implemented strategies range between ten to thirty percent (Raps, 2005). In contrast, Cândido and Santos (2015, p. 237) argue that "it is often claimed that 50 to 90 percent of strategic initiatives fail.........they are controversial". However, the different claims made by the authors of previous studies are sound proof to argue that the success rate of strategic planning and implementation is not adequate. Therefore, both for academic researcher and business practitioner, it is interesting to investigate the reasons behind the failure of strategic planning and implementation at micro level. Here, the micro level means the barriers in terms of what, how and why questions. As for example: competencies of strategist (what?), the role of management or people's actions (how?), and strategic failure (why?).

Nevertheless, the knowledge of barriers in strategic planning and implementation will help an organization to develop a clear path in accomplishing organizational objectives, hence, contribute in improving the firm's performance. Specifically, this research attempts to address a few questions: What are the barriers in strategic planning and implementation? What is the process of strategic planning and implementation in practice (i.e. what does a strategic planner think and do in practice in the process of strategic planning and implementation)? Why do strategic planning and implementation fail in practice? By addressing these research questions the study aims to meet the following two objectives:

- To identify the critical factors facilitating and/or disrupting the strategic planning and implementation.
- To present a framework for effective and/or efficient strategic planning, implementation and monitoring.

The research concept of this study is presented in Figure 1. As shown in the figure the research aims to present a framework for effective and/or efficient strategic planning, implementation and monitoring. Methodologically the research adopts a thematic analysis following an inductive approach to reach a common understanding of current practice of strategic planning, implementation and monitoring, hence to develop the proposed framework. In addition, barriers of strategic planning and implementation, potential solutions in overcoming the identified barriers are discussed. Also, the competencies of a good strategic planner are explored and discussed. The data required for the study were collected through an online survey conducted among the managers of Finnish small and medium size enterprises (SMEs).

The study is organized in different sections. First, the concept of strategy is presented as a theoretical background considering resource based view (RBV) and industry organization (IO) theories as the theoretical lens. Second, the research methodology is described which includes the sample population, methods of data collection and the procedure of data analysis. Third, the research findings and results are presented. Finally, the concluding remarks are made, including critical analysis of results and findings, theoretical and managerial implication, also the research limitation and future research possibilities are discussed.

Figure 1. Research concept
Source: *Own presentation of author.*

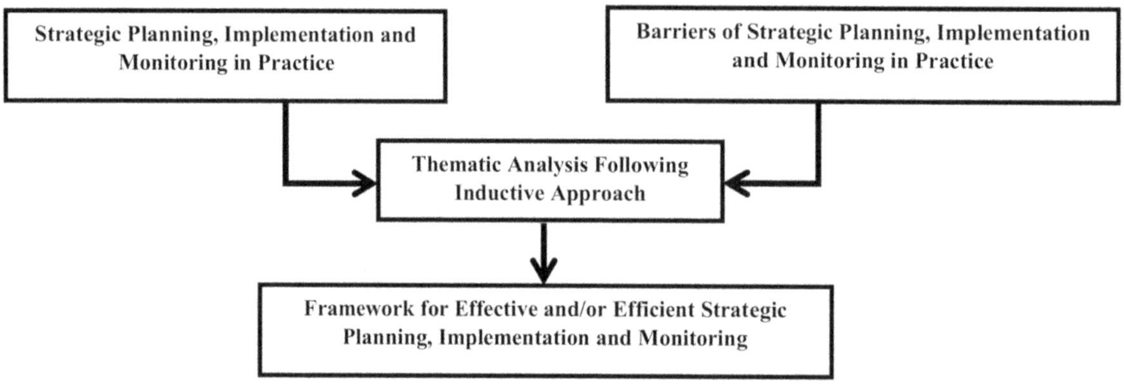

THEORETICAL BACKGROUND

The Concept of Strategy: A Perspective to RBV and IO theory

A firm operates and serves the surrounding environment (Emery & Trist, 1965) which is highly dynamic and complex. Therefore, a firm should scan and understand the consequence of the changing business environment (Moon & Ruona, 2015) not only to make better utilization of resources and capabilities, but also to gain and sustain competitive advantage through strategic plan and actions. This indicates that internal and external environments are crucial in shaping a firm's strategy. Therefore, in this study to build the concept of strategy the focus lies on the RBV and IO theories. On the basis of these theories the strategy process can be seen both as emergent and intended. Following the work of Mintzberg & Waters (1985) emergent strategy is defined as the behavior or actions taken by a firm to overcome the environmental threats over time i.e. outside in approach, while intended strategy refers to the behavior or actions taken by a firm on the basis of its resources and capabilities i.e. inside out approach (see Figure 2). As shown in Figure 2, the main argument of this section is that the RBV and IO theories are complementary to each other in shaping a firm's strategy.

Irrespective of business sector and size, a firm needs to have a clear strategy to survive and grow in the changing business environment. The term strategy is related to every aspect of business; it is related to day-to-day activities carried by a firm. In the literature RBV and IO theories have been widely accepted not only as a core concept to differentiate the performance difference between competing firms but also as the foundations for a firm's strategy. This is because of two reasons. First, IO theory is precisely concerned with the opportunities and threats' streaming from the environment and asserts that the industry forces in which a firm operates are very important for a firm's growth and survival. Basing on I/O theory Porter (1985) has argued that a firm's strategic position in the market is determined by five forces: threats of new entrant, threats of substitute product, bargaining power of buyer, bargaining power of supplier, and rivalry among the existing firms. Thus, by identifying competitive intensity, profitability, and industry attractiveness the Porter's five force model not only helps to evaluate the present strategic strength but also supports in shaping the future strategic move. Second, with the development of RBV there is shift from industry to firm specific effects in identifying the sources of sustainable competitive advantage among the strategy researchers (Spanos & Lioukas, 2001). Likewise, the proponent of RBV (e.g. Peteraf, 1993; Barney, 1991; and many more) argues that a firm's competitiveness is the result of the firm's capabilities in mobilizing organizational resources. It is due to the fact that a firm competes in the marketplace on the basis of its resources and capabilities (Peteraf & Bergen, 2003). Besides this fact, organizational resources and capabilities are the foundations of the firm's strategy (Feurer & Charbaghi, 1995; Grant, 1991) and are highly influenced by the firm's operating environment.

Figure 2. The concept of strategy: A perspective to RBV and IO theories
Source: *Own presentation of author.*

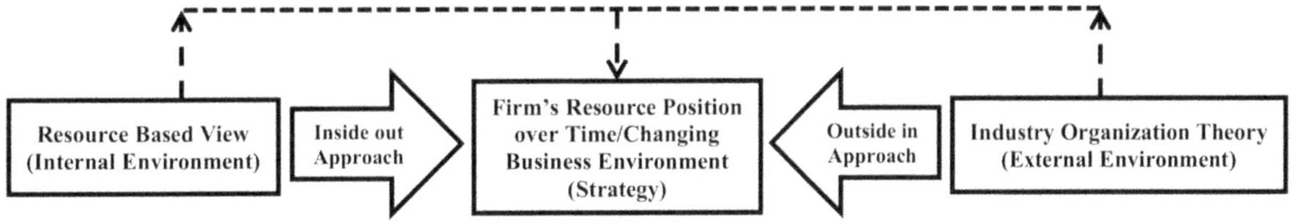

Thus, from the above discussion, it can be concluded that firm's resources, capabilities, and the external environment are critical to strategic planning and implementation. In business practices, strategies act as a connecting bridge between the firm and its operating environment. In other words a firm utilizes its resources and capabilities to overcome the challenges posed by operating environment through strategic actions. This indicates that through strategic actions a firm utilizes its resources and capabilities to overcome the challenges posed by the operating environment. According to Raduan, Jegak, Haslinda, and Alimin (2009, p. 412) "I/O perspective offered strategic management a systematic model for assessing external competition within an industry.......... RBV is indeed crucial as it can be used as a conceptual guideline for business organization in particular to enhance their competitive advantage". In this notion researchers like Drnevich & Kriauciunas (2011) have argued in the support of complementary nature of RBV and IO theories. Indeed, the core notion of strategy as a fit between internal competencies and external environment incorporates the RBV and IO perspective. Here, the strategic fit means the strategic choice made by a firm which assures the best possible use of resources and capabilities in regard to the external business environment. Through the assumption of RBV a firm can make an assessment of its strength and weakness while from Porter's five forces (a perspective to IO theory) a firm can identify its opportunities and threats in the industry (Spanos & Lioukas, 2001). In a similar manner, Mahoney & Pandian (1992) also support the view that RBV and industrial organization research are complementary. This shows that IO theory has implication on RBV, RBV has implication to firm's strategic posture, showing that IO theory and RBV are mutually inclusive theories that have significant impact on firm's strategic planning and implementation (see Figure 2).

RESEARCH METHODOLOGY

Study Population, Sample, and Data Collection

The data required for this study was collected from Finnish SMEs through an online survey. There were 36 respondents who participated in the survey, representing 7 micro enterprises (1-9 employees), 19 small enterprises (10-49 employees), and 10 medium size enterprises (50-249 employees). Among the 36 respondents there were 4 managers, 17 directors, 14 owners and one person working on other positions in the company. In a similar manner, these respondents correspond to 3 primary industries (industry making use of natural resources and includes agriculture, forestry and fishing, mining, and extraction of oil and gas sector), 8 secondary industries (industries using raw materials supplied by the primary sector), and 25 tertiary industries (industries involved in the service sector).

Development of the Survey Instrument: Why Open Ended Questions?

In order to address the research question and to meet the research objective, the study adopted open ended questions to collect the data. This is because of a few reasons. First, the open ended question provides an opportunity for respondents to express their opinion on the subject matter without any restrictions and to the point (Jackson & Trochim, 2002). Second, while answering respondents can think, edit and present exactly what they want in response to the asked questions. This view is supported by Roberts et al., (2014) and argues that open ended survey provides an opportunity to respondents to express their opinion in own thinking. Likewise, open ended questions are clear questions without any suggestion to the answer but results in clear answers (Popping, 2015). Third, "in comparison to interviews or focus groups, open-ended survey questions can offer greater anonymity to respondents and often elicit more honest responses" (Jackson & Trochim, 2002, p.307). However, to avoid ambiguity in data analysis the respondents were asked to be as precise as possible in answering the survey questions. There were eight open ended questions in total as listed as follows:

1. What are the barriers that your organization is facing in strategic planning?
2. In your opinion how to overcome the barriers of strategic planning?
3. What are the barriers that your organization is facing in strategic implementation?
4. In your opinion how to overcome the barriers of strategic implementation?
5. Do you use any strategic planning tool? If yes what are they?
6. In your opinion what are the competencies of a good strategic planner?
7. Could you please describe your organization's strategic planning process?
8. Could you please describe your organization's strategic implementation and monitoring process?

Data Analysis: Why Thematic Analysis Through an Inductive Approach?

This study was designed to explore the common practices of strategic planning and implementation among the organizations which participated in the survey. Furthermore, the research was aimed at developing a framework for effective and/or efficient strategic planning and implementation rather than assessing the impact of strategic planning and implementation on organizational success (see Figure 1). Therefore, the qualitative method was selected. Likewise, the research adopts a data driven inductive approach (observation – pattern analysis – theory) supported with thematic analysis. Under this approach identification of codes and themes are made fully relying on the information available in the collected data, while here, in the case of this research data are written text from respondent. Therefore, it is reasonable to say that an inductive approach is appropriate in validating the content of thematic analysis (Jackson & Trochim, 2002).

Thematic analysis is a qualitative research "method for identifying, analysing, and reporting patterns (themes) within data. It minimally organises and describes your data set in (rich) detail. However, it also often goes further than this, and interprets various aspects of the research topic" (Braun & Clarke, 2006, p. 79). This means the thematic analysis helps to not only understand the current practices, but also to make consistent interpretation from the data by detecting and identifying factors or variables that influence respondent's behavior, actions and thoughts (Alhojailan, 2012). Furthermore, in comparison to other qualitative methods (e.g. grounded theory and hermeneutic phenomenology) thematic analysis needs a low level of interpretive complexity (Vaismoradi, Turunen & Bondas, 2013). These are the few reasons to use thematic analysis for data analysis and interpretation.

In order to conduct a thematic analysis, the study followed the six phase model proposed by Braun & Clarke (2006). According to the authors to make a meaningful interpretation from qualitative data, thematic analysis needs to be performed in six phases as mentioned below:

1. **Familiarizing with the Data:** This means the collected data needs to be read and re-read to become more and more familiar with the data, hence to notice key ideas in the collected data. In other words, it helps a researcher to identify, examine, list and note down underlying ideas to make meaningful sense from the collected data.

2. **Initial Code Generation:** It is the process of breaking data into distinct idea or events, labelling it with a name, and finally sorting and organizing the raw data into different concepts. It's a back-forth process that forms the basis for theme development.
3. **Searching for Themes:** This means categorizing codes into different themes and collecting relevant data on each identified theme. Here, themes are the patterned responses, something important to the data set in responding the particular research question.
4. **Reviewing of Themes:** This means the checking of themes in relation to the coded extracts and the entire dataset to generate a thematic map. Here, thematic maps are the display of attributes which are related to a specific topic, theme or subject area.
5. **Defining and Naming of Themes:** This means a comprehensive refinement of developed themes in understanding the collected data, generating clear definition (i.e. what does each theme mean? What does it represent? What ideas have been captured?) and developing names for each theme.
6. **Report Producing:** This is the final step in a thematic analysis where the identified themes are classified into explicit categories for final analysis, which are then interpreted in regard to research questions and supported with literature (previous research) to produce the final report.

Reliability and Validity: Why Intercoder Reliability?

To establish the reliability and validity, raw data were coded independently by the respective author of the present study, which was then brought to two external persons for discussion to find common agreement. This approach of establishing reliability and validity is often known as "intercoder reliability or inter-rater reliability" (Vaismoradi et al., 2013, p. 403) and has been common in practice among researchers (e.g. Carroll, Kaltenthaler, FitzGerald, Boland & Dickson, 2011; Thomas et al., 2004). However, peer checking of intercoder reliability has been subjected to criticism because "one researcher merely trains another to think as she or he does when looking at a fragment of text" (Vaismoradi et al., 2013, p. 403). Furthermore, some authors say that this may lead to the problem of objectivity, as all the intercoder may not look the data with similar subjectivity. Therefore, following the suggestion made by Miles & Huberman (1994) the validation of codes and themes was done in two stages: early and late stages of the data analysis. Also, Roberts et al., (2014) have suggested defining the dimensions for coding before searching for codes, even it is beneficial to draw some examples as a guide of reference. Therefore, discussion was held before and after during the process of coding and theme making. In both stages similar procedure was followed. According to Sykes (1990, 1991) in qualitative research the reliability and validity can be ensured by developing consistent and meaningful results and findings through a careful documentation of the cases. As the study was following an inductive approach, analytic themes were developed from the survey data. Approximately 60% of the developed codes and themes were identical between the author and external person, remaining codes and themes were determined by mutual agreement. Thus, the reliability and validity were established.

RESULT AND FINDINGS

Use of Strategic Planning Tools

In response to the use of strategic planning tools, the study found nine distinct tools that have been commonly used in strategic planning. They are: Strength-weakness-opportunities-threat (SWOT) analysis[1], Porter's five force model, Balance score card (BSC), Value chain analysis, Strategy canvas, Critical success factor analysis, Situation-task-action-result (STAR) analysis, Blue ocean strategy, and Boston consulting group (BCG) analysis. Somehow, this finding is consistent with the previous study made by Glaister and Falshaw (1999) in the United Kingdom, who found that what-if analysis, analysis of key success factors, financial analysis of competitors, SWOT analysis, and core capability analysis were the top strategic planning tools adopted by service and manufacturing companies in the United Kingdom. Likewise, a study conducted among SMEs in Western Australia, Singapore, Hong Kong, and Malaysia showed that SWOT, PEST, financial ratio analysis, and budgeting are the frequently used strategic planning tools (Frost, 2003). This shows that there is an increasing trend and/or regional difference in the use of planning tools in strategic planning. Table 1 summarizes currently used strategic planning tools by Finnish SMEs (current study sample).

As shown in the Table 1, majority of the respondent claimed that they use SWOT analysis as a strategic planning tool, followed by BSC, Porter's five force model, and Value chain analysis respectively. Few respondents outlined that they hire external consultants for guidance in strategic planning. However, seven respondents claimed that they do not use any of the strategic planning tools. According to Recklies (2008) in the process of strategic planning it is very important to:

1. Understand the business, organizational strategy and assumptions behind the strategy,
2. Have innovative ways of strategic planning,
3. Have organization specific strategic planning process i.e. "customized strategic planning.

Table 1. Use of strategic planning tools

Strategic Planning Tools	Number of SMEs Using This Tool
SWOT	18
Porter's five force model	3
BSC	4
Value chain analysis	3
Strategy canvas	1
Critical success factor analysis	1
STAR model	1
Blue ocean strategy	1
BCG analysis	1
No strategic planning tool at all	7

Source: Own presentation of author.

Furthermore, the author claims that these approaches help to ensure improvements on planning cycle, cultural context, strategic analysis, agenda setting, strategy finding, as well as strategy implementation and strategy execution" (Recklies, 2008, p. 5). Indeed, the organizational culture, type of business and the operating environment has a significant impact on the use of strategic planning tools and process. This might be the reason for diversification in the use of strategic planning tools.

Current Practice of Strategic Planning, Implementation, and Monitoring

Strategic planning is found to be a comprehensive action plan for organizational growth, which is based on systematic mapping of changes in the business environment and its impact on organization, supported with internal targets and means (resources). Likewise, strategic implementation and monitoring is found to be the act of execution of the chosen strategic plan and actions, supported with a control procedure which is equipped with procedures and explanation for not only to overcome what if scenario but also to motivate managers and employees through incentives. The process of strategic implementation and monitoring through constant feedback on employee and organizational performance helps in ensuring that the chosen strategic paths are towards the intended organizational goals and objectives. From the analysis of collected data, it was perceived that in practice it is very difficult to separate the process of strategic planning, implementation and monitoring as they go hand in hand. Strategic planning, implementation and monitoring are almost overlapping. In either case the respondents highlighted the importance of a strong commitment from both the management and employees.

There was found to be a lot of diversity in the state of current practice of strategic planning. For example, few respondents claimed that the strategy is being made by top managers and transferred to the lower level employee (top down approach) while others claimed that the process of strategic planning starts from lower level employee's opinion and finally approved by top managers (bottom up approach). In a similar manner the process of strategic planning was found to be both formal and non-formal. The process approach[1] of strategic planning was found to be the most common approach among the SMEs considered in this research. The time horizon of strategic planning was found to be in different range, for example, one year, three years, five years, three to five years, and even one of the respondents claimed for 10 year long plans. There found to be regular updates of strategic plans depending upon the changes in the business environment and organization's internal circumstances, usually updates are made two to four times a year. During the process of strategic planning, strategic seminars and workshops involving top management, lower level employees and network partners were also found to be common in the research group. This showed that strategies are the result of discussion between chief executive officer, entrepreneur, employee and network partner. However, to design an effective strategic plan (competitive strategy) it is important to know the customer's business in detail before the meetings and discussions; the focus should be towards the enhancement of customer's business rather than improving the features of product/service. The respondents assured that the customer details can be gained through critical factor analysis, i.e. customer analysis, market analysis, competitor analysis, company analysis, environment analysis. Besides these findings, key approaches of strategic planning were found to be as mentioned as below:

1. Target – Process – Audition – Inspection of the result – Approval/rejection of the strategic plan.
2. Idea – Target – Options – Strategic decision/plan.

3. Scenario planning – Cost effective strategies for resource allocation – What if analysis (situations) - Target – Process – Audition – Inspection of the result – Approval/rejection of the strategic plan.
4. Systematic mapping of changes in the business environment and its impact on organization – Strategic updates – New strategies and action plans.
5. Mirror the different scenarios against the current states of the organization – Project the ambition scenarios – Action plan.
6. Preliminary work with executives and external consultant – Internal assessment – Strategy in actions.
7. Internal assessment by chief executive and technology officer on a regular basis – Discussion with external co-operating partner – Action plan.
8. Executive group meeting followed by a team meeting (e.g. quality team) and personnel meeting with different groups – Advancement of strategy – Five year plan frameworks.
9. Annual planning – Detail discussion in a wide group – Drawing action plan with 2 to 4 priorities – Developing measurement tools for control and monitoring – Action plan.

Likewise, the process of strategic implementation and monitoring was also found to be diversified. The process of strategic implementation is mainly concerned with monitoring of daily organizational activities and projecting near future scenario, but not too far into the future. The board of director, chief executive officer and owner are in-charge of implementation and responsible for controlling the strategic actions and plans. It's an executive group work. On the other hand strategic implementation through the immediate supervisors of the unit was also found to be common in practice among the research group. Follow up plans (monitoring) in each stage of strategic implementation, a process approach was common practice in implementing and monitoring of strategic (organizational) activities. Likewise, for obtaining higher efficiency clear instructions are provided to the employee, and if information turns out to be imperfect, the employee has the authority to move to "Plan B" as much as it is possible and practical. However, information needs to be exchanged constantly between the employee and his/her direct supervisors. Monitoring activities are carried out in different ways. For example, every next week office meeting, monthly follow-up and annual evaluation or by the end of the project, even after a couple of years; therefore, strategy is not a separate piece from the operations but one thing being constantly perceived. Surprisingly, there found to be no clear monitoring and milestones in some cases, all what is being done is to serve the strategy. Besides these findings, key approaches of strategic implementation and monitoring were found to be as mentioned below:

1. Office meeting, every two weeks - Follow-up of the developed key performance indicator – Reporting to the board for feedback and comments.
2. Target – Release – Monitoring according to the strategy – Inspections of results – Re-evaluation of formulated strategies.
3. Management by objectives - Performance evaluations – Working models of the executive and steering group.
4. Monthly follow-up of the key strategic performance indicators - Collection of feedback – Correcting process and actions.
5. Collect constant feedback and information from the customers - Based on the feedback constantly go through the quality checking of product and organization's operations procedure – Make correcting decisions.

6. Every main goal needs to have a separate "group" and "a leader" - The executive group participates and monitors the strategies being implemented and succeeded.

Strategic Planning Barriers and their Mitigation

Barriers of strategic planning are the obstacles/constraints/hindrances that prevent effective strategic planning. Here, the effective strategic planning refers to the clear and concise strategy that clarifies mission, vision and organizational objectives. It mainly answers the four key questions: What to do? When to do? How to do? and Why to do? Likewise, the mitigation of strategic planning barriers refers to the actions that are taken in-order to eliminate or reduce the frequency and severity of different barriers in the process of strategic planning. The process or act of mitigation helps a strategic planner to reduce the impact of barriers in strategic planning. In practice there were found to be different kinds of barriers to strategic planning and ways of mitigation. Table 2 summarizes the identified barriers of strategic planning with examples of each type and their potential solutions in practice.

The Table 2, is made with careful evaluation of the respondent's answers, the barriers of strategic planning were categorized into seven distinct groups. The cited examples of each identified barrier and its potential solution is derived from responses. The research does not claim that the identified barriers of strategic planning represent all issues faced by the manager/strategic planner of Finnish SMEs. However, the research believes that it provides a good estimate of the problems faced by Finnish SMEs in the practice of strategic management.

Table 2. Barriers to strategic planning, examples and potential solution

Strategic planning Barriers	Examples to Strategic Planning Barriers	Potential Solutions to Strategic Planning Barriers in Practice
Lack of proper communication	• Unstructured planning process. • Limited involvement of lower level employee in strategic planning. • No appropriate methods for collecting feedback and comments.	• By employing more systematic strategic planning process. • By providing regular training and strategy workshop to employee. • By encouraging and providing opportunities for lower level employee to participate in strategic planning process.
Lack of time, planning expertise and skills	• Inefficient use of time in planning. • Incompetence of management. • Organizational tasks are not prioritized. • Lack of experienced strategic planner. • Unrealistic projections and milestones. • Weak implementation plan.	• By scheduling and prioritizing strategic planning to a higher level than the current level. • By employing organizational change and redistributing tasks among the employee. • By providing regular training and strategy workshop to employee. • By hiring an external expert into the work of the management or the board. • By developing network partners and co-operating activities.
Lack of sufficient finance	• The challenges of funding strategic change. • Limited capital. • Unwillingness of the management/ shareholder to increase the investment (because development of the markets is obscure in the target of 3-5 years).	• By negotiating and settling with banks, funding agencies and the shareholders; the government should ensure lending to the SMEs. • By increasing the funding and capital, believing in one's own vision and coping with the circumstances.

continued on following page

Table 2. Continued

Strategic planning Barriers	Examples to Strategic Planning Barriers	Potential Solutions to Strategic Planning Barriers in Practice
Environmental barriers	• Difficulty in predicting the market situation (e.g. customer demand, competitor's move). • Changing competitors, e.g. big companies buying the small companies. • The competing technologies are constantly changing in the market. • Too many variables need to be considered at the time of strategic planning. • Rapidly and constantly changing business environment. • Uncertainties related to the future (e.g. development of the industry, taxation). • Unavailability of reliable and trustworthy information to make decisions.	• By incorporating flexibility in strategic planning. • By making a specific adjustment to the core market, customer satisfaction and developing the service. • By creating clear vision and objective. • By strengthening the participation of the sales team in strategic planning. • Sometimes the decisions have to be made by instincts, even a bad decision is better than no decision at all i.e. one has to believe in own vision and experience. • Acquisition of foreign customers, weighting of the different business sectors also would be an advantage. • By developing a systematic process of strategic planning. • By encouraging government and authorities in making rapid decisions. • By developing network partners and co-operating activities.
Lack of industry knowledge	• Insufficient or lack of broader knowledge of the industry, e.g. recent development, current and future potential competitors, future movement and opportunities, history of industry sales (past, present and future expectations), marketing trend within the industry etc.	• By hiring an external expert into the work of the management or the board. • By developing network partners and co-operating activities.
Lack of long term vision	• The focus is too much in the presence. • Unclear ambition of the management/owner. • There is no vision in which a strategy can base on. • Narrow perspective, the expected result is too general and broad.	• By conducting open discussion between management, owner and employee. • By developing a clear vision based on what the company (and the owner) wants. • With an agile and continuing strategic review.
Too much bureaucracy	• The delays caused by permissions and bureaucracy. • Difficulties of getting stakeholders along. • Strict rules and regulations.	• The chain of command should be more personal and flexible. • By providing certain rights to operational and line managers in decision making or even to the lower level employee.

Source: Own presentation of author.

Strategic Implementation Barriers and their Mitigation

Barriers of strategic implementation are obstacles/constraints/hindrances that prevent efficient strategic implementation. Here, efficient strategic implementation refers to the situation where strategies can be implemented successfully and efficiently as planned. Strategic implementation is mainly concerned with putting planned strategies into actions, so that the desired mission, vision and organizational objectives can be achieved without interruption. As described in the case of strategic planning (the above section) the mitigation of strategic implementation barriers refers to actions that are taken in-order to eliminate or reduce the frequency and severity of different barriers in the process of strategic implementation. The process or act of mitigation helps a strategic planner to reduce the impact of barriers in strategic implementation. In practices there found to be different kinds of barriers to strategic implementation and ways of mitigation. Table 3 summarizes the identified barriers of strategic implementation with examples of each type and their potential solutions in practice.

Table 3. Barriers to strategic implementation, examples and potential solution

Strategic Implementation Barriers	Examples to Strategic Implementation Barriers	Potential Solutions to Strategic Implementation Barriers in Practice
Poor implementation plan	• The strategy remains as a loose dream, no clear milestones and roadmap. • The varying commitment from the distribution and network partner to strategic choice. • Strategy remains in the upper level, the management knows it better not communicated to operational level.	• By improving management practices. • By appointing the right person in the right job with the right skills. • By validating the rules of the game. • By providing regular training and strategic workshops.
Resistance to change	• The rigidity of the trade union. • Employee attitude, it has always been done this way. • Inefficient control of resources. • Limited skills and knowledge.	• By dismantling the rigidities of trade union through discussion. • By emphasizing and promoting internal communication. • By anticipating each employee in the process of strategic planning. • By motivating and encouraging employee in implementing strategy. • By providing regular training and strategic workshops.
Time and resource management	• Too many strategic implementations at the same time. • Neither enough resources, nor money and time (this also depends on the situation of business environment). • Limited knowledge of time and resource management. • Difficulty of finding the most appropriate and reliable network partner nearby.	• By adopting the principle of one thing at a time. • By hiring a qualified, experienced development manager possessing a higher education degree. The person should also be a bit salesman. • By providing regular training and strategic workshops. • By doing things in a smarter way. • By networking and expanding the number of business partners.
Financial	• Poor economic situation. • Lack of capital. • Limited financial resources and funding.	• Improving the result by rationalizing and prioritizing. • By getting capital from outside. • By attracting foreign customers from active industries. • By optimizing operating cost for better financial access.
Customer loyalty	• The customers' slippage to technology that is so called universally trendy.	• With strong marketing, so called teaching signaling/ messaging. • By persistency, never give up.
Competitors move	• Difficulty of taking into account all the variables of strategic planning beforehand. • Compulsive movements in real time.	• By strengthening the participation of the sales team/ department in the monitoring of the stakeholders' strategy. • By networking and expanding the amount of business partners.
Limited employee participation	• Lack of employee motivation. • Limited involvement of employee in strategic planning/building activities.	• By rewarding the employee for their achievement and good participation, a hard piece (justice) • By providing regular training and strategic workshops.

Source: Own presentation of author.

After a careful evaluation of the respondent's answers, the barriers of strategic implementation were categorized into seven distinct groups, as shown in Table 3. The cited examples of each identified barriers and potential solutions are derived from responses. The research does not claim that the identified barriers of strategic implementation represent all issues faced by the manager/strategic planner of Finnish SMEs. However, the research believes that it provides a good estimate of the problems faced by Finnish SMEs in the practice of strategic management.

Identified Competencies of a Good Strategic Planner

Strategic leader provides a clear direction for an organization and its employee. For a successful strategic planning, implementation and control, a strategic leader not only needs to micromanage strategies, but also needs to act as an umbrella under which the management develops appropriate strategies for organizational growth and success. Most importantly a strategic leader should be able to answer what, why and how questions in creating organizational mission, vision, and objectives. Strategic leaders are action takers who have the capability to make a choice between different alternatives and have the capability to motivate employees in reaching the organizational goals and objectives. In business practices, through strategies, a strategic leader acts as a connecting bridge between internal operations of an organization and external changing business environment. Therefore, a strategic leader needs to be opportunistic and curious in utilizing organizational resources and improve productivity. Likewise, awareness, responsibility, bureaucratic and leadership capability, decision making ability, initiative was found to be the competencies of a good strategic leader. Besides these competencies, there were found to be few more characteristics that make a strategic leader to be good. The Table 4 summarizes each group with examples derived from respondents answer.

As shown in Table 4, there found to be eight different competencies of a good strategic planner. These competencies were identified after a careful evaluation of the responses. However, these competencies are not the complete list to describe a good strategic planner, but provide a brief idea about the competencies held by a manager/strategic planner in Finnish SMEs.

DISCUSSION

Strategic planning, implementation and monitoring have been equally emphasized in literature as enablers of organizational success. Likewise, each of these processes have been defined and discussed separately, but in practice they are complementary in nature, meaning the processes of strategic planning, implementation and monitoring are mutually inclusive events which go side by side. However, for effective planning, efficient implementation and monitoring of strategies the competencies of a strategic leader count a lot. This study confirmed that Finnish SMEs do practice both the top down and the bottom up approach in the process of strategic planning, at least in the research sample. In comparison to the top down approach of strategic planning the bottom up approach might be beneficial not only in strategic planning but also in strategic implementation because it ensures greater participation and motivation to a lower level employee. After all, people working at the operational level are keys to efficient strategic implementation. In this vein, Rodomska (2014) has argued that the implementation of strategies fails with resistance from employees, as identified in this study. Therefore, the employee participation and open communication/discussion between higher and lower level employee has remained an important aspect of systematic strategic planning, implementation and monitoring. Through a study in Latin America Brenes, Mena, and Molina (2008) has confirmed that successful companies give priorities for systematic strategic planning. This means the systematic strategic approach in strategic management ensures a better understanding of the organization's mission, vision and objectives. This helps not only in effective strategic planning, but also at efficient implementation and monitoring of identified strategies. This is consistent with the argument made by Balogun & Johnson (2004). According to these authors

Table 4. Competencies of a good strategic planner and examples

Competencies of a Good Strategic Planner	Examples to Identified Competencies of a Good Strategic Planner
Business intelligence (entrepreneurship)	• Ability to understand the overall situation. • Knowledge of markets and economy. • Understanding of profitability and causality of business. • Realize the meaning of customer. • Experience from the industry. • Ability to understand the big picture, i.e. sees the forest from the trees. • Knowledge of target market and technology. • Knowledge of customer and the customer of the customers.
Creative	• Experience from the industry. • Thinking capability. • Ability to answer, what, why and how questions. • Ability to perceive what if situation. • Curiosity to use several perspectives and different tools of management.
Responsiveness	• Capability to analyze the company and surrounding society and react to the changes. • Ability to gain trust and develop relationship with different stake holders.
Courageous	• Ability to translate strategies into actions constantly, i.e. "the radar tuned on", means there is possible to fine-tune the strategy according to the circumstances. • Risk taking ability.
Patience	• Capability to listen the stories from employee, customer and different organizational stakeholders. • Concentration capability, thinking and planning ability.
Analytical capability	• Ability to understand human nature and ability to include stakeholders' commitment to the planning. • Ability to summarize the actions and plans.
Future orientation	• Ability to see the future, i.e. "the gift of a sight". • One must have vision about the direction of the industry. • There has to be a clear target from the owner that the strategy is defined for. • Result oriented, i.e. accurate in going through the process.
Analytical leadership skills	• Capability to evaluate present situation and view the future. • Realistic expectation for the future. • Ability to translate strategy into action (easy to dismount of strategy).

Source: Own presentation of author.

two way communication between top and lower level employee allows better interpretation, adaption and implementation of strategies at the operational level.

In the study sample it was found that majority of Finnish SMEs do make strategic plans for one year, three years, five years, and three to five years, and even one of the respondents claimed to have plans for 10 years. Likewise, the Finnish SMEs make the process of monitoring of strategic implementation in every two weeks, monthly, annually or by the end of the project. The current practice of strategic planning, implementation and monitoring in Finnish SMEs shows that strategy is not a separate piece from operations but these are things being perceived constantly. By examining the current practice of strategic management among Finnish SMEs, the study has identified not only the barriers of strategic planning, implementation and monitoring but also the potential ways to overcome those identified barriers have been presented (see Tables 2 and 3 for details). Likewise, the study has also highlighted the competencies of a good strategic planner (see Table 4 for details). The distinctive barriers of strategic planning, implementation and monitoring were similar to those identified in the previous researches (e.g. Al-Ghamdi, 1998; Beer & Eisenstat, 2000; O'Regan & Ghobadian, 2002; Nazemi, Asadi, & Asadi, 2015). Brenes et al. (2008) have also discussed factors leading to successful implementation of strategy which are similar to the findings of this study. The current practice of strategic planning, implementation and monitoring among Finnish SMEs found to be similar to the five step strategic planning process as discussed by Dye and Sibony (2007). According to the authors those five steps are: identify the strategic issues which might have potential impact on future performance, conduct the strategic discussion within the right group of people, adapt the planning cycle according to the needs, develop implementation and monitoring processes, and finally deploy right people in the right task. Likewise, the competencies of a good strategic planner were similar to those discussed in earlier research (e.g. Schoemaker & Krupp, 2015). However, there is some difference in the identified barriers in the present study and previous studies; this difference might be due to the fact that these studies have been conducted at different time frame, geographical location and industry structure.

MANAGERIAL IMPLICATIONS

From the point of competencies of a good strategic leader, the research could find implied possibilities to overcome the barriers of strategic planning, implementation and monitoring. In other words, by developing good competencies a strategic leader can reduce the barriers of strategic planning, implementation and monitoring beforehand. Likewise, the study concludes that strategic planning, implementation and monitoring are the operational function that guarantees better utilization of the firm's resources and capabilities which are being directed by the competencies that a strategic leader holds. The competencies of a strategic leader enable not only effective strategic planning, but also enable efficient implementation of strategies and monitoring, hence, the better utilization of the organization's resources and capabilities to reach the desired mission, vision and organizational objectives. However, for better utilization of resources, capabilities, and to overcome the internal and external environmental barriers it is necessary to follow some common principle in the process of strategic planning, implementation and monitoring. In this vein Beinhocker and Kaplan (2002) have emphasized redesigning the process of annual strategic planning so that the real time strategy can be made. Therefore, a framework has been presented which shows how past experiences and future expectations can be correlated through present actions, i.e. strategic planning, implementation and monitoring (see Figure 3). Indeed, the knowledge of how to deploy resources in strategic implementation leads to sustained strategic advantage by virtue of which the process of strategic implementation goes spontaneous (Barney, 2001).

In the sample of the study the author tried to understand how a firm establishes strategies, directions and action plans to reach the common goals and organizational objectives in practice. Also, the emphasis was given to the actions and practices conducted by managers in mitigating internal and external threats. After analyzing existing strategic management practices in the study sample, an extended strategic framework has been presented to overcome the barriers of planning, implementation and monitoring. The presented framework is entirely based on respondent's opinion and the author's understanding in the process of data analysis (see section: current practice of strategic planning, implementation and monitoring). However, the framework is believed to provide better strategic choice, so that the firm becomes more adaptive to changing internal and external operating environment. Either one way or another the majority of respondent's view of strategic planning, implementation and monitoring is closely related to plan-do-check-act (PDAC) cycle (also called Deming's cycle), which is a strategic management concept. The planning phases are related to collections and analysis of data and information from past experience and converting it into present actions to identify the areas of competitive advantage through strategic

Figure 3. Framework for effective and/or efficient strategic planning, implementation and monitoring
Source: *Own presentation of author.*

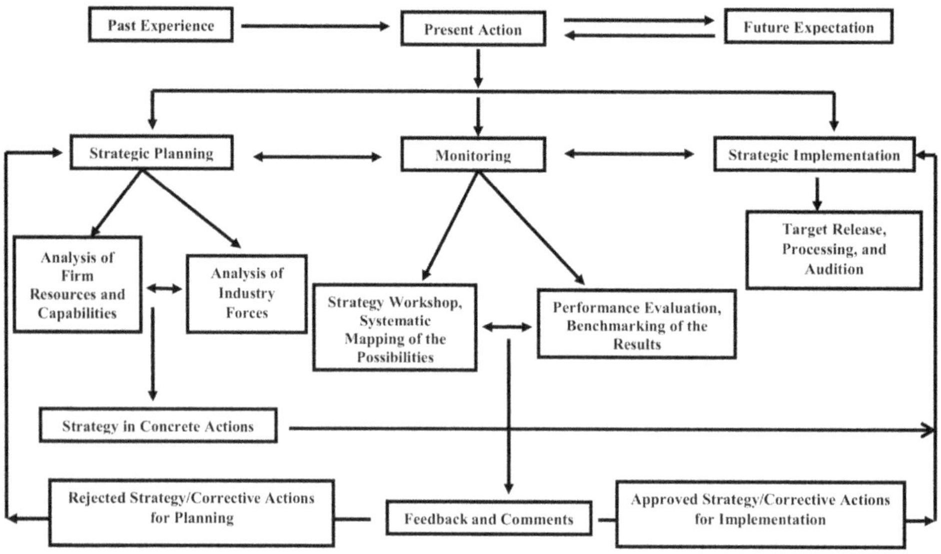

actions and plans. Based on the firm's capabilities and industry forces a firm should develop different alternatives, so that the strategies can be matched with the present needs and future expectations. The strategic planning can be monitored either through a strategic workshop or systematic mapping of the different alternatives (see Figure 3). In the "*do*" phase developed strategies are taken into practice, representing strategic implementation in the framework; here the monitoring can be carried out by evaluating performance and benchmarking of the results. However, there should be close co-ordination between the process of monitoring both in planning and implementation phase as shown in the framework (see Figure 3). This is due to the fact that in practice it's really difficult to determine when the planning ends and the process of strategic implementation starts. Check and act phase represents the process of monitoring where results are collected and matched with the action plan so that the future corrective actions can be developed, if necessary. Feedback and comments gained through the process of monitoring can be one of the most effective ways of overcoming the barriers of strategic planning and implementation because it helps to map present actions and future expectation based on past experiences. However, the future expectation signals a strategic planner in designing an alternative course of actions and decisions. A similar thought has been proposed by Gates (2010, p. xi) and says projection of "future scenarios allow organizations to explore multiple potential futures and generate robust strategies and early warning signs to understand how the future is unfolding".

The RBV asserts that firm's strategies are the result of the assets base of the firm and its competitors while Porters five forces asserts that firm's strategies are the result of constraints arising from broader industry and public policy environment. Likewise, the turbulence in the business environment is the measure of resource transfer between different stakeholders acting within the firm's operating environment. In business practices, barriers in strategic planning, implementation and monitoring are the result of imbalance between external industry forces, the organizational resources and capabilities. In a similar manner, by analyzing firm's resources, capabilities and industry forces, a strategic leader can prepare their minds for real time strategy making as argued by Beinhocker & Kaplan (2002). This shows that the firm's resources and industry forces are the two key parameters in synthesizing the firm's strategy (see Figure 3) in real time. Here, the real time strategy is defined as the act of making purposeful choices in identifying and exploiting a niche in the market through present actions and reach future goals, guided by past experiences, supported with organizational resources and capabilities. Based on the proposed framework, the study suggests that the process of strategic planning, implementation and monitoring are guided mainly by three factors: past experiences, present actions and future expectations. Here, the argument is that there is one-way communication between past experience and present actions while there is two way communications between present actions and future expectations in the process of strategic planning, implementation and monitoring (see Figure 3). This is because of two reasons. First, the past experience guides/influences/impacts the present action taken by a firm and present action guides the future expectations. Second, to reach the organizational objectives the present actions need to be modified or altered according to the changing business environment (e.g. competitors move, internal situation, changes in the business landscape, changes in tax policies, and rules and regulations from the government). This is in line with the argument made by Fitzsimmons (2006), according to the author for better understanding and wiser judgement in turbulent times a strategist should be able to make an assessment of uncertainty and potential risk, because uncertainty in business environment provides both opportunities and threats (Takala & Usitalo, 2012). Furthermore, Fitzsimmons (2006, p. 135) argues that "intuition and judgment of decision makers will always be vital to strategy" as it helps in choosing right alternatives from different options. In this study it is called competencies of a good strategic leader (see Table 4). Thus, by presenting a framework the study has tried to help managers and strategic planners to overcome some of the practical issues of strategic planning, implementation and monitoring.

FUTURE RESEARCH DIRECTIONS

Although many scholars have tried to identify the barriers of strategic planning, implementation and monitoring, few scholars have paid attention to the ways of mitigating those barriers. Also, there have been few attempts to understand the competencies of a good strategic leader from a practice point of view. This study has tried to address these issues. However, the research was limited to the qualitative study considering only 36 SMEs. Hence, the results and findings cannot be generalized, though it provides meaningful results and findings for both strategic practitioners and academic researchers. Therefore, future studies considering a larger sample size, comprising a wider group of industries has been recommended, so that the results and findings of the present study can be compared, generalized, and validated. Additionally, to avoid the tautology in research finding and refine the concept developed in this study, future research could adopt a longitudinal case considering past experience, present actions and future expectations as shown in the framework above (see Figure 3).

CONCLUSION

It is very important for a manager and strategic leader to know the barriers of strategic planning, implementation and monitoring. This will help to improve organizational performance. During the turbulent business environment the strategy needs to be updated all the time, which makes it very challenging to maintain effective and/or efficient strategic planning, implementation and monitoring. Therefore, to understand the barriers of strategic planning, implementation and monitoring, this study has highlighted the current practice of strategic planning, implementation and monitoring processes followed by Finnish SMEs. From a practical view point potential solutions in overcoming the identified barriers have been proposed. The research believes that results and findings of this study will help managers of SMEs and strategic planners in better practice of strategic management, hence to improve organizational performance in a changing business environment. Likewise, the competencies of a strategic leader or manager count a lot in the process of strategic management. In this regards the present study has identified a few competencies of a good strategic planner. By developing the identified competencies, a strategic leader or manager can enhance their practice of strategic management. In addition, the proposed framework provides a practical guideline in strategic planning, implementation and monitoring, especially to SMEs. However, internal assessment of resources and capabilities, and its systematic mapping with the external business environment is crucial not only for better strategic planning, but also very important for the successful implementation of developed strategies.

REFERENCES

Al-Ghamdi, S. M. (1998). Obstacles to successful implementation of strategic decisions: The British experience. *European Business Review*, *98*(6), 322–327. doi:10.1108/09555349810241590

Alhojailan, M. I. (2012). Thematic analysis: A critical review of its process and evaluation. *West East Journal of Social Sciences*, *1*(1), 39–47.

Azevedo, C. L., Almeida, J. P. A., van Sinderen, M., & Pires, L. F. (2015). Towards capturing strategic planning in EA. In *Enterprise Distributed Object Computing Conference (EDOC), 2015 IEEE 19th International*. doi:10.1109/EDOC.2015.31

Balogun, J., & Johnson, G. (2004). Organizational restructuring and middle manager sensemaking. *Academy of Management Journal*, *47*(4), 523–549. doi:10.2307/20159600

Barney, J. (1991). Firm resources and sustained competitive advantage. *Journal of Management*, *17*(1), 99–120. doi:10.1177/014920639101700108

Barney, J. B. (2001). Is the resource-based "view" a useful perspective for strategic management research? Yes. *Academy of Management Review*, *26*(1), 41–56.

Beer, M., & Eisenstat, R. A. (2000). The silent killers of strategy implementation and learning. *Sloan Management Review*, *41*(4), 29–40.

Beinhocker, E. D., & Kaplan, S. (2002). Tired of strategic planning. *The McKinsey Quarterly*, *2*, 48–57.

Braun, V., & Clarke, V. (2006). Using thematic analysis in psychology. *Qualitative Research in Psychology*, *3*(2), 77–101. doi:10.1191/1478088706qp063oa

Brenes, E. R., Mena, M., & Molina, G. E. (2008). Key success factors for strategy implementation in Latin America. *Journal of Business Research*, *61*(6), 590–598. doi:10.1016/j.jbusres.2007.06.033

Bryson, J. M. (1988). A strategic planning process for public and non-profit organizations. *Long Range Planning*, *21*(1), 73–81. doi:10.1016/0024-6301(88)90061-1

Cândido, C. J., & Santos, S. P. (2015). Strategy implementation: What is the failure rate? *Journal of Management & Organization*, *21*(02), 237–262. doi:10.1017/jmo.2014.77

Carroll, C., Kaltenthaler, E., FitzGerald, P., Boland, A., & Dickson, R. (2011). A thematic analysis of the strengths and weaknesses of manufacturers submissions to the NICE Single Technology Assessment (STA) process. *Health Policy (Amsterdam)*, *102*(2), 136–144. doi:10.1016/j.healthpol.2011.06.002 PMID:21763025

Drnevich, P. L., & Kriauciunas, A. P. (2011). Clarifying the conditions and limits of the contributions of ordinary and dynamic capabilities to relative firm performance. *Strategic Management Journal*, *32*(3), 254–279. doi:10.1002/smj.882

Drucker, P. F. (1954). *The practice of management: A study of the most important function in American society*. New York, NY: Harper & Brothers.

Dye, R., & Sibony, O. (2007). How to improve strategic planning. *The McKinsey Quarterly*, *3*, 40–49.

Emery, F. E., & Trist, E. L. (1965). The causal texture of organizational environments. *Human Relations*, *18*(1), 21–32. doi:10.1177/001872676501800103

Feurer, R., & Chaharbaghi, K. (1995). Strategy development: Past, present and future. *Management Decision*, *33*(6), 11–21. doi:10.1108/00251749510087614

Fitzsimmons, M. (2006). The problem of uncertainty in strategic planning. *Survival*, *48*(4), 131–146. doi:10.1080/00396330601062808

Franken, A., Edwards, C., & Lambert, R. (2009). Executing strategic change: Understanding the critical management elements that lead to success. *California Management Review*, *51*(3), 49–73. doi:10.2307/41166493

Frost, F. A. (2003). The use of strategic tools by small and medium-sized enterprises: An Australasian study. *Strategic Change*, *12*(1), 49–62. doi:10.1002/jsc.607

Gates, L. P. (2010). *Strategic planning with critical success factors and future scenarios: An integrated strategic planning framework* (Technical Report No. CMU/SEI-2010-TR-037). Carnegie Mellon University. Retrieved September 1, 2016 from Software Engineering Institute web site: http://resources.sei.cmu.edu/library/asset-view.cfm?assetID=9673

Glaister, K. W., & Falshaw, J. R. (1999). Strategic planning: Still going strong? *Long Range Planning*, *32*(1), 107–116. doi:10.1016/S0024-6301(98)00131-9

Grant, R. M. (1991). The resource-based theory of competitive advantage: Implications for strategy formulation. *California Management Review*, *33*(3), 114–135. doi:10.2307/41166664

Higgs, M., & Rowland, D. (2005). All changes great and small: Exploring approaches to change and its leadership. *Journal of Change Management*, *5*(2), 121–151. doi:10.1080/14697010500082902

Jackson, K. M., & Trochim, W. M. (2002). Concept mapping as an alternative approach for the analysis of open-ended survey responses. *Organizational Research Methods*, *5*(4), 307–336. doi:10.1177/109442802237114

Mahoney, J. T., & Pandian, J. R. (1992). The resource-based view within the conversation of strategic management. *Strategic Management Journal*, *13*(5), 363–380. doi:10.1002/smj.4250130505

Mankins, M. C., & Steele, R. (2005). Turning great strategy into great performance. *Harvard Business Review*, *83*(7/8), 64–72. PMID:16028817

Miles, M. B., & Huberman, A. M. (1994). *Qualitative data analysis: An expanded sourcebook*. SAGE Publications.

Mintzberg, H., & Waters, J. A. (1985). Of strategies, deliberate and emergent. *Strategic Management Journal*, *6*(3), 257–272. doi:10.1002/smj.4250060306

Mintzberg, H., Ahlstrand, B., & Lampel, J. B. (2009). *Strategy safari: Your complete guide through the wilds of strategic management*. Edinburg Gate, UK: Pearson Education Limited.

Moon, H., & Ruona, W. (2015). Towards a deeper understanding of strategic learning. *Leadership and Organization Development Journal*, *36*(6), 657–674. doi:10.1108/LODJ-10-2013-0139

Naujoks, H. (2010). Strategic management in turbulent times. In L. Stadtler, A. Schmitt, P. Klarner, & T. Straub (Eds.), *More than Bricks in the Wall: Organizational Perspectives for Sustainable Success* (pp. 104–111). Springer Gabler. doi:10.1007/978-3-8349-8945-1_11

Nazemi, S., Asadi, S. T., & Asadi, S. T. (2015). Barriers to strategic planning implementation; case of: Mashhad electricity distribution company. *Procedia: Social and Behavioral Sciences*, *207*, 2–9. doi:10.1016/j.sbspro.2015.10.142

ORegan, N., & Ghobadian, A. (2002). Effective strategic planning in small and medium sized firms. *Management Decision*, *40*(7), 663–671. doi:10.1108/00251740210438490

Peteraf, M. A. (1993). The cornerstones of competitive advantage: A resource-based view. *Strategic Management Journal*, *14*(3), 179–191. doi:10.1002/smj.4250140303

Peteraf, M. A., & Bergen, M. E. (2003). Scanning dynamic competitive landscapes: A market-based and resource-based framework. *Strategic Management Journal*, *24*(10), 1027–1041. doi:10.1002/smj.325

Popping, R. (2015). Analyzing open-ended questions by means of text analysis procedures. *Bulletin of Sociological Methodology. Bulletin de Methodologie Sociologique*, *128*(1), 23–39. doi:10.1177/0759106315597389

Porter, M. E. (1985). *Competitive advantage: Creating and sustaining superior performance*. New York, NY: FreePress.

Radomska, J. (2014). Linking the main obstacles to the strategy implementation with the companys performance. *Procedia: Social and Behavioral Sciences*, *150*, 263–270. doi:10.1016/j.sbspro.2014.09.054

Raduan, C. R., Jegak, U., Haslinda, A., & Alimin, I. I. (2009). Management, strategic management theories and the linkage with organizational competitive advantage from the resource-based view. *European Journal of Soil Science*, *11*(3), 402–418.

Ralston, P. M., Blackhurst, J., Cantor, D. E., & Crum, M. R. (2015). A structure–conduct–performance perspective of how strategic supply chain integration affects firm performance. *Journal of Supply Chain Management*, *51*(2), 47–64. doi:10.1111/jscm.12064

Raps, A. (2005). Strategy implementation – an insurmountable obstacle? In Handbook of Business Strategy. Emerald Group Publishing Limited.

Recklies, O. (2008). Problems and barriers to strategic planning. *Economics and Organization of Future Enterprise*, *1*(1), 3–11. doi:10.2478/v10061-008-0001-2

Roberts, M. E., Stewart, B. M., Tingley, D., Lucas, C., Leder-Luis, J., Gadarian, S. K., & Rand, D. G. (2014). Structural topic models for open-ended survey responses. *American Journal of Political Science*, *58*(4), 1064–1082. doi:10.1111/ajps.12103

Schoemaker, P. J., & Krupp, S. (2015). Overcoming barriers to integrating strategy and leadership. *Strategy and Leadership*, *43*(2), 23–32. doi:10.1108/SL-01-2015-0001

Scholes, K., & Johnson, G. (2005). *Exploring corporate strategy*. Edinburg Gate, UK: Prentice Hall International.

Sirmon, D. G., Hitt, M. A., & Ireland, R. D. (2007). Managing firm resources in dynamic environments to create value: Looking inside the black box. *Academy of Management Review*, *32*(1), 273–292. doi:10.5465/AMR.2007.23466005

Spanos, Y. E., & Lioukas, S. (2001). An examination into the causal logic of rent generation: Contrasting Porters competitive strategy framework and the resource-based perspective. *Strategic Management Journal*, *22*(10), 907–934. doi:10.1002/smj.174

Sykes, W. (1990). Validity and reliability in qualitative market research: A review of the literature. *Journal of the Market Research Society*, *32*(3), 289–328.

Sykes, W. (1991). Taking stock: Issues from the literature on validity and reliability in qualitative research. *Journal of the Market Research Society*, *33*(1), 3–12.

Takala, J., & Usitalo, T. (2012). Resilient and proactive utilization of opportunities and uncertainties in service business.*Proceedings of the University ofVaasa, Finland.* University of Vaasa.

Thomas, J., Harden, A., Oakley, A., Oliver, S., Sutcliffe, K., Rees, R., & Kavanagh, J. (2004). Integrating qualitative research with trials in systematic reviews. *BMJ (Clinical Research Ed.)*, *328*(7446), 1010–1012. doi:10.1136/bmj.328.7446.1010 PMID:15105329

Vaismoradi, M., Turunen, H., & Bondas, T. (2013). Content analysis and thematic analysis: Implications for conducting a qualitative descriptive study. *Nursing & Health Sciences*, *15*(3), 398–405. doi:10.1111/nhs.12048 PMID:23480423

Waterman, R. H. Jr, Peters, T. J., & Phillips, J. R. (1980). Structure is not organization. *Business Horizons*, *23*(3), 14–26. doi:10.1016/0007-6813(80)90027-0

ADDITIONAL READING

Aaltonen, P., & Ikävalko, H. (2002). Implementing strategies successfully. *Integrated Manufacturing Systems*, *13*(6), 415–418. doi:10.1108/09576060210436669

Dincer, O., Tatoglu, E., & Glaister, K. W. (2006). The strategic planning process: Evidence from Turkish firms. *Management Research News*, *29*(4), 206–219. doi:10.1108/1409170610665068

Okumus, F. (2003). A framework to implement strategies in organizations. *Management Decision*, *41*(9), 871–882. doi:10.1108/00251740310499555

Powell, T. C. (1992). Strategic planning as competitive advantage. *Strategic Management Journal*, *13*(7), 551–558. doi:10.1002/smj.4250130707

KEY TERMS AND DEFINITIONS

Barriers of Strategic Planning and Implementation: Obstacles/constraints/hindrances preventing effective strategic planning and implementation.

Environmental Turbulence: The measure of resource transfer between different stakeholders acting within a firm's operating environment.

Inductive Approach: A research approach which begins with the observation followed by a pattern analysis and finally a theory or result and findings are produced.

Industry Organization (IO) Theory: A theory of the firm precisely concerned with the opportunities and threats' streaming from the environment. It asserts that the industry forces in which a firm operates are very important for a firm's growth and survival.

Open Ended Survey Questions: An unstructured but clear question that provides better opportunities for respondents to express their views and opinion on the subject matter. It also provides greater anonymity to respondents and often elicits more honest and clear responses.

Resource Based View (RBV): A theory of the firm which claims that a firm's competitiveness is the result of the firm's capabilities in mobilizing organizational resources. RBV asserts that firm's strategies are the result of the asset base of the firm and its competitors and provides guidelines for organizations to gain competitive advantage.

Strategic Implementation: An act of execution of the chosen strategic plan and actions, supported with a control procedure. It mainly concerned with putting planned strategies into actions, so that the desired mission, vision and organizational objectives can be achieved without interruption.

Strategic Planning: Act of developing clear and concise strategy that clarifies mission, vision and organizational objectives by answering four key questions: What to do? When to do? How to do? and Why to do?

Thematic Analysis: A qualitative research method for identifying, detecting and analyzing factors or variables that influence respondent's behavior, actions and thoughts in regards to a particular course of actions. Thus, it helps in developing and reporting themes/pattern analysis within the collected data.

ENDNOTES

1 Editors' note: See chapter 16 for a review of instruments and tools of the situational analysis including SWOT and compare with the ideas suggested in this chapter.

Chapter 12
Knowledge Management Optimization through IT and E-Business Utilization:
A Qualitative Study on Serbian SMEs

Aleksandar M. Damnjanović
Higher School of Academic Studies "DOSITEJ", Serbia

ABSTRACT

This chapter, which is the first part of fundamental research related to optimization of SMEs knowledge management, elaborates relevant key optimization criteria. First section deals with optimizing electronic forms of communication, web site material, management effectiveness, customer management and procedures for electronic responses, channels of communication, SME cooperation mechanisms. In the second part analyzed is knowledge management optimization, modernization of SMEs, the level of use of modern information, development of employees and partners, financing the development of IT employees, the budget for information technology, innovative knowledge management in budget planning, the benefits of electronic payment cards, communication of employees and management and IT innovation. Exploration of these questions further examines impact of new technologies, IT in optimizing the development of competitive capacity and advantages from SME technological inputs and outputs. Finally, the chapter offers appropriate conclusions and recommendations.

INTRODUCTION

At the EU - Western Balkans Summit held in mid-2003 in Thessaloniki, the various Western Balkans countries (Albania, Bosnia and Herzegovina, the Former Yugoslav Republic of Macedonia, Kosovo, Serbia and Montenegro) formally adopted the European Charter for Small Enterprises. This decision was made within the framework of the Lisbon agenda to improve cooperation between the member states of the European Union and the countries of the Western Balkans (Eur-Lex, 2016). It reflects the commitment of these countries to small and medium enterprises (SME) as a major source of new jobs,

DOI: 10.4018/978-1-5225-1949-2.ch012

innovation, economic dynamics and social inclusion. Since its adoption the Charter has become the main platform for pursuing optimal policies for the development of small enterprises and entrepreneurship in the whole region (Ožegović & Sajfert, 2009).

In addition to the European Charter, in 2010 the European Commission also adopted the Strategy for smart, sustainable and inclusive economic growth – or briefly Europe 2020, (European Commission, 2010). This strategy builds on the Lisbon Strategy, which defines the main principles for formulating an optimal policy for the development of SMEs. All objectives of the strategy are aimed at increasing employment, strengthening research and innovation, education, reduction of greenhouse gas emissions and boosting energy efficiency and fighting poverty. The strategy defines these three main operational objectives:

1. Smart growth implying optimization of economy based on knowledge and innovation;
2. Sustainable growth implying stimulation of optimized economic competitiveness, efficient use of resources, while taking into consideration the conditions of globalization;
3. Inclusive growth fostering growth of employment and education, social protection systems and combating poverty.

From 2009 onward, monitoring of the implementation of this strategy and the progress of SMEs reform policies in the Western Balkans has been effectively carried out. The economic environment in Western Balkan countries is therefore based on the progress toward market economies and private sector development. However, besides some minor improvements of economic trends in the region, the trend of negative business development continues in most of these countries, which requires further efforts towards optimizing key factors of production.

According to leading economists and relevant researches cited in the chapter, the three reform areas that can assist the Western Balkan countries and allow them to quickly overcome the crisis, are as follows: continued fiscal consolidation, improvement of business climate and optimization of knowledge and skills necessary for economic development. These three reform areas can contribute significantly to economic growth in the Western Balkans. Fiscal issues and business climate are once again emphasized very extensively in the June 2016 issue of the World Bank Group's Global Economic Prospects (World Bank, 2016). Consequently, this chapter explores the significance of the third pillar as knowledge management optimization through IT and e-business utilization. This factor is analyzed in the context of competitiveness.

Within this framework optimization of competitiveness through IT and electronic business is considered among the most important issues in many countries and companies (Njeguš, 2009). The struggle for competitiveness is not just between companies on a national and regional level but also between countries at international level. Measuring the competitiveness of the national economy is of particular importance, given that it contributes to the advancement of business and overall economic growth (Milošević, 2014). If competitiveness in the global market is poor, then the national economy suffers, leading to protectionism, non-transparent government subsidies and barriers to enterprise development (Jokić, 2010). Optimization through IT and e-business is an effort to counter such trends (Jevtić et al., 2011) which motivates the author to explore this phenomenon deeper.

Since 2005 the World Economic Forum has based its analysis of the competitiveness on the Global Competitiveness Index (World Economic Forum, 2016). This is a very comprehensive index that includes micro and macroeconomic foundations of national competitiveness. In order to rank the national economy, this methodology takes into account the key factors and systematizes them in institutions, infrastructure, macroeconomic environment, health, primary education and higher education, vocational training, goods market efficiency, labor market efficiency, financial market development, technological readiness, market size, business sophistication and innovation, which are all the key optimization factors in IT and e-business.

Comparing the Balkan countries with US and developed European countries in terms of optimizing competitiveness, it is evident that they are lagging far behind in reducing the existing difference compared to these countries. This could have negative implications in terms of approaching EU membership and participation in European and world integration trends. In order to reduce this lag, it is necessary for national economic policies in the Balkans to reduce high operating costs and tax levies, huge public debt, high unemployment, resolve difficulties in the liquidation of enterprises and create a positive business climate that would optimize competitiveness and innovation in the global market.

The Internet as the largest and most popular global computer network, also known as an information highway intended for data exchange, is underutilized in the Balkan countries for optimization of competitiveness (Šarac, Jevremović, & Radovanović, 2015). As it will be elaborated later the author advocates establishing computer networks between all relevant scientific and business institutions, the use of e-business and computers as well as technology of packet transmission for the mutual exchange of information (Živadinović, 2010). This technology allows the sharing of information in smaller packets and then sending them to specific computers. The Balkan countries could therefore advance in the optimization of business competitiveness with the application of this technology. This will allow connections and exchange of data between different business computer networks. Closed computer network of dedicated professionals grows into open-ended networks in order to optimize competitiveness of enterprises. On a commercial basis internet facilities thus became a factor in optimizing competitiveness of enterprises. The Internet is thus becoming a network of a large number of networked computers around the world that are interconnected by optic cables, telephone lines, satellite links and other forms of communication connection, which operate on the basis of specific optimized business protocols. This network enables seamless two-way communication between all participants in the optimized business information traffic (OECD, 2004).

In the context of research in this chapter Internet is a computer network infrastructure that is built by a specific standard used by all users, connected to each other by using *TransmissionControlProtocol/ InternetProtocol (*TCP/IP), as well as individual business networks that form a single logical network in which all share a common addressing scheme. In today's information society the Internet is a global communications medium (Milosavljević, 2013) and as such it is a factor optimization in terms of the author's research. Networks within and outside the company have their own services for business optimization. Business services are a set of functions packaged as a single entity and set up somewhere on the Internet to serve *ad hoc* other optimized programs and networks. The competitiveness of services and optimization in this regard can be grouped into: public services, basic business services, commercial search services, security services and system business services.

Web is one of the most important Internet platforms for business optimization. It allows access to multimedia and distributed databases in developing competitiveness and optimization of the company management functions (Ranđić & Lekić, 2009). Using text, graphics, multimedia, and related files, it forms a vast network of easily accessible data. In order to review and study the issues of competitiveness and optimization of business operations with the use of IT and e-business, the chapter will study, analyze and elaborate on the following key criteria formulated from research questions: Optimizing electronic forms of communication and web site, the degree of management effectiveness (Mašić, 2010), optimizing customer management (Damnjanović & Milisavljević, 2009) and related web site material, the development of management systems and procedures for electronic responses and appeals, channels of internal communication, cooperation mechanisms between organizational units of the company, optimizing knowledge management, modernization of enterprises and employees. The author will further analyze the level of use of modern information, development of employees and partners, financing and budget for development of employees and technology in the information sector, financial investments in information technology, innovative knowledge management in budget planning, the benefits of electronic payment cards, communication of employees and company management, innovation in the field of information technology, progress arising from the use of new technologies, the level of commitment of management to introduce new information technologies and employee motivation to manage new technologies.

Exploration of these criteria through desktop and research questions for empirical research covers in detail the role and impact of new technologies, especially information technology in optimizing the competitive advantages (Wang, 2016) of small and medium-sized enterprises. Research eventually assesses the degree of optimization of technological inputs and outputs in the development of competitive capacity of SMEs and offers appropriate conclusions.

BRIEF LITERATURE REVIEW

Besides already cited sources, academic literature comprises a whole range of different approaches to the topic of SME development and to the impact of modern technologies on strategic competitiveness whereby each approach reveals important aspects of the issues (Bešić, Nikolić, & Damnjanović, 2010). Scientific literature differs in volume, methods of analyses and interpretation. Many scientists deal with impact of entrepreneurship on economic growth. They especially follow impact of highly-developed enterprises on economic growth and impact of the newly-founded companies – i.e. entrepreneurial activities in founding (Reynolds, 2000). On the other hand, a number of researches are oriented towards the significance of environmental factors which have a negative impact on the development of companies. For example, a similar analysis was conducted in eighty countries, aiming to identify the limitation factors for companies (Ayyagari, Demirgüç-Kunt & Maksimovic, 2006). Research results of this particular analysis conclude that lack of financial resources and political instability of the country directly influences growth of companies, while fiscal policy and legislation are not as much important. Literature also deals with the creation of business environment (Vemić, 2008) which provides business facilities and guarantees to the entrepreneurs and financiers for fulfillment of their rights and contracts (Beck, Demirgüc-Kunt, & Maksimovic, 2008). These analyses of business environment show, among other things, that less bureaucratized business environment works in favor of the economic growth (Altenburg & Drahenfels, 2006).

EMPIRICAL DATA ON KNOWLEDGE MANAGEMENT OPTIMIZATION THROUGH IT AND E-BUSINESS UTILIZATION

Main Subject, Starting Points, and Objectives of the Research

Subject of research in this chapter is application of information technology (IT) and E-business in optimization of competitive advantage of SMEs. Fast changes in places where SMEs conduct their activities lead to the constant finding of answers in relation to challenges that they are exposed to (Bešić, Sajfert, & Damnjanovic, 2009). A solution to this question is found in the advancement of strategic competitive features, through application of IT and E-business.

New business forms are made due to the quick development of information and communication technologies (ICT), (Veljović, Vulović, & Damnjanović, 2009), which become basis of modern business. In order to survive at the market, SMEs must conform to the new challenges and rules imposed by modern political, social, economic and technological environment. Because of the competition challenges, advantages of modern IT (Obradović et al., 2011) must be recognized by top management.

On the basis of previous elaboration the main objective of the chapter is to examine application of IT and E-business in optimization of competitive advantage of SMEs for improvement of overall performances of organization.

Target group treated in this chapter involves SMEs which represent a significant part of the Serbian economy and have the greatest impact on the development of this economy.

Research starts with a hypothesis that with application of IT (Njeguš, 2010) and E-business optimization SMEs can significantly create and improve competitive advantage through e-commerce skills provided in learning centers, web trade knowledge and skills as was previously accomplished in the project implemented by the Central Ostrobothnia Polytechnic between the 1st of September 2000 and the 11th of November 2002 (European Commission, 2004).

Main Hypothesis of the Research

Through theoretical, empirical and desk research, hypothesis set in this work is related to testing the impact of ICT and E-business on competitive advantage of SMEs.

Main Hypothesis: Optimization of SMEs business through IT and E-business had a positive impact on competitiveness in the researched period.

RESEARCH METHODOLOGY

Methodological-Theoretical Framework of the Research

Statistical analysis of basic and control variables is used in this chapter for testing the hypothesis and results of empirical research. With regard to theoretical worthiness of statistical methods used in the research, methods involved three basic approaches of the author: *descriptive-statistical analysis* of the observed data, *statistical testing* of dependency of variables that are considered as important for overall innovative and technological performances and export motivation of SMEs on the one hand, and in relation to the size of an enterprise on the other hand. *Pearson's test of independence* for modalities of two signs in the sample set is also used. For the formal testing of all hypotheses model of *binary logistic regression* was used, since the dependent variable is of a dichotomous type.

The basic method of *descriptive-statistical analysis* of the observed data represents the first step. In this way the author will define the so-called dominant characteristics of the observed set, i.e. calculate the basic parameters, the mean value and standard deviation for each of the hypotheses (variables) within the sample.

In the next phase of the statistical analysis of the observed data the author will perform *statistical testing* of dependable variables, derived on the basis of individual responses by the SME respondents on the one hand, and in relation to the size of an enterprise on the other hand. In this case the author uses the well-known *Pearson's X2 test for independence* for modalities of two signs in the sample set. Number of methods for statistical testing - based on the so-called Pearson's χ^2 statistics - represents one of the oldest methods for statistical conclusions. However - as the readers will observe in the research – this group of statistical tests is widely applied today in the statistical practice and various fields of social sciences.

The basic unit for statistical analysis is each individual enterprise, and the author will realize the X^2 independence test on the basis of positive answers - *from the sample including 137 SMEs* – on the suitable survey questions, i.e. total number of positive values of the observed variables. In case that so-called *Zero hypothesis* (H_0) for independence of the observed signs is true, i.e. when values of the observed variables are not dependent on the size of SMEs, then the expected variable frequencies related to each individual hypothesis (from 1A to 3B) will be proportional to the frequencies derived from sample stratification and based on the size of the company.

For that purpose, if f_i' denotes the expected frequencies, and f_i empirical ones, i.e. total number of positive answers per each variable – then, the realized value of so-called X^2 statistics is obtained through standard statistical procedure, by the following pattern:

$$\chi^2 = \sum_{i=1}^{r} \frac{\left(f_i - f_i'\right)^2}{f_i'}, \qquad (1)$$

where f_i and f_i' represent empirical and expected frequencies; r – represents total number of different modalities (number of variables, i.e. survey questions). By its definition, it is clear that statistics represents measure for adequate deviation of empirical and expected frequencies. Therefore, "small: values of X^2 statistics will be beneficial for the zero hypothesis H_0 for independence of the respondents' answers in relation to the size of an enterprise. On the other hand, "high" X^2 values give grounds for the rejection of this hypothesis and adoption of the opposite one – that structure of the answers directly depends on the size of the enterprises. More precise formulation for the testing rules, i.e. decision on adoption or rejection of the null hypothesis may be formulated in this way:

In the Table of χ^2 distribution included is the critical value $c_{r-1;1-\alpha}$ which represents "borderline" for adoption, i.e. rejection of H_0 hypothesis. In this way, the following *test rule* is applicable:

H_0 adopted for $\chi^2 \leq c_{r-1;1-\alpha}$ H_0 rejected for $\chi^2 > c_{r-1;1-\alpha}$

Research hypotheses are as follows:

Hypothesis 1A: SME investments, innovations and new technologies had a positive impact on their competitive abilities and exports in the research period.

Hypothesis 2A: SMEs strongly require modernization and introduction of innovations based on IT utilization.

Hypothesis 3A: Majority SMEs hold and plan certain financial resources intended for the development of IT in their environment.

Hypothesis 1B: SME investments into innovations and new information technologies had a positive impact on the internal performances of the company in the research period.

Hypothesis 2B: SME investments into innovations and new information technologies had a positive impact on the relationship with customers in the research period.

Hypothesis 3B: SME investments into innovations and new information technologies had a positive impact on the performances of electronic communication and environmental business in the research period.

By testing hypotheses through empirical and desk research the author tests the positive impact of business optimization on competitiveness. Hypotheses are tested with the following questions followed by presentation of research results.

Eventual positive impact on competitiveness is explored, where the use suitable survey questions are as follows:

1. Have innovations in IT field been monitored?
2. Has the achieved advancement resulting from the use of new technologies been monitored and assessed?
3. Have modern information technologies been used for communication between the employees and the management?
4. Have modern information technologies been sufficiently used within information management?
5. Has knowledge management been optimized for the purpose of modernization of the company and development of employees?
6. Has knowledge management been optimized within organization?
7. Have cooperation mechanisms between organizational units of the company been optimized?
8. Have channels of internal communication been optimized?
9. Has knowledge been managed innovatively within the budget planning?

The above stated hypotheses are tested with empirical research and statistical analysis. A sample of 137 SMEs was formed, of which 103 were small and 34 medium enterprises.

Survey questions are formulated in such a way that they can confirm accuracy of the hypothesis. Answers of the surveyed respondents are coded with 0 (negative answer) and with 1 (positive answer). Answers derived in this way are perceived as dichotomous chance variables, with two values: 0 and/or 1.

RESEARCH RESULTS

The author now presents a suitable conclusion from the defined hypotheses, supposing a positive impact on competitiveness with optimization of business. Formally, the author introduces adequate dichotomous variables, which show (negative or positive) answers of the respondents. On the basis of the respondents' answers, variable (survey question) takes one out of two values:

- Yes (value – 1)
- No (value – 0)

Data gained in this way are quantified with especially introduced dichotomous (0/1) variables, or so-called indicators of the observed feature within the sample. In the next section, Table 1 presents the research questions used by the author.

Table 2 now provides an evaluation performed by the author.

Table 1. Research questions

QUESTIONS	1	2	3	4	5	Mean
1. Have innovations in IT field been monitored?	0%	4.5%	32%	18%	45.5%	4.04%
2. Has the achieved advancement resulting from the use of new technologies been monitored and assessed?	0%	9%	32%	23%	36%	3.86%
3. Have modern information technologies been used for communication between the employees and the management?	9%	4%	14%	23%	50%	4.00%
4. Have modern information technologies been sufficiently used within information management?	0%	9%	36%	23%	32%	3.77%
5. Has knowledge management been optimized for the purpose of modernization of the company and development of employees?	0%	0%	23%	27%	50%	4.27%
6. Has knowledge management been optimized within organization?	0%	0%	23%	45.5%	31.5%	4.09%
7. Have cooperation mechanisms between organizational units of the company been optimized?	0%	4.5%	18%	50%	27.5%	4.00%
8. Have channels of internal communication been optimized?	0%	18%	9%	36.5%	36.5%	3.91%
9. Has knowledge been managed innovatively within the budget planning?	4.50%	18%	36%	23%	18.50%	3.32%

Source: Author formulated questions.

Table 2. Evaluation of responses to questions

EVALUATION									
QUESTION	1	2	3	4	5	6	7	8	9
EVALUATION									
1	0%	0%	9%	0%	0%	0%	0%	0%	4,50%
2	4.5%	9%	4%	9%	0%	0%	4.5%	18%	18%
3	32%	32%	14%	36%	23%	23%	18%	9%	36%
4	18%	23%	23%	23%	27%	45.5%	50%	36.5%	23%
5	45.5%	36%	50%	32%	50%	31.5%	27.5%	36.5%	18,50%
MEAN VALUE	4,04%	3,86%	4,00%	3,77%	4,27%	4,09%	4,00%	3,91%	3,32%

Source: The projection of the author

The validity of hypotheses which supposes that investments in SMEs innovations and new information technologies in previously explored period had a positive impact on relationship with the consumers is now examined. Part of the interview survey concerning this hypothesis had five questions, i.e. adequate dichotomous variables, whose observed values, along with the rest of descriptive-statistic indexes are displayed in Table 3.

Variable X_3 has the biggest observed value; it presents positive attitudes of the SME respondents about keeping proactive and regular relations with the customers. The majority of respondents, 83.58% of them, answered this question confirmatively. Similarly, variable X_4 shows nearly 80% of positive answers about IT budget planning. As for variable X_1, more than three quarters of the surveyed SME sample positively answered the question about active policy for informing customers. It is important also to highlight the value of variable $X_{5,}$ on the basis of which the author concludes that more than 70% of the respondents sufficiently use modern information technologies for the purpose of collection of information. Variable X_2 shows that almost two thirds of SME respondents affirm use of oriented web presentations. In the next section, Table 3 presents the statistical analysis.

Contrary to previous variables, variable X_2 decidedly has the smallest observed value, i.e. surprisingly small number of SME respondents, only 28.36% of them, said that they actually possess user-oriented web-presentations suitable for each customer individually.

The results of X^2 testing of the above described variables, observed as the hypothesis modalities, in relation to the size of SMEs, are given in Table 2. The realized value $X^2 = 12.67$ is slightly smaller than in the majority of the previously obtained values. This is, however, a consequence of the relatively small number of the degrees of freedom ($n = 4$), because X^2 statistically exceeds the critical values of X^2 test – at both levels of significance: $p <0.01$ and $p <0.05$. Therefore, with less than 1% of risk, the author finds that there is a statistically significant relationship of the characteristics expressed in dichotomous variables (X_1, X_2,..., X_5) to the size of SMEs. In the next section, Table 4 presents the realized frequency values.

In other words, a positive impact on relations with customers, realized by investments and innovations in new information technologies, is directly dependent on the size of SMEs. Clearly, the readers can observe "dominance" of the medium-sized enterprises which expressed more positive attitudes in

Table 3. Statistical analysis of responses to questions

Variables	Description (survey questions)	SMEs number	Proportion (p)	Standard deviation
X_1	IT innovations - monitored	112	0.8175	0.3877
X_2	Achieved advancement arising from IT utilization - monitored and assessed	58	0.4234	0.4959
X_3	IT in communication between employees and management - used	83	0.6058	0.4905
X_4	Modern IT in information knowledge – used	53	0.3869	0.4870
X_5	Knowledge management for modernization of the company and development of employees - optimized	98	0.7153	0.4513
X_6	Knowledge management within organization – optimized	55	0.4015	0.4902
X_7	Cooperation mechanisms between organizational units - optimized	124	0.9051	0.2931
X_8	Channels of internal communication – optimized	110	0.8029	0.3978
X_9	Knowledge within budget planning managed - innovatively	32	0.2336	0.4231

Source: Calculation and projection of the author

Table 4. Realized frequency values by SME size and X^2 statistics of the independence test

Variables	Small Enterprises $f_i f_i'$		Medium-sized Enterprises $f_i f_i'$		Σ
X_1	80	76.98	32	35.02	112
X_2	27	39.12	31	18.88	58
X_3	58	57.05	25	25.95	83
X_4	25	35.09	28	17.91	53
X_5	68	64.89	30	33.11	98
X_6	25	36.42	30	18.58	55
X_7	92	82.10	32	41.90	124
X_8	80	72.83	30	37.17	110
X_9	14	21.02	18	10.98	32
Σ	247	247	132	132	**X^2=12.67****

Source: Calculation and projection of the author

relation to the majority of survey questions of the hypothesis. This disproportion of affirmative responses is especially high among the X_1, X_3 and X_4 variables that were previously analyzed. An insight into the formulation of survey questions that correspond to these three variables clearly shows that responsible management system, communication with customers and developing user-friendly web-sites can be particularly related to the medium-sized enterprises.

DISCUSSION OF RESEARCH RESULTS

Research starts with the premise that innovative business models of small and medium-sized enterprises based on introduction of new technologies and communication may rapidly improve internal knowledge and abilities, enhance technical and technological capacities, create better approach to new ideas, products. In this way, a long-term, sustainable strategic competitiveness and advantage of SMEs is ensured - which is the goal of introduction of new technologies, knowledge and methods into business practice of SMEs.

In the previous section the author has conducted detailed statistical analysis of these hypotheses, beginning with the basic descriptive-statistical methods (determination of the total number of positive answers, their proportions and standard deviations), and then checking of dependency in relation to the size of SMEs. The overall statistical analysis was based on the application of relevant theoretical facts and on modern approaches within the theoretical-statistical analysis of related issues. Therefore, the author considers the results and conclusions as completely authoritative. At this stage the research, the author will try to connect all important results and statistical indicators which are mostly related to the validity of 1A-3B Hypotheses. The author provides their summary now, with suitable interpretation.

Tables in the chapter generally show mean values of all important results gained through previous statistical analysis. The results are unified and grouped per each Hypothesis (1A-3B), and they are displayed with vertical indents. Inside are mean values of total number of positive answers per each hypothesis individually. It should be highlighted that differences among values are not high: they vary within the interval of approximately 78.23 respondents (Hypothesis 3B) and up to somewhat more than

89 of the examined SMEs (Hypothesis 1B). Similarly, proportions of these answers fluctuate inside a relatively small interval. However, persuasively the highest average of positive answers was obtained in observing Hypothesis 1B, i.e. the majority of respondents - at an average 65.04% - believe that innovations and new IT had positive impact on the internal performances of the company in the research period. Inversely, minimal number of positive SME responses were "equally" given for Hypotheses 3A and 3B, somewhat above 58%. It means that a minimal number of respondents confirmed planning of certain financial resources intended for development of IT, i.e. innovations and that new IT had a positive impact on internal performances of SMEs in the observed period.

However, it is important to emphasize that all of the observed proportional values of positive responses *exceed half the observed sample*, i.e. majority of the respondents generally responded to questions positively. This fact is stimulating, because it demonstrates that majority of SMEs has a positive attitude towards innovations and investments into new IT. In the next section, Table 5 presents the mean values for basic statistical indicators.

Readers will observe now the second segment of statistical research which is testing dependency of hypotheses, observed in relation to the size of the surveyed SMEs. This testing was done strictly formally, with the well-known X^2 test for independence of two signs (characteristic of the basic set, i.e. its sample). All realized values of X^2 statistics are displayed also in Table 5, along with the corresponding number of the degree of freedom and statistical indicator – *contingency coefficient* of two (dependent) signs. This coefficient is defined as *the dependency measure of the observed signs*, i.e. it quantitatively (in percentage) presents degree of dependency among them. Hypothesis 3A has the highest value of the contingency coefficient and of X^2 statistics. Derived value of 51.47% indicates the percentage of the surveyed SMEs, whose planning of financial sources for the development of IT is directly dependent on size of the SME.

Table 5. Summary presentation of mean values for basic statistical indicators

Mean value	HYPOTHESES					
	1A	2A	3A	1B	2B	3B
Number of SMEs:	85.70	83.83	81.00	89.11	79.88	78.23
Proportion (p) :	0.6256	0.6111	0.5857	0.6504	0.5961	0.5838
Standard deviation:	0.4301	0.4413	0.4404	0.4298	0.4413	0.4495
χ^2 *- statistics:*	46.52**	15.51	48.29**	37.65**	28.24**	44.69**
Degrees of freedom (n):	9	10	16	8	7	12
Significance Level (a)	4.82(-7)	0.1145	4.28(-5)	8.74(-6)	1.99(-4)	1.64(-5)
Contingency Coefficient	50.76%	32.20%	51.47%	46.83%	41.72%	50.00%
RP Coefficient	3.4273	2.8523	2.3400	3.1987	2.2623	2.0244
Regression Coefficient	0.4393	0.4328	0.2950	0.5180	0.6105	0.3827
Relative errors of evaluation	3.49%	5.70%	10.27%	4.45%	9.27%	7.27%
Cumulative relative error	3.28%	4.06%	5.87%	3.30%	12.32%	6.46%
Logistic Model	98.09%	98.16%	97.57%	98.75%	93.72%	98.44%
Linear Model	97.9%	98.0%	97.4%	98.5%	94.8%	98.2%

Source: Calculation and projection of the author

Inversely, the only hypothesis that does not confirm statistical dependency in relation to the number of employees is Hypothesis 2A. This hypothesis supposes that modernization and innovations based on utilization of modern IT are necessary for the companies. Naturally, this sort of need is supposed to be equally represented in all SMEs, regardless of their size and number of employees. Thus, the obtained results are in accordance with this obvious fact. As for other hypotheses, dependency in relation to this sign of the observed SMEs is existent. However, statistically presented with contingency coefficient, this dependency is not significantly high - it is between 40% and 50%.

The author will now briefly describe the last segment of this statistical research in relation to the formation and analysis of the corresponding regressive models of dynamics of dichotomous variables as quantitative indicators of the SMEs positive responses. The highest mean value has the RP coefficient of the first Hypothesis 1A – it is 3.4273. On the contrary, minimal value "belongs" to the RP-coefficient of the Hypothesis 3B.

One of the important gross statistic measures refers to the assessed values of the regression coefficient that the author has calculated. These coefficients point out to the degree and level of positive impact of regressive, dichotomous variables on the character of the sample observed with each individual hypothesis. The highest significance level has regressive coefficients of the Hypothesis 2B, which at an average value to 0.6105. This means that with the increase of positive SME attitudes in relation to the questions regarding this Hypothesis *the highest positive effect of investments into innovations and new information technologies in relation to the consumers* will be obtained. Level of such increase is quantitatively presented with given values of the regression coefficient - which, at an average, amounts to 61.05%! As for other hypotheses, mean values are somewhat smaller, bust almost all their mean values exceed 30% level of positive effects. Exception is only Hypothesis 3A – which has the smallest mean value of regression coefficients (only 0.2950). This hypothesis was already analyzed in the previous section of our exposition.

It is necessary to emphasize here perhaps the most important fact that the author formally demonstrated positivity of all obtained, assessed values of the regression coefficients. In this way, positive impact of configured qualities and characteristics of the SME sample on the level and degree of investments into new information technologies has been displayed – which *confirms the validity of all above-formulated hypotheses.*

The author will now reflect and compare the quantity level of the obtained regression models for all six formerly formulated hypotheses. Presented were the so-called *relative errors of evaluation of the regression coefficient* individually, as well as *cumulative relative error* of the regressive model, defined as relation between mean squared error and corresponding observed mean logistic function – for each Hypothesis: 1A-3B. Thus, it presents one more adequate quantitative indicator of the quality of the logistic models that were obtained formerly (Papić, 2011). Cumulative relative error of logistic models is lowest in case of Hypotheses 1A and 1B – it amounts only to 3.30%. On the other hand, the highest cumulative relative error of logistic-regressive model has Hypothesis 2B – it is 12.32%. Models with lower relative error correspond to the higher values of determination coefficient and *vice versa.* In case of Hypothesis 2B this gives basis for "suspicion" into correctness of the choice of the mentioned model. In case of the mentioned Hypothesis 2B higher value of determination coefficient for linear model is marked (R^2=94.8%) in relation to the corresponding logistic model, where - R^2=93.72%. Therefore, it is only here that an inadequate "selection" of the regressive model is observed.

In all other cases, determination coefficients for the logistic model, though by values close to the corresponding coefficients for the linear models, are insignificantly higher. In other words they provide a basis for the correctness of the selection of the logistic-regressive model as an adequate theoretical model which in the best way interprets the obtained empirical data.

FUTURE RESEARCH DIRECTIONS

The subject of knowledge management optimization through it and e-business utilization has been elaborated in this chapter. The author tried to prove the importance and necessity of investments into application of modern technologies as innovative input for advancement of overall performances of the company. Analyzed were knowledge and ability of the employed and management, formation of new services and communication with customers, profit for the employed, shareholders and suppliers, external communication with the community and global world to achieve increased competitive and comparative advantages. Conducted research covered six hypotheses within the group of general and internal aspects of relation between competitiveness and performances, through utilization of new information technologies. Obtained results offer a range of possibilities for further theoretical and practical research, which paves the way for strategic competitiveness and increased efficiency of the key SMEs activities.

Within this framework optimization of competitiveness through IT and electronic business is considered among the most important issues in many countries and companies. The struggle for competitiveness is not just between companies on a national and regional level but also between countries at international level. Measuring the competitiveness of the national economy is of particular importance, given that it contributes to the advancement of business and overall economic growth. If competitiveness in the global market is poor, then the national economy suffers, leading to protectionism, non-transparent government subsidies and barriers to enterprise development. This also remains for future research.

The remaining key issues for the finalization of this research project for SMEs include optimization of communication with the customers, web business, mechanisms of suggestions and appeals by customers and suppliers, communication with partners, optimal assessment and financing of the employee' development, IT developmental financing. This is also recommended for future research.

Fast development of information technologies influences creation of new business forms and becomes ground for infrastructure of modern business. Information technologies impose new rules and challenges on the businessmen, to which they must accommodate in order to survive at the market. Modern management is supposed to identify all advantages of modern technologies that influence advancement and creation of new values, which helps SMEs strategic advantage to overpower competition and future research can be useful in this area.

Dynamics and levels of modern markets' competitiveness force SMEs to be flexible, innovative and to react fast on the requirements of the customers. Future research needs to explore how modern technologies can play a crucial role in creation of strategic advantages of SMEs, so that they become agile, connected and able to quickly exchange all information internally as well as externally. An important segment for the success of SMEs is adequate adjustment of possibilities offered by information technologies and future research can reflect more deeply how to practically improve business activities, functions and processes.

CONCLUSION

The desk research of theoretical grounds, references and empirical results influenced the possible limitations as well as potential innovative research arising from this chapter. According to the structure of this work the following main conclusions describe the empirically obtained results and findings.

The introductory part precisely defines scientific and social context of the problem through description of social and economic actuality of the research, as well as modern business and technological trends. The concept of modern SMEs, with emphasis on technology to improve business conditions and results as a prerequisite for planning and formulation of strategy, represent sound background for the social significance and validity of the chosen topic and conducted research.

Selection of literature relates to ICT with sustainable competitive advantage of SMEs, reflecting also on strategic management, resource theories, theories of sustainable competitive advantage. This approach allowed testing of hypotheses within this theoretical framework. Information technologies and activities related to them can be realized in an SME provided that suitable platform is prepared beforehand. Literature review given in this chapter points out to the fact that a company is able to achieve competitive advantage through utilization of IT and knowledge.

General results of the chapter refer to the justification of research in relation to the impact of modern technologies on competitive capacities of SMEs, on the level of innovative and technological input and their further impact. Key factors include internal advancement of overall companies' functions, creation of better communication with customers through modern information technologies, companies' incentive to invest into innovative products, exports and processes of electronic communication with wider community, institutions and global market.

It can be concluded that development of SMEs presents a basic factor of social-economic development for any modern economy. Fast growth in number of SMEs and construction of the system of developmental stimulation, are important achievements of Serbian economy in the observed period from 2000 to 2011 when the first surveys were done by the author and in the years thereafter. Comparative developmental analysis of SME development in Serbia reveals lagging compared to the EU average particularly considering sales and profit per employee. World economic crisis has deepened the problems which Serbian economy usually has to face with, and it had restrictive influence on the development of SMEs. Recession effects (reduction of external and internal demand and investments, risk increase and investment costs, fear from failure), had a negative impact on business and business environment generally. Reduction of local consumption was coupled with faster growth of imports than the export. Although export of high-tech products grows fast, Serbian economy still depends on the export of low-tech technology. Application of modern technology in the function of competitive trading of modern enterprises shows that in terms of equipment and utilization of computers Serbian SMEs still lag behind the average of the EU-28 and other highly developed countries, while in terms of Internet utilization they are lagging considerably behind the majority of the EU member states. Internet is mostly used for bank and financial services, while electronic business (buying and selling) is still insufficiently present. Therefore, level of development and economic structure of Serbia is usually more comparable to the surrounding countries: Croatia, Bosnia and Herzegovina, Albania, Montenegro and Macedonia and for this reason proposals and conclusions are related to the development of these countries and to their economic structure and innovativeness within the field of ICT.

It further derives from the chapter findings that lack of innovations in SMEs is related to non-formulated vision, poor management structure, absence of growth strategy, deficiency of resources and growing exposure to global factors and competition. SME advantages lie in flexibility, motivation of management and entrepreneurial values. Practice of electronic business as an innovative concept of management is spreading. It introduces radical changes in business functions by implementation of information technologies and systems into daily work of SMEs. ICT as networked computer applications are also becoming a crucial factor enabling changes in business processes. According to the findings of this work, it may be concluded that economic competitiveness of the Western Balkan countries and of Serbia as part of the region generally depends on the level of utilization, readiness and valorization of modern ICT. Significant lagging of these countries, especially of Serbia are visible, which indirectly influenced productivity and caused a low level of current innovations related to products, services, business processes and models. Revival of productivity in the Western Balkan countries in the last few years is related to the adoption and expansion of new technologies and advance of production and process innovations. Transfer of technologies from more advanced countries to the economy of these countries and Serbia would directly contribute to the growth of productivity and competitiveness. Companies should constantly improve the quality of products and adopt the cheapest methods of production. Global challenges require bigger participation of SMEs in international supply chains in order to survive, provided that they are able to adjust themselves to the new technologies, especially to ICT. From the aspect of improvement of companies' capacity this work emphasizes that ICT offers an open platform, decentralized for communication, cooperation, innovations, improvement of productivity and economic growth. ICT promotes the possibility of closer integration into global economy and interaction which boosts general prosperity in the observed region. Since services which support ICT are omnipresent and necessary in everyday life and economy, it is possible to speak of digital era of Internet economy (European Commission, 2016). Capacity of economy, of governmental services and society to use an opportunity and get acquainted with challenges of various fields (environment, education, health service, demography) requires the use of ICT and connection to IT basic networks on the Internet. Since Internet can also function as a catalyst for further enhancement of developmental goals through policies which facilitate convergence, its use stimulates creativity, boosts confidence and broadens possibilities for global economic, social and cultural development. Corporate leaders want to find new values. In the last fifteen years they succeeded in raising productivity of the transaction processes by combining technological investments and re-engineering processes. It is also underlined that ICT are helpful for the SME performances, particularly when combined with complementary intangible assets[1] and abilities of employees. E-business is growing, but adoption of more complex business applications is slower. Relatively small number of companies comprehensively adjusted their business concepts, value chains, organization, communication with suppliers and customers to the new technological solutions. Thus, for instance, internal integration of the system of electronic orders with other functions such as marketing or external integration with suppliers and customers is rare. Considering that the first phase of ICT use is related to the readiness of companies to introduce ICT tools, there is an issue of gap between bigger companies SMEs, so that it is necessary for the wider community to encourage the use of these technologies in smaller companies.

Information technologies are therefore an important subject for consideration by researchers and theoreticians. Most distinguished theoretical works were cited, pointing out the importance of human capital in creation of sustainable competitive advantage of companies in relation to modern information technologies. In order that employees and managers may be effective agents for application of information technologies and tools, they are supposed to have necessary competences related to information technologies. Many resource theories advocate fusion of the human resources factor and information technologies in order to create sustainable competitive advantage of the companies. Therefore, investments into ICT contribute to the deepening of capital and therefore this can assist in raising productivity, particularly through multifactor productivity growth. This may help to reduce transaction costs and to boost innovations. However, this requires development of IT skills. Success of technological and organizational innovations depends on the capacities and skills of individuals at all management levels to absorb changes and to interpret them properly and quickly. An important obstacle for this accomplishment is related to the process of ageing of European and also Serbian workforce.

It is concluded that competences of both the employed and managers in the field of ICT, their knowledge and their computer skills are key factors for the enhancement of companies' competitiveness generally, especially for SMEs. Companies, including SMEs, use also other methods to improve competences and skills of the employed and of managers, such as: practice, learning through transfer of good practice, exchange visits to companies, dialogue with customers and suppliers, meetings related to the development of personnel, rotation at the workplace. These are all essentially based on "learning by doing" methods. Such training leads to unspoken, implied competences and skills that may create a basis for the companies' competitiveness, but they are not easily recognizable with standard assessment, i.e. with University degrees. Literature on business and management issues used in this research confirms the concept of competence development and its significance. This concept comes from psychology, and the term has been used in education since 1920. Each company may develop a base of competences through various measures. It may also outsource suitable competences out of the company, and develop human resources already existing in the organization. These investments will be effective only if used in the way which meets market requirements.

For Serbia, competitiveness inevitably has an export dimension, considering limited market size. Lack of innovativeness in utilization of new technologies and in creation of new products, high operating costs, insufficient knowledge on market requirements and unsatisfactory productivity caused a huge deficit in Serbian foreign trade. Accordingly, strengthening of SME productivity and competitiveness must become the most important direction in future development, along with modern education in new technological competences, innovativeness, technological development, as well as development of clusters, in order that the potentials may become more visible leading to achievement of maximal results.

Attainment of a certain level of international competitiveness is one of the prerequisites for the accession into the EU, since countries must meet the condition of effective market economy and competitive enterprises should be able to stand the pressure of the common market. Therefore, increase of competitiveness will play a crucial part in long-term economic development, but also in integration into European and world economy. The main goal in development of national competitiveness, according to the findings of this chapter, is permanent advancement of long-term, sustainable growth of productivity through IT and e-business utilization. Modern technologies and their fast implementation change traditional business models. SME management is therefore supposed to think strategically about how to prepare organizations for these observed challenges of the new environment: combination of growing internet technologies, increased computerization, faster digital communication, talent and resource management, new methods of organizational structure, visualization of information technologies that create new methods for consumption of goods and services. Also, according to the findings of this chapter, entrepreneurs should consider the suitable business models, with whose application and transformation SMEs will practically be able to capitalize and enhance advances in the market. Finally, conclusions from this work advocate establishing computer networks between all the relevant scientific and business institutions, the use of e-business and computers as well as technology of packet transmission for the mutual exchange of information. What is now required is further understanding of the implications of the author's findings and more refined researches to discriminate among alternative IT and e-business utilization hypotheses on SMEs.

REFERENCES

Altenburg, T., & Drachenfels, C. (2006). *Business environment reforms: Why it is necessary to rethink priorities and Strategies*. German Development Institute. Retrieved July 27th, 2016 from German Development Institute web site: https://www.die-gdi.de/uploads/media/Altenburg_Drachenfels_2008_Business_environment_reforms.pdf

Ayyagari, M., Demirgüç-Kunt, A., & Maksimovic, V. (2006). *How Important Are Financing Constraints? The Role of Finance in the Business Environment*. World Bank Policy Research Working Paper, No. 3820.

Beck, T. H. L., Demirgüc-Kunt, A., & Maksimovic, V. (2008). Financing patterns around the world: Are small firms different? *Journal of Financial Economics*, *89*(3), 467–487. doi:10.1016/j.jfineco.2007.10.005

Bešić, C., Nikolić, M. & Damnjanović, A. (2010). *Strategijski menadžment*. Čačak: Fakultet tehničkih nauka.

Bešić, C., Sajfert, Z. & Damnjanović, A. (2009). *Menadžment promena*. Čačak: Fakultet tehničkih nauka.

Damnjanović, A., & Milisavljević, V. (2009). *Ciljni marketing*.Paper presented at 12th Internatinal Conference Quality and Reliability Management(ICDQM '09). Čačak.

European Commission. (2004). Retrieved July 27, 2016, from EC web site: http://ec.europa.eu/regional_policy/en/projects/finland/e-commerce-for-smes

European Commission. (2010). *EUROPE 2020: A strategy for smart, sustainable and inclusive growth*. Retrieved July 27, 2016 from http://publications.europa.eu/resource/cellar/6a915e39-0aab-491c-8881-147ec91fe88a.0004.02/DOC_1

European Commission. (2016). *The importance of the digital economy*. Retrieved July 27, 2016, from EC web site: https://ec.europa.eu/growth/sectors/digital-economy/importance_en

Eur-Lex. (2016). Retrieved July 28, 2016, from Eur-Lex, Access to European Union law web site: http://eur-lex.europa.eu/legal-content/EN/TXT/?uri=CELEX:52004DC0064

Jevtić, B., Damnjanović, A., Matić, R., & Grozdanić, R. (2011). Izazovi industrijske politike u IKT sektoru. *Ekonomski vidici,16*(3), 572-588.

Jokić, D. (2010). *Preduzetništvo. Užice: Naučno istraživački centar (NIC)*. Užice.

Mašić, B. (2010). *Menadžment*. Beograd: Univerzitet Singidunum.

Milosavljević, M., Veinović, M., & Grubor, G. (2013). *Informatika*. Beograd: Univerzitet Singidunum.

Milošević, D. (2014). *Strategija rasta i razvoja preduzeća*. Beograd: Visoka škola za poslovnu ekonomiju i preduzetništvo.

Njeguš, A. (2009). *Poslovni informacioni sistemi*. Beograd: Univerzitet Singidunum.

Njeguš A. (2010). *Informacioni sistemi u turističkom poslovanju*. Beograd: Fakultet za turistički i hotelijerski menadžment, Univerzitet Singidunum.

Obradović, S., Jevtić, B., Grozdanić, R., Damnjanović, A., & Vulović, R. (2011). *Uloga IKT u koordinaciji intermodalnog transporta.*Paper presented at14thInternational Conference Quality and Reliability Management (ICDQM '11), Čačak.

OECD. (2004). *ICT, E-business and SMEs*. Retrieved July 27th, 2016 from OECD web site: http://www.oecd.org/internet/ieconomy/34228733.pdf

Ožegović, L., & Sajfert Z. (2009). *Preduzetništvo*. Novi Sad: Fakultet za ekonomiju i inženjerski menadžment.

Ranđić, D., & Lekić, S. (2009). *Menadžment*. Beograd: Beogradska poslovna škola.

Reynolds, P. D. (2000). National Panel Study of U.S. Business Start-ups: Background and Methodology. In J. A. Katz (Ed.), *Databases for the Study of Entrepreneurship* (pp. 153–227). Amsterdam: JAI Press. doi:10.1016/S1074-7540(00)04006-X

Šarac, M., Jevremović, A., & Radovanović, D. (2015). *Internet marketing*. Beograd: Fakultet za informatiku i računarstvo, Univerzitet Singidunum.

Veljović, A., Vulović, R. & Damnjanović, A. (2009). *Informaciono komunikacione tehnologije u menadžmentu*. Čačak: Tehnički fakultet.

Vemić, M., (2008). *Nauka o privrednom sistemu i ekonomskoj politici u tranziciji*. Novi Sad: Cekom books.

Wang, S. (2016). *Enterprise Management and Performance Evaluation of E-business in SMEs: Based on Human Resource Management Perspective*. Retrieved July 27th, 2016 from SERSC web site: http://www.sersc.org/journals/IJUNESST/vol9_no3/31.pdf

World Economic Forum. (2016). *Global Competitiveness Report*. Retrieved July 27th, 2016 from World Economic Forum web site: https://www.weforum.org/reports/global-competitiveness-report-2014-2015

World Bank. (2016). *Global Economic Prospects June 2016: Divergences and Risks*. Washington, DC: World Bank. Washington, DC: World Bank. Retrieved September 7th, 2016 from World Bank web site: http://www.worldbank.org/en/publication/global-economic-prospects#data

Živadinović, J., (2010). *Poslovna informatika*. Beograd: Visoka škola za poslovnu ekonomiju i preduzetništvo.

ADDITIONAL READING

Barringer, B., & Ireland, D. (2008). *Entrepreneurship successfully launching new ventures*. New Jersey: Pearson Education.

Damnjanović, A., Jovanović, M., Šljivić, S., Grozdanić, R., & Jevtić, B. (2011). Innovations in electronic commerce as a competitive advantage of SMEs. In *Institutional change as a determinant of economic development of Serbia in 2011*. Kragujevac: Faculty of Economics.

Grozdanić, R. (2006). *Ekonomija preduzetništva*. Čačak: Tehnički fakultet.

Kotler, P., Wong, V., Saunders, J., & Armstrong, G. (2007). *Principles of Marketing*. Belgrade: MATE.

Samuelson, P., & Nordhaus, W. (2009). *Economics*. Belgrade: MATE.

KEY TERMS AND DEFINITIONS

Communication: The process of reaching common understanding, in which SME managers and stakeholders not only share business information, news, ideas and feelings but also develop an exchange meaning.

Electronic Business: All electronically brokered information exchanges, both within an SME and with external partners strengthening a whole range of business activities.

Information Management: Implementation of to gather, communicate it with internal and external stakeholders, and it to empower SME for faster and more effective.

Information Technologies: The use of electronic devices, storage, networking and other physical devices, infrastructure and applications to create, process, store, secure and exchange different forms of electronic data.

Knowledge Management: The holistic management of an SME's knowledge portfolio aimed at creating value and achieving operational and strategic objectives.

Optimization: Achievement of best results in knowledge management with the application of IT and e-business tools.

Web Business: The conduct of business-to-consumer, business-to-business, and more recently, consumer-to-consumer business through the web and web applications.

ENDNOTES

1 Editors' note: For a more detailed discussion of this point see, for example, chapter 15 on optimization of intangibles in SMEs.

Chapter 13
Entrepreneurship and Enterprise Value Creation in Support of Smart, Sustainable, and Inclusive Growth in the European Union

Julia S. Stefanova
Economic Research Institute at the Bulgarian Academy of Sciences, Bulgaria

Zachary Wenner
Fulbright Commission, Bulgaria

ABSTRACT

In 2010, the European Union codified "Smart, Sustainable, and Inclusive Growth" (SSIG) as part of the Europe 2020 Strategy. Introduced in the context of the global financial crisis, SSIG seeks to restore economic resiliency and foster a more equitable, innovative and environmentally sustainable European economy. The research evaluates the processes of entrepreneurship through which small and medium enterprises (SMEs) contribute to SSIG, including through "innovation-driven entrepreneurship," "sustainable entrepreneurship" and "social entrepreneurship." The research additionally assesses the creation of an enabling regulatory and fiscal environment for these modes of value creation, their associated socio-economic outcomes and their monitoring and evaluation practices.

INTRODUCTION

In adherence to traditional enterprise models, investors hold ownership rights while enterprises pursue the principle objective of profit maximization. As a result of the global financial crisis, however, the notion that a narrow focus on financial returns is the most efficient method to ensure sustainable revenue growth and market reach has become increasingly tenuous. In recent years, several of the concepts underlying traditional enterprise models have progressed to integrate a more multi-dimensional view of

DOI: 10.4018/978-1-5225-1949-2.ch013

value creation, including social, financial and environmental dimensions (Emerson, 2003). Global and localized drivers have catalyzed this shift, including: a focus on long term value creation as a risk mitigation strategy; constrained and sometimes depleting public resources, which have encouraged adoption of private sector social and sustainable models and practices; and an increasing body of scholarship which demonstrates the financially material nature of global trends such as demographic changes and natural resource scarcity (Wilson, Silva & Richardson, 2015).

In particular, the European Union (EU) has become increasingly specialized in its regulatory and financial support for small and medium-sized enterprises (SMEs). This specialization relates to both continuing efforts to foster economic development and increasing consideration of the extra-financial value of such growth. Through several of its 2014-2020 Multi-annual programming instruments, the Europe 2020 Strategy delineates SMEs based on their distinct contributions to Smart, Sustainable and Inclusive Growth (SSIG). The research analyzes the SME subsets that align to and advance smart, sustainable and inclusive growth. The SME categories are not absolutes, nor are their associated modes of growth and value creation. Rather, the categorizations are general attempts to align aspirational modes of development with commensurate SME interventionist strategies. Of course, enterprises often possess the characteristics of multiple SME subsets, and may, in turn, contribute to various strands of SSIG. The chapter further proposes a methodology to evaluate SMEs' *ex ante* potential to advance SSIG and to measure the *ex post* materialization of that potential as a result of EU intervention. This methodology incorporates theories of entrepreneurship, value creation and the measurement of SMEs' socio-economic outcomes.

Authors identify the role of entrepreneurship and innovation at the macroeconomic and microeconomic levels. *The macroeconomic research* strand delineates the main factors advancing SME entrepreneurship and innovation, the characteristics of the SME ecosystem and the relationship between entrepreneurship, innovation and economic growth. With respect to the *microeconomic factors*, the research identifies the performance metrics related to SME entrepreneurship, innovation and value creation.

The reserch utilizes two main approaches: theoretical analysis of SME subsets and comparative evaluation of various strands of entrepreneurship within EU member states. Thus the chapter expects to make a contribution to the theoretical and practical development of an efficient system of resources allocation that provides incentives and financial infrastructure for SMEs to optimize their financial and extra-financial value and generate diversified and resilient growth for the European economy.

Incorporating a theoretical methodological review and practical comparative analysis, the present chapter provides evidence with respect to two key assumptions:

Assumption 1: Entrepreneurship and innovation are main drivers for enterprise value creation (microeconomic aspect) and economic growth (macroeconomic aspect).

Assumption 2: The majority of European SMEs have the potential to more effectively leverage entrepreneurial and innovative approaches to value creation, which incentivizes EU intervention at the institutional level through various financial instruments and programmes.

METHODOLOGICAL BACKGROUND

The research on SME entrepreneurship and innovation and their impact on firm performance and economic growth is set against the backdrop of the global financial crisis and the pressing need to address economic stagnation. SMEs are the principal economic growth drivers of European development and critical to the realization of the Europe 2020 Strategy (European Commission, 2010). Thus, adherence to the principles of Smart, Sustainable and Inclusive Growth (SSIG) depends on the ability of the EU to establish an enabling environment for SMEs' development. An extensive body of academic research (see Table 1) recognizes the relationship between entrepreneurship and economic growth. The key contributions relate to enterprise formation, innovation in products/services/processes and the entrance into new markets through knowledge and capital accumulation.

The outcomes of SME entrepreneurship include employment growth, increased taxation revenues and enhanced economic output. According to Kritikos (2010), entrepreneurship produces such outcomes through the development of new technologies, products or process innovations. These developments drive competition among enterprises, while lowering prices and bolstering product diversification for consumers. For Geroski (1989), fluid, dynamic and constant enterprise formations and exits positively relate to productivity. Portela (2013) established a positive relationship between entrepreneurship and economic growth based on the European Social Survey, which incorporated data from nineteen EU countries. The research identified trust in a country's institutional framework as an important determinant driving economic growth due, among other aspects, to its facilitation of risk-taking behavior among economic agents. Therefore, the framework linking entrepreneurship to economic growth encompasses three levels: micro-individual (skills and attitudes), firm (innovativeness and firm performance) and macroeconomic (increased competition and economic growth). An extensive body of academic research recognizes the relationship between entrepreneurship activity and firm performance (see Table 2).

With respect to the financial impact of entrepreneurship, the literature (see Table 2) indicates a positive relationship with firm performance. SMEs operating in environments characterized by high levels of uncertainty and competition tend to adopt more entrepreneurial strategies – encouraging innovation, risk taking and allowing for first-mover advantages. Following is a review of major academic research establishing a link between innovation and economic growth (see Table 3).

Table 1. Relationship between entrepreneurship and economic growth

Criteria of Relationship	Academic Sources
Entrepreneurship as a catalyst of economic growth	Schumpeter (1934)
Entrepreneurship as a catalyst of economic growth through accumulation of economic knowledge and innovation	Keilbach and Audretsch (2005)
Entrepreneurship as a cause of high economic growth in the US in the 1990s and 2000s	Hubbard (2008)
Entrepreneurship and its impact on economic growth	Smith and Culbertson (2010)

Source: Own presentation of the authors.

Table 2. Relationship between entrepreneurship activity (EA) and firm performance

Criteria of Relationship	Academic Sources
Positive relationship between EA and firm performance	Rauch, Wiklund, Lumpkin, and Freese (2009)
Positive relationship between entrepreneurial behavior in social enterprises (including nonprofits) and revenues	Anderrson (2011)
Direct positive influence of EA on firm profitability (ROA, ROS, free cash flow)	Silva (2012)

Source: Own presentation of the authors.

Table 3. Relationship between R&D (innovation) and economic growth

Criteria of Relationship	Academic Sources
Positive relationship confirmed using international panel data	Frantzen (2000), Griffith, Redding and Van Reenen (2004)
Positive effects of R&D spillovers on Total Factor Productivity (TFP) from industrialized to developing countries	Coe, Helpman and Hoffmaister (1995)
Domestic R&D and FDI increase productivity and added value growth	Savvides and Zachariadis (2003)
Positive relationship between innovation (patent stock) and per capita GDP in OECD and non-OECD countries	Ulku (2004)
Positive long-term effects of R&D on economic growth	Aghion and Howitt (1998).
Positive relationship between R&D and TFP in the US	Scherer (1982), Zachariadis (2003)

Source: Own presentation of the authors.

The academic research establishing a link between innovation and economic growth (see Table 3) provides support for the role of endogenous growth models (Romer, 1986). Technological innovation is created in R&D processes by leveraging human capital and the existing knowledge stock. Ulku (2004) found that a 1% increase in innovation (in terms of patent data and R&D expenditures) raises GDP per capita income by around 0.05% in both OECD and non-OECD countries. Moreover, according to research by Frontier Economics (2013), venture financed companies account for less than 6% of total private sector employment in Europe, yet they generate around 12% of all industrial innovation while their spending on R&D amounts to 8% of all industrial spending on R&D. The total economic value of the patents granted to these firms in Europe over a five-year period is over EUR 350 billion.

The traditional explanation for the positive relationship between innovation and firm perfomance (see Table 4) can be found in Schumpeter's work (1934). Schumpeter argued that innovative new products face limited competition and allow firms to realize relatively high profits. The fact that innovative firms persistently earn higher profits is a result of improved capabilities based on increased knowledge and economies of scale. Leveraging data from the Australian stock exchange, Bosworth and Rogers (1998) found that R&D activities and intangible assets have a positive relationship with the market value of firms.

The above methodological review of the relationship between entrepreneurship and innovation, and this relationships' impact on enterprise value and economic growth, provides evidence in support of *Assumption 1.*

Table 4. Relationship between R&D (innovation) and firm performance

Criteria of Relationship	Academic Source
Continuous innovation activity as a key source of long term firm success	Rosenbusch, Brinckmann and Bausch (2011)
Innovative firms, particularly in the high-tech sectors, tend to realize higher employment growth, and cause positive spillovers for other firms	Stam and Wennberg (2009)
Firms which engage in technological innovation experience higher growth in employment, net revenue, labor productivity, and market share	Kannebley, Sekkel and Araujo (2010)
Innovation as a key factor for firm success and sustainability	Jimenez and Sanz-Valle (2011); Wolfe (1994)
Positive effect of innovations on operating profit	Geroski, Machin, and Van Reenen (1993)
Significant impact of product innovation on firm performance	Artz, Norman, Hatfield and Cardinal (2010)

Source: Own presentation of the authors.

ENTREPRENEURSHIP AND SMART GROWTH

Globalization has continued to accelerate the transition from managerial to entrepreneurial economies. The EU has sought to facilitate this shift, while also targeting pre-defined modes of entrepreneurship and SME value creation in pursuit of SSIG. According to the European Commission (2010), SME growth, entrepreneurship and the formation of a modern European economy are interdependent. The creation of an enabling environment for SME entrepreneurship is dependent on formal and informal institutions, which according to various authors, may be evaluated at the macro-economic, regional, industrial level and micro-individual firm levels. Following is a classification compiled by the authors of this chapter.

Macro-level factors of SME entrepreneurship relate to GDP growth, employment, gross domestic investment, cost levels, inflation and interest rates (Highfield & Smiley, 1987). Public sector activities -- including direct spending and the leveraging of political capital -- in the areas of education, infrastructure and health are also catalytic factors. For Murphy, Schleifer and Vishny (1999), so too are the protection of property rights and the provision of balanced public and private sector incentive structures. State regulation of market entry reduces enterprise formation and depresses levels of entrepreneurship (Ciccone & Papaioannou, 2006). For Micco and Pages (2006), rigid labor market regulations negatively impact market entry and impose limitations on firm entrepreneurship. According to Djankov, Glanser, McLiesh, Ramalho and Schleifer (2008), administrative burdens associated with taxes similarly diminish SME entrepreneurship.

Regional factors of SME entrepreneurship include: transportation costs, localization and mobility of human capital, extensive R&D activity, availability of financial capital (Bartik, 1989), population density and employment, income growth, supply of B2B services, standard of living, technology and knowledge transfer, and rates of new firm formation. For Ejermo (2009), a principal determinant of regional innovation is quality-adjusted patent data, which he recognized for its link to regional R&D activity. For example, 90% of regional variation in economic growth (as measured by total factor productivity) is associated with the accumulation of knowledge stock (patent activity) and regional enterprise formation. A research by Lueckgen (2004) of entrepreneurial activity showed that the German regions with the highest number of nascent entrepreneurs are Cologne, Western Saxony/Leipzig, Munich, Emscher-Lippe and Middle Mecklenburg/Rostock. Wach (2012) undertook research of 109 SMEs in two provinces in Southern Poland within the Regional Entrepreneurship Monitor project and found strong correlations between availability of regional capital, quality of entrepreneurial infrastructure, rate of firm formation and the development of SMEs.

Industrial level factors advancing entrepreneurship encompass the potential for profit generation, existence of entry/ exit barriers, demand/ supply forces, and agglomeration of production structures (Reynolds, 1992). Profits, industry growth and size are positively related to enterprise formation according to the Swedish Entrepreneurship Forum (2016). Entry and exit of firms can serve as a proxy for the entrepreneurial environment at the industry level; however, for Baumol, Panzar, and Willig (1982), this fluidity may be influenced by internal (organizational and management) or external factors (including threat to entry). Therefore, net entry is a more accurate proxy for expansion of entrepreneurial activities. In technologically advanced industries, the entry of foreign firms may serve to enhance entrepreneurial activities.

Finally, micro individual firm-level variables of entrepreneurial activity include human capital – which Evans and Leighton (1989) characterized as a prerequisite factor – as well as equity availability and social networks. The latter two variables additionally foster lower transaction costs (Williamson, 1971).

According to Fonseca, Lopez-Garcia and Pissarides (2001), high start-up costs and high taxes reduce levels of entrepreneurship. For Glaeser and Kerr (2009), higher proportions of small and independent firms increase entrepreneurship activity due, among other aspects, to their access to skilled labor and finance, as well as larger profit margins and ability to leverage knowledge drawn from internal and external networks.

Entrepreneurship Impact: Measurement and Evaluation

The Global Entrepreneurship Monitor (GEM) project provides homogeneous cross-country measures of entrepreneurship activity, while also evaluating the relationship between these measures and economic development. It additionally publishes the Total Entrepreneurship Activity (TEA) Index as an annual measure of individual self-employment and enterprise formation.

The entrepreneurship ecosystem is strongest in the innovation-driven economies, such as Switzerland, the Netherlands and Luxembourg, which generate higher per capita income, jobs and social value. In 2015 (GEM Consortium, 2016), the EU countries with the highest level of TEA were Estonia, Luxembourg, Ireland, The Netherlands, Portugal and Slovak Republic. According to a comparative study of entrepreneurial activity in the EU and US (Swedish Entrepreneurship Forum, 2016), TEA is considerably higher in the US, while the lowest levels of TEA are found in Germany, France and the countries most severely hit by the global financial crisis – namely, Portugal, Italy, Greece and Spain.

In addition to GEM, EU entrepreneurship activity at regional level is measured using the Regional Entrepreneurship and Development (REDI) Index. Initially, this index was constructed to measure contextual features across the Spanish regions but was subsequently extended to encompass regional level entrepreneurship in 125 European regions by incorporating individual and specific institutional variables at regional level (Szerb, Zoltan, Komlosi & Argiles, 2015). The results from the 2015 survey indicate that the 10 regions with the highest entrepreneurial activity in Europe are two Danish, four Swedish, two UK, one French and one Irish region. The Danish Hovedstaden region, centered by Copenhagen, stands out as the most entrepreneurial region in the EU. According to the index, there is a close connection between entrepreneurship and regional development. Thus, larger, more developed regions with higher per capita GDP generally rank ahead of less developed regions -- even within the same country. The bottom 10 places of least entrepreneurial European regions are occupied by four Hungarian, four Romanian and two Greek regions from the 24 surveyed EU member states.

INNOVATION-DRIVEN ENTREPRENEURSHIP

Globalized economies increasingly rely on innovation-driven SMEs i.e. those that not only maximize economic output through conventional channels but form new opportunities for firm value creation through the production of novel goods or services. Innovation capacity continues to play a central role in growth dynamics – not only in areas related to high technology products but also in terms of "hyper-local" innovation that addresses regional challenges and links communities to regional, national or international networks. While increased trade integration within Europe and at the global level has allowed for economies of scale and specialization, trans-border production processes have also placed greater strain on SMEs to pursue sector-specific innovations in which they have a competitive advantage, while simultaneously identifying and cultivating the requisite human capital.

The EU's restructuring of its production and innovation structures targets various stages of the SME development cycle. In addition to direct SME intervention, the EU cultivates an innovation ecosystem – from educational access to the provision of venture capital. In terms of SME segmentation, the EU identifies innovation-driven SMEs by both their operational models and outcomes, including trans-border orientation, high risk to reward ratios, and the degree or method of change (*e.g.*, radical changes to product, processes, services and marketing).

Schumpeter (1934) identified several intertwined forms of innovation, all of which rely on creative disruption and divergence from traditional production patterns. The Oslo Manual (OECD, 2005) similarly subcategorizes forms of innovation based on its manifestation at various levels of the logic and results chain. These include the leveraging of technological products and processes, organizational and marketing structures. According to Nevens, Summer and Uttal (1990), innovation is of particular importance in technology driven industries characterized by the brevity of product life cycles.

The Innovation-Driven Entrepreneurship Environment

The innovation environment is particularly sensitive to the adequacy of the legal framework for the protection of intellectual property rights, the scope of tax-based incentives for innovation, competition policies, market regulations, development of financial markets, efficiency of trade and investment. In addition, stable macroeconomic conditions and low interest rates directly affect levels of innovation activity.

The structure of the national innovation system includes the anchors of knowledge generation (*e.g.*, human capital and R&D expenditure), diffusion capacity (*e.g.*, the existence of linkages among various knowledge transfer networks), the absorption capacity of companies (*e.g.*, the propensity and resources to engage in innovation), demand for innovations in the economy and innovation governance (*e.g.*, fiscal stimuli, funds for innovation and favorable legislation). The NUI Galway and Whitaker University (2015) introduced the "triple helix" concept to the evaluation of innovation environment. This approach centers on the linkages between universities, public research organizations, the private sector, and government entities at various points of the innovation process. The development of national innovation systems requires effective partnership between these entities.

In the EU the innovation environment of EU member states is measured using the Innovation Union Scorecard (IUS) and Community Innovation Surveys. The European Commission publishes an annual IUS for each member state based on indicators and data of national statistics agencies. At the EU level the most innovative countries (European Commission, 2014) Finland, Sweden and the Netherlands share features within their national innovation systems, including the prominence of business R&D activities (especially patent applications), educational performance and effective resource mobilization. Finland has the highest R&D intensity in the EU (3.55% of GDP, compared to 2.07% for EU-28 and 2.79% for USA) and is a top performer in terms of business R&D spending (2.44% of GDP).

The active government support of knowledge-based high-growth enterprises is presupposed by their contribution to and significant impact on economic growth, employment and innovation and takes the form of improved access to financing (including venture capital), innovation policy measures and export boosting incentives (Vemic, 2013). Taking Finland as an example and EU outperformer, the European Structural and Investment Funds are an important source of funding for research and innovation activities. Between 2007-2013, 29.3% of these funds were earmarked for this purpose. The Finnish government has accelerated the reforms in the national research and innovation systems through provision of various tax incentives for SMEs, particularly those in the start-up stage, to increase the size of the domestic venture capital market. The adopted Action Plan for Research and Innovations in 2012 emphasized the need to increase the number of innovative SMEs with high growth potential, enhance the quality and internationalization of Finnish SMEs, accelerate the process of public-private partnerships in industrial cluster formation, and foster specialization through the Innovative Cities Programme.

EU countries demonstrating lower levels of innovation include Romania (R&D intensity of 0.49% of GDP), Bulgaria (0.64% of GDP) and Slovak Republic (0.82% of GDP). These countries share such features as: low levels of economic competitiveness, fragmentation of the public R&D systems, insufficient number of patent applications, no active collaborative links between public research centers and the private sector, inadequate levels of public spending on innovations, weak commercialization of research results and problematic enforcement of intellectual property legislation.

Innovation-Driven Entrepreneurship Impact: Measurement and Evaluation

The measures of innovation encompasses input innovation indicators, process (organization) and performance (output) indicators. There are aslo attempts at using composite innovation measures gauging the innovation potential of SMEs along the entire production chain.

Input Indicators of firm innovation include R&D intensity, R&D personnel, number of issued patents (Paces & Griliches, 1984), R&D resources allocated for external technology (*e.g.*, acquisition of patents, trademarks or licenses), and expenditures for new product marketing. In a similar vein, resources for training related to introduction of new or changed products are treated as inputs (Rodgers, 1998). Possible weaknesses of input indicators are that they do not account for process efficiency or economic and qualitative value, nor do they measure the technological complexity of goods or services.

Process innovation indicators generally reflect the organizational processes, management systems and design of the firm's innovation system (Howells, 1995). While the effects of the introduction of new products and services can be directly measured in monetary terms, those of *process innovations* are not directly tangible and are more difficult to gauge. According to Loefsten (2014), a firm's innovation strategy should align to its organizational structure in order to propel sustained company growth. Rapid innovation occurs within firms with clearly established innovation strategies, which is well understood by management and is aligned to other business areas. The model of organizational innovation is based on a resource-based theory of firm performance (Penrose, 1959), in which knowledge is an intangible resource resonating throughout the organization, which contributes to new processes and structures, directly impacting future performance.

The value of innovation activities is often drawn from *output indicators*, including profit, revenue growth, share performance, market capitalization and productivity. Alternative output measures encompass production of new or improved products, patents, patent quotes, trademarks, and designs. However, patents do not directly correspond to commercially viable innovations because enterprises often choose to internalize those that are commercially sensitive. Costs influence patent preference, thus creating situations in which patents are withheld from commercial application. Moreover, the technological level and the economic value of patents are extremely heterogeneous and not all innovations are patented.

A *composite innovation index*, which combines input, process and output indicators, captures the various factors of performance and provides a comprehensive evaluation of the firm's levels of innovation (Carayannis, 2010). Utilizing correlation analysis, Carayannis established that R&D expenditures correlate with the percentage of firm sales and that innovation correlates with venture capital provided for innovative needs. The number of patents correlates only with revenues and expected sales.

At the microeconomic level, various studies have established that SMEs and large corporations (LCs) innovate in a divergent way (Vossen, 2010). The relative strength of LCs is associated with ample resources (Nooteboom, 1994) to finance R&D (including internal funds). LCs have high output and synergy potential, diversified portfolio of R&D projects, well-developed marketing channels and realize scale and scope economies in the R&D process due to higher productivity, division of labor and spreading of risks. Alternatively, the innovative strengths of SMEs lie in the high motivation of management and employees, efficient and fast communication, specialized knowledge and skills, and high flexibility. Most empirical findings suggest that SMEs conduct R&D more efficiently than LCs (Vossen, 2010) and more effectively take advantage of knowledge spillovers from corporate R&D. According to Cohen and Klepper (1992), LCs are better at large-scale fundamental and technological innovations with higher average economic value while SMEs predominantly undertake product innovations driven by their flexibility and market proximity. In one LC for example, at 3M, the R&D process is organized in three levels: 1) basic (fundamental) science-based research into new materials in cooperation with universities and research centers; 2) technological and process R&D; and 3) R&D focused on product development. These differences in the innovation patterns of SMEs and LCs point toward the potential for collaborative open innovations between them by integrating SMEs into various innovation networks and systems. However, Stuart (2000) demonstrated that older and larger organizations derive fewer benefits from these forms of cooperation than their smaller start-up counterparts.

SOCIAL ENTREPRENEURSHIP AND INCLUSIVE GROWTH

Within Europe, individuals at risk of financial poverty increased by 10 million between 2009 and 2012. Lower growth and depressed productivity levels, along with a deteriorating social environment, has not only increased unemployment, it has reduced the capacity of government and civil sector organizations to support individuals, families and communities during jobless periods. Demographic shifts, such as a rapidly ageing population, have compounded these trends, while wealth inequality has magnified these issues for specific segments of the population, particularly minority groups. Social entrepreneurship, embodied in the SME form of social enterprise (SE), seeks to mitigate and stem these trends, reducing social exclusion, increasing access to social services and effectively allocating scarce resources to empower all segments of the population, often through collective bargaining power.

The founder of the Ashoka Fellowship, William Drayton (Leviner, Churchfield & Wells, 2012), first coined the phrase "social entrepreneurship" as a process through which an entrepreneur explicitly seeks new solutions to entrenched social issues.

SE are not exclusively motivated by financial interests; rather, they are driven by the objective to provide goods or services to the community, particularly minority populations (*i.e.,* a social purpose). This process further enhances the generation of collective externalities, which may be leveraged for communal wellbeing.

SEs are multi-stakeholder enterprises which combine the ownership interests and claims of diverse groups of stakeholders, while also developing trust and social capital through their organizational model. In doing so, SEs may lower transaction costs through the notion, and dynamics, of trust. Social capital in these organizations improves production factors (Laville, 2010) and, in some cases, the return on financial capital. SEs often impose limits to the distribution of profits and tend to operate at the convergence of the public, private and civil sectors to address the capacity shortfalls and leverage the resources of each (Nyssens, 2006).

Dees (2009) first described the macroeconomic perspective of SEs in distinguishing between "adaptive" and "allocative" efficiency. The former is a dynamic concept, which refers to the ability of organizations to address problems through a process of continual adaptation. SEs are reliant on supportive institutional environment (*e.g.*, legal, cultural, organizational and environmental institutions) to promote the still relatively new model of SE and mobilize commensurate levels of social and political capital. The microeconomic aspect of SEs refers to the localized generation of sustainable and innovative solutions to social issues. These enterprise forms contribute to social cohesion and employment through novel organization and production methods, which are among the main goals of the Europe 2020 Strategy. In traditional management theory, the firm centers on value capture. In contrast, SEs facilitate diffuse value *creation*. For SEs, business models based on intellectual property protections and the creation of competitive advantage have been supplanted by open source mechanisms in which solutions are mutually developed and shared by many stakeholders (*i.e.,* a multi-stakeholder model).

The Social Entrepreneurship Environment

Framework conditions for the development of social entrepreneurship include the presence of highly qualified human resources, beneficial and conducive legal and institutional environment, stable political context and socially inclusive system.

On the supply side, social entrepreneurship is driven by the following factors:

- Financial support in the form of grants for early stage development, prizes for social innovations, debt and equity instruments, "patient" capital, crowdfunding, social impact bonds and venture philanthropy;
- Business networks such as the development of social incubators, accelerator programs, mentoring and coaching and professional services (*e.g.,* legal, marketing, fiscal, accounting, governance and human resources);
- Development of skills for social entrepreneurship and innovation through courses such as training for social entrepreneurs, university programmes, and mobility schemes.

On the demand side, social entrepreneurship is influenced by:

- Public procurement and commissioning of socially innovative goods and services;
- Supporting private demand through tax incentives and subsidies;
- Development of the national knowledge base and the measuring of impact and outcomes;
- Implementation of national cluster policies and support for R&D cooperation in the social sphere.

Influenced by favorable and well-developed ecosystems, SEs have the potential to generate significant social, economic and environmental impact and to contribute to the achievement of Europe 2020 Strategy goals. These goals include the creation of socially sustainable jobs, the facilitation of social integration and improved social standards. As a result, according to a Global Entrepreneurship Monitor survey (GEM Consortium, 2009), the share of the population employed by SEs in Belgium is 4.1%, 7.5% in Finland, 3.1% in Germany, 5.7% in UK, and 3.3% in Italy. Approximately one in four businesses founded in Europe is a SE. The social economy in Europe is estimated to engage as many as 15 million employees, equivalent to about 7% of the working population in the EU. The greatest number of SEs is found in the UK, followed by Germany, Spain, France and Finland.

Similarly, according to a global comparative study of social entrepreneurship (GEM Consortium, 2016), the average rate of social entrepreneurship activity (SEA) among 58 surveyed economies is 3.2%. The innovation-driven economies with the highest SEA in the EU are Luxembourg (over 6%), Finland, Ireland, Estonia and Slovak Republic. At the other end of the spectrum are Germany, Spain and Greece (less than 2% SEA). The survey shows that innovative SEs globally are 1.6% (lowest number in Bulgaria and Greece, highest - in Luxembourg and Sweden).

Social Entrepreneurship Impact: Measurement and Evaluation

Social impact measurement (SIM) broadly refers to a set of processes intended to evaluate the social value and impact generated by SEs. The conception of "social impact" includes both social value creation (Emerson, Wachowicz & Chun, 2000) and "social return" (Clark and Eustace, 2009). According to Clifford (2014), "social impact" encompasses the results from an economic activity, refers to multi-stakeholder effects, and includes both positive and negative outcomes (Wainwright, 2002).

SEs can leverage SIM to set targets, monitor and improve performance and mobilize funding. Because SEs balance an array of outcomes while attempting to align incentives among multiple stakeholders, SIM is often regarded as an imperfect process still in the initial stages of development (Epstein & McFarlan, 2011). To date, there is neither commonly agreed definition of SIM, nor a shared understanding of the general aim of SIM. Mulgan (2010) proposes about 10 ways to measure social value and Rockefeller Foundation has developed a catalogue of over 20 different approaches to measuring the social impact.

Global initiatives for SIM include those of the OECD, the EU and G8. The OECD (2015) recommends establishing adequate data infrastructure and development of streamlined and standardized processes to more effectively monitor the evolution of the social impact investment market. At the EU level, the European Commission (2014) has drawn up guidelines for SIM based on the criteria for the European Social Entrepreneurship Fund passport. According to the document, a unique set of SIM indicators may not be preferable. Instead, The Commission recommended the adoption of a social impact measuring process with mandatory characteristics defining measurement disclosure. The choice of frameworks and indicators should be selected by the SE in accordance with multiple stakeholders, and include validation by an independent audit. Finally, G8 (2014) has elaborated guidelines which further expand the European Commission framework and enlarge the definition of social impact to include environmental goals of SEs. The long-term vision of G8 is the development of a more formalized impact measurement standard and reporting system that enhances the availability of impact data.

At the microeconomic level and in adherence to a *stakeholder approach to measuring social impact,* the process should satisfy the information needs of the principal stakeholders of the SE. For this reason, a variety of SIM indicators is necessary in order to meet the demand of various stakeholders (Harlock, 2013), including social accounting and auditing, social return on investment, and cost-benefit analysis. This approach is in conformity with the global social impact framework process.

SUSTAINABLE ENTREPRENEURSHIP AND SUSTAINABLE GROWTH

Without disentangling economic growth from environmental degradation, current consumption and production patterns will continue to preclude the realization of a sustainable future. Current economic systems promote resource inefficiency, while failing to price these missteps at their true costs. This type of resource inefficient development will have deleterious effects on the economy, particularly as it faces increasing costs for essential raw materials, absence of security of supply and price volatility. Sustainable entrepreneurship seeks to restructure EU economic growth to promote rather than degrade the environment. It integrates the long-term social, economic and environmental impact of venture opportunities. According to Cohen, Smith and Mitchell (2008), entrepreneurship is in itself a means to further sustainable objectives.

The World Commission on Environment and Development ("*The Brundtland Commission*") (1987) first introduced the concept of sustainable entrepreneurship. The commission outlined the need to strike a balance between society, economy and ecology (the so called "triple-bottom line")[1]. For Gibbs (2009), sustainable entrepreneurship includes a process of creative destruction in combination with the development of sustainable social, economic and environmental systems, which align the incentives of various internal and external stakeholders (De Palma & Dobes, 2010).

The Sustainable Entrepreneurship Environment

The macro-level factors of sustainable entrepreneurship encompass the environment within which sustainable entrepreneurs operate. These factors include institutional development, the efficiency of existing public policies and the responsiveness of firms to shifting regulatory requirements. Hall, Daneke and Lenox (2010) identify these conditions as prerequisite to the balancing of economic, social and environmental objectives. Alternatively, the micro-level aspects of sustainable entrepreneurship refer to the creation of sustainable value chains and the cultivation of sustainable business models (Parrish, 2010).

At the EU level, environmental sustainability is evaluated by the extent to which the natural environment is contributing to overall national competitiveness. The EU member states leading in environmental sustainability include Sweden, Finland and Latvia. These countries all demonstrate commitments to existing EU environmental legislation, ratification of international environmental treaties and application of environmentally sustainable practices by their SMEs. On the other hand, the EU countries exhibiting lower levels of sustainability include Hungary, Malta and Romania (WEF, 2014). These countries are expected to complement future economic growth with efforts to protect and enhance the natural environment.

Sustainable Entrepreneurship Impact: Measurement and Evaluation

Many of the measurement and evaluation efforts regarding sustainable entrepreneurship arise from various international initiatives. In 1987, "*The Brundtland Commission*" brought the environmental issue high on the political agenda by stressing the link between economic, social and environmental challenges. In 2000, the UN adopted the Millennium Declaration according to which countries undertake measures for eliminating poverty, promoting social equity, democracy and environmental sustainability. In order to operationalize the conception of sustainable development, the UN codified the Millennium Development Goals (MDG) (UN, 2000) and later adopted Agenda 2030, addressing the economic, social and environmental aspects of sustainable development in a balanced manner (UN, 2015). Research by Seelos and Mair (2005a) established that of 74 enterprises within the Schwab Foundation network, 68% have a direct impact on MDGs while 8% have an indirect impact. Among the social enterprises directly impacting the MDGs, the majority are working to reduce poverty, empower women and conserve natural resources. Furthermore, SEs often impact more than one MDG.

Economic, social and environmental aspects are also part of the EU Sustainable Development Strategy, which was enacted in 2001. Progress toward the objectives of the Strategy is assessed using a set of social development indicators (SDI). The Europe 2020 Strategy advances the EU Sustainable Development Strategy by contributing to increased coherence and integration of the three dimensions - economic, social and environmental - in EU polices. Key indicators for social and economic development are real GDP per capita, eco-innovation, R&D expenditures and employment. At EU level the latest data from Eurostat for 2015 show that leading EU member states according to GDP per capita are Norway, Luxembourg and Ireland, while at the lower end stand Bulgaria and Romania whose GDP per capita represents only about 30% of the EU average. Regarding total employment rate indicator, best performers are Sweden, the Netherlands and Finland (exceeding the EU-28 average rate of 70%), while the worst performing countries are Greece and Italy, respectively. Last but not least, the most eco-innovative countries are Denmark, Sweden and Finland, while according to this indicator EU "laggards" are Poland, Bulgaria and Cyprus.

EUROPEAN UNION SOLUTIONS TO BOOSTING SMART, SUSTAINABLE AND INCLUSIVE ENTERPRISE VALUE CREATION

Based on the comparative analysis of innovative, inclusive and sustainable SME development in the EU, an array of disparities can be identified between the leading northern and north-eastern EU member states and the southern and central-eastern European ones. Boosting innovative, inclusive and sustainable SME growth in the lagging countries will require increased levels of investments in R&D, accelerated adoption of ICT, improved business environments, extensive collaboration between research centers, businesses and universities, and better education and social security systems. The comparative analysis among EU countries substantiates *Assumption 2* of the authors, which indicates that the majority of European SMEs are insufficiently entrepreneurial and innovative, and this necessitates EU institutional level intervention for enhancing their potential through various financial support measures and programmes.

The *"Europe 2020" Strategy* (European Commission, 2010) prioritizes research and innovation (R&I) as a means to promote smart, economically sustainable and socially inclusive economic growth. The financial and economic crisis negatively impacted the progress toward some of the Europe 2020 targets such as employment, R&D and poverty reduction.

In view of promoting smart economic growth, the *Innovation Union Initiative* advances the Single Market objective, framework conditions for business innovation through single EU patent, EU-level patent court, improved intellectual property protection and access to capital for innovative SMEs. The Initiative also seeks to strengthen the linkages between educational institutions, businesses and innovation centers. *The Framework Programme for Research and Innovation - Horizon 2020* is intended to provide capital access in support of SMEs that engage in R&I. The framework promotes an integrated approach to SME development through building a society and economy based on knowledge and innovation across the EU in view of attaining the R&D target of 3% of GDP by 2020. The *Industrial Leadership pillar of the Programme* aims to speed up development of technologies and innovations that will support innovative SMEs growth and international presence.

Furthermore, t*he Initiative for Industrial Policy in a Globalized Era of the Europe 2020 Strategy* establishes a framework for a modern industrial policy that directly bolsters SME entrepreneurship and competition. At the national level, member states are encouraged to improve the business environment for innovative SMEs, reduce administrative burdens and boost the quality of commercial law. The EU still continues to underperform in terms of building smart, innovation-driven and knowledge-based economies as compared to other advanced economies (WEF, 2014). The innovation gap between the EU, the US and Japan is notably due to levels of ICT expenditures, triad patent applications and population with tertiary education.

To enhance SME competitiveness at the EU level, for the programming period 2014-2020, the European Commission adopted the *Competitiveness of Small and Medium-sized Enterprises Programme* (COSME). It seeks to foster a more entrepreneurial environment and supports SMEs in areas such as capital access and cross-border market expansion. The venture capital financial instruments under the COSME programme complement "Horizon 2020." By 2017, the Programme is projected to provide more than 100.000 SMEs with various forms of credit, venture capital or guarantees worth EUR 10.76 bln.

The evaluations (European Commission, 2012) of Competitiveness and Innovation Programme (2007 - 2013) showed that the Programme supported the economic growth of SMEs. A key effect of the Programme was to generate multiplier effects and to attract public and private funding for projects in the field of venture capital in the EU. According to a survey of SMEs which were beneficiaries of the guarantee and loan financial facilities under the Programme, the following has been established:

- The majority of European SMEs utilizing funding under the guarantee facility recognized the positive effects of the funding (76%) and indicated that they have created more than 10 jobs post funding (over 47% of SMEs);
- Over 83% of SMEs utilized the financial facility for early stage development and 61% of those using the guarantee facility realized product innovation;
- About 48% of SMEs which leveraged the financial facility for early stage development achieved over 21% growth in turnovers over the previous three years, while over 94% indicated that the scheme had a positive effect on their long-term growth.

Despite these positive effects, the evaluation of the indicators of the Competitiveness and Innovation Programme acknowledged significant weaknesses, including low entrepreneurial spirit in Europe, where less than 50% of the population indicated a preference for self-employment, unfavorable business environment and administrative burdens that restrict entrepreneurship and capital access.

A survey by the author (Stefanova, 2014) of innovative Bulgarian SMEs showed that that the majority of Bulgarian SMEs predominantly use own capital and retained earnings to finance their operations. About 20% of Bulgarian SMEs have used financial resources under European Programmes such as the Operational Programme "Competitiveness" and the Joint European Resources for Micro and Medium Enterprises (JEREMIE) (Executive Agency for SME Encouragement in Bulgaria, 2011). The main benefits that these SMEs have obtained in this process include establishment of new contacts, expansion into new markets, and exchange of experience with companies in the same or other sectors. Difficulties that Bulgarian SMEs highlighted include: burdensome application procedures, unclear programme requirements for co-financing, and slow and difficult utilization of amounts exceeding the terms of the contracts. Approximately 5% of Bulgarian SMEs have developed products that are subject to protection under intellectual property rights. Obstacles in patenting trademarks include the extremely low level of intellectual property protection in Bulgaria (especially for "small players") and violation of the terms for publication by the patent office.

In support of inclusive economic growth, the Single Market Act of 2012 included the Social Business Initiative – an instrument established under the *EU Programme on Social Change and Innovation (2007-2013)*– to support social enterprise development. The Social Business Initiative identified three strands of action in support of SEs in EU: improving access to finance, increasing the visibility of SEs and optimizing the legal environment in which these enterprises operate. The *European Programme for Employment and Social Innovation 2014-2020* recognizes social innovation as a means for solving entrenched social issues and supports SEs with demonstrable long-term impact. The social economy and social entrepreneurship constitute an integral part of the EU social market, which, in aggregate, foster social convergence and inclusive labor markets. The Programme bolsters the inflow of socially-driven capital through the Microfinance and Social Entrepreneurship Axis by providing equity, quasi-equity, loan instruments and grants to SEs of up to EUR 500,000.

Finally, the European Commission developed an implementation mechanism for sustainable economic growth within the 2001 *European Strategy for Sustainable Development* (ESSD). This mechanism espouses a mode of economic development in which growth, social progress and preservation of the environment are mutually reinforcing. All EU policies include evaluation of expected economic, social and environmental costs and benefits in order to facilitate benchmarking and comparison. Moreover, publicly quoted companies are encouraged to publish "triple bottom line" statements in their annual reports to evaluate performance against economic, environmental and social criteria as part of their corporate social responsibility. The ESSD and The Europe 2020 Strategy are complementary. The Europe 2020 Strategy is a dynamic mid-term strategy, while the ESSD is a long-term framework document. The European Commission continues to monitor progress of the ESSD based on a pre-defined Sustainable Development Indicators in coordination with Eurostat.

FUTURE RESEARCH DIRECTIONS

Future research may be directed to financial instruments for SME subsets, particularly venture capital and social entrepreneurship funds. SMEs, especially those in their seed and early expansion stages, encounter both internal and external barriers[2] to capital access, some of which are cyclical and some structural (Wilson et al., 2015). Three EU Programme funds include venture capital (VC) as a focal investment vehicle, namely JEREMIE, COSME and Horizon 2020. The VC instruments could be delineated and analyzed based on features such as targeted SME subsets and return expectations.

An additional research area within the domain of the topic would be an empirical evaluation of the determinants which make up the "ecosystem" for various SME subsets through a national or comparative regional approach. These determinants would encompass access to finance, depth of the capital markets, level of venture capital and philanthropic activity, B2B services, macroeconomic policy instruments, entrepreneurial culture and the social environment.

Prospective research could also empirically analyze the effects of entrepreneurship and innovation on a regional or national basis with particular reference to EU Programme effectiveness on the long-term growth and SME development. Moreover, research efforts may identify and compare the drivers and barriers facing social enterprises in underperforming EU member states.

CONCLUSION

Theoretical and empirical evidence confirms the existence of positive relationships between entrepreneurship, innovation, enterprise value creation and economic growth. Furthermore, the above comparative analysis of SME subsets indicates significant divergence across EU countries. If a Smart, Sustainable and Inclusive Growth model is made on the basis of the theoretical and the practical approach, proposed by the research, it may be able to articulate country-specific measures at institutional and market level, needed to overcome the competitiveness and innovation divide in EU and with renewed impetus to advance the goals of SSIG.

The EU continues to restructure its SME interventions to foster value creation and entrepreneurship in support of SSIG. The provision of support to these subsets of enterprises and the pursuit of fully integrated financial markets may serve to promote innovation, redress the recessionary aftereffects of the financial crisis and secure sustainable economic growth for Europe's 23 million SMEs. However, the extent to which the above policy interventions succeed in their socioeconomic vision relies not only on the development of SME growth markets or the realization of SSIG. Ultimately, their efficacy depends on the long-term health of European SMEs and their ability to support EU citizens across all segments of society in a sustainable, innovative and equitable way across Europe.

REFERENCES

G8. (2014). *Measuring impact.* Retrieved February 20, 2016, from http://www.socialimpactinvestment.org/reports/Measuring%20Impact%20WG%20paper%20FINAL.pdf

Aghion, P., & Howitt, P. (1998). *Endogenous growth theory*. Cambridge, MA: The MIT Press.

Andersson, F. O. (2011). Organisational capacity and entrepreneurial behaviour in nonprofit organisations: An empirical exploration. *Voluntary Sector Review*, *2*(1), 43–56. doi:10.1332/204080511X560611

Artz, K. W., Norman, P. M., Hatfield, D. E., & Cardinal, L. B. (2010). A longitudinal study of the impact of R&D, patents, and product innovation on firm performance. *Journal of Product Innovation Management*, *27*(5), 725–740. doi:10.1111/j.1540-5885.2010.00747.x

Bartik, T. (1989). Small business start-ups in the United States: Estaimates of the effects of characteristics of states. *Southern Economic Journal*, *55*(4), 34–56. doi:10.2307/1059479

Baumol, W., Panzar, J., & Willig, R. (1982). *Contestable markets and the theory of industrial structure.* New York: Harcourt Brace Jovanovich, Inc.

Bosworth, D., & Rogers, M. (1998). *Research and development, intangible qssets and the performance of large Australian companies*. Melbourne: Melbourne Institute.

Carayannis, L., & Provance, M. (2010). Measuring firm innovativeness. Towards composite innovation index built on firm innovative posture, propensity and performance attributes. *International Journal of Innovation and Regional Development*, *1*(1), 90–107. doi:10.1504/IJIRD.2008.016861

Ciccone, A., & Papaioannou, E. (2006). *Red tape an delayed entry*. London: CEPR.

Clifford, J. (2014). *Impact evaluation by social enterprises: measuring the un-measurable?* Retrieved April 20, 2016, from http://www.oecd.org/cfe/leed/10th-fplg-meeting.htm

Coe, D. T., Helpman, E., & Hoffmaister, W. A. (1995). *North-south R&D spillovers.* NBER Working Paper, No. 5048. Cambridge, MA: National Bureau of Economic Research.

Cohen, B., Smith, B., & Mitchell, R. (2008). Toward a sustainable conceptualization of dependent variables in entrepreneurship research. *Business Strategy and the Environment*, *17*(2), 107–119. doi:10.1002/bse.505

Cohen, W. M., & Klepper, S. (1992). The anatomy of R&D intensity distributions. *The American Economic Review*, *82*(4), 773–799.

De Palma, R., & Dobes, V. (2010). An integrated approach towards sustainable entrepreneurship. Experience from the TEST project in transitional economies. *Journal of Cleaner Production*, *18*(18), 1807–1821. doi:10.1016/j.jclepro.2010.07.025

Dees, J. G. (2009). *Social ventures as learning laboratories.* Retrieved March 15, 2016, from https://centers.fuqua.duke.edu/case/wp-content/uploads/sites/7/2015/02/Article_Dees_SocialVenturesasLearningLaboratories_2009.pdf

Djankov, S., Glanser, T., McLiesh, C., Ramalho, R., & Schleifer, A. (2008). *The effect of corporate taxes on investment and entrepreneurship. NBER WP 13765.* Cambridge: NBER. doi:10.3386/w13756

Ejermo, O. (2009). Regional innovation measured by patent data – does quality matter? *Industry and Innovation*, *16*(2), 141–165. doi:10.1080/13662710902764246

Emerson, J. (2003). The blended value proposition. *California Management Review*, *45*(4), 35–51. doi:10.2307/41166187

Emerson, J., Wachowicz, J., & Chun, S. (2000). *Social return on investment: exploring aspects of value creation in the non-profit sector.* San Francisco: The Roberts Foundation.

Epstein, M. J., & McFarlan, F. W. (2011). Measuring the efficiency and effectiveness of a nonprofit's performance. *Strategic Finance*, *93*(4), 27–34.

European Commission. (2010). *Europe 2020. A European strategy for smart, sustainable and inclusive growth.* Retrieved August 16, 2016, from http://ec.europa.eu/eu2020/pdf/COMPLET%20EN%20BARROSO%20%20%20007%20-%20Europe%202020%20-%20EN%20version.pdf

European Commission. (2012). *Entrepreneurship and innovation programme.* EIP performance report. Retrieved August 1, 2016, from http://ec.europa.eu/cip/files/docs/eip_performance_report_january_2012_en.pdf

European Commission. (2014). *Proposed approach to social impact measurement in European Commission legislation and in practice relating to EuSEFs and the EuSI.* Brussels: GECES.

European Commission. (2014). *Research and innovation performance in the EU. Innovation union progress at country level.* Retrieved July 26, 2016, from http://ec.europa.eu/research/innovation-union/pdf/state-of-the-union/2014/iuc_progress_report_2014.pdf

Evans, D., & Leighton, S. (1989). Some empirical aspects of entrepreneurship. *The American Economic Review*, *79*(3), 519–535.

Executive Agency for SME Encouragement in Bulgaria. (2011). *Analysis of the state and the factors for the development of SMEs in Bulgaria*. Sofia: Executive Agency for SME Encouragement in Bulgaria.

Fonseca, R., Lopez-Garcia, P., & Pissarides, C. (2001). Entrepreneurship, start-up costs and employment. *European Economic Review*, *45*(4-6), 692–705. doi:10.1016/S0014-2921(01)00131-3

Frantzen, D. (2000). R&D, human capital and international technology spillovers: A cross country analysis. *The Scandinavian Journal of Economics*, *102*(1), 57–75. doi:10.1111/1467-9442.00184

Frontier Economics. (2013). *Exploring the impact of private equity on economic growth in Europe*. London: Frontier Economics Ltd.

GEM Consortium. (2009). *Global entrepreneurship monitor 2009. Global report*. Retrieved May 10, 2016, from http://www.babson.edu/Academics/centers/blank-center/global-research/gem/Documents/gem-2009-global-report.pdf

GEM Consortium. (2016). *Global entrepreneurship monitor 2015/2016. Global report*. Retrieved August 4, 2016, from http://www.gemconsortium.org/report/49480

Geroski, P., Machin, S., & Van Reenen, J. (1993). The profitability of innovating firms. *The Rand Journal of Economics*, *24*(2), 198–211. doi:10.2307/2555757

Geroski, P. A. (1989). Entry, innovation, and productivity growth. *The Review of Economics and Statistics*, *71*(4), 572–578. doi:10.2307/1928098

Gibbs, D. (2009). Sustainability entrepreneurs, ecopreneurs and the development of a sustainable economy. *Greener Management International*, *55*, 63–78.

Glaeser, E., & Kerr, W. (2009). Local industrial conditions and entrepreneurship: How much of the spatial distribution can we explain? *Journal of Economics & Management Strategy*, *18*(3), 623–663. doi:10.1111/j.1530-9134.2009.00225.x

Griffith, R., Redding, S., & Van Reenen, J. (2004). Mapping the two faces of R&D productivity growth in a panel of OECD industries. *The Review of Economics and Statistics*, *86*(4), 883–895. doi:10.1162/0034653043125194

Hall, J. K., Daneke, G. A., & Lenox, M. J. (2010). Sustainable development and entrepreneurship: Past contributions and future directions. *Journal of Business Venturing*, *25*(5), 439–448. doi:10.1016/j.jbusvent.2010.01.002

Harlock, J. (2013). *Impact measurement practice in the UK third sector: a review of emerging evidence*. Birmingham, UK: University of Birmingham.

Highfield, R., & Smiley, R. (1987). New business start-ups and economic activity: An empirical investigation. *International Journal of Industrial Organization*, *5*, 51–66. doi:10.1016/0167-7187(87)90006-3

Howells, J. (1995). A socio-cognitive approach to innovation. *Research Policy*, *24*(6), 883–894. doi:10.1016/0048-7333(94)00804-3

Hubbard, R. G. (2008). *Nondestructive creation. Entrepreneurship and management research in the study of growth.* Retrieved August 11, 2016, from http://www.sciencedirect.com/science/article/pii/S0161893808000355

Jimenez, J. D., & Sanz-Valle, R. (2011). Innovation, organizational learning and performance. *Journal of Business Research*, *64*(4), 408–417. doi:10.1016/j.jbusres.2010.09.010

Kannebley, S., Sekkel, J. V., & Araujo, B. C. (2010). Economic performance of Brazilian manufacturing firms: A counterfactual analysis of innovation impacts. *Small Business Economics*, *34*(3), 339–354. doi:10.1007/s11187-008-9118-x

Keilbach, M., & Audretsch, D. (2005). *The knowledge spillover theory of entrepreneurship.* Retrieved August 10, 2016, from http://web.mit.edu/iandeseminar/Papers/Fall2005/audretschkeilbach.pdf

Kritikos, A. (2010). *Entrepreneurs and their impact on jobs and economic growth.* Retrieved August 1, 2016, from http://wol.iza.org/articles/entrepreneurs-and-their-impact-on-jobs-and-economic-growth.pdf

Laville, J. (2010). *The social enterprise: towards a theoretical socio-economic approach.* Retrieved January 12, 2016, from http://www.ssc.wisc.edu/~wright/Social%20Economy%20PDFs/THE%20EMERGENCE%20OF%20SOCIAL%20ENTERPRISE/TEOSE-17.pdf

Leviner, N., Churchfield, L., & Wells, D. (2012). *Understanding the impact of social entrepreneurs: Ashoka's answer to the challenge of measuring effectiveness.* Retrieved January 16, 2016, from http://www.nj.gov/state/programs/pdf/faith-based-social-understanding-impact.pdf

Loefsten, H. (2014). Product innovation processes and the trade-off between product innovation performance and business performance. *European Journal of Innovation Management*, *17*(1), 61–84. doi:10.1108/EJIM-04-2013-0034

Lueckgen, I. (2004). *Nascent entrepreneurs in German regions. Evidence from the Regional Entrepreneurship Monitor Germany.* Retrieved August 2, 2016, from http://repec.iza.org/dp1394.pdf

Micco, A., & Pages, C. (2006). *The economic effects of employment protection: evidence from international industry-level data.* Discussion paper 2633. Bonn: IZA.

Mulgan, G. (2010). *Measuring social value.* Retrieved May 12, 2016, from http://www.ssireview.org/articles/entry/measuring_social_value

Murphy, K., Schleifer, A., & Vishny, R. (1999). The allocation of talent: The implications for growth. *The Quarterly Journal of Economics*, *106*(2), 503–530. doi:10.2307/2937945

Nevens, T., Summer, G. L., & Uttal, B. (1990). Commercializing technology: What the best companies do. *Harvard Business Review*, (May–June), 154–163.

Nooteboom, B. (1994). Innovation and diffusion in small firms: Theory and evidence. *Small Business Economics*, *6*(5), 327–347. doi:10.1007/BF01065137

NUI Galaway & Whitaker University. (2015). *Macro and firm level innovation measurements and indicators.* Retrieved May 25, 2016, from http://galwaydashboard.ie/publications/white-paper-measurement-2015.pdf

Nyssens, M. (2006). *Social enterprise at the crossroads of market, public and civil society*. London: Routledge.

OECD. (2005). *Oslo Manual*. Retrieved May 28, 2016, from https://www.oecd.org/sti/inno/2367580.pdf

OECD. (2015). *Building the evidence base. Social impact investment. Preliminary version*. Retrieved May 16, 2016, from http://www.oecd.org/sti/ind/social-impact-investment.pdf

Paces, A., & Griliches, Z. (1984). Patents and R&D at the firm level: a first look. In Z. Griliches (Ed.), *R&D, patents, and productivity* (pp. 55–72). Chicago: University of Chicago Press.

Parrish, D. (2010). Sustainability-driven entrepreneurship: Principles of organization design. *Journal of Business Venturing*, *25*(5), 510–523. doi:10.1016/j.jbusvent.2009.05.005

Penrose, E. (1959). *The theory of the growth of the firm*. Oxford, UK: Oxford University Press.

Portela, M. (2013). *Entrepreneurship and economic growth: macroeconomic analysis and effects of social capital in EU*. Retrieved August 1, 2016, from http://www.intechopen.com/books/entrepreneurship-born-made-and-educated/entrepreneurship-and-economic-growth-macroeconomic-analysis-and-effects-of-social-capital-in-the-eu

Rauch, A., Wiklund, J., Lumpkin, G. T., & Freese, M. (2009). Entrepreneurial orientation and business performance: An assessment of past research and suggestion for the future. *Entrepreneurship. Theory into Practice*, *33*(3), 761–781.

Reynolds, P. (1992). Predicting new-firm births: interactions of organizational and human populations. In D. Sexton & J. Kasarda (Eds.), *Entrepreneurship in the 1990s* (pp. 59–71). Boston: PWS-Kent Publishing.

Rodgers, M. (1998). *The definition and measurement of innovation*. Working paper 10. Melbourne: Melbourne Institute of Applied Economic and Social Research.

Romer, M. (1986). Increasing returns and long run growth. *Journal of Political Economy*, *94*(5), 1002–1037. doi:10.1086/261420

Rosenbusch, N., Brinckmann, J., & Bausch, A. (2011). Is innovation always beneficial? A meta-analysis of the relationship between innovation and performance in SMEs. *Journal of Business Venturing*, *26*(4), 441–457. doi:10.1016/j.jbusvent.2009.12.002

Savvides, A., & Zachariadis, M. (2003). *International technology diffusion and TFP growth*. Oklahoma State University.

Scherer, F. M. (1982). Inter-industry technology flows and productivity growth. *The Review of Economics and Statistics*, *64*(4), 627–634. doi:10.2307/1923947

Schumpeter, J. A. (1934). *The theory of economic development. An inquiry into profits, capital, credit, interest, and the business cycle*. Cambridge, MA: Harvard University Press.

Seelos, C., & Mair, J. (2005a). Social entrepreneurship: Creating new business models to serve the poor. *Business Horizons*, *48*(3), 241–246. doi:10.1016/j.bushor.2004.11.006

Silva, I. (2012). *Entrepreneurial and innovative behavior in Spanish SMEs. Essay on profitability, growth and export intensity* (Doctoral dissertation). Retrieved August 12, 2016, from https://ddd.uab.cat/pub/tesis/2013/hdl_10803_117454/imds1de1.pdf

Smith, M., & Culbertson, S. (2010). *Enhancing entrepreneurship: The role of goal orientation and self-efficacy.* Retrieved August 8, 2016, from http://jca.sagepub.com/content/early/2011/01/13/1069072710385543.abstract

Stam, E., & Wennberg, K. (2009). The roles of R&D in new firm growth. *Small Business Economics, 33*(1), 77–89. doi:10.1007/s11187-009-9183-9

Stefanova, J. (2014). *Venture capital in the EU: prospects of Bulgarian companies to access new forms of financing*. Veliko Turnovo: Faber Publishing House.

Stuart, T. E. (2000). Inter-organizational alliances and the performance of firms: A study of growth and innovation rates in a high technology industry. *Strategic Management Journal, 21*(8), 791–811. doi:10.1002/1097-0266(200008)21:8<791::AID-SMJ121>3.0.CO;2-K

Swedish Entrepreneurship Forum. (2016). *The Entrepreneurial challenge. A comparative study of entrepreneurial dynamics in China, Europe and the US.* Retrieved August 4, 2016, from http://entreprenorskapsforum.se/wp-content/uploads/2016/06/GEM_Internationell_Rapport_2016_Webb.pdf

Szerb, L., Zoltan, A., Komlosi, E., & Argiles, R. (2015). *Measuring entrepreneurial systems: The regional entrepreneurship and development index.* Retrieved August 1, 2016, from http://ec.europa.eu/regional_policy/sources/docgener/studies/pdf/regional_entrepreneurship_development_index.pdf

Ulku, H. (2004). *R&D, innovation and economic growth. An emprical evidence.* IMF Working Papers 04/185. Washington, DC: IMF.

United Nations. (2000). *Millenium declaration.* Retrieved August, 16, 2016, from http://www.un.org/millennium/declaration/ares552e.htm

United Nations. (2015). *Transforming our world: the 2030 agenda for sustainable development.* Retrieved August, 16, 2016, from http://www.un.org/ga/search/view_doc.asp?symbol=A/RES/70/1&Lang=E

Vemic, M. (2013). Evaluation of knowledge-based high-growth enterprises in Serbia's economic development. *Online Journal of Applied Knowledge Management, 1*(2), 1–14.

Vossen, R. (2010). *Combining small and large firm advantages in innovation: theory and evidence.* Retrieved August 1, 2016, from http://www.rug.nl/research/portal/files/3183201/98b21.pdf

Wach, K. (2012). *Impact of the regional business environment on the development of SMEs in Southern Poland.* Retrieved August 3, 2016, from http://www.businessenvironment.org/dyn/be/docs/272/The%20Impact%20of%20Business%20Environment%20on%20Small%20and%20.pdf

Wainwright, S. (2002). *Measuring impact: A guide to resources.* London: NCVO.

Williamson, O. (1971). The verical integration of production: Market failure considerations. *The American Economic Review, 61*(2), 112–123.

Wilson, K. E., Silva, F., & Richardson, D. (2015). *Social impact investment: building the evidence base.* Retrieved December 14, 2015, from http://www.oecd.org/sti/ind/social-impact-investment.pdf

Wolfe, R. (1994). Organizational innovation: Review, critique and suggested research directions. *Journal of Management Studies*, *31*(3), 405–431. doi:10.1111/j.1467-6486.1994.tb00624.x

World Commission on Environment and Development ("The Brundtland Commission"). (1997). *Our Common Future*. Retrieved May 30, 2016, from http://www.un-documents.net/our-common-future.pdf

World Economic Forum. (2014). *The Europe 2020 competitiveness report. Insight report. Building a more competitive Europe.* Retrieved July 25, 2016, from http://www.weforum.org/Europe2020

Zachariadis, M. (2003). R&D, innovation, and technological progress: A test of the Schumpeterian framework without scale effects. *The Canadian Journal of Economics. Revue Canadienne dEconomique*, *36*(3), 566–686. doi:10.1111/1540-5982.t01-2-00003

ADDITIONAL READING

Alter, S. K. (2006). Social enterprise models and their mission and money relationships. In A. Nicholls (Ed.), *Social entrepreneurship: new models of sustainable social change* (pp. 205–232). Oxford: Oxford University Press.

Audretsch, D. B. (1995b). *Innovation and industry evolution*. Cambridge, Massachusetts: MIT Press.

Austin, J., Stevenson, H., & Wei-Skillern, J. (2006). Social and commercial entrepreneurship: same, different, or both? *Entrepreneurship theory and practice*, 30(1), 1-22.

Baldwin, J., & Johnson, J. (1999). Entry, innovation and firm growth. In Z. Acs (Ed.), Are small firms important? Their role and impact (51-79). Dordrecht, Boston, MA and London: Kluwer Academic Publishers. doi:10.1007/978-1-4615-5173-7_4

Baumol, W. (1990). Entrepreneurship: Productive, unproductive and destructive. *Journal of Political Economy*, *98*(5), 893–921. doi:10.1086/261712

Borzaga, C., & Solari, L. (2001). Management challenges for social enterprises. In C. Borzaga & J. Defourny (Eds.), *The Emergence of Social Enterprises* (pp. 333–349). New York: Routledge.

Certo, S. T., & Miller, T. (2008). Social entrepreneurship: Key issues and concepts. *Business Horizons*, *51*(4), 267–271. doi:10.1016/j.bushor.2008.02.009

Choi, D. I., & Gray, E. R. (2008). The venture development processes of sustainable entrepreneurs. *Management Research News*, *31*(8), 558–569. doi:10.1108/01409170810892127

Coleman, J. S. (1990). *Foundations of social theory*. Cambridge: Harvard University Press.

Coombs, R., Narandren, P., & Richard, A. (1996). A literature-based innovation output indicator. *Research Policy*, *25*(4), 403–413. doi:10.1016/0048-7333(95)00842-X

Daft, R. L. (1978). A dual-core model of organizational innovation. *Academy of Management Journal*, *21*(2), 193–210. doi:10.2307/255754

Damanpour, F. (1991). Organizational innovation. A meta-analysis of effects of determinants and moderators. *Academy of Management Journal*, *34*(3), 555–590. doi:10.2307/256406

Dart, R., Clow, E., & Armstrong, A. (2010). Meaningful difficulties in the mapping of social enterprises. *Social Enterprise Journal*, *6*(3), 186–193. doi:10.1108/17508611011088797

Dees, J. G. (1998). Enterprising nonprofits. *Harvard Business Review*, *76*(1), 54–67. PMID:10176919

Dees, J. G., & Economy, P. (2001). Social entrepreneurship. In J.G.Dees, J.Emerson & P.Economy (Eds.). Enterprising non-profits: a toolkit for social entrepreneurship (1-18). Canada and USA: Wiley.

Gibbon, J., & Dey, C. (2011). Developments in social impact measurement in the third sector: Scaling up or dumbing down?'. *Social and Environmental Accountability Journal*, *31*(1), 63–67. doi:10.1080/0969160X.2011.556399

Gompers, P., & Lerner, J. (2001). The venture capital revolution. *The Journal of Economic Perspectives*, *15*(2), 145–168. doi:10.1257/jep.15.2.145

Hellman, T., & Puri, M. (2000). The interaction between product market and financing strategy: The role of venture capital. *Review of Financial Studies*, *13*(4), 959–984. doi:10.1093/rfs/13.4.959

Highfield, R., & Smiley, R. (1987). New business start-ups and economic activity: An empirical investigation. *International Journal of Industrial Organization*, *5*, 15–28. doi:10.1016/0167-7187(87)90006-3

Justo, R., & De Castro, O. J. (2008). Indicators of entrepreneurship activity: Some methodological contributions. *International Journal of Entrepreneurship and Small Business*, *6*(4), 604–621. doi:10.1504/IJESB.2008.019504

Keogh, P. D., & Polonsky, M. J. (1998). Environmental commitment: A basis for environmental entrepreneurship? *Journal of Organizational Change Management*, *11*(1), 38–49. doi:10.1108/09534819810369563

Kolodinsky, J., Stewart, C., & Bullard, A. (2006). Measuring economic and social impacts of membership in a community development financial institution. *Journal of Family and Economic Issues*, *27*(1), 27–47. doi:10.1007/s10834-005-9002-7

Kortum, S., & Lerner, J. (2000). Assessing the contribution of venture capital to innovation. *The Rand Journal of Economics*, *31*(4), 647–692. doi:10.2307/2696354

Linnanen, L. (2002). An insiders experiences with environmental entrepreneurship. *Green Management International*, *38*(38), 71–80. doi:10.9774/GLEAF.3062.2002.su.00008

Mair, J., Robinson, J., & Hockerts, K. (2005). *Social entrepreneurship*. New York: Palgrave Macmillan.

Martinelli, F. (2012). Social innovation or social exclusion? Innovating social services in the context of a retrenching welfare state. In F. Hans-Werner, H. Josef, & H. Jürgen (Eds.), *Challenge Social Innovation* (pp. 169–180). Berlin: Springer-Verlag. doi:10.1007/978-3-642-32879-4_11

Nicholls, A. (2010). The institutionalization of social investment: the interplay of investment logics and investor rationalities. *Journal of social entrepreneurship*, 1(1), 70-100.

Parasuraman, A., & Zeren, L. M. (1983). R&D's relationship with profits and sales. *Research Management, 26*(1), 25–28.

Pastakia, A. (1998). Grassroots ecopreneurs: Change agents for a sustainable society. *Journal of Organizational Change Management, 11*(2), 157–173. doi:10.1108/09534819810212142

Peredo, A. M., & McLean, M. (2006). Social entrepreneurship: A critical review of the concept. *Journal of World Business, 41*(1), 56–65. doi:10.1016/j.jwb.2005.10.007

Pol, E., & Ville, S. (2009). Social innovation: Buzz word or enduring term. *Journal of Socio-Economics, 38*(6), 878–885. doi:10.1016/j.socec.2009.02.011

Prahalad, C. K., & Hammond, A. (2002). Serving the world's poor, profitably. *Harvard Business Review, 80*(9), 48–57. PMID:12227146

Putnam, R. D. (1993). The prosperous community: Social capital and public life. *The American Prospect, 13*, 35–42.

Romer, P. M. (1990). Endogenous technical change. *Journal of Political Economy, 98*(5, Part 2), 71–102. doi:10.1086/261725

Sadowski, M., & Roth, A. (1999). Technology leadership can pay off. *Research and Technology Management, 42*(6), 32–33.

Schaltegger, S. (2002). A framework for ecopreneurship: Leading pioneers and environmental managers to ecopreneurship. *Green Management International, 38*(2), 45–58. doi:10.9774/GLEAF.3062.2002.su.00006

Schaltegger, S., & Wagner, M. (2011). Sustainable entrepreneurship and sustainability innovation: Categories and interactions. *Business Strategy and the Environment, 20*(4), 222–237. doi:10.1002/bse.682

Schaper, M. (2005). Understanding the green entrepreneur. In M. Schaper (Ed.), *Making ecopreneurs: developing sustainable entrepreneurship* (pp. 3–12). Aldershot: Ashgate Publishing Limited.

Skiltere, D., & Jesilevska, S. (2013). Building the system of innovation capability indicators: Case of Latvia. *Eurasian Journal of Business and Economics, 6*(12), 113–118.

Spencer, J. W. (2003). Firms knowledge sharing strategies in the global innovation system: Empirical evidence from the flat panel display industry. *Strategic Management Journal, 24*(3), 217–233. doi:10.1002/smj.290

Staber, U. (1997). An ecological perspective on entrepreneurship in industrial districts. *Entrepreneurship and Regional Development, 24*(1), 37–48.

Walley, E. E., & Taylor, D. W. (2002). Opportunists, champions, mavericks: A typology of green entrepreneurs. *Green Management International, 38*(38), 31–43. doi:10.9774/GLEAF.3062.2002.su.00005

Walley, L., Taylor, D., & Greig, K. (2010). Beyond the visionary champion: testing a typology of green entrepreneurs. In M. Schaper (Ed.), *Making ecopreneurs: developing sustainable entrepreneurship* (pp. 59–74). Farnham, Surrey, England: Gower Publishing Ltd.

Zahra, S. A., Gedajlovic, E., Neubaum, D. O., & Shulman, J. M. (2009). A typology of social entrepreneurs: Motives, search processes and ethical challenges. *Journal of Business Venturing*, *24*(5), 519–532. doi:10.1016/j.jbusvent.2008.04.007

KEY TERMS AND DEFINITIONS

Inclusive Growth: Growth that reduces poverty and inequality, raises employment rates through modernized labor markets and welfare systems, and ensures benefits for a large part of a country's labor force. It requires investment in human capital, structural transformations, effective social protection, non-discrimination and social inclusion.

Innovation: Successful commercial or social exploitation of new ideas where the idea is successfully brought to market. In a social context it leads to reduced social costs or improved access to social services through effective alternatives to existing arrangements.

Social Enterprise: An operator in the social economy whose main objective is to have a social impact rather than make a profit for their owners or shareholders. It provides goods and services for the market in an entrepreneurial and innovative manner by involving multiple stakeholders affected by its commercial activities and uses its profits primarily to achieve social objectives.

Social Investment: The provision and use of repayable finance to generate social as well as financial returns. Social returns are improved outcomes for society and range substantially – from, for example, improved individual health to employability amongst disadvantaged groups, to provision of community goods and services.

Social Impact: The effects of an organization's activities on the well-being of society members with short-term and long-term outcomes and a particular focus on attaining systematic social improvement.

Sustainable Development: "Development that meets the needs of the present without compromising the ability of future generations to meet their own needs". (The World Commission on Environment and Development, 1987).

Value Creation: Process of generation of financial and non-financial value based on firm's business model through transformation of tangible and intangible (financial, manufactured, intellectual, human, social, natural) assets to produce short-term, medium- and long-term outputs, outcomes, and impact for the benefit of the organization, its stakeholders, society at large, and the environment. Innovation and entrepreneurship are central to the process of value creation which require use of measures for its evaluation that reflect the full costs and benefits created by the firm.

ENDNOTES

1 Editors' note: For a different angle of view on tripple bottom line, see also chapter 17, on integrating sustainable development into portfolio management through application of open innovation.

2 Editors' note: For a more detailed discussion of this point see, for example, chapter 11 on overcoming the barriers of strategic planning in SMEs.

Chapter 14
Process Management for SMEs:
Barriers, Enablers, and Benefits

Lidia Sanchez-Ruiz
University of Cantabria, Spain

Beatriz Blanco
University of Cantabria, Spain

ABSTRACT

More than 90% of companies in Spain, and other European countries, are SMEs. Therefore it is imperative to support and develop their competitiveness due to their contribution to the national economy. Thus, process management is said to be a useful means to improve efficiency and achieve a competitive advantage. Taking this into consideration the aim of this study is to identify and rank the barriers, enablers and benefits that SMEs encounter when implementing process management and to propose different initiatives to ease the implementation process. In order to fulfil this objective, a survey was conducted among SMEs that have already implemented process management so that their experience could help other SMEs. Finally, a ranking of barriers, enablers and benefits is obtained and some suggestions for managers are proposed.

INTRODUCTION

Currently businesses operate in a highly competitive environment which has become tougher as a result of the economic crisis which started at the end of 2007. Thus, not only do companies have to offer their customers what they need, but they also have to do it in the most efficient way. Companies are obliged to overhaul their strategy in order to find new perspectives to efficiently manage their resources and achieve a competitive advantage. As a result, survival is closely related to competitiveness. It follows logically that the higher the company competitiveness the higher the chances of success. Companies have a wide range of philosophies at their disposal to improve their efficiency in order to achieve this. Among them, process management (PM) will be analyzed in this study, as it has been considered an empowering philosophy of quality and efficiency, the two paramount concepts in today's environment.

DOI: 10.4018/978-1-5225-1949-2.ch014

On the other hand, due to the relevance that small and medium enterprises (SMEs) have in the economic environment of most developed and developing countries, this study will be focused on them. Overall, the aim of this chapter is first to identify and rank the barriers, enablers and benefits that SMEs encounter when implementing PM and to propose different initiatives to ease the implementation process. Having said that, the authors organized the chapter as follows. The first section integrates the theoretical framework and a literature review. The second section describes the data and methodology. The third section includes the results and, finally, future research and conclusions are stated in section four.

THEORETICAL FRAMEWORK

Literature Review

PM has been widely studied in literature as Sidorova and Isik (2010) and Houy, Fettke and Loos (2010) concluded in their studies (see Table 1). In fact, both studies highlighted the dramatic increase in the number of articles published related to PM topics during the analyzed periods.

However, that increase was not equally distributed among countries (see Table 2) and more research is needed in certain areas. That was the case of Spain, which was placed in a rather unfavorable position with only four published papers about PM. Although the results may be biased because the used databases were mainly Anglo-Saxon, the difference was large enough to be taken into consideration.

Given this huge difference and the absence of literature reviews on PM focused on the Spanish literature, a review using Dialnet database, which is the largest repository for Spanish-written publications, was carried out by the authors (Sanchez & Blanco, 2014). Their study concluded that practitioners are interested in the PM concept. However, research is in the early stages of development.

Furthermore, a thematic analysis showed that, as in the international case, the implementation of PM systems is a recurring theme. However, some research gaps were detected on topics such as main PM factors (barriers, enablers, motivations, and benefits), human resources management, corporate culture and innovation.

Table 1. Main characteristics of the international PM reviews

TITLE	Business process research: a cross-disciplinary review	Empirical research in business process management-analysis of an emerging field of research
AUTHOR	Sidorova, A.; Isik, O.	Houy, C.; Fettke, P.; Loos, P.
YEAR	2010	2010
DATABASES USED	EBSCOHost	Two databases (not specified)
METHODOLOGY	Latent Semantic Analysis (LSA)	Scientometric methods
No PAPERS REVIWED	2700	355
PERIOD	09/1927 – 12/2008	1991-2010
ORIENTATION	Theory	Empirical papers

Source: Authors

Table 2. Number of PM papers published by country

Country	Total number of articles	First article published in the year...
United Kingdom	105	1993
United States	97	1994
...	...	...
...	...	...
Spain	**4**	**2004**

Source: (Houy et al., 2010, p.632)

Overall, it is considered that the PM field, despite being a relatively mature field in the Anglo-Saxon literature, still presents many opportunities for development in the Spanish-speaking environment, especially in the above mentioned topics.

The Spanish business framework has a significant share of SMEs, greater than 99%, as in other European countries (INE, 2016). Consequently, guaranteeing the survival and efficiency of SMEs is essential due to their contribution to the national economy. Thus, PM could be the means to achieve this goal.

Overall, based on both national and international reviews carried out, a conceptual review of the PM concept together with its main barriers, enablers and benefits was done. In the following lines, the main results are summarized. It is important to highlight that the obtained results are not only referred to SMEs because authors did not find studies devoted specifically and exclusively to SMEs. Thus, the relevance of these kinds of studies is reinforced.

The Concept of Process Management

As stated before, there is an umpteen number of studies focused on PM. Hence, there are also many different definitions. By way of example, in the following lines, some PM definitions will be stated.

Elzinga, Horak, Chung-Yee and Bruner (1995, p.119) offered a general definition of PM by characterizing it as "Any structured approach used to analyse and continually improve fundamental activities, such as manufacturing, marketing, communications, and other major elements of a company's operation". Additionally, Armistead and Rowland (1996) stated that "It is the management of the organization by the consideration of their business processes".

Other authors described PM from a technological perspective. This is the case of van der Aalst (2004) who defined it as "A generic software approach that is driven by explicit process designs to enact and manage operational business processes". Similarly, Reijers (2006, p.390) stated that PM is "A piece of generic software that supports activities such as the modelling, analysis and enactment of business processes" (p.135).

In this study the authors consider PM from a wider perspective. This point of view coincides with more recent definitions such as the one of Smith and Fingar (2007). According to these authors, "Not only does process management encompass the discovery, design and deployment of business processes, but also the executive, administrative and supervisory control over them to ensure that they remain compliant with business objectives for the delight of customers" (p.4). Equally, Ko, Lee and Lee (2009) define PM as "supporting business processes using methods, techniques and software to design, enact, control and analyse operational processes involving humans, organizations, applications, documents and other sources of information" (p.745).

Overall, it may be concluded that PM is a way of managing the company by focusing the attention in the efficient management of its processes.

Process Management Barriers

A barrier could be defined as any factor that obstructs, hinders or limits the PM implementation initiative. Before surveying managers about their specific experiences, authors carried out a literature review. Table 3 summaries the results of the review.

Process Management Enablers

Enablers could be understood as any factor that eases or simplifies the PM implementation initiative. As in the previous case, a rigorous literature review was done whose main results are included in Table 4.

The high number of enablers found shows the relevance of the topic as well as the existing concern related to ways of empowering PM.

Table 3. PM barriers

Barriers	Authors
Resistance to change (employees)	(Ravesteyn & Batenburg 2010; Dawe, 1996; Paper, Rodger & Pendharker, 2001; Lee & Dale 1998; Hill & Collins 1998).
Traditional structure of the company (functional oriented)	(Wahlich, 2004) in (Kohlbacher, 2010; Paper 1998).
Lack of understanding of the "PM" concept	(Armistead & Pritchard, 1999; Corallo, Margherita, Scalvenzi & Storelli, 2010).
Wrong implementation strategy	(Paper & Chang, 2005; Armistead & Pritchard, 1999).
Technology is not lined up with the aims of PM	(Paper & Chang, 2005).
Lack of management commitment	(Holland & Kumar 1995; Feltes & Karuppan, 1995; Davenport & Short, 1990 ; Neubauer, 2009)
Incorrect identification of processes on which work is going to be done	(Holland & Kumar, 1995).
Communication problems	(Melan, 1989).
Lack of time	(Lee & Dale).
Lack of long term commitment	(Bergey, Smith, Tiley & Wirderman, 1999) in (Paper & Chang, 2005).
A legal system out of control	(Paper & Chang, 2005).
Lack of training resources	(Paper & Chang, 2005).
Employees chained to old technologies	(Paper & Chang, 2005).

Source: Authors

Table 4. PM enablers

Enablers	Authors
Use of Information and Communication Technologies	(Kirchmer, 2009) in (Kohlbacher, 2010); Ravesteyn & Batenburg, 2010; Paper, Rodger & Pendharker, 2001; O'Neill & Sohal, 1999; Sandhu & Gunasekaran, 2004; McAdam & McCormack, 2001; Corallo et al., 2010; Ongaro, 2004).
Staff involvement	(Oakland & Sohal, 1996; Waldman, 1994; Wang, 2004) in (Hung, 2006); Ravesteyn & Batenburg, 2010; Cooper & Markus; Berrington & Oblich, 1995; Bashein, Markus & Riley, 1994; Hinterhuber, 1995; Bruss & Ross, 1993; Arendt, Landis & Meister, 1995) in (Zairi, 1999); Dawe, 1996; Jackson, 1997; (Feltes & Karuppan, 1995; Paper, 1998; Melan, 1989; Ittner & Larcker, 1997; Sandhu & Gunasekaran, 2004; Mir, Mir & Mosca, 2002; Caron, Jarvenpaa & Stodard, 1994.
High management support	(Bashein, Markus & Riley, 1994; Hammer & Stanton 1995; Rastogi, 1994; Talwar, 1993) in (Zairi, 1999); Towers, 1996; Holland & Kumar, 1995, p.83; Stanton, Hammer & Power, 1993; (Paper & Rodger, 2001; Paper, 1999; Paper & Dickinson, 1997; Paper & Mok, 2003; Paper, Rodger & Pendharkar, 2000) in (Paper & Chang, 2005); Paper et al., 2001; Ravesteyn & Batenburg, 2010; Feltes & Karuppan, 1995; Jackson, 1997; Paper, 1998; Clemmer, 1994; Melan, 1989; Ittner & Larcker, 1997; Sandhu & Gunasekaran, 2004; Lee & Dale, 1998; Sentanin, Almada Santos & Chiappetta Jabbour, 2008
Existence of an open culture (that motivates the generation of new ideas and tolerates failure)	Stanton et al., 1993; (Paper & Dickinson, 1997) in (Paper & Chang, 2005); Clemmer, 1994; Melan, 1989; Lee & Dale, 1998; (Winklhofer, 2002) in (Paper & Chang, 2005; Zairi & Sinclair, 1995; Cooper & Markus, 1995; Savvas, 2005.
Correct identification of the processes	Holland & Kumar, 1995; Ravesteyn & Batenburg, 2010.
High previous process orientation level	Reijers, 2006; Ravesteyn & Batenburg, 2010; Paper, 1998.
Motivate working in multifunctional teams	Paper & Chang, 2005; Cooper & Markus, 1995; Paper, 1998; Towers, 1996; Dawe, 1996; Ittner & Larcker, 1997; Paper et al., 2001; Sandhu & Gunasekaran, 2004; Ongaro, 2004.
Integrate process improvement objectives in strategic objectives	Ravesteyn & Batenburg, 2010; (Hammer, 1990) in (Zairi, 1999); Davenport & Short, 1990; Zairi, 1997; Sentanin et al., 2008; Corallo et al., 2010; Ongaro, 2004; Neubauer, 2009.
Correct development of the process quality, control and maintenance	Ravesteyn & Batenburg, 2010.
Having previous experience about implementing that kind of practices	Ravesteyn & Batenburg, 2010.
Common work method	Ravesteyn & Batenburg, 2010; Paper, 1998; Paper et al., 2001.
People included in the project implementation should be adequate	Ravesteyn & Batenburg, 2010.
Training courses about PM	Paper & Chang, 2005; Jackson, 1997; (Berrington & Oblich, 1995; Bashein et al., 1994; Arendt et al., 1995; Worsley, 1994) in (Zairi, 1999); Paper, 1998; Clemmer, 1994; Towers, 1996; Dawe, 1996; Ittner & Larcker, 1997; Davenport & Short, 1990; Paper et al., 2001.
Design and development of a custom business process map	Paper & Chang, 2005; Paper et al., 2001.
The promotion and development of creativity among employees increases the likelihood of success	Cooper & Markus, 1995.
Open communication	(Arendt et al., 1995; Hammer & Stanton, 1995; Carr and Johansson, 1995) in (Zairi, 1999); Zairi & Sinclair, 1995; Jackson, 1997; Dawe, 1996, p.86; Ittner & Larcker, 1997; Ravesteyn & Batenburg, 2010; Paper et al., 2001; McAdam & McCormack, 2001.
Existence of rewards systems that motivate employees	(Feltes & Karuppan, 1995; Hinterhuber, 1995; Dawe, 1996; Towers, 1996; Ostroff, 1992) in (Zairi, 1999); Feltes & Karuppan, 1995; Furey, 1993; Dawe, 1996.
Empowerment	(Bashein et al., 1994; Hinterhuber, 1995; Dawe, 1996; Thomas, 1994; Cooper, 2000) in (Zairi, 1999); Paper, 1998; Dawe, 1996; Ittner & Larcker, 1997.

continued on following page

Table 4. Continued

Enablers	Authors
Appropriate definition of jobs and allocation of responsibilities	(Talwar, 1993) in (Zairi, 1999).
Having enough resources	(Furey, 1993; Bruss & Ross, 1993; Boyle, 1995; Bashein et al., 1994) in (Zairi, 1999); (Furey, 1993).
Having external information (about competitors, benchmarking...)	(Carr & Johansson, 1995) in (Zairi, 1999).
Customer requirements and expectations must be defined and measured	(Jackson, 1997; Rastogi, 1994; Hall, Rosenthal & Wade, 1990) in (Zairi, 1999); Jackson, 1997; Paper, 1998; Zairi, 1997; Ongaro, 2004.
Having help from an external consultancy	(Rastogi, 1994) in (Zairi, 1999); Harrison & Pratt, 1993; Clemmer, 1994.
Integrate PM in other wider improvement systems	(Zairi 1999); (Feltes, Karuppan 1995).
Manage the change towards PM with all the stakeholders	Stanton et al., 1993.
Appointing process managers	Ongaro, 2004.
Leadership	Ongaro, 2004.

Source: Authors

Process Management Benefits

Finally, another review was made in order to identify the main benefits obtained after implementing PM. Table 5 summaries the results of the review, reinforcing the idea of PM as a useful mean to achieve better results.

It could be seen that the topic of PM benefits arouses much interest among researchers. Additionally, the variety of benefits is really broad, affecting different areas and levels of the company.

DATA AND METHOD

Motivation and Objective

Based on the relevance of SMEs for the Spanish economy and the role of PM as a means to empower their competitiveness, the aim of this study is to identify and rank the barriers, enablers and benefits that SMEs encountered when implementing PM.

Although it is true that there are numerous studies along literature that analyzed those factors related to PM, the authors found a gap to cover as those studies did not analyze the relevance of each factor. Therefore, there was no evidence whether all factors (barriers, enablers, benefits) were equally important or if there were differences among them. Thus, the authors consider that this research may be of interest to develop the topic.

Table 5. PM benefits

Benefits	Authors
Reduce the use of resources increasing efficiency	Salgueiro, 1999; Pérez Fernández de Velasco, 2004.
Cost reduction	Hammer, 2007; Kohlbacher, 2010; Küng & Hagen, 2007; Zairi & Sinclair, 1995; Salgueiro, 1999; Pérez Fernández de Velasco, 2004; Paper et al., 2001; Skrinjar, Bosilj-Vuksic & Indihar-Stemberger, 2008.
It's much easier to measure and reduce launch times	Salgueiro, 1999; Pérez Fernández de Velasco, 2004.
Measurements are easily stated	Salgueiro, 1999; Armistead & Machin, 1998.
Objectives are better established	Salgueiro, 1999.
Barriers and problems are easier identified	Salgueiro, 1999.
Mistakes are reduced and it is easier to prevent them	Salgueiro, 1999; Küng & Hagen, 2007; Zairi & Sinclair, 1995; Reijers, 2006, p.389.
A wider perspective of the company is offered	Salgueiro, 1999; Armistead & Machin, 1998.
Company is focused on the customer	Salgueiro, 1999; Pérez Fernández de Velasco, 2004; Clemmer, 1994; Ongaro, 2004.
It is possible to analyse the value added by every activity of the company	Salgueiro, 1999.
Total quality is improved in every activity of the company	Salgueiro, 1999.
Relationships and communication improve	Salgueiro, 1999; Armistead & Machin, 1998.
It gives a global and wider perspective of the company (value chain) and its internal links	Pérez Fernández de Velasco, 2004; Salgueiro, 1999.
It helps to take important decisions as it enables the identification of constraints and barriers in order to achieve the established objectives.	Pérez Fernández de Velasco, 2004; Salgueiro, 1999; (Schmelzer & Sesselmann, 2006) in (Kohlbacher, 2010).
Clearly allocating responsibilities makes self-evaluation a common practice in the company together with the process improvement co-responsibility	Pérez Fernández de Velasco, 2004; Palmberg, 2010.
It facilitates the implementation of wider systems such as Lean Management, Total and Quality Management	Ravesteyn & Batenburg, 2010.
Relationships between departments improve	Pérez Fernández de Velasco, 2004; Silvestro & Westley, 2002.
Product and process quality improvements	Hammer, 2007; (Hirzel, 2008) in (Kohlbacher, 2010); Melan, 1989; Paper, 1998; Reijers, 2006; Ongaro, 2004; (Hinterhuber, 1995) in (Kohlbacher, 2010,); (Bullita, 2006) in (Kohlbacher, 2010); (Schmelzer & Sesselmann, 2006) in (Kohlbacher, 2010); (Setti & Stückl, 2006) in (Kohlbacher, 2010); Küng & Hagen, 2007.
Time reductions	Hammer, 2007; (Hirzel, 2008) in (Kohlbacher, 2010); (Zairi & Sinclair, 1995); (Classe & Mundle, 1997) in (Kohlbacher, 2010); Hertz, Johansson & de Jager, 2001; (Mittermaier & Braun, 2004) in (Kohlbacher, 2010); (Schmelzer & Sesselmann, 2006) in (Kohlbacher, 2010); (Wahlich, 2004) in (Kohlbacher, 2010); Paper, 1998; Reijers, 2006; Ongaro, 2004; Skrinjar et al., 2008; (Setti & Stückl, 2006) in (Kohlbacher, 2010).
Internal/External customer satisfaction increases	(Hinterhuber, 1995) in (Kohlbacher, 2010); (Hirzel, 2008) in (Kohlbacher, 2010); (Bullita, 2006) in (Kohlbacher, 2010); Hertz et al., 2001; (Schima, 2004) in (Kohlbacher, 2010); (Wahlich, 2004) in (Kohlbacher, 2010); Zairi & Sinclair, 1995; Ongaro, 2004; (Skrinjar et al., 2008; (Wahlich, 2004) in (Kohlbacher, 2010).
Productivity increases	(Hinterhuber, 1995) in (Kohlbacher, 2010); Melan, 1989; Armistead & Machin, 1998; (Bullita, 2006) in (Kohlbacher, 2010); Küng & Hagen, 2007; Zairi & Sinclair, 1995; Feltes & Karuppan, 1995; Armistead & Machin, 1998.

continued on following page

Table 5. Continued

Benefits	Authors
Value added increases by eliminating those activities that are not competitive and strengthening the main abilities	(Hinterhuber, 1995) in (Kohlbacher, 2010); Skrinjar et al., 2008.
Property ambiguity is eliminated	
A clear description of the process boundaries and the relationship among processes are established	Melan, 1989.
Sales increase (physical units)	(Bullita, 2006) in (Kohlbacher, 2010); (Wahlich, 2004) in (Kohlbacher, 2010).
Clear definition of process that is used as a learning tool for new employees	Melan, 1989; Zairi & Sinclair, 1995
Clarity, regarding the responsibility structure, is improved	(Pauls, 2006) in (Kohlbacher, 2010); Küng & Hagen, 2007.
Income increase	(Bullita, 2006) in (Kohlbacher, 2010); (Wahlich, 2004) in (Kohlbacher, 2010); Ittner & Larcker, 1997; (Schmelzer & Sesselmann, 2006) in (Kohlbacher, 2010); Hammer, 2007.
More proactive company	Melan, 1989.
Stock is reduced	Hertz et al., 2001; (Schmelzer & Sesselmann, 2006) in (Kohlbacher, 2010).
Delivery reliability increases	Hertz et al., 2001; (Mittermaier & Braun, 2004) in (Kohlbacher, 2010); (Schmelzer & Sesselmann, 2006) in (Kohlbacher, 2010); (Setti & Stückl, 2006) in (Kohlbacher, 2010)
Company competitiveness increases	Küng & Hagen, 2007.
Systematic search for potential improvements	(Pauls, 2006) in (Kohlbacher, 2010); Pérez Fernández de Velasco, 2004.
Transparency increases	(Schima, 2004) in (Kohlbacher, 2010).
Employees versatility increases	(Schima, 2004) in (Kohlbacher, 2010); Paper, 1998.
Market share increases	(Wahlich, 2004) in (Kohlbacher, 2010); Zairi & Sinclair, 1995.
Financial improvements	(Wahlich, 2004) in (Kohlbacher, 2010); Ittner & Larcker, 1997; Zairi & Sinclair, 1995.
Absenteeism is reduced	Middel, op de Weegh & Gieskes, 2007.
Company flexibility increases	Reijers, 2006; Skrinjar et al., 2008.
Relationships with suppliers are improved	Authors.

Source: Authors

Study Design

In order to fulfill the above mentioned objective, a survey was designed and conducted among companies from Cantabria, a region in the north of Spain. The following stages were followed in order to complete the study:

- **Literature Review:** From the literature review authors could discover what had been written on the topic, as well as to identify potential items to be included in the empirical study. Results of the literature review have been included in this chapter's *theoretical framework*.
- **Survey Design and Validation:** An initial draft of the survey was designed. It was shown to experts in order to validate its content and structure. Experts were academics (including Professors of Business and Management), practitioners (upper and mid-level managers) from manufacturing

and service companies, consultants and quality managers. Their main suggestions were about vocabulary issues, inclusion or exclusion of some items and adaptation of certain items so they were appropriate for both, manufacturing and service companies.

- **Final Survey Design:** After collecting all the information needed, including the experts' opinions, the final survey was designed. Companies had to value how important the barriers, enablers and benefits included in the survey were when they implemented PM.
- **Identification of Target Population:** First, in order to identify the target population, all Cantabrian companies over 20 employees were asked whether or not they practised PM. The fact of establishing this minimum number of employees is due to the fact that, based on authors' experience, companies under 20 employees do not usually implement PM as the effort needed might not be compensated. This idea was later reinforced during the survey and it could be verified that the lowest percentage of companies (around 12%) that practised PM occurred among the smallest SMEs (between 20 and 50 employees).

Companies were identified using the directory of companies offered by ICANE, Cantabrian Statistical Institute. The research technical record is shown in Table 6.

Finally, 299 responses were received which represents a response rate of 37%. Among those companies 168 companies affirmed practising PM and were sent the complete survey.

- **Data Collection:** After sending the survey, 96 valid responses were received. Among them, 77 were SMEs with less than 250 employees (European Commission, 2016). The results of this study are based on the responses of those 77 SMEs.

Methodology: Rasch Measurement Theory

Data analysis includes a descriptive analysis, which was done by using the SPSS software, and Rasch analysis, which was done by using the Winsteps 3.92 software (Linacre, 2016). Since the use of Rasch Measurement Theory (RMT) is not generalised in the business and management field, the following lines will describe the methodology and its main applications.

In the Business and Management area it is very common to use Likert scales to value a group of items that, in theory, integrate the construct that we are interested in. However, the scores obtained from Likert scales cannot be considered measurements because they are ordinal scores and do not have an additive structure, a characteristic that only interval variables have.

Table 6. Technical record

Characteristics	Survey
Population	Cantabrian companies with more than 20 employees (808)
Geographical scope	Autonomous Community of Cantabria
Unit of analysis	Company
Response rate	37% (299 responses)

Source: Authors

RMT, initially developed by George Rasch (1960), solves this problem by transforming ordinal variables into interval variables. RMT is based in three principles: unidimensionality (all the items are referred to the same construct or latent variable); invariance (results are independent from the samples of subjects and items used); and additivity.

When it comes to transforming the ordinal variables into interval variables, the Theory is based in the following statement where Rasch explained that "persons who are more able/more developed have a greater likelihood of correctly answering all the items in the observation schedule. And, easier items are more likely to be answered/reached correctly by all persons" (as cited in Bond & Fox, 2007, p.28).

Then, the mathematical model is derived from the logistic function that relates the increasing probability of response to all items with the persons' ability (Gonzalez Montesinos, 2008).

From this statement, a new concept is defined: distance. It is the difference between the ability of the person (β) and the difficulty of the item (δ). Three situations may happen depending on the result of this difference:

- Si β-δ>0: probability of the subject answering correctly to the item is more than 50%
- Si β-δ<0: probability of the subject answering correctly to the item is less than 50%
- Si β-δ=0: probability of the subject answering correctly to the item equals 50%

The mathematical expression of the model derives from this idea. Bond and Fox (2007), Algumalai, Curtis and Hnugi (2005) and von Davier and Carstensen (2007) discuss it in their approach.

With regard to data treatment, several software packages are available, for instance, Facets or Winsteps. In this study the 3.92 Winsteps software is used (Linacre, 2016).

Applications of the Methodology

In this section applications of the RMT which will be used in this chapter are briefly explained:

1. **Dimensionality Analysis:** Dimensionality is a necessary condition to use RMT so it must be checked. As well as methodological reasons, this checking is interesting when new constructs are defined because it allows affirming that their definition is correct.
2. **Global and Individual Reliability and Validity of Measures:** A measure is valid when it measures what is supposed to be measuring. And a measure is reliable when it has a high ability of reproducibility.
3. **Classification of Persons and Items:** Rasch methodology offers a hierarchical classification of the items that integrate the construct depending on the importance given by the persons. Equally, persons can also be ordered according to their level of ability.

RESULTS

In order to rank the different factors (barriers, enablers and benefits) related to PM, RMT was used. Before proceeding with the classification, some validation checks must be done. Specifically:

- Dimensionality analysis: it must be checked that the defined constructs in each case are unidimensional, this is, all the items are measuring the same reality;
- Global reliability and validity measures;
- Individual reliability and validity of measures: sometimes items, although measuring the same reality, have an unexpected behaviour and (based on the invariance principle) should be taken out from the sample in order to avoid validity problems. The same could happen with subjects (in this case company).

Process Management Barriers

First, it is important to validate the construct defined after the panel of experts which, in this case, was integrated by 10 items.

Construct dimensionality analysis (see Table 7) showed that the construct met the requirements established in the methodology manual (Linacre, 2012) and, as a result, it could be considered to be unidimensional. The two requirements are (Linacre, 2012):

1. The eigenvalue of the unexplained variance of the first contrast has to be lower or equal to 3.
2. The percentage of explained variance by the first contrast has to be lower than the percentage of raw variance explained by items.

Global validity and reliability results are also positive (see Table 8). As noted, the values obtained are consistent with those required (Linacre, 2012), both for the case of the means (MNSQ valid in the range (0.5 to 1.5) as the standard variances (ZSTD close to 0) and for all measures, both individuals and items. In terms of reliability (the optimum value is 1, but values over 0.70 are accepted), it can be seen that it is of an acceptable level for the question persons and excellent for items.

Finally, when the data are complete, the correlation (Pearson correlation) should be 1 for person and -1 for items (Linacre, 2012). In this research, it is acceptable in both cases, persons and items. Therefore, it is concluded that measurements are valid and reliable.

Finally, regarding individual reliability, all items' measures were reliable. However, some problems were identified with 9 companies which were taken out from the sample. Therefore, in this case, the final sample was integrated by 68 companies.

Table 7. Variance components screen plot for items (barriers)

	Empirical			**Model**
Total Raw Variance in observations	19.1	100.0%		100.0%
Raw variance explained by measures	9.13	47.7%		48.4%
Raw variance explained by persons	2.35	12.3%		12.5%
Raw variance explained by items	6.7	35.4%		35.9%
Raw unexplained variance (total)	10.0	52.3%	100.0%	51.6%
Unexplained variance 1st contrast	2.14	11.2%	21.4%	

Source: Authors

Table 8. Reliability and validity of measures (barriers)

	INFIT		OUTFIT		RELIABILITY	CORRELATION
	MNSQ	ZSTD	MNSQ	ZSTD		
Persons	0.88	-0.3	0.85	-0.3	0.73	0.93
Items	1.00	0.0	1.05	0.3	0.96	-1.00

Source: Authors

Once the validation process has been done, Figure 1 represents the final results obtained with respect to barriers hierarchy. On the left side of the vertical line (representing the construct) companies are denoted by an identification code. Companies located at the top (E221, E267, E096, E171, E264,) are the ones that have valued the selected barriers higher. On the right side of the vertical line, items are located. In contrast with what happens to companies, the most important barrier is the one located at the bottom (GPP5.3) and the less important one is item GPP5.9. The complete list could be seen in Table 9, where barriers are classified from highest to lowest importance. The first column indicates the

Figure 1. Barriers variable map
Source: Authors

```
MEASURE                                    PERSON - MAP - ITEM
                                               <more>|<rare>
   2                                                 +
                                                     |
                                                     |
                                                     |
                                                     |
                                                     |
                                                     |
                                             E177    |
                                                     |  GPP6.9
                                                     |T
   1                                                 +
                                                    T|  GPP6.14
                          E012  E135  E257  E277     |
                                            E051     |
                                E060  E171  E231     |S
                                            E131    S|
                    E098  E187  E199  E214  E216     |  GPP6.13
              E014  E037  E105  E210  E255  E267     |  GPP6.10  GPP6.7
              E020  E069  E097  E117  E118  E155     |  GPP6.11  GPP6.8
                          E002  E103  E208  E248     |
   0          E032  E043  E111  E127  E157  E243   M+M  GPP6.4
        E057  E104  E108  E110  E153  E279  E281     |
                                E070  E152  E221     |  GPP6.15  GPP6.3    GPP6.5
        E096  E145  E162  E163  E172  E207  E292     |  GPP6.12
                                E009  E109  E125     |  GPP6.1
                          E047  E129  E217  E285    S|  GPP6.16  GPP6.6
              E068  E099  E120  E138  E148  E233     |S
                                            E018     |
                                                     |
                                            E045     |
  -1                                        E089    T+
                                                     |T
                                            E264     |
                                                     |  GPP6.2
                                                     |
                                                     |
                                                     |
                                                     |
                                                     |
                                                     |
  -2                                                 +
                                               <less>|<freq>
```

Table 9. Rank of PM barriers

Code	Name
GPP5.3	Resistance to change (employees, unions)
GPP5.2	Traditional structure of the company (functional oriented)
GPP5.1.	Lack of understanding of the "PM" concept
GPP5.8	Lack of time
GPP5.7	Lack of management commitment in the long term
GPP5.10	Lots of financial resources were needed
GPP5.6	Technology is not lined up with the aims of PM
GPP5.5	Lack of training
GPP5.4	Wrong implementation strategy
GPP5.9	Lack of high management support

Source: Authors

code of the barrier, where GPP stand for "PM in Spanish", whereas the second column includes the whole name of the barrier.

Table 9 shows that the main three barriers are "resistance to change", "traditional structure of the company" and "lack of understanding of PM concept". Although all the results are going to be analyzed deeper in the following section, it may be seen that the success of PM highly depends on human resources issues. Thus, the relationship between PM and human resources should be analyzed deeper in future research projects.

Process Management Enablers

As in the previous case, before obtaining the final classification, some validation checks were conducted. First, construct dimensionality analysis (see Table 10) showed that the construct met the requirements (Linacre, 2012):

- The eigenvalue of the unexplained variance of the first contrast is lower than 3, in this case is 2.4;
- The percentage of explained variance by the first contrast (9%) is lower than the percentage of raw variance explained by items (23.4%).

Therefore, the dimensionality analysis showed that the construct met the requirements and, as a result, it could be considered to be unidimensional.

Global validity and reliability results are also positive (Table 11). MNSQ values are in the valid range (0.5 to 1.5), whereas ZSTD are close to 0. In terms of reliability (the optimum value is 1, but values over 0.70 are accepted), it can be seen that it is of an acceptable level for question persons and excellent for items. Finally, when the data are complete, the correlation should be 1 for person and -1 for items (Linacre, 2012). In our case, it is acceptable in both cases. Therefore, it is concluded that measurements are valid and reliable.

Table 10. Variance components screen plot for items (enablers)

	Empirical			Model
Total Raw Variance in observations	23.73	100.0%		100.0%
Raw variance explained by measures	10.73	40.2%		39.9%
Raw variance explained by persons	4.42	16.6%		16.5%
Raw variance explained by items	6.30	23.6%		23.5%
Raw unexplained variance (total)	16.0	59.8%	100.0%	60.1%
Unexplained variance 1st contrast	2.40	9.0%	15.0%	

Source: Authors

Table 11. Reliability and validity of measures (enablers)

	INFIT		OUTFIT		RELIABILITY	CORRELATION
	MNSQ	ZSTD	MNSQ	ZSTD		
Persons	0.93	-0.4	0.92	-0.3	0.75	1.00
Items	0.99	-0.2	1.03	0.0	0.96	-1.00

Source: Authors

Finally, regarding, individual reliability, all items' measures were valid and reliable. However, some problems were identified with 6 companies which were taken out from the sample. Therefore, the final sample was integrated by 71 companies.

Figure 2 represents the final results obtained with respect to enablers. Results referred to item importance, which can be seen on the left hand side of the vertical line in Figure 2, have been summarized in Table 12.

"High management support", "integration of process improvement objectives in strategic objectives" and "open communication" are the three main enablers according to companies' experience. On the other side, "Existence of rewards systems that motivate employees" and "Manage the change towards PM with all the stakeholders" are the least important enablers. The authors consider that by empowering these enablers, some of the previous barriers could be reduced.

Process Management Benefits

The defined construct "PM benefits" was not unidimensional so RMT could not be applied. As a result, a descriptive analysis was done. Table 13 shows the main benefits obtained by surveyed companies when implementing PM.

As it could be seen from Table 13 there is a wide range of benefits derived from PM implementation. According to companies' opinion, the most important benefits are both internal and external. Thus, "transparency incrementation", "easiness to establish measurements" and "improvement of the relationship between departments" can be highlighted. Whereas, "competitiveness increase", "increase of external customer satisfaction" or "more proactive company" could be understood as external benefits.

Figure 2. Enablers process map
Source: Authors

```
  MEASURE                               PERSON - MAP - ITEM
                                            <more>|<rare>
    2                                             +
                                                  |
                                                  |
                                                  |
                                                  |T
                                                  |
                                                  |  GPP5.9
    1                                             +
                                                  |
                                                  |S
                                                 T|  GPP5.4   GPP5.5   GPP5.6
                                                  |  GPP5.10  GPP5.7
                                      E221  E267  |
                                E096  E171  E264  |
    0               E110  E118  E145  E153  E187  +M
                                            E045 S|
                          E002  E068  E162  E216  |
              E047  E051  E177  E208  E233  E257  |
                          E057  E060  E135  E157  |
                          E032  E065  E131  E294  |S GPP5.1   GPP5.8
              E012  E104  E111  E207  E214  E292 M|
   -1                     E089  E097  E138  E277  +  GPP5.2
        E066  E069  E103  E108  E120  E163  E231  |  GPP5.3
                          E098  E099  E125  E246  |
                                      E105  E152  |T
                    E014  E070  E117  E127  E248  |
                                E009  E279  E281 S|
                                      E018  E199  |
   -2                                             +
                                                  |
                                                  |
                                            E155 T|
                                                  |
                                            E255  |
                                                  |
   -3                                             +
                                      E148  E243  |
                                                  |
                                                  |
                                                  |
                                                  |
                                                  |
   -4                                 E037  E285  +
                                            <less>|<freq>
```

Emerging Solutions and Recommendations for SME Process Management

Based on the previous results, in this section, the authors are going to propose some initiatives that SMEs should start in order to implement PM in an efficient way.

Resistance to change is the main barrier faced by companies when implementing PM. This is a worrying result taking into consideration that human resources are a key element to guarantee PM success. In fact, many factors related to human resources management: employees' involvement, open communication, existence of an open culture and working in multifunctional teams have been identified as key PM enablers.

Thus companies should work in order to reduce resistance to change, for instance through developing a training policy. In this way a change in employees' mindset could be achieved. Employees may be less reluctant to implement PM if they understand the objectives and the potential benefits. On many occasions, employees reject new management systems because they think that they are a way of control encouraged by managers.

Table 12. Rank of PM enablers

Code	Name of the Enablers
GPP6.2	High management support
GPP6.6	Integrate process improvement objectives in strategic objectives
GPP6.16	Open communication
GPP6.1	Use of Information and Communication Technologies
GPP6.12	Having help from an external consultancy
GPP6.3	Existence of an open culture (that motivates the generation of new ideas and that tolerates failure)
GPP6.5	Motivate working in multifunctional teams, eliminating departmental barriers
GPP6.15	Common work method
GPP6.4	Training courses about PM
GPP6.11	Having external information (about competitors, benchmarking...)
GPP6.8	Low average age
GPP6.7	Having previous experience about implementing that kind of practices
GPP6.10	Having time
GPP6.13	Integrate PM in other wider systems such as Total Quality Management or Lean Management
GPP6.14	Manage the change towards PM with all the stakeholders
GPP6.9	Existence of rewards systems that motivate employees

Source: Authors

It is true that, due to their size and characteristics, SMEs may have more problems than big companies in allocating time and resources to training. In fact, this problem could be worse as company size is reduced. Big companies can usually devote resources to offer their employees training sessions in the company. It is true that this may be an expensive practice for SMEs, however other possibilities exist. For instance, it could be interesting to integrate a cluster with other SMEs from the same area or from the same sector in order to offer common training sessions. Additionally, some local governments and public institutions offer financial assistance for training. The second problem, related to time allocation, could be solved by using online learning. By using the opportunities new technologies offer (webinars), employees could have access to training when work load is not very high.

Traditional structure of the company is the second main barrier[1]. This is probably related to the functional structure that many companies still have. This structure hinders the relationship between different departments that sometimes seem to be competitors instead of working through a common objective. Changing this attitude is not easy and, while it persists, the implementation of new management systems is strongly impeded. In this case a good communication strategy may be developed so that departments start to work together. It should be highlighted that departmental structure does not have to disappear completely when implementing PM, but both structures have to be combined as many authors suggest.

Lack of understanding of the PM concept is the third most important barrier. It may directly affect the previous barrier `resistance to change´ which due to a lack of understanding may lead to mistrust and resistance. In this case companies have a problem which, again, could be reduced by training employees. Through training, firstly, a common language of work is established encouraging that all employees understand, participate and benefit from PM initiatives. Employees have to understand why

Table 13. Rank of PM benefits

	- IMPORTANCE +				
	1	2	3	4	5
Transparency increased	3.90%	7.79%	25.97%	35.06%	27.27%
Company competitiveness increased	12.99%	5.19%	16.88%	41.56%	23.38%
Measurement are easily stated	6.49%	5.19%	24.68%	40.26%	23.38%
Relationship between departments improved	10.39%	5.19%	27.27%	35.06%	22.08%
Clear definition of process that was used as a learning tool for new employees	16.88%	6.49%	19.48%	35.06%	22.08%
It facilitates the implementation of wider systems such as Lean Management and Total Quality Control	21.05%	13.16%	18.42%	26.32%	21.05%
External customer satisfaction increased	14.29%	1.30%	22.08%	42.86%	19.48%
More proactive company	9.09%	5.19%	24.68%	41.56%	19.48%
Response time was reduced	18.18%	7.79%	16.88%	37.66%	19.48%
PM facilities the decision management process	10.39%	2.60%	35.06%	33.77%	18.18%
Employees versatility increased	16.88%	12.99%	25.97%	25.97%	18.18%
Innovation and continuous improvement initiatives increased	6.49%	2.60%	32.47%	41.56%	16.88%
Cycle time was reduced	22.08%	10.39%	27.27%	23.38%	16.88%
Work in teams is improved	9.09%	3.90%	29.87%	42.86%	14.29%
Overproduction was reduced	23.38%	9.09%	28.57%	24.68%	14.29%
Quality defects were reduced	14.29%	9.09%	16.88%	46.75%	12.99%
Cost reduction	23.38%	14.29%	23.38%	25.97%	12.99%
Communication improved	5.19%	5.19%	35.06%	42.86%	11.69%
Productivity increased	22.08%	3.90%	29.87%	32.47%	11.69%
Unnecessary motion was reduced	20.78%	18.18%	23.38%	28.57%	9.09%
Unnecessary transport was reduced	33.77%	18.18%	15.58%	23.38%	9.09%
Relationship with suppliers is improved	23.38%	11.69%	28.57%	28.57%	7.79%
Stock reduction	36.36%	18.18%	19.48%	18.18%	7.79%
Market share increased	23.38%	19.48%	31.17%	19.48%	6.49%
Overproduction was reduced	45.45%	20.78%	14.29%	12.99%	6.49%
Sales increased (physical units)	41.56%	15.58%	24.68%	11.69%	6.49%
Waiting times were reduced	22.08%	14.29%	37.66%	20.78%	5.19%
Financial improvements	22.08%	25.97%	27.27%	22.08%	2.60%
Absenteeism was reduced	44.16%	29.87%	16.88%	9.09%	0.00%

Source: Authors

PM is beneficial and why it is necessary in the current business environment. As a result, the whole implementation process would be easier.

Lack of time is an important barrier as well. Sometimes the problem lies, at least in part, in the fact that PM is understood as a non-core activity and no time is allocated to it. Nonetheless, as it was highlighted above, PM is a complex and time-consuming process, especially during the first implementation stages when everything is to be done. This barrier is probably worse for SMEs than for big companies. Some big companies often have a person, or even a department, responsible for implementing new philosophies. In SMEs, however, the person responsible for leading the process (if there is any) is usually responsible for other departments and tasks. Probably the best way of facing this problem is by creating a formal group of people responsible for implementation. By creating the group, PM implementation becomes a core activity, occupying the place it deserves. This group would have to allocate some hours of the week to establish objectives and plan how to implement PM.

The following barrier in order of importance is lack of management commitment in the long term. This result contrasts with the fact that lack of high management support is the least important barrier in this study and the main enabler. The only difference between both barriers is the period of time. It may be concluded that, whereas surveyed companies counted on manager's support at the beginning of the project, it was reduced as time progressed.

In this sense, managers should be aware that they play a key role during the whole implementation process. Their attitude may be even more important in the case of SMEs because managers usually work more closely with employees. Although it is true that in the initial stages they should invest more time as they have to convince and motivate all the employees together with allocating resources, their role continues in the following stages as facilitators and leaders. Additionally, many SMEs are family companies so that the manager is the founder and becomes a natural leader. Therefore, his role is even more important.

Concerning the item "*lots of financial resources were needed*", its position on the rank should be taken into consideration. According to previous studies, it was concluded that the implementation of PM initiatives does not require a significant economic effort, but time investment. However, the intermediate position of this item could be caused by two different possibilities: whether companies do not understand correctly the efforts needed to implement PM, or the economic effort is higher than it was stated by previous literature. Therefore, more research should be done in this respect.

Taking into consideration that results refer to SMEs and due to, as it was said before, they do not have the same resources as big companies, it may be concluded that they need more financial resources as they might have to hire some services that big companies possess. For instance, big companies may have full-time employees devoted to PM implementation whereas SMEs might have to hire an external advisor to start with the implementation.

Lack of high management support, wrong implementation strategy and lack of training are the least important barriers. This may be considered a very positive result. However, some comments should be made with reference to this result.

Training is a key mechanism that can help managers to reduce resistance to change or improve understanding of problems[2]. However, the fact that surveyed companies considered that enough training is provided leads the authors to conclude that the content or structure is not well designed, otherwise there would be understanding problems. Therefore, the relevance of offering good training must be highlighted.

Regarding the barrier "*lack of high management support*", some nuances should be highlighted. Firstly, the fact that high management support is the least important barrier could change if the study was conducted among companies that do not practice PM. Second, results may also be biased because the survey was mainly answered by upper and middle managers.

Regarding the results related to PM enablers, it may be concluded that they are quite coherent with those obtained in the previous section. First, the relevance of high management support is again highlighted. The second most important enabler is "*integrate process improvement objectives in strategic objectives*". This result clearly reinforces the idea of allocating time and resources to PM. Obviously, PM implementation is not about sporadic actions but continuous projects that involve a lot of people.

There is also a group of enablers that highlight the importance of information along the PM implementation initiative. Thus, the existence of an open communication and culture and having external information are considered to be key enablers. These results are directly related with the benefits derived from PM. Hence, if information is properly managed, transparency will be increased, measurements will be easily stated, and relationships between departments will improve. Therefore, PM is also about information management. Companies should adopt an open philosophy with regard to information management.

Among the least important enablers, the "*existence of rewards systems that motivate employees*" should be emphasised. The authors considered that this result is very interesting as, initially, may be seen as contradictory with previous ones. Why are rewards not important if employees' involvement is a key element? The authors think that the involvement should not be forced by giving rewards. Employees should be trained about PM and should experience their benefits. This would probably be the best way to guarantee PM sustainability; otherwise, it could fail if rewards stop. It would be interesting to analyse whether SMEs have more problems to give rewards to employees because they might have less available resources than big companies.

Finally, concerning the benefits rank obtained after the analysis, several ideas should be highlighted. First, there is a high variety of benefits that range from increasing transparency to reducing unnecessary motion. Thus, PM could positively affect every department or activity in the company. It is a transversal tool. Secondly, after transparency incrementation, the increase of company competitiveness appears as the second most important benefit. Therefore, authors' initial statement (PM is useful for increasing competitiveness) is reinforced.

It could also be seen that PM helps companies to increase external customer satisfaction by reducing response time and reducing quality problems. Thus, companies are able to better adapt to the current environment where time has become a key competitive factor and customers are more exigent. PM seems to be also useful for employees: their versatility increases, the clear definition of processes is used as a learning tool and work in groups is increased. Therefore, PM is beneficial for many stakeholders. Overall, based on previous results, SMEs can see how to benefit from PM and which factors they should take into consideration when implementing this management philosophy.

FUTURE RESEARCH DIRECTIONS

Despite the rigorous approach, the study had two main limitations. First, due to the geographical scope of the study, results should be treated with caution. Nonetheless, the authors consider that the sample size is big enough so that results could be considered significant. Secondly, as it has been already stated, since the surveyed companies are already committed to PM, results may be slightly biased. Thus, in the future, it would be interesting to include companies that have not implemented PM yet or companies that failed.

Replicating the study in other regions, even in other countries, would allow analyzing whether cultural differences influence PM initiatives. It would also be interesting to analyze whether there are different perceptions depending on the characteristics of the company (sector, location). Authors have already started to work in this direction by replicating the study in another region (Cardiff, Wales). RMT will be employed again, specifically a DIF (differential item functioning analysis) will be done.

Moreover, in order to obtain a more general perspective of the PM phenomenon, it would be useful to carry out a deeper research surveying not only managers but also medium managers and employees. In this case, surveys could be combined with in-depth interviews.

Based on the results obtained in this study, it would be interesting to go one step further and develop new studies with SMEs. For instance, it would be useful to employ *action research* as a means to increase cooperation between researchers and SMEs. Since the lack of understanding and lack of time are two main barriers SMEs face, researchers could act as facilitators; at the same time, researchers would benefit from being an active part during the implementation.

It is very important to continue working on the topic "SMEs adopting PM" because there is not enough research about this topic. This gap was detected after doing the literature review as most of the studies were based on big companies.

From the methodological point of view, it would be interesting to combine RMT with other methodological tools. For instance, structural equation modeling could be a very useful instrument in order to analyze whether there are relationships between the different constructs.

At this point it would be interesting to classify the different factors (barriers, enablers and benefits) into thematic fields (human resources, organizational, operations, customer satisfaction and so on) in order to check whether the influence of the different groups of factors is the same or not.

Moreover, it is paramount to develop new research trying to establish whether there is a positive relationship between PM implementation and organizational performance. It is true that some effects could already be identified as PM benefits, nonetheless a more quantitative approach should be adopted.

CONCLUSION

From a theoretical point of view, this study designs, develops and validates a new construct integrated by several PM factors. Design and validation processes have been rigorously developed by asking a panel of experts and using RMT.

It also adds value to the field because, as it is not based on an individual company, a general rank of the main PM barriers, enablers and benefits could be obtained, covering an existing gap in literature. From a practical point of view, in the discussion section, this chapter highlights several issues which should be considered by managers.

Overall, the authors consider that PM should be promoted among SMEs as it is a management philosophy that improves business efficiency. This could be seen from the benefits that companies admitted obtaining. The high variety of benefits shows that PM is a holistic approach that advocates adopting a value chain approach. In fact, this idea can be transferred out of the company's limits, cooperating with customers and suppliers, and not only building more efficient companies, but also more efficient supply chains.

REFERENCES

Alagumalai, S., Curtis, D. D., & Hnugi, N. (2005). *Applied Rasch Measurement: A Book of Exemplars*. Springer.

Arendt, C., Landis, R. & Meister, T. (1995). The human side of change - part 4. *IEE Solutions*, 22-27.

Armistead, C., & Pritchard, J. P. (1999). Business process management - lessons from European business. *Business Process Management Journal*, *5*(1), 10–35. doi:10.1108/14637159910249144

Armistead, C. G., & Rowland, P. (1996). *Managing business processes: BPR and beyond*. New York, NY: Wiley.

Bashein, B., Markus, M., & Riley, P. (1994). Precondition for BPR success and how to prevent failures. *Information Systems Management*, *11*(2), 7–13. doi:10.1080/10580539408964630

Bergey, J., Smith, D., Tiley, S., Weiderman, N. & Woods, S. (1999). Why reengineering projects fail. *Carnegiee Mellon Software Engineering Institute - Product Line Practice Initiative,1*, 1-30.

Berrington, C., & Oblich, R. (1995). Translating reengineering into bottom-line results. *Industrial Engineering (American Institute of Industrial Engineers)*, 24–27.

Bond, F. G., & Fox, C. M. (2007). *Applying the Rasch Model. Fundamental Measurement in the Human Sciences* (2nd ed.). Lawrence Erlbaum Associates, Inc.

Bullita, C. (2006). Geschäftsprozessmanagement bei Siemens Medical Solutions. In A. Geschäftsgebiet, H. J. Schmelzer, & W. Sesselmann (Eds.), *Geschäftsprozessmanagement in der Praxis* (pp. 475–489). Munich: Hanser.

Caron, J. R., Jarvenpaa, S. L., & Stodard, D. B. (1994). Business reengineering at CIGNA Corporation: Experiences and lessons learned from the first five years. *Management Information Systems Quarterly*, *18*(3), 233–250. doi:10.2307/249617

Classe, D., & Mundle, J. (1997). Geschäftsprozessgestatung im Rahmen des CIP-Prozasses bei Bosch. In S.-P. Riekhof (Ed.), *H.C* (pp. 211–224). Stuttgart: Beschleunigung von Geschäftsprozessen.

Clemmer, J. (1994). Process re-engineering and process improvement - not an either/or choice. *CMA Magazine,* 36-39.

Cooper, R., & Markus, M. L. (1995). Human Reengineering. *Sloan Management Review*, *36*(4), 39. PMID:10139766

Corallo, A., Margherita, A., Scalvenzi, M., & Storelli, D. (2010). Building a Process-Based Organization: The Design Roadmap at Superjet International. *Knowledge and Process Management*, *17*(2), 49–61. doi:10.1002/kpm.340

Dawe, R. L. (1996). Systems are people too. *Transportation & Distribution*, *37*(1), 86.

Elzinga, D.J., Horak, T., Lee, C-Y., & Bruner, C. (1995). Business process management: survey and methodology. *Engineering Management, IEEE Transactions on,42*(2), 119-128.

European Commission. (2016). *Entrepreneurship and Small and medium-sized enterprises (SMEs). Business-friendly environment. What is an SME?* Retrieved 5th August, 2016 from EU web site: http://ec.europa.eu/growth/smes/business-friendly-environment/sme-definition/

Feltes, P., & Karuppan, C. (1995). Reengineering: Getting down to the business of doing business. *Industrial Management (Des Plaines)*, *37*(4), 3–12.

González Montesinos, M. J. (2008). *El Análisis de Reactivos con el Modelo Rasch.* Manual Técnico A. Instituto Nacional para la Evaluación de la Educación, Universidad de Sonora.

Jackson, N. (1997). Business Process Re-engineering '96. *Management Services, 41*(2), 34.

Hammer, M. (2007). The Process Audit. *Harvard Business Review, 85*(4), 111. PMID:17432158

Hammer, M., & Stanton, S. (1995). *The reengineering revolution.* New York: HarperCollins.

Hertz, S., Johansson, J. K., & de Jager, F. (2001). Customer-oriented cost cutting: Process management at Volvo. *Supply Chain Management, 6*(3/4), 128–142. doi:10.1108/13598540110399174

Hill, F. M., & Collins, L. K. (1998). The positioning of BPR and TQM in long-term organisational change strategies. *TQM Journal, 10*(6), 438–446. doi:10.1108/09544789810239182

Hinterhuber, H. (1995). Business process management: The European Approach. *Business Change and Re-engineering, 2*(4), 63–73.

Hirzel, M. (2008). Erfolgsfaktor Prozessmanagement. In M. Hirzel, F. Kühn, & I. Gaida (Eds.), *Prozessmanagement in der Praxis* (pp. 11–22). Gabler.

Holland, D., & Kumar, S. (1995). Getting past the obstacles to successful reengineering. *Business Horizons, 38*(3), 79–85. doi:10.1007/978-3-8349-9436-3_1

Houy, C., Fettke, P., & Loos, P. (2010). Empirical research in business process management. Analysis of an emerging field. *Business Process Management Journal, 16*(4), 619–661. doi:10.1108/14637151011065946

Hung, R. Y. Y. (2006). Business Process Management as competitive advantage: A review and empirical study. *Total Quality Management & Business Excellence, 17*(1), 21–40. doi:10.1080/14783360500249836

INE. (2016). *Instituto Nacional de Estadística (Spanish Statistical Institute).* Retrieved 5th August, 2016 from INE web site: http://www.ine.es/

Ittner, C. D., & Larcker, D. F. (1997). The performance effects of process management techniques. *Management Science, 43*(4), 522–534. doi:10.1287/mnsc.43.4.522

Kirchmer, M. (2009). *High Performance through Process Excellence.* Berlin: Springer.

Kohlbacher, M. (2010). The effects of process orientation: A literature review. *Business Process Management Journal, 16*(1), 135–152. doi:10.1108/14637151011017985

Küng, P., & Hagen, C. (2007). The fruits of Business Process Management: An experience report from a Swiss bank. *Business Process Management Journal, 13*(4), 477–487. doi:10.1108/14637150710763522

Lee, R. G., & Dale, B. G. (1998). Business process management: A review and evaluation. *Business Process Management Journal, 4*(3), 214–225. doi:10.1108/14637159810224322

Linacre, J. M. (2012). Winsteps® Rasch measurement computer program User's Guide. Beaverton, OR: Winsteps.com.

Linacre, J. M. (2016). Winsteps® Rasch measurement computer program. Beaverton, OR: Winsteps.com.

McAdam, R., & McCormack, D. (2001). Integrating business processes for global alignment and supply chain management. *Business Process Management Journal*, *7*(2), 113–130. doi:10.1108/14637150110389696

Melan, E. H. (1989). Process Management: A Unifying Framework For Improvement. *National Productivity Review*, *8*(4), 395–406. doi:10.1002/npr.4040080407

Middel, R., Weegh, S. O. D., & Gieskes, J. (2007). Continuous improvement in The Netherlands: A survey-based study into current practices. *International Journal of Technology Management*, *37*(3-4), 259–271. doi:10.1504/IJTM.2007.012262

Mir, A., Mir, R., & Mosca, J. B. (2002). The new age employee: An exploration of changing employee-organization relations. *Public Personnel Management*, *31*(2), 187–200. doi:10.1177/009102600203100205

Mittermaier, G., & Braun, M. (2004). Geschäftsprozessmanagement bei Infineon. In H. Ellringmann & H. J. Schmelzer (Eds.), *Geschäftsprozessmanagement inside* (pp. 1–40). Munich: Hanser.

Neubauer, T. (2009). An empirical study about the status of business process management. *Business Process Management Journal*, *15*(2), 166–183. doi:10.1108/14637150910949434

Oakland, J. S., & Sohal, A. (1996). *Total Quality Management: Text with Cases*. Melbourne: Butterworth-Heinemann.

ONeill, P., & Sohal, A. S. (1999). Business Process Reengineering A review of recent literature. *Technovation*, *19*(9), 571–581. doi:10.1016/S0166-4972(99)00059-0

Ongaro, E. (2004). Process management in the public sector: The experience of one-stop shops in Italy. *International Journal of Public Sector Management*, *17*(1), 81–107. doi:10.1108/09513550410515592

Palmberg, K. (2010). Experiences of implementing process management: A multiple-case study. *Business Process Management Journal*, *16*(1), 93–113. doi:10.1108/14637151011017967

Paper, D. (1998). BPR: Creating the conditions for success. *Long Range Planning*, *31*(3), 426–435. doi:10.1016/S0024-6301(98)80009-5

Paper, D. (1999). The enterprise transformation paradigm: The case of Honeywells industrial automation and control unit. *Journal of Information Technology Cases and Applications*, *1*(1), 4–23. doi:10.1080/15228053.1999.10855927

Paper, D., & Chang, R. (2005). The state of business process reengineering: A search for success factors. *Total Quality Management & Business Excellence*, *16*(1), 121–133. doi:10.1080/1478336042000309907

Paper, D., & Dickinson, S. (1997). A comprehensive process improvement methodology: experiences at Caterpillar's Mossville engine centre. In M. Khosrowpour & J. Liebowitz (Eds.), *Cases on Information Technology Management in Modern Organizations*.

Paper, D., & Mok, W. (2003). The relation between BPR and ERP systems: A failed project. *Annals of Cases on Information Technology*, *5*, 45–62. doi:10.4018/978-1-59140-061-5.ch003

Paper, D., & Rodger, J. A. (2001). Managing radical transformation: a holistic paradigm. Academic Press.

Paper, D., Rodger, J.A., & Pendharkar, P.C. (2000). *Development and initial testing of a theoretical model of transformation*. Academic Press.

Paper, D., Rodger, J. A., & Pendharkar, P. C. (2001). A BPR case study at Honeywell. *Business Process Management Journal, 7*(2), 85–99. doi:10.1108/14637150110389416

Pérez Fernández de Velasco, J. A. (2004). *Gestión por procesos: cómo utilizar ISO 9001:2000 para mejorar la gestión de la organización*. Madrid: ESIC.

Rasch, G. (1960). *Probabilistic models for some intelligence and attainment tests*. Chicago: The University of Chicado Press.

Rastogi, P. (1994). Nature, significance and rationale of business process reengineering. *Productivity, 35*(3), 467–476.

Ravesteyn, P., & Batenburg, R. (2010). Surveying the critical success factors of BPM-systems implementation. *Business Process Management Journal, 16*(3), 492–506. doi:10.1108/14637151011049467

Reijers, H. A. (2006). Implementing BPM systems: The role of process orientation. *Business Process Management Journal, 12*(4), 389–409. doi:10.1108/14637150610678041

Salgueiro, A. (1999). *Cómo Mejorar los Procesos y la Productividad*. Madrid: AENOR.

Sanchez, L., & Blanco, B. (2014). La gestión por procesos. Un campo por explorar. *Dirección y Organización, 54*, 54–71.

Sandhu, M., & Gunasekaran, A. (2004). Business Process Development in project-based industry. *Business Process Management Journal, 10*(6), 673–693. doi:10.1108/14637150410567875

Savvas, A. (2005, March). Cultural resistance main cause of BPM project failure. *Computer Weekly*.

Schima, M. (2004). Geschäftsprozessmanagement bei der Kurz Gruppe. In H. Ellringmann & H. J. Schmelzer (Eds.), *Geschäftsprozessmanagement inside* (pp. 1–40). Munich: Hanser.

Schmelzer, H. J., & Sesselmann, W. (2006). *Geschäftsprozessmanagement in der Praxis*. Munich: Carl Hanser.

Sentanin, O. F., Almada Santos, F. C., & Chiappetta Jabbour, C. J. (2008). Business process management in a Brazilian public research centre. *Business Process Management Journal, 14*(4), 483–496. doi:10.1108/14637150810888037

Setti, C., & Stückl, R. (2006). Geschäftsprozessmanagement in der Praxis. In H. J. Schmelzer & W. Sesselmann (Eds.), *Geschäftprozessmanagement in der Praxis* (pp. 461–474). Munich: Hanser.

Sidorova, A., & Isik, O. (2010). Business process research: A cross-disciplinary review. *Business Process Management Journal, 16*(4), 566–597. doi:10.1108/14637151011065928

Silvestro, R., & Westley, C. (2002). Challenging the paradigm of the process enterprise: A case-study analysis of BPR implementation. *Omega, 30*(3), 215–225. doi:10.1016/S0305-0483(02)00028-2

Skrinjar, R., Bosilj-Vuksic, V., & Indihar-Stemberger, M. (2008). The impact of business process management orientation on financial and non-financial performance. *Business Process Management Journal, 14*(5), 738–754. doi:10.1108/14637150810903084

Smith, H., & Fingar, P. (2007). *Business Process Management: The Third Wave*. Meghan-Kiffer Press.

Stanton, S., Hammer, M., & Power, B. (1993). Reengineering: Getting everyone on board. *I.T. Magazine*, *25*(4), 22.

Talwar, R. (1993). Business re-engineering—a strategy-driven approach. *Long Range Planning*, *26*(6), 22–40. doi:10.1016/0024-6301(93)90204-S

Towers, S. (1996). Re-engineering: Middle managers are the key asset. *Management Services*, *40*(12), 17.

Van der Aalst, W. M. P. (2004). Business Process Management: A personal view. *Business Process Management Journal*, *10*(2), 135. doi:10.1108/bpmj.2004.15710baa.001

Von Davier, M., & Carstensen, C. H. (2007). Multivariate and Mixture Distribution Rasch Models. Extensions and Applications. New York: Springer Science + Business Media, LLC. doi:10.1007/978-0-387-49839-3

Wahlich, S. M. (2004). Prozessorientatierte Organisation bei Vaillant Hepworth. In H. Ellringmann & H. J. Schmelzer (Eds.), *Geschäftsprozessmanagement inside* (pp. 1–40). Munich: Hanser, Munich.

Winklhofer, H. (2002). Information Systems project management during organizational change. *Engineering Management Journal*, *14*(2), 33–37. doi:10.1080/10429247.2002.11415160

Zairi, M. A. M. (1999). BPR implementation process: An analysis of key success and failure factors. *Business Process Management Journal*, *5*(1), 87–112. doi:10.1108/14637159910249108

Zairi, M., & Sinclair, D. (1995). Business process re-engineering and process management A survey of current practice and future trends in integrated management. *Business Process Re-Engineering & Management Journal*, *1*(1), 8–30. doi:10.1108/14637159510798248

ADDITIONAL READING

Bateman, N., & Arthur, D. (2002). Process improvement programmes: A model for assessing sustainability. *International Journal of Operations & Production Management*, *22*(5/6), 515–526. doi:10.1108/01443570210425156

Bateman, N., & Rich, N. (2003). Companies perceptions of inhibitors and enablers for process improvement activities. *International Journal of Operations & Production Management*, *23*(2), 185–199. doi:10.1108/01443570310458447

Bawden, R., & Zuber-Skerritt, O. (2002). The concept of process management. *The Learning Organization*, *9*(3), 132–138. doi:10.1108/09696470210428859

Beheshti, H. M., & Lollar, J. G. (2003). An empirical study of US SMEs using TQM. *TQM & Business Excellence*, *14*(8), 839–847. doi:10.1080/1478336032000090798

Bessant, J. (1998). Developing continuous improvement capability. *International Journal of Innovation Management*, *2*(4), 409–429. doi:10.1142/S1363919698000183

Bessant, J., Caffyn, S., Gilbert, J., Harding, R., & Webb, S. (1994). Rediscovering continuous improvement. *Technovation*, *14*(1), 17–29. doi:10.1016/0166-4972(94)90067-1

Bessant, J., & Francis, D. (1999). Developing strategic continuous improvement capability. *International Journal of Operations & Production Management*, *19*(11), 1106–1119. doi:10.1108/01443579910291032

Bessant, J., Caffyn, S., & Gallagher, M. (2001). An evolutionary model of continuous improvement behavior. *Technovation*, *21*(2), 67–77. doi:10.1016/S0166-4972(00)00023-7

Goh, T. N., Low, P. C., Tsui, K. L., & Xie, M. (2003). Impact of Six Sigma implementation on stock price performance. *Total Quality Management and Business Excellence*, *14*(7), 753–763. doi:10.1080/1478336032000090969

Hekmatpanah, M., Sadroddin, M., Shahbaz, S., Mokhtari, F., & Fadavinia, F. (2008). Six Sigma process and its impact on the organizational productivity. *World Academy of Science. Engineering and Technology*, *43*, 365–369.

Kinney, M. R., & Wempe, W. F. (2002). Further evidence on the extent and origins of JITs profitability effects. *Accounting Review*, *77*(1), 203–225. doi:10.2308/accr.2002.77.1.203

Sanchez, L. (2014). *Implementation of control and continuous improvement techniques in business process.* (Unpublished doctoral dissertation). University of Cantabria.

Sanchez, L., & Blanco, B. (2012). Business Process Management in Project-Based Companies: A New Methodology. *International Journal of Advances in Management and Economics*, *1*(3), 35–41.

Sanchez, L., & Blanco, B. (2015). Three decades of continuous improvement. *Total Quality Management and Business Excellence*, *25*(9-10), 986–1001. doi:10.1080/14783363.2013.856547

Sanchez, L., & Blanco, B. (2016a). Process management implementation level. A case study from Cantabria. *International Letters of Social and Humanistic Sciences*, *67*, 70–77. doi:10.18052/www.scipress.com/ILSHS.67.70

Sanchez, L., & Blanco, B. (2016b). How do companies implement process management? The case of Cantabrian companies. *Business. Management and Economics Research*, *2*(1), 1–9.

Von Brocke, J., & Sinnl, T. (2011). Culture in business process management: A literature review. *Business Process Management Journal*, *17*(2), 357–377. doi:10.1108/14637151111122383

Womack, J.P. & Jones, D.T. (2005). *Lean Thinking. Cómo utilizar el pensamiento lean para eliminar los despilfarros y crear valor en la empresa.* Barcelona: Ediciones Gestión 2000.

KEY TERMS AND DEFINITIONS

Barrier: Factor that hinders or makes a process more difficult.

Benefit: A positive consequence of a fact.

Competitive Advantage: Skill, knowledge or equipment a company possesses thanks to which a company is better positioned than its competitors.

Enabler: Factor that eases or simplifies a process.

Process: A sequence of activities that transform inputs into outputs.

Process Management: A management philosophy focused on the efficient management of processes aimed at improving internal efficiency and satisfying the customer.

Rasch Measurement Theory: Useful methodology to work with categorical variables. It is based on three principles: unidimensionality, invariance and additivity.

SME: Small and Medium Enterprises.

ENDNOTES

[1] Editors' note: For more specific angles of view on this point see and compare, for example, chapter 11 on overcoming the barriers to strategic planning in SMEs.

[2] Editors' note: For a more complete treatment, see chapter 16 where a similar discussion is given in connection with change management.

Chapter 15
Development of an Optimization Tool for Intangibles in SMEs:
A Case Study from Serbia with a Pilot Research in the Prestige by Milka Company

Olja M. Arsenijević
Union - Nikola Tesla University, Serbia

Drago Orčić
University of Belgrade, Serbia

Edita Kastratović
Faculty of Business Economics and Entrepreneurship, Serbia

ABSTRACT

The purpose of this chapter is to present research findings and address the conclusion that intellectual capital is present, measurable and can be optimized in SMEs. The theoretical part of the chapter presents a review of tools for measuring intellectual capital. By using this concept, SMEs recognize the importance of intellectual capital and accept methods by which they can convert their businesses into knowledge based companies. This approach has evolved very dynamically as changes in business development and intellectual capital continue to develop. The chapter also provides the newly created tool for optimizing intellectual potential in SMEs, tested through a pilot research in an SME, the "Prestige by Milka" company. As it can be seen from the chapter, neglecting intangibles frequently results in suboptimal business development. It is expected that the conclusions and recommendations of this study will reveal possible directions for further development of optimization potential and intellectual capital in SMEs.

DOI: 10.4018/978-1-5225-1949-2.ch015

INTRODUCTION

Intellectual capital has become a subject of serious study in the last twenty years. Values not enumerated in the classic financial statements began to be noticed. Scientists and practitioners are seriously researching the concept of intellectual capital management in order to increase its value. They examine the methods for measuring intellectual capital in companies in order to establish measurable targets for stakeholders.

Structural changes in running a business, caused by computerization, globalization and humanization of management processes determine the leading role of intellectual capital in a new economy. For example, the practice of open innovation can produce a leverage effect, focusing the intellectual effort of business entities on the creation of new high technologies. The organization of the "*flow of mental energy*" is also created when using crowd sourcing through powerful software to enable many professionals, often motivated by personal interests to quickly find effective solutions for problems in the company.

Clearly, intellectual capital affects the daily work of the organization, speed and reliability of business processes, but also determines the quality of goods and services. Business success means both material (i.e. profit, growth, income) and intangible (i.e. brand, customer loyalty) results of the organization.

Authors identified these two key questions which can provide more insight into the area of intellectual capital and which represent the core aim of this chapter:

1. How to manage intangible resources, which according to their characteristics are completely different from the tangible resources and are not subject to the basic, currently in effect, economic rules?
2. How to manage intellectual capital if you do not have an adequate system for monitoring and measurement of intangible resources?

It is understood that an intangible resource cannot be evaluated in a traditional way, although the necessity of its evaluation cannot be questioned. Obviously, the financial measures must be complemented with new standards of intellectual capital that will enable the measurement, evaluation, and management of intellectual capital which is in the age of knowledge an essential resource of any company, particularly of small and medium-sized enterprises.

Intellectual capital is an increasingly important part of the overall value of the company. Although not shown in the balance sheet, it is a fact that it has significant value. In fact, invisible capital value often exceeds the value of the visible and stated capital.

Accounting mechanisms have been developed to measure exclusively tangible assets. However, in many companies from advanced economies, intellectual capital reaches up to 90% of the shares of a company. Investors even trade in intangible capital, without having real insight into its value. The challenge facing modern management is certainly finding an accurate model for measuring intellectual capital. Regardless of the importance of intellectual capital, its measurement is still not fully developed.

Therefore, methods for measuring intellectual capital are emerging for the needs of modern organizations whose primary value is knowledge. These are organizations in which knowledge is the most important resource of successful business, which are aware of the existence of an invisible value organization and they want to manage it.

In scientific research circles intellectual capital is discussed by many authors including Brooking (1996), Edvinsson (2002), Sveiby (2007), and Stewart (1997). In the last ten years many new researches also appeared on the role of intellectual capital in the market of the developing and transition countries such as Taiwan, Russia, Malaysia, South Africa and Western Balkans.

In order to secure competitive advantage and innovation potential, the Intellectual Capital Statement project - Made in Europe has started involving 25 European companies, 6 business associations, research institutes and several independent experts. The project has directives for measurement and strategic management of intellectual capital of companies. In its approach the European Directive consequently divides intellectual capital into three categories (InCaS, 2008):

1. **Human Capital:** Where every employee adds to organization's added value, including qualifications, skills, experience, leadership capacity, communication skills and motivation;
2. **Structural Capital:** Which implies organizational structures and processes of the organization. E.g. corporate culture, communication and interpersonal relations, development of new products and services, industrial innovation;
3. **Relational Capital:** One that reflects the relationship among shareholders, owners, business partners, investors, customers, suppliers and others.

There seems to be a broad consensus on the above classification on human capital, organizational capital and relational capital (Bjurstrom & Roberts, 2007; Guthrie, Klauda & Morrison, 2012). However, attention has gradually shifted from categories to the dynamic aspects and multifaceted nature of intellectual capital (Kianto, 2007), because these intellectual capital subcategories cannot be rigidly separated in a meaningful manner (Nielsen & Dane–Nielsen, 2010). According to them, it is not possible to identify a priori the features and functions of a company's intellectual capital because they depend on the original combination that is set up in the specific company context (Mouritsen, 2004).

Similarly, one of the most important goals of EU directives is to support SMEs in Europe, which are actively engaged in scientific research for building intellectual capital. Raising the level of information about intellectual capital enables solving problems when seeking research and innovation project funds in Europe. Authors of the chapter argue that insufficient financing was conditioned by an absence of information on intellectual capital and will consider this criterion in their case study.

BACKGROUND WITH LITERATURE REVIEW

Many definitions and classifications associated with it have been proposed for understanding intellectual capital it and where it is located within a company, especially in the early stages of research (Catasus & Chaminade, 2007). In the Danish Guideline for Intellectual Capital Statements the concepts of intellectual capital and knowledge management are intertwined as these intrinsic aspects are given a body from which management can be conducted (Rimmel, 2004). Here intellectual capital is considered as a phenomenon that allows an "activation" of intangible resources, i.e. the knowledge resources connected to employees, customers, technologies and processes (Rimmel, 2004).

The importance of intellectual capital is confirmed by the multitude of other empirical research. For example, Pulić (2000) published the results of research conducted in the period 1992-1999, on a random sample of 30 companies in Croatia. His results showed that there is a causal link between the level of intellectual capital and the market value of the organization. Bontis, Keow and Richardson (2000) did a research on a sample of 107 Malaysian companies. The main result of his study reveals significant positive correlation between intellectual capital and enterprise performance, regardless of the industry. The impact of individual components of intellectual capital on the performance of the company is different in different industries (Bontis, Keow & Richardson, 2000). In a sample of more than five thousand French companies, analyzed in the period 1998-2000 Kremp and Mairesse (2004) have come up with results that by increasing intensity of knowledge management by 1%, the level of innovation in the organization increases by 4% and productivity by 3%. Taiwanese researches on 4254 public companies, covering 1992-2002 have shown that the level of intellectual capital affects the profitability and commercial expenses (Chen, Cheng & Hwang 2005). Another Taiwanese study on 81 public companies has shown that there is high complementary impact of intellectual capital on the market value of a company. This effect was significantly higher in high-tech companies (Tseng & Goo, 2005). Non-linear impact on the profitability of companies was shown in a survey conducted in 2003 on 297 Taiwanese companies (Huang & Liu, 2005). Furthermore, from 1997 to 2001 research was conducted on all IT companies from Taiwan. The results showed that individual components of intellectual capital show strong linear impact on the performance of companies, and influences performances also indirectly through other components of IC (Wang & Chang, 2005). The fact that the level of intellectual capital is directly related to the cost-effectiveness of companies and its market value and with no impact on productivity is proved by the study done on 80 public Taiwanese companies (Shiu, 2006). Regarding the impact of IC on the type of innovation activities, it can be said that it is very strong and intense. The bigger the IC impact the mpre radical innovation is. This was confirmed by a research carried out on 93 US companies (Subramaniam &Youndt, 2005). Russian researches also tried to determine the impact of IC on the performance of their companies in these empirical studies:

- Research conducted from 2002-2006 on 19 big Russian companies, has confirmed that there is a direct link between the added value, created on the basis of IC and return on assets - IC factors that influence the value-added enterprises identified (Bayburina & Golovko, 2008);
- According to a survey on 401 companies in the Perm region 2005-2007, there is a positive relationship between intellectual capital and the growth rate of the company's revenues, depending on the industry. Molodchik, Shakina and Bykova in 2012 published the results of a survey carried out during the five years on 332 European trade companies, which shows that IC factors can both increase and decrease the value of the company in the short and long term where state and industry and size of companies played a major role.
- Based on a survey of 50 firms listed on Nairobi Securities Exchange, the study "Intellectual Capital and Performance of Firms Listed on Nairobi Securities Exchange" examines the relationship between intellectual capital and organizational performance (Kariuki, Obonyo, & Ogutu, 2015).

More recently the global economic crisis attracted particular interest to relationship between company investment behavior and intellectual capital. According to Guevara & Bounfour (2013), during a downturn companies are challenged to find the best ways to reallocate their resources. Intangibles are then considered priorities since they are '*a stream for future benefits*' (Patel & Narain, 2008). An extensive body of managerial literature further considers the intangible-intensive strategies of companies. It is studied in the papers by Williams & Nones (2009), Martínez Torres (2014) and Williams & Lee (2009). Many recent studies such as those by Clarkson, Richardson and Vasivari (2011) and Bocquet, Bas and Poussing (2013) also deal with proactive and reactive strategic behavior in changing environments. However, the analysis of proactive and reactive decisions in intangibles is still underdeveloped in literature. This impedes deeper understanding of the role of intangible intensive strategies in companies during economic crises.

As stated by Dahlmann & Brammer (2011) companies do not tend to change their strategies considerably regarding intangibles. This implies that intangible intensive strategies chosen before an economic crisis either provides companies with sustainable competitive advantages or aggravates the situation during a recession. These findings were established in surveys carried out by Wilson and Eilertsen (2010). However, as discussed by Kunc and Bhandari (2011) some companies follow reactive strategies by cutting a substantial part of expenses associated with their intangible portfolio. Thus, proactive strategies along with reactive ones are of particular interest when considering economic distress.

Consequently, this chapter attempts to fill a gap in the literature dedicated to the analysis of proactive company strategies in intangibles. In putting forward a hypothesis that intangibles create competitive advantages, intangible-intensive strategies are also proposed by the authors to deal with consequences of the global economic crisis from 2008-2009.

There is a cluster of other papers that also consider different elements of strategic company investments in relationship to intellectual capital: in innovations by Archibugi, Filippetti and Frenz (2013), in human resources, by Crook Russell T. (2011), Unger, Rauch, and Frese (2011), in marketing by and Hassan Hamid and Bohairy (2010).

Most of these studies were based on surveys and reveal the attitudes of decision makers towards external economic shocks and outcomes of these attitudes towards company performance. Undoubtedly, behavioral factors are some of the most relevant in turbulent conditions. These researches do not, however, challenge the objective capacity of different intangible-intensive strategies to keep a company performing well in an economic crisis.

Furthermore, some idiosyncratic assets must support each strategic company position (Nickerson & Silverman, 1997). In fact, intangible assets are among those peculiar assets that can provide a company with competitive advantages and allow it to outperform rivals. Hence they become the core strategic resources for a business. They enable an organization to differentiate itself from rivals and consequently create sustainable value (Lev, 2001; Kristandl & Bontis, 2007).

Intangible-intensive strategy dealing with these idiosyncratic assets is the main counterpart of the strategic behavior of any company. The author agrees with Bottani (2010) that these common features form a strategic profile of a cluster of companies.

The research approach of this chapter also suggests the study of a company profile in overcoming its idiosyncratic characteristics. For the purpose of this study, the strategic profile as a unit of observation is considered and not a company which is a research approach also applied one earlier study (Cho, Lee, Hwa, & Hwang 2012).

The concept of intellectual capital is closely related to the creation, sharing and management of knowledge within companies.

RESEARCH DESIGN AND METHODOLOGY

Challenges of Measuring Intellectual Capital

During the measurement process, different elements that give life to intellectual capital have to be identified, i.e. the intangible resources, the interactions among them and value creation. It is with reference to all these aspects that the measurement takes place. Indicators should be consistent with the mobilization of resources aimed at achieving targets and financial success (Grasenick & Low, 2004) and the choice of indicators should to be associated with specific elements is not easy. Therefore, there is the risk of designing indicators that only partially represent the elements that make up the complex web of intellectual capital. The need to combine financial and non-financial indicators derives from the inability of the former to adequately represent the complex web of intangible resources. The financial measures are symptomatic: they measure only the final outcomes, not allowing for the identification of the causes that generate the results. Eccles (1991), among others, argues that using non-financial measures allows to capture the causes of the company's success and for this author leading indicators "drive" the performance of lagging indicators.

The measurement of intellectual capital for internal management purposes ensures the availability of information in order to support and guide the decision-making process towards value creation through an efficient and effective management of intellectual capital (Sveiby, 2007). In particular, the measurement process should be mainly focused on the relations among the intangible assets that make up the complex system of intellectual capital, i.e. the real source of value creation. This perspective would capture the dynamic nature of the phenomenon under investigation, allowing managers to monitor the way in which actions on intellectual capital contribute to the overall corporate performance. Thus, the measurement for internal management purposes produces knowledge on intellectual capital. It is not a mere representation of the past, but it allows for the identification of problems and risks in the present (Montemari & Nielsen, 2012). Moreover, information attained from measures can be considered as a stimulus to make changes in the future, creating goals and weighing alternative courses of action (Mouritsen, 2004). Measurement and management are, therefore, closely related terms.

Over the past 15 years, academics and practitioners have developed many definitions, measures and frameworks. However, they use static modes of intellectual capital measurement which do not fit with the real essence of intellectual capital. Therefore, a growing number of contributions, such as from O'Donnell, Henriksen and Voelpel (2006), have argued that the existing frameworks for intellectual capital measurement are not able to fully explain the value creation process triggered by intangible resources (Dumay & Cuganesen, 2011). The relationships among intangible resources, and the value creation are not explicitly handled in the *Intellectual Capital Measurement Systems* (ICMS). Most systems are based on categories that have been identified as one of the main problems for intellectual capital measurement (Mouritsen, 2009). In fact, categories create separations: intangible resources are mainly extracted from the context in which they are deployed and are then measured "*on hold*", not "*in action*". In this way, categories in fact hinder the full consideration of dynamic aspects of intellectual capital.

Indicators for Measuring Intellectual Capital

The purpose of measuring intellectual capital is to discover the factors that create value and are not covered by traditional measurement systems. Thus, each indicator should contribute to revealing the hidden organizational processes[1] that create value. In order for the intellectual capital to generate the maximum value it needs to be made visible. The use of indicators for this purpose is largely different from traditional measurements which are basically looking for numbers.

Intellectual capital indicators are used to detect and define intangible resources of an organization. Their role is to make visible the invisible, illustrate it and measure it. From a management perspective, indicators are used to detect whether an entire organization is on the right path, i.e. whether the decisions of individuals are consistent with the objectives and strategies of the company. This means that a good indicator can always be adapted to new and unexpected demands.

In search of meaningful indicators, it is necessary first of all to identify the intellectual capital of the organization, to locate it and find out how it works, i.e. how it creates value. There are no general rules when it comes to intellectual capital. The content of intellectual capital and the way in which it is transformed into added value depends on the organization and what it wants to achieve. The results of measuring intellectual capital cannot be expressed in numbers at any price.

When defining appropriate indicators it is important to determine the number of indicators for one influential factor. Other existing reports (i.e. management, personnel, marketing) can be used as indicators of intellectual capital but in this case it is important that these indicators be viewed from a new perspective. Since indicators are a basis of measuring intellectual capital, measurement accuracy and usability of the intellectual capital report largely depend on the defined indicators.

According to Roberts (1999) the basic technical criteria for indicators are as follows (p. 1-22):

- **Precision:** The goal is to make as small a fluctuation as possible, i.e. the range of results. Imprecision of indicators may have dramatic influence on the organization or individuals. If they are inaccurate, they cast doubt on the measuring system and cause dissatisfaction of employees or managers. Therefore, accuracy is also an important motivator in the organization.
- **Objectivity:** For an indicator to provide objective data, it must exclude any kind of bias, although in some situations, certain subjective assessment is the only one possible.
- **Timeliness:** The indicator does not reveal anything if its measurement is not done at the right time. As with precision, timeliness and indicators used to measure it can be a good motivator because if, for example, an award earned by measuring indicators is untimely, the motivation will be lesser. If indicators are determined at the right time, there will be enough time to act on the basis of the given results.
- **Simplicity:** Indicators must be expressed in an understandable way for all, to avoid confusion, increase accuracy of information and to avoid ambiguity in their use or operation. SME needs to clearly define what and how is being measured by the indicator and formulate it in a language which is common in an organization.

According to another approach used by the firm Buckman Laboratories the following elements are assessed as indicators (Buckman, 2016):

- Percentage of the company that is effectively committed in dealing with customers;
- Percentage of income that is invested in the system of knowledge transfer;
- Number of employees with a college degree;
- Percentage of sales of new products (less than five years) in relation to total sales.

The FIVA (Framework of Intangible Valuation Areas) model enables an organization to identify potential indicators of business performance, to connect them with the influences of their intangible value, and then measures them to control and assess their impact on achievement of strategic requirements of knowledge management. FIVA provides a methodology for the efficient use of business resources and knowledge, to support the needs and opportunities for the organization to reach a specific market advantage (Green & Ryan, 2005).

Methods for Measuring Intellectual Capital

If the methods for evaluating intellectual capital are classified into methods of measurement and methods of assessment, one can agree that methods for measuring mainly deal with *internal* measuring and reporting on performance in order to gain management insight that helps companies to improve their business. On the other hand, methods for assessing intellectual capital are primarily aimed at determining the economic value of the company and its intangibility. These approaches mainly deal with the *external* view of the company and are designed to help analysts and investors to evaluate the financial worth of an organization.

Without a measuring system and continuous monitoring of business performance and the results achieved, there is no proper orientation in the global market. In this regard, measuring and recording are an integral part of every business.

If the company does not have a clear picture of the situation in which it is located, it has no opportunity to improve its business.

Measuring intangible assets of intellectual capital has become a necessity in every company. In the last few decades a lot has been done in the process of finding ways to accommodate and reflect intangible assets in the financial statements, which greatly lose on relevance when they omit the most important and most valuable assets in the economy of knowledge.

In a modern economy, there obviously appears a clear need and necessity for reports on intellectual capital as an addition to traditional financial statements.

The inventors of intellectual capital came to the conclusion that knowledge should not be placed on the balance and mix with the experience of over 500 years old traditional accounting which is solely intended for reporting on business performance, i.e. financial capital.

Basically there are two systems or two categories according to which measurement is classified:

1. Measuring methods which use indicators to provide management with a better insight into the intangible resources of their company;
2. Assessment methods which try to assign a monetary value to intangible resources within the company.

Some of the existing methods for measuring intellectual capital identified by the author, are as follows:

- Intangible Assets Monitor (IAM);
- Danish Guidelines;
- Meritum Guidelines;
- The balance of knowledge – which is the most current.

To date, more than 30 other methods for measuring the efficiency of intellectual capital have been registered, but the problem is that, out of these methods, majority appear only once as a method, when they are used by individual authors, and in many cases they are not really used in practice.

Therefore, officially, there is still no generally accepted method for measuring intellectual capital. The author's tool for defining the value of intangible capital tries to integrate both methods. Therefore, the authors' tool for measuring the value of intangible capital covers all key segments of the company such as:

- Structural capital: i.e. internal and external capital
- Human capital

Definitions of human, structural and relational capital are conveniently provided in the Key Terms and Definitions section of the chapter.

MAIN COMPONENTS AND CHARACTERISTICS OF THE PROPOSED OPTIMIZATION TOOL FOR INTANGIBLES IN SMEs

In line with the previous theoretical analysis, the subject of this empirical work is the analysis of results obtained using the author's tool to measure intangible resources and capital in a company. The main task is to establish the value of the intangible resources of the company, with the tools presented. The authors present the newly created tool for optimizing intellectual potential in SMEs, verified through pilot research in the case study of company " Prestige By Milka ".

The company "Prestige By Milka" has began operating in 2002 in Subotica. Until 2000 it operated as a small hairdressing salon with only 2 employees. Upon arrival of Dragana Milankovic as company CEO the business started to develop and modernize. Already in 2004 the company has 7 and today 11 permanent employees and highly skilled workers. In addition to providing hairdressing services, in 2007 the company expanded its offering with the sale of cosmetics for hair care and signed a cooperation agreement with the well-known international company L'Oreal. In 2011 it invests in new office space of 170m2 and starts to operate from there. In 2012 the company expands its business by opening a new salon on the second location in the city. Thanks to the quality of service and professional way of doing business, the number of clients is constantly increasing so that the company currently serves 2,214 regular clients who are satisfied with the service. Thanks to its expertise and quality of cooperation with L'Oreal the company owner and manager meanwhile became ambassador for L'OREAL in the Serbian market, an international presenter and coach of L'OREAL, and from 2015 an HCF president of the club of high fashion in Serbia.

After research, based on the results and proposed recommendations for improving management of business the company has taken concrete actions according to prescribed guidelines. Teams were formed and persons appointed responsible for specific tasks relating to improvement of business. Already after a short time observed were positive results primarily reflected through the improvement of the atmosphere among employees. The findings, concrete guidelines, defined tasks and instructions that have contributed significantly to entrepreneurship and creativity of employees became visible, which resulted in greater cohesion among employees. The main challenges are the lack of experience of the employees in the management of the change process and daily tasks and duties. On that basis authors came to the conclusion that it is advisable the company should hire a mid-career consultant to manage the implementation and evaluation of business recommendations in the process. The value of company intangible resources of was determined by measuring these groups of internal and external resources combining measurements of assessors and information obtained through questionnaires.

The values of internal resources measured by the assessors included:

- *Effectiveness of management;*
- *Organizational processes;*

- *Business processes;*
- *Business management systems.*

The values of internal resources measured through questionnaires included:

- *Organization;*
- *Management;*
- *System;*
- *Competitiveness;*
- *Innovation;*
- *Codification.*

The values of the external resources included:

- *Key suppliers;*
- *Clients;*
- *Value of goods and services.*

Finally, human resources in the company were measured by observing the following parameters:

- *Number of employees;*
- *Number of newly employed in the reporting period;*
- *Number of workers who left the company in the reporting period;*
- *Average age of employees;*
- *Average time spent in the company;*
- *Number of employees by level of professional competence;*
- *Number of training received by level of professional competence;*
- *Employee creativity;*
- *Personal development of employees;*
- *Teamwork of employees;*
- *Innovativeness of employees;*
- *Employee initiative.*

Measurement results are expressed in numerical, percentage and financial values.

SPECIFIC OBJECTIVES OF THE CASE STUDY RESEARCH IN SERBIA

The new economy based on knowledge sets new demands on all companies. They have to transform their work and adapt to the new conditions.

The significance of this research lies in the fact that it seeks to point out new directions and offer possible solutions to measure the value of intangible capital in small and medium-sized enterprises. In this research, one SME is analyzed for a case study, because SMEs should be the main drivers of renewal of the economic system and. Authors believe that SMEs have hidden potential of intangible values that are not adequately exploited.

In direct conversation with respondents prior to the completion of the questionnaire, the conclusion is that there is an awareness of the need to identify the value of intangible resources in small and medium-sized enterprises, but respondents did not have an opportunity to meet with adequate tools for their measurement.

The objectives of developing an instrument to measure the intangible value in an SME are:

1. Assessment of the identified dimensions of intellectual potential in the organization, by analysis of the research results;
2. Consideration of the identified dimensions;
3. Creating a roadmap for development of intellectual resources in the organization;
4. Conversion of identified dimensions, through tool indicators, into financial indicators.

In the context of the case study, it was also necessary to complete the following tasks:

1. Examine whether there is a dependence on each of the identified dimensions of individual independent variables (gender, level of education, size of the organization, ownership structure);
2. Based on the results obtained in cases where it was determined that there is the dependence of identified dimensions of the independent variables, determine reference values for each subset of the basic set which features a unique value of independent features;
3. On the basis of these results make a reduction in the number of variables that correspond to each of the dimensions;
4. Bearing in mind the dimensions identified, their standard reference values, set an integral instrument for assessing the intellectual potential and capital in an organization.

Research Hypotheses

Starting from the proposed tasks and objectives of the research, the following hypotheses of the research are set:

Hypothesis (H1): It is possible to identify n dimensions that characterize the intellectual potential and capital in an SME.

Hypothesis (H2): It is possible to standardize the instrument for assessing each of the n-identified dimensions of intellectual resources and capital in an SME.

Hypothesis (H3): It is possible to measure, monitor and optimize the elements which affect the magnification of the value of the intellectual resources and capital in an SME.

Research Methods

Precise definition of the research problem and topic is an important factor, given the breadth of the field of observation.

While working on this chapter, scientific and professional literature, professional journals and literature available online were used. In this study two main methods were used for data collection and analysis: method of theoretical analysis and survey method. The method of theoretical analysis was used to study the secondary data.

The results of this analysis were used primarily as a guide for the design and implementation of the empirical research.

As the main research method the survey method was used in order to collect the desired data on the appropriate SME interviewees in the field. The data was examined using statistical methods and techniques in order to view and interpret the results, and based on them draw the appropriate conclusions.

In the process of research the qualitative and the quantitative approach were combined. The data obtained in the survey will be analyzed:

1. **Quantitatively:** With the appropriate mathematical and statistical methods;
2. **Qualitatively:** With the methods of induction, analysis, synthesis, description.

The techniques and instruments were selected within the descriptive research method which is appropriate to the object and purpose of this research.

Research Results

The results of the value of intangible resources of the company "Prestige By Milka", were reached through the application of tools by which the critical value of intangible assets of company was measured. Through individual interviews authors estimated the core values of the company's key business. Using the tool for measuring the value of intangible capital, all key segments of the company structural and human capital were covered.

The interviews involved the director or owner of the company, managers, supervisors and workers.

Interviews were conducted by an expert, independent consultant-interviewer. Interviews were performed individually - "one on one". Measurements covered the mentioned key elements of business functioning.

The measured results are reported with the typical numeric rating scale with values from one to five having the following characteristics:

5. **Excellent:** Full and efficient implementation with lasting results;
4. **Appropriate:** A satisfactory level of implementation, but not perfect;
3. **Border:** Implemented but incomplete, the results are mediocre and non-permanent;
2. **Incorrect:** Much more needs to be improved;
1. **Nothing:** Nothing has been done.

The sum of the results obtained is a measure of the value of intangible resources of the company. The result indicates the value of existing intangible resources of the company, but also pinpoints the areas that can be improved. Presentation of the obtained values of intangible assets of the company in the form of tables is provided by authors. Table 1 includes general information on the surveyed SME.

Table 2 provides results of assessment of intangible values of human capital.

Especially interesting is the low score of innovation among employees, which is not the result of real incapacity to be innovative, but of power culture in the organization. Table 3 provides results of assessment of intangible values of structural capital.

Table 4 provides the second set of results of assessment of intangible values of structural capital.

Table 1. About the surveyed company

COMPANY NAME	"Prestige By Milka"
HEAD OFFICE	Novi Sad
COMPANY NUMBER	55593213
TAX ID	101657408
DATE OF ESTABLISHMENT	04 07 2002
TELEPHONE NUMBER	021 45 13 13, 063 537 535
CONTACT PERSON	DRAGANA MILANKOVIĆ
DATE	21. 05.2015.
OBSERVING PERIOD	01.01.2014-01.01.2015.

Source: Presentation of authors.

Table 2. Assessment of intangible values of human capital

INTANGIBLE VALUE OF HUMAN CAPITAL	Notice			
		Current year	Previous	Difference
1	2	3	4	5
INTERNAL RESOURCES			N/A	N/A
ESTIMATED VALUE OF IC HUMAN RESOURCES				
1. NUMBER OF EMPLOYEES		11		
2. NEWLY EMPLOYED IN THE REPORTING PERIOD		2		
3. LEFT THE COMPANY IN THE REPORTING PERIOD		1		
2. AVERAGE AGE OF EMPLOYEES		31		
3. AVERAGE YEAR SPENT IN COMPANY		7		
NUMBER OF EMPLOYEES BY LEVEL OF PROFESSIONAL COMPETENCE				
1. I. LEVEL - MANAGER		3		
2. II. LEVEL - STYLIST - MENTOR		2		
3. III. LEVEL - STYLIST		3		
4. IV. LEVEL – HAIR-DRESSER		2		
5. V. LEVEL – EXTRA WORKER		1		
6. VI. LEVEL -		0		
NUMBER OF TRAININGS HELD BY LEVEL OF PROFESSIONAL COMPETENCE				
1. I. LEVEL - MANAGER		6		
2. II. LEVEL – STYLIST - MENTOR		4		
3. III. LEVEL - STYLIST		4		
4. IV. LEVEL – HAIR DRESSER		3		
5. V. LEVEL – EXTRA WORKER		3		
6. VI. LEVEL -		0		

continued on following page

Table 2. Continued

INTANGIBLE VALUE OF HUMAN CAPITAL	Notice			
		Current year	**Previous**	**Difference**
INTANGIBLE VALUE OF HUMAN CAPITAL	Notice	Amount in %		
		Current y.	Prev.	Difference
1	2	3	4	5
EXTERNAL RESOURCES			N/A	N/A
ESTIMATED VALUE OF IK HUMAN RESOURCES Questionnaire		**78%**		
1. EMPLOYEE CREATIVITY		80%		
2. PERSONAL DEVELOPMENT		95%		
3. TEAMWORK OF EMPLOYEES		83%		
4. INNOVATIVENESS OF EMPLOYEES		65%		
5. INITIATIVE OF EMPLOYEES		65%		

Source: Author calculations.

Table 3. Assessment of intangible values of structural capital 1

INTANGIBLE VALUE OF STRUCTURAL CAPITAL	Notice	amount in %		
		Current y.	**Prev.**	**Difference**
1	2	3	4	5
INTERNAL RESOURCES			N/A	N/A
ESTIMATED VALUE OF IC - Questionnaire		**71.94%**		
ORGANIZATION		**70.00%**		
1. CLEARLY DEFINED VISION		80.00%		
2. DEFINED STRATEGY		65.00%		
3. CLEARLY DEFINED TASKS		65.00%		
MANAGEMENT		**75.00%**		
1. MANAGEMENT COMPETENCE		80.00%		
2. MANAGEMENT SUPPORT		70.00%		
3. MANAGEMENT INNOVATIVENESS		75.00%		
SYSTEM		**65.00%**		
1. SYSTEM EFFICIENCY		75.00%		
2. SYSTEM SUSTAINABILITY		70.00%		
3. DECISION MAKING SYSTEM		50.00%		
COMPETITIVENESS		**88.33%**		
1. COMPETENCE		95.00%		
2. SKILLS		90.00%		
3. IMPLEMENTATION		80.00%		
INNOVATIVENESS		**65.00%**		

continued on following page

Table 3. Continued

INTANGIBLE VALUE OF STRUCTURAL CAPITAL	Notice	amount in %		
		Current y.	Prev.	Difference
1. INNOVATIVE IDEAS		65.00%		
2. INNOVATION DEVELOPMENT		75.00%		
3. CREATIVITY LEVEL		55.00%		
CODIFICATION		**68.33%**		
1. DEFINED STANDARDS		80.00%		
2. TRANSFER OF KNOWLEDGE		70.00%		
3. STORAGE		55.00%		

Source: Author calculations.

Table 4. Assessment of intangible values of structural capital 2

INTANGIBLE VALUE OF STRUCTURAL CAPITAL	Notice	Amount in %		
		Current y.	Prev.	Difference
1	2	3	4	5
INTERNAL RESOURCES			N/A	N/A
ESTIMATED VALUE OF IC - EVALUATOR		**71.43%**		
EFFECTIVENESS OF MANAGEMENT		**82.86%**		
1. PLANS, VISION, MISSION...		60.00%		
2. EFFECTIVENESS OF MANAGEMENT		80.00%		
3. BUSINESS INDICATORS		80.00%		
4. COMPANY VALUE		80.00%		
5. MANAGER STANDARDS		100.00%		
6. MANAGERS COMPETENCE		100.00%		
7. MENTORING		80.00%		
ORGANIZATIONAL PROCESSES		**94.29%**		
1. BUSINESS TASKS		100.00%		
2. TEAMWORK		100.00%		
3. COMPETENCE OF EMPLOYEES		100.00%		
4. DEVELOPMENT		100.00%		
5. RESPONSIBILITIES		80.00%		
6. COMMUNICATION		100.00%		
7. PRIZES AND AWARDS		80.00%		
BUSINESS PROCESSES		**77.14%**		
1. RELATIONSHIP WITH CUSTOMERS		100.00%		
2. QUALITY CONTROL		80.00%		
3. RELATIONSHIP WITH SUPPLIERS		60.00%		
4. INNOVATIVENESS		60.00%		

continued on following page

Table 4. Continued

INTANGIBLE VALUE OF STRUCTURAL CAPITAL	Notice	Amount in %		
		Current y.	Prev.	Difference
5. CODIFICATION		60.00%		
6. METICULOUSNESS		100.00%		
7. QUALITY IMPROVEMENT		80.00%		
BUSINESS MANAGEMENT SYSTEMS		**31.43%**		
1. BUSINESS PLAN		20.00%		
2. DEVELOPMENT		20.00%		
3. DISTRIBUTION OF RESOURCES		20.00%		
4. MONITORING BUSINESS PROCESSES		20.00%		
5. EMPLOYEE INVOLVEMENT IN PLANNING		20.00%		
6. SYSTEM FOR BUSINESS PROCESSES IMPROVING		60.00%		
7. STIMULATION		60.00%		

Source: Author calculations.

The company did not clearly define vision, mission, goals, which in this segment is the lowest score. The same goes for relationship with suppliers. With expertise of the owner of the company and the slow fluctuation of employees, quality standards, knowledge, skills, were built in the company, which measurement recorded, but it was not codified and standardized through processes, records, manuals, standardized training. The measurement has shown that there is no plan of company development, which inevitably leads to stagnation and degradation of business results.

Table 5 provides the third set of results of assessment of intangible values of structural capital.

Table 5. Assessment of intangible values of structural capital 3

INTANGIBLE VALUE OF STRUCTURAL CAPITAL	Notice			
		Current y.	Prev.	Difference
1	2	3	4	5
EXTERNAL RESOURCES			N/A	N/A
ESTIMATED VALUE OF IC				
SUPPLIERS				
1. NUMBER OF KEY SUPPLIERS		3		
2. TOTAL NUMBER OF INVOICES BY KEY SUPPLIERS		112		
3. AVERAGE CHARGE BY INVOICE		244.58 €		
4 COST OF MATERIALS		27,392.44 €		
CLIENTS				
1. TOTAL NUMBER OF CLIENTS		2,214		
3. TOTAL NUMBER OF CUSTOMERS WHO USED SERVICES 1+ TIMES in%				

continued on following page

Table 5. Continued

INTANGIBLE VALUE OF STRUCTURAL CAPITAL	Notice			
		Current y.	Prev.	Difference
3.a. MONTHLY		8.67%		
3.b. QUARTERLY		19.38%		
3.c. SEMI ANNUALLY		24.62%		
3.d. ANNUALLY		47.34%		
TRANSPORT OF GOODS AND SERVICES				
1. TOTAL NUMBER OF WORKING ORDERS		11,226		
2. AVERAGE SERVICE PRICE		14.67 €		
3. AVERAGE INCOME		12.23 €		
4. TOTAL INPUT		164,733.84 €		
5. VALUE REALIZED		**137,341.40 €**		

Source: Author calculations.

Internal intangible value in the company "Prestige By Milka" evaluated by assessors is 71.43% out of 100%. The percentage share of intangible resources in total revenue of "Prestige By Milka" is 83.37% out of possible 100%. The percentage share of intangible resources in total revenue achieving IS 83.37%.

Table 6 provides an author's analysis of results of measurement of intangible values.

Overall assessment of intangible values of human resources in the company '"Prestige By Milka" is 78% out of 100%. Overall assessment of internal intangible values in the company "Prestige By Milka" through evaluation form is 71.94% out of 100%.

Table 6. Analysis of the results of measurement of intangible values

EVALUATION OF INTANGIBLE VALUES OF HUMAN RESOURCES IN THE COMPANY	
EMPLOYEE CREATIVITY	80%
EMPLOYEE'S PERSONAL DEVELOPMENT	95%
TEAMWORK OF EMPLOYEES	83%
EMPLOYEE'S INNOVATIVENESS	65%
EMPLOYEE INITIATIVE	83%
INTERNAL ASSESSMENT OF INTANGIBLE VALUES IN THE COMPANY THROUGH EVALUATION FORM	
ORGANIZATION	70%
MANAGEMENT	75%
SYSTEM	65%
COMPETITIVENESS	88,3%
INNOVATIVENESS	65%
CODIFICATION	68%

continued on following page

Table 6. Continued

INTERNAL ASSESSMENT OF INTANGIBLE VALUES IN THE COMPANY EVALUATED BY ASSESSORS	
EFFECTIVENESS OF MANAGEMENT	82,86%
ORGANIZATIONAL PROCESSES	94,29%
BUSINESS PROCESSES	77,14%
BUSINESS MANAGEMENT SYSTEMS	31,43%

Source: Author calculations.

FUTURE RESEARCH DIRECTIONS

After conducting the pilot studies, it is necessary to align the indicators in the instrument to be as accurate as possible, and then perform measurements on a larger sample of companies in order to be able to perform comparisons of different industries and provide guidance for the practical implementation of the optimization tool for measuring intangible assets in SMEs.

What is not covered by measurement in this case study, which opens like a completely new field of research, is the entrepreneurial capital. Research considering entrepreneurial capital and its impact on business performance on small business is still developing. Entrepreneurial capital consists of the competence and commitment related to entrepreneurial activities in the organization (Erikson, 2002). Despite different interpretations, it has been found that entrepreneurial capital impacts innovation (Hughes M. Hughes P. & Morgan, 2007). Obviously entrepreneurial firms are more successful in innovation. Even at the societal level, it seems to the authors that entrepreneurship is essential for producing more innovation. An organization with high entrepreneurial capital will be more innovative through having electronic and mental infrastructures that enable and support knowledge codification, collection, memorizing, socializing, and search, and thereby aid innovation. For all these reasons, in subsequent measurements, it is necessary to include entrepreneurial capital. It is therefore necessary to extend the tools, create appropriate indicators and to measure entrepreneurial capital as an integral part of the intellectual capital of SMEs.

In the analysis of data obtained in this pilot case study, it was noticed that the part referring to the measurement of mutual trust in the organization is missing, and through special interviews with entrepreneurs and employees authors came to the conclusion that this is one of the most important categories within an organizational culture.

CONCLUSION

After the pilot measurement of intellectual potential of the "Prestige By Milka" company performed with the tool presented in this chapter, a few low ratings are observable in the segment of external and internal resources: the low score of innovation and initiative employees, unclear mission, vision and goals of the company, inadequate relationship with suppliers, as well as the lack of codification of knowledge with the aim of preserving it and for further transmission.

On the basis of the measurement of intellectual potential several recommendations were given to improve operations and optimize the intellectual potential in this SME.

Low score of innovation among employees is not the result of real inability of employees to be innovative but it is the result of a culture of power in the organization. This approach has resulted in rapid initial development and growth of the company, but for further progress it is essential for that culture of power to evolve into the culture of the task. Therefore, it will increase the initiative of employees and contribute to greater productivity and better business results.

To improve business results, it is suggested that the owner, along with 2-3 employees form a team that would clearly define the vision, mission, and short and long term business goals of the company and inform other employees.

The current relationship with suppliers is reflected in the fact that they dictate the rules of cooperation because the company activities rely on one key supplier. In order to improve relationships with suppliers it is proposed to incorporate a larger number of suppliers and establish standard of cooperation. In order to promote innovation the observed company needs to take the following five clear steps: 1. Create an innovative team 2. Develop a long-term and short-term plan for innovation 3. Define annual budget for innovation 4. Define and assign responsibility to the team 5. Create a mechanism for evaluation. It is also suggested that, with the help of consultants, the company codifies all existing knowledge and turns it into a permanent tangible value of the company through standardized trainings, manuals, codices, rule books and procedures.

Finally, it is necessary to develop a business plan of short and long-term development of the company. It is important to note that there are quality resources and opportunities for business improvement through development and investment.

The conclusion based on measurements is that existing resources are not properly utilized and it is possible to significantly improve business results by better managing resources. The proposal is to analyze the utilization of the existing resources, and the strategy and implementation to be created on the basis of the measurement results.

Involving employees in planning and setting business objectives will contribute to better understanding and consequently better employee engagement to achieve higher business results.

Permanently and systematically dealing with the improvement of business processes would significantly impact on business performance. For now, the observed company is engaged in an ad hoc approach.

The case study shows that the future of SMEs can be reflected in the fact that they must be grounded in knowledge. It is also important that entrepreneurs recognize the importance of intellectual capital and to know the methods by which their companies can turn into knowledge-based enterprises.

Authors tried to contribute to earlier findings that the intellectual capital is present, measurable, and can be improved in SMEs. After analyzing and processing the obtained data, it can be concluded that it is possible to measure, monitor and improve the elements that influence the possible increase and value of intellectual potential and capital in an SME which was the single main objective of this chapter.

REFERENCES

Archibugi, D., Filippetti, A., & Frenz, M. (2013). Economic crisis and innovation: Is destruction prevailing over accumulation? *Research Policy*, *42*(2), 303–314. doi:10.1016/j.respol.2012.07.002

Bjurstrom, E., & Roberts, H. (2007). *The Principle of Connectivity: Networked Asets, Strategic Capabilities and Bundled Outcomes. In Intellectual Capital Revisited: Paradoxes in the Knowledge Intensive Organization* (pp. 45–60). Cheltenham, UK: Edvard Elgar.

Байбурина Э.Р. Головко, Т.В. (2008). Эмпирическое исследование интеллектуальной стоимости крупных российских компаний и факторов ее роста. Корпоративные финансы. N° 2 (6). (pp. 5–19).

Bocquet, R., Bas, C., Mothe, C., & Poussing, N. (2013). Are firms with different CSR profiles equally innovative? Empirical analysis with survey data. *European Management Journal*, *31*(6), 642–654. doi:10.1016/j.emj.2012.07.001

Bontis, N., Keow, W. C. C., & Richardson, S. (2000). Intellectual capital and business performance in Malaysian industries. *Journal of Intellectual Capital*, *1*(1), 85–100. doi:10.1108/14691930010324188

Bottani, E. (2010). Profile and enablers of agile companies: An empirical investigation. *International Journal of Production Economics*, *125*(2), 251–261. doi:10.1016/j.ijpe.2010.02.016

Brooking, A. (1996). *Intellectual Capital*. London: Cengage Learning.

Buckman Laboratories. (2016). Retrieved August 16th, 2016, from Buckman web site: http://www.buckman.com/en/about-us/history

Catasus, B. & Chaminade. (2007). Intellectual capital: Paradoxes and expansion. In *Intellectual capital revisited: paradoxes in the knowledge intensive organization*. Edvard Elgar.

Chen, M. C., Cheng, S. J., & Hwang, Y. (2005). An empirical investigation of the relationship between intellectual capital and firms market value and financial performance. *Journal of Intellectual Capital*, *6*(2), 159–176. doi:10.1108/14691930510592771

Cho, D. W., Lee, Y. H., Hwa, A., & Hwang, M. K. (2012). A framework for measuring the performance of service supply chain management. *Computers & Industrial Engineering*, *62*(3), 801–818. doi:10.1016/j.cie.2011.11.014

Clarkson, P., Li, Y., Richardson, G., & Vasivari, F. (2011). Does it really pay to be green? Determinants and consequences of proactive environmental strategies. *Journal of Accounting and Public Policy*, *30*(2), 122–144. doi:10.1016/j.jaccpubpol.2010.09.013

Crook Russell, T. (2011). Does Human Capital Matter? Strategic Human Resource Management. *Human Resource Management Journal*, *21*, 93–104.

Dahlmann, F., & Brammer, S. (2011). Exploring and Explaining Patterns of Adaptation and Selection in Corporate Environmental Strategy in the USA. *Organization Studies*, *32*(4), 527–553. doi:10.1177/0170840611403670

Dumay, J., & Cuganesan, S. (2011). Making sense of intellectual capital complexity: Measuring through narrative. *Journal of Human Resource Costing & Accounting*, *15*(1), 24–49. doi:10.1108/14013381111125305

Eccles, R. (1991). Analyzing cognitive maps to help structure issues or problems. *European Journal of Operational Research*, *159*, 673–686.

Edvinson, L. (2002). *Corporate Longitude: What you need to know to navigate the knowledge economy*. Financial Times Prentice Hall.

Erikson, T. (2002). Entrepreneurial capital: The emerging ventures most important asset and competitive advantage. *Journal of Business Venturing*, *17*(3), 275–290. doi:10.1016/S0883-9026(00)00062-8

Grasenick, K., & Low, J. (2004). Shaken, not stirred: Defining and connecting indicators for the measurement and valuation of intangibles. *Journal of Intellectual Capital*, *5*(2), 268–281. doi:10.1108/14691930410533696

Green, A., & Ryan, J. C. H. (2005). A framework of intangible valuation areas (FIVA): Aligning business strategy and intangible assets. *Journal of Intellectual Capital*, *6*(1), 43–52. doi:10.1108/14691930510574654

Guevara, D., & Bounfour, A. (2013). Monitoring intellectual capital: A case study of a large company during the recent economic crisis. *Knowledge Management Research & Practice*, *11*(2), 196–207. doi:10.1057/kmrp.2013.12

Guthrie, J. T., Klauda, S. L., & Morrison, D. A. (2012). Motivation, achievement and classroom contexts for information book reading. Retrieved from http//www.corilearning.com/researchpublications/

Hassan,, S.B., Hamid, M.S.B., & Bohairy, H. A. (2010). Perception of destination branding measures: A case study of Alexandria destination marketing organizations. *International Journal of Euro-Mediterranean Studies*, *3*(2), 271–288.

Huang, C. J., & Liu, C. J. (2005). Exploration for the relationship between innovation, IT and performance. *Journal of Intellectual Capital*, *6*(2), 237–252. doi:10.1108/14691930510592825

Hughes, M., Hughes, P., & Morgan, R. (2007). Exploitative Learning and Entrepreneurial Orientation Alignment in Emerging Young Firms: Implications for Market and Response Performance. *British Journal of Management*, *18*(4), 359–375. doi:10.1111/j.1467-8551.2007.00519.x

InCaS. (2008). *Intellectual Capital Statement for Europe. A new approach to make SMEs successful.* Retrieved August 16th, 2016, from InCaS web site: http://www.psych.lse.ac.uk/incas/page99/page99.html

Kariuki, A. K., Obonyo, P., & Ogutu, M. (2015). Intellectual Capital and Performance of Firms Listed on Nairobi Securities Exchange. *1st DBA-Africa Management Review International Conference* (pp. 118-142).

Kianto, A. (2007). What do we really mean by dynamic intellectual capital*? International Journal of Learning and Intellectual Capital*, *4*(4), 342–356. doi:10.1504/IJLIC.2007.016332

Kremp, E., & Mairesse, J. (2004). *Knowledge Management, Innovation and Productivity: A Firm Level Exploration Based on French Manufacturing CIS3 Data*. NBER Working Paper No. 10237. Retrieved from http://www.nber.org/papers/w10237.pdf

Kristandl, G., & Bontis, N. (2007). Constructing a definition for intangibles using resourced based view of the firm. *Management Decision*, *45*(9), 1510–1524. doi:10.1108/00251740710828744

Kunc, M. & Bhandari R. (2011). Strategic development processes during economic and financial crisis. *Management Decision*, *49*(8), 1343 – 1353.

Lev, B. (2001). *Intangibles: Management, measurement, and reporting*. New York: Brookings Institution Press.

Martínez-Torres, M. R. (2014). Identification of intangible assets in knowledge based organizations using concept mapping techniques. *R & D Management*, *44*(1), 42–52. doi:10.1111/radm.12037

Molodchik, M., Shakina, E., & Bykova, A. (2012). Intellectual capital transformation evaluating model. *Journal of Intellectual Capital, 13*(4), 444–461. doi:10.1108/14691931211276089

Montemari, M., & Nielsen, C. (2012). Value creation maps. In Business Models: Networking, Innovating and Globalizing. BookBoon.com.

Mouritsen, J. (2004). Measuring and intervening: How do we theorize intellectual capital management? *Journal of Intellectual Capital, 5*(2), 257–267. doi:10.1108/14691930410533687

Mouritsen, J. (2009). Classification, measurement and the ontology of intellectual capital entities. *Journal of Human Resource Costing & Accounting, 13*(2), 154–162. doi:10.1108/14013380910968665

Rimmel, G. (2004). *Danish Guidelines on Intellectual Capital Reporting*. Retrieved May 6, 2016, from Gothenburg University web site: http://gupea.ub.gu.se/bitstream/2077/3058/1/SNEE2004_article_The_Danish_Guidelines_on_Intellectual_Capital_Reporting1.PDF

Nickerson, J., & Silverman, B. (1998). Intellectual Capital Management Strategy: The Foundation of Successful New Business Generation. *Journal of Knowledge Management, 1*(4), 320–331. doi:10.1108/EUM0000000004603

Nielsen, C. & Dane-Nielsen, H. (2010). The emergent properties of intellectual capital: A conceptual offering. *Journal of Human Resource Costing &Accounting, 14.*

ODonnell, D., Henriksen, L. B., & Voelpel, S. C. (2006). Guest editorial: Becoming critical on intellectual capital. *Journal of Intellectual Capital, 7*(1), 5–11. doi:10.1108/14691930610639732

Patel, N., & Narain, U. (2008). *Can intangible assets act as a cushion in a downturn?* Retrieved December 16, 2015, from Watson Wyatt web site: http://www.watsonwyatt.com/pubs/directions/media/2009_EU_12783_ Directions_CP05_Intangibles.pdf

Pulic, A. (2000). *VAIC – an accounting tool for IC management*. Retrieved 11, March, 2016, from VAIC web site: http://www.vaic-on.net/start.html

Roberts, H. J. E. (1999). The Control of Intangibles in the Knowledge-intensive Firm.*22nd Annual Congress of the European Accounting Association*, Bordeaux, France.

Shakina, E., & Barajas, A. (2015). An intangible-intensive profile of companies: Protection during the economic crisis 2008-2009. *Management WP BRP, 33*, 1–22.

Shiu, H. (2006). The Application of the Value Added Intellectual Coefficient to Measure Corporate Performance: Evidence from Technological Firms. *International Journal of Management, 23*(2), 356–365.

Stewart, T. A. (1997). *Intellectual capital: The New Wealth of Organizations*. New York: Doubleday.

Subramaniam, M., & Youndt, M. A. (2005). The influence of intellectual capital on the types of innovative capabilities. *Academy of Management Journal, 48*(3), 450–463. doi:10.5465/AMJ.2005.17407911

Sveiby, K. E. (2007). *Methods for Measuring Intangible Assets*. Retrieved 10, March, 2016, from Sveiby web site: http://www.sveiby.com/articles/ IntangibleMethods.html

Tseng, C., & Goo, Y. (2005). Intellectual capital and corporate value in an emerging economy: Empirical study of Taiwanese manufacturers. *R & D Management*, *35*(2), 187–201. doi:10.1111/j.1467-9310.2005.00382.x

Unger, J. M., Rauch, A., Frese, M., & Rosenbusch, N. (2011). Human capital and entrepreneurial success: Ameta-analytical review. *Journal of Business Venturing*, *26*(3), 341–358. doi:10.1016/j.jbusvent.2009.09.004

Wang, W. Y., & Chang, C. (2006). Intellectual capital and performance in causal models: Evidence from the information technology industry in Taiwan. *Journal of Intellectual Capital*, *6*(2), 222–236. doi:10.1108/14691930510592816

Williams, C., & Lee, S. H. (2009). Exploring the internal and external venturing of large R&D-intensive firms. *R & D Management*, *39*(3), 231–246. doi:10.1111/j.1467-9310.2009.00553.x

Williams, C., & Nones, B. (2009). R&D subsidiary isolation in knowledge intensive industries: Evidence from Austria. *R & D Management*, *39*(2), 111–123. doi:10.1111/j.1467-9310.2009.00544.x

Wilson, J. W., & Eilertsen, S. (2010). How Did Strategic Planning Help During The Economic Crisis? *Strategy and Leadership*, *38*(2), 1087–8572. doi:10.1108/10878571011029000

ADDITIONAL READING

Amar, A. D. (2002). *Managing Knowledge Workers: Unleashing Innovation and Productivity*. Quorum Books.

Andriessen, D. (2004). *Making sense of intellectual capital: designing a method for the valuation of intangibles*. Burlington, USA: Elsevier Butterworth-Heinemann.

Bukovich, W. & Williams, R. (1999). *The knowledge management fieldbook*. Prentice-Hall, Financial.

Caplan, R., & Norton, D. (1996). *The Balanced Scorecard: Translating Strategy into Action*. Harvard Business Review Press.

Caplan, R. (2010). *Conceptual Foundations of the Balanced Scorecard*, Working Paper. HBS.

Chatzkel, J. L. (2003). *Knowledge Capital: How Knowledge-Based Enterprises Really Get Built*. Oxford University Press.

Copeland, T., Koller, T., & Murrin, J. (2000). Valuation: Measuring and Managing the Value of Companies. 3rd ed. John Wiley & Sons. Mc Kinsey and Company.

Czerniawska, F. (1999). *Management Consultancy in the 21st Century*. Macmillan. doi:10.1007/978-1-349-14873-8

Delong, D. W. (2004). *Lost Knowledge: Confronting the Threat of an Aging Workforce*. Oxford University Press. doi:10.1093/acprof:oso/9780195170979.001.0001

Firer, S., & Williams, S. M. (2003). Intellectual capital and traditional measures of corporate performance. *Journal of Intellectual Capital*, *4*(3), 348–360. doi:10.1108/14691930310487806

Fruin, W. M. (1997). *Knowledge Works: Managing Intellectual Capital at Toshiba*. Oxford University Press.

Hirschey, M. (1982). Intangible Capital Aspects of Advertising and R&D Expenditures. *The Journal of Industrial Economics*, *30*(4), 375–390. doi:10.2307/2097924

Hislop, D. (2009). *Knowledge Management in Organizations: A Critical Introduction*. Oxford University Press. IC Rating. Retrieved December 16, 2015, from IC rating website: http://www.icrating.com

Klein, D. (2011). *The Strategic Management of Intellectual Capital*. Transferred to Taylor & Francis.

Roos, G., Pike, S., & Fernstrom, L. (2005). *Managing Intellectual Capital in Practice*. Elsevier.

Stern, J. M. (2001). *The EVA Challenge: Implementing Value Added Change in an Organization*. New York, NY: Wiley.

Stewart, T. (2002). *The Wealth of Knowledge: Intellectual Capital and the Twenty-first Century Organization*, Financial Times.

Sullivan, P. (2001). *Profiting from Intellectual Capital: Extracting Value from Innovation*. Wiley.

Sveiby, K. E. *The Intangible Assets Monitor*. Retrieved December 16, 2015 from Sveiby web site: http://www.sveiby.com/articles/companymonitor.html

Swartz, N. P., & Firer, S. (2005). Board structure and intellectual capital performance in South Africa, Meditari. *Accountability in Research*, *13*(2), 145–166.

Tan, H. P., Plowman, D., & Hancock, P. (2007). Intellectual capital and financial returns of companies. *Journal of Intellectual Capital*, *8*(1), 76–95. doi:10.1108/14691930710715079

Teece, D. J. (2002). *Managing Intellectual Capital: Organizational, Strategic, and Policy Dimensions*. Oxford University Press. doi:10.1093/0198295421.001.0001

KEY TERMS AND DEFINITIONS

Human Capital: The value every employee adds to organization's added value, including qualifications, skills, experience, leadership capacity, communication skills and motivation.

Intellectual Capital: All knowledge within an organization which has the potential to create value when applied in line with the mission, vision and goals organization and refers to the intangible capital owned by companies: human capital, structural capital and network capital.

Measuring Intellectual Capital: Essential and very important in order to compare different companies, to estimate their real value or even to control their improvement year by year. Used to improve the way in which companies manage intellectual resources that generate value maximizing advantages for the company.

Medium Sized Enterprise: An enterprise that has between 50 employees and 249 employees and has either an annual turnover not exceeding €50m or an annual Balance Sheet total not exceeding €43m.

Network (Relational) Capital: One that reflects the relationship among shareholders, owners, business partners, investors, customers, suppliers and others.

Small Enterprise: An enterprise that has fewer than 50 employees and has either an annual turnover or an annual Balance Sheet total not exceeding €10m.

Structural Capital: Organizational structures and processes of the organization such as corporate culture, communication and interpersonal relations, development of new products and services and industrial innovation.

ENDNOTES

1 Editors' note: For the more process minded, see and compare with chapter 14 discussion on process management in SMEs.

APPENDIX: CASE SUPPORT MATERIAL

Questions and Answers

1. What is the overall problem presented in this case?

The aim of this case study is to contribute to the knowledge that the intellectual capital is present, measurable, and can be optimized in SMEs. In line with the previous theoretical analysis, the subject of this empirical work is the analysis of results obtained using the author's tool to measure intangible resources and capital in the company „ Prestige By Milka “. The main task is to establish the value of the intangible resources of the company with the tools presented. The authors presented the newly created tool for optimizing intellectual potential in SMEs, verified through a pilot research in the company „Prestige By Milka “.

2. What are the factors affecting the problem(s) related to this case?

The main challenges are the lack of experience of the employees in the management of the change. The company did not clearly define vision, mission, goals, which in this segment is the lowest score. The same goes for relationship with suppliers. With expertise of the owner of the company and the slow fluctuation of employees, quality standards, knowledge, skills, were built in the company, which measurement recorded, but it was not codified and standardized through processes, records, manuals, standardized training processes and daily tasks and duties.

3. Discuss managerial, organizational, and technological issues and resources related to this case.

Low score of innovation among employees is not the result of real inability of employees to be innovative but it is the result of a culture of power in the organization. This approach has resulted in rapid initial development and growth of the company, but for further progress it is essential for that culture of power to evolve into the culture of the task. Therefore, it will increase the initiative of employees and contribute to greater productivity and better business results. The current relationship with suppliers is reflected in the fact that they dictate the rules of cooperation because the company activities rely on one key supplier.

4. What role do different players (decision-makers) play in the overall planning, implementation and management of the information technology applications?

The owner of the company makes all decisions in the process of planning and implementation, as well as in connection with the procurement and use of information technology. Culture of power and roles is dominant. There is no authorization and delegation of responsibilities.

5. What are the possible alternatives and pros and cons of each alternative facing the organization in dealing with the problem(s) related to the case?

In the analysis of data obtained in the pilot case study, it was observed that the part referring to the measurement of mutual trust in the organization is missing, and through special interviews with entrepreneurs and employees, authors came to the conclusion that this is one of the most important categories within an organizational culture, as evidenced by the number of references.

6. What are some of the emerging technologies that should be considered in solving the problem(s) related to the case?

After conducting the pilot study, it is necessary to align the indicators in the instrument to be as accurate as possible, and then perform measurements on a larger sample of companies in order to be able to do comparisons of different industries and provide guidance for the optimization tool for measuring intangible assets in SMEs.

7. What is the final solution that can be recommended to the management of the organization described in the case? Provide your arguments in support of the recommended solution.

On the basis of the measurement of intellectual potential several recommendations were given to improve operations and optimize the intellectual potential in this SME.

To improve business results, it is suggested that the owner, along with 2-3 employees forms a team entrusted to clearly define the vision, mission short and long term business goals of the company. Consequently the team would also inform other employees.

In order to improve relationships with suppliers it is proposed to incorporate a larger number of suppliers and establish standard of cooperation. In order to promote innovation the company needs to take the following steps:

- Create an innovative team
- Develop a long-term and short-term plan for innovation
- Define annual budget for innovation
- Define and assign responsibility to the team
- Create a mechanism for evaluation.

It is also suggested that, with the help of consultants, the company codifies all existing knowledge and turns it into a permanent tangible value of the company through standardized trainings, manuals, codices, rule books and procedures. Permanently and systematically dealing with the improvement of business processes, would significantly impact on business performance. For now, the company is engaged in an ad hoc approach. Additionally, it is necessary to develop a business plan of short and long-term development of the company. It is important to note that there are quality resources and opportunities for business improvement through development and investment.

The conclusion based on measurements is that existing resources are not properly utilized and it is possible to significantly improve business results by better managing resources. The proposal is to analyze the utilization of the existing resources, and the strategy and implementation to be created on the basis of the measurement results. Involving employees in planning and setting business objectives will contribute to better understanding and consequently better employee engagement to achieve higher business results.

Epilogue and Lessons Learned

As emphasized, after conducting the pilot study, it is necessary to align the indicators in the instrument to be as accurate as possible, and then perform measurements on a larger sample of companies in order to be able to do comparisons of different industries and provide guidance for the optimization tool for measuring intangible assets in SMEs.

After analyzing and processing the obtained data, it can be concluded that it is possible to measure, monitor and improve the elements that influence on increasing of the value of intellectual potential and capital in an SME. The future of SMEs can also be reflected in the fact that it must be grounded in knowledge. It is also important that entrepreneurs recognize the importance of intellectual capital and to know the methods by which their companies can turn into a knowledge-based enterprise. Authors believe that the main aim of this case study was achieved and that it contributes to the knowledge which testifies that the intellectual capital is present, measurable, and can be improved in SMEs.

Chapter 16
Structural Change Management for Sustainable SME Development:
Applying Classical Management Tools

Robert M. Molnar
Technical College of Applied Sciences in Zrenjanin, Serbia

ABSTRACT

This chapter addresses several approaches in investigation of the SME life cycles, particularly in respect to the structure, functions and behavior of these businesses. Once SMEs have been established they start to change both their structure and behavior. In fact, it could be stated that SMEs operate in a permanently changing environment. There are a lot of factors which influence these changes and optimizing their management is a great challenge for most SMEs. As a result of these factors and suboptimal SME management practices survival rate of new established SMEs is low. For researchers and practitioners it is therefore crucial to discover how to increase long-term survival chances of SMEs through relevant structural change. In order to contribute to this goal the chapter will consider the conceptual and research framework of the structure and practical management behavior of SMEs in change management processes. The chapter also provides suggestions on how to use practical management tools such as the PEST, SWOT and TOWS matrices and ways of their analysis.

INTRODUCTION

Structural changes inevitably occur with the development of human society, both on a personal and on a global level. These structural changes are being reflected in value systems and put small, flexible organizations in a more favourable position compared to larger ones in terms of adaptability. Market competition between companies also forces small and medium size enterprises (hereinafter: SMEs) to be both innovative and productive. Speed of changes that take place in human society, even in markets, increased the competitive advantage of innovative and flexible companies, so they put at a disadvantage

DOI: 10.4018/978-1-5225-1949-2.ch016

those companies which are weighed down by large structures and bureaucracy. With this in mind, perhaps it is not surprising that the share of small enterprises in the total number of companies is in constant increase. In addition to their number, small companies have a significant role in employment generation which is higher when compared with large ones and what is also very important, small companies could be the driving force for local economic development, particularly out of urban areas (Birch, 1981).

However, a small number and scope of their business functions result in the fact that SMEs individually have little impact on their environment and the power to modify it according to their needs. But, for the same reasons SMEs have an ability to react quickly to the changes happening in their environment. Conversely, large enterprises strive to establish stability in their operations, as well as keep up their structure, opening the door for the development of a structured bureaucratic organization. This inevitably hinders their flexibility and innovation which are prerequisites for success in market competition (Molnar, 2010).

In the second half of the twentieth century, large enterprises and multinational companies were the main factor of economic growth and development. However, at the same time in both developed and developing economies, SMEs became an increasingly important economic and social factor. With their innovations and willingness to take a risk, entrepreneurs in these SMEs emerged as a potential force that changes the existing situation and contributes to overall progress. The willingness of societies to support such individuals often predetermined their own overall progress (Penezić & Jakopin, 2002).

Enterprise growth eventually develops and changes its organizational structure and its goals, as well as ways of achieving them. The initial phase in the life cycle of SMEs is often chaotic, uncertain and risky period in their lives (Stanković, 1998). Those companies which survive this period in many cases avoid re-exposure to risks and uncertainties with which they are faced. They sustain their businesses in a smaller framework, not only relating to the number of employees, but also to other business factors. In this way, owners of SMEs themselves become a limiting factor in their own SME life cycle.

Nevertheless, the number and rate of creation of "small" companies among the overall business entities are increasing day by day. SMEs emerge on a daily basis, but they also disappear on a daily basis. Whether this is due to their inability to overcome all the "children's diseases" on their development path or simply the law of the market in overall economic environment remains for further research. This is definitely a question that requires an answer. There are many ways in which small businesses are trying to survive, but modest overall resources that are at their disposal, are certainly one of the limiting factors of their survival and eventual further development.

In terms of ruthless struggle for market survival, the problems faced by SMEs are not just problems of individuals - entrepreneurs and managers. But rather, these problems have a wider social-economic dimension. An SME as an open organizational system is in intensive relationship with the environment, and its dependence on stakeholders is at an extremely high level. Since an SME does not have the power to change the environment, the only strategy is to adjust to it (Molnar, 2010).

As a result, and strongly dependent on success of SME management, numerous statistical evidence shows that less than a half of established SMEs survive for more than five years (EUROSTAT, 2016). Evidence seems more favorable in developed countries mostly due to their entrepreneurial culture level as well as organized systemic support provided by institutions of financial and non-financial SME support (SBA, 2014; Australian Government, 2013). Statistical SME performance data also reveals that failure to secure the survival is one of the most fundamental SME problems as well as opportunities.

For researchers and practitioners it is therefore crucial to discover how to increase long-term survival chances of SMEs. In order to contribute to this goal this chapter will implement the conceptual research framework on the structure and practical behavior of SMEs. Many researches have been conducted in area of business failure. Some of them emphasize problems with finances (Hagen, 2009), while others report non-financial problems such as poor marketing, bad business planning, lack of skills, poor leadership as crucial ones (Molnar, 2010; Wagner, 2013). But, somehow *unmet need for structural changes* in an SME has never been listed among most frequent reasons for unsuccessful business or finally for its failure. This is the main research gap and one of the intentions of this chapter is an attempt to go beyond it.

METHODOLOGY IMPLEMENTED

This chapter presents a conceptual framework of structural change management in sustainable SME development. The main result of this chapter is presented as a model of an *SME growth in order to achieve sustainable development.* This model has been developed as a synthesis of previously developed models and concepts in the relevant field that have been analyzed by broad literature research. High level of abstraction has been implemented as well in order to develop the model as it is. The literature sources for this research could be divided in two groups:

1. Primary, where researches in relevant field has been conducted on a broad SME base, and
2. Secondary, where relevant concepts has been developed.

BACKGROUND WITH LITERATURE REVIEW

Once SMEs have been established they start to change both their own structure and behavior. So, it could be highlighted that SMEs operate in permanent changes which originate both internally and externally. There are a lot of factors which influence these changes and optimizing their management is a great challenge for most SMEs.

The life cycle curve determination is one of the important tasks in investigating SME durability (Adižes, 1994). There are behavioral similarities among SMEs behaviour in respect to the position on the life cycle curve. At the same time, management approach and practice in how to achieve business goals has strong inluence on SME future determination in terms of success. Once an SME starts to grow, it undergoes permanent changes. Sound management should be able to deal with problems that arise due to change. Some of them can be foreseen, but some not. In dealing with problems solving different conflicts arises. The causes of conlifct have a complex nature that is as complex as SME business operation itself. But conflicts are expected and have to be carrefully managed in order to keep or turn them into the constructive ones (Adizes, 1992). A proper SME management tries to optimize together different interests of various stakeholders into an SME (partners, management, employees, to mention just a few) with the goals and strategic interest of the SME as a business.

The author would like to point out several other life cycle issues. For example, what is truly the driving force for an SME in its life cycle adventure? Is proper management enough to ensure a bright business future for an SME? One of the very important factors which shouldn't be neglected is *innovation*. It could be described as a specific objective of entrepreneurs to use change as an opportunity for the realization of various production and service activities. Therefore, entrepreneurs look at change as something desirable and healthy. They even look for them and use them as opportunities for practicing a pro-active approach in their everyday business (Drucker, 1991).

Following an SME on its life-cycle "travel" and bearing in mind previous remarks a question arises. Is it possible to develop/change an SME without changing its elements, particularly people within it? This question is tightly related with entrepreneur and SME management structure as well. Learning organization is an imperative in modern management. By an SME development there have been numerous changes undertaken, not just in "material" world but in "psychological" one too. If it is not achieved, that will be a big obstacle for SME future development (Sengi, 2003). According to this theory, there are 5 features of a learning organization:

1. Systems thinking,
2. Shared vision,
3. Challenging of mental models,
4. Team learning, and
5. Personal mastery, all of which are of tremendous importance for SMEs.

Besides the life cycle approach there are other theories the author has used extensively in his SME research (Molnar, 2010). First, the *system approach* which should be implemented in all efforts to achieve long term survival of an SME. The theory suggests that different physical, biological as well as social/organizational and other systems have certain similarities in their behaviors and have some common characteristics on higher levels of abstraction. The fact that an SME is an organizational system means that it performs and behaves on the highest level of complexity among all other known systems such as technical and biological ones (Boulding, 1956). Consequently, SME management efforts include more or less mistakes before a concrete activity even started. The other fact that is also important for this chapter relates to the behavior of a system, where it is not a sum of the behaviour of its parts, but rather it a product of their interactions (Ackoff, 1999). Beside these, any man made system can be classified as purposeful one. Relating to SMEs which are organizational systems, their members are purposeful individuals who intentionally formulate their objectives (Ackoff & Emery, 1972).

The second major theory closest to the author's views is implementation of laws of physics in the organization. This approach was previously called "Organizational Physics" (Sisney, 2013). This comprehensive approach incorporated in itself most of the previously developed approaches/theories, mentioned before with well known laws of physics. In this theory there are classified "Six Laws of Organizational Physics" which deal with an organization performance and can help to improve it. For an entrepreneur or manager, it is crucial to understand implications of each law. The first one, states that "An organization is a complex adaptive system", originates from systems theory and suggests that in every analysis, intervention and reaction, the organization should be considered in its totality. The second law, states that "An organization is subject to the first law of thermodynamics" suggesting that every organization has a finite amount of energy and if it strives to increase then it must obtain it from the outside". The third law states that "An organization is subject to the second law of thermodynamics" where due to entropy every system falls apart over time. So, organization first uses its energy to manage and counter disintegrating forces, and survival could be used to drive integration with its environment. The fourth law, states that "An organization must shape and respond to its environment and act as a whole system, including its parts and sub-parts". This implies that every organization is like a chaotic system which has patterns or forces throughout it. The fifth law states that "An organization is subject to the conditions of its environment" which is the driving principle of evolution suggesting that not the strongest or most intelligent companies will survive but those which are best adapted to the environment. Finally, the sixth law states that "An organization is subject to the laws of motion" putting into correlation Newton's laws of motion (inertia, acceleration and action-reaction) with the principles of organizational change and momentum.

One of the latest researches in the field of organizational change implies that uncertainty may not have an inevitable negative influence on organizational development (Zhang, 2016). In this study, a co-evolutionary model of organizational change has been developed. This model also emphasizes the interaction between institutional-level factors and organizational-level change as bridged by top management perceptions of organizational uncertainty where its dynamics affects organizational change.

Highlights of Types of Change Management

But, what is really the driving force behind the attained structural changes in SMEs? There are some processes which should be properly managed in order to achieve certain levels of change. Changes in an SME could be treated on two levels by two types of management programs (Bourda, 2013):

1. Systematic organization implying a wide change initiative that involves an organization through a wide transformation effort, particularly changes in its structures, and
2. Specific internal change management or change control program that involves providing tools and processes to control daily operational or project - specific changes, including the people side of change.

As a result of highlighted changes in structure, people and relationships, the behavior of SME naturally evolves and changes as well.

It is crucial for managers to identify change, modify or adapt it and consequently use a model for SME transformation to help an SME survive or even develop. There are at least two leading models in change management practice which the author would like to highlight.

One of the first developed and applied change management models for understanding organizational development refers to the three-stage process of change, namely the *Unfreeze–Change–Freeze* stages (Lewin, 1947). This model takes into the consideration psychological aspects of behavior modification in these three different stages. In the *Unfreezing* stage there is low level of resistance to change, mostly because of recognizing and accepting the need for change. It is reasonable to expect uncertainty. In the next stage named *Change* it is necessary to develop new attitudes to encourage behavior necessary for change to occur. In this stage, people and employees need to understand the benefits of change. In the final stage of change, named *Freezing,* processes should stabilize and there is a need for supporting and reinforcing new change conditions. It is important to acknowledge people's efforts that will reinforce their belief in future changes.

The next major model widened the stages that should be taken into consideration when dealing with change management. The eight-stage change model describes change process for managing change in organizations, mostly in large ones (Kotter, 2012), from which SMEs can learn. The eight steps are as follows:

1. **Establishing a Sense of Urgency:** People have to see the need to change;
2. **Creating a Guiding Coalition:** Create a group with authority and credibility to lead the change efforts;

3. **Developing a Vision and Strategy:** Create a vision and develop strategies with clear aim and way forward;
4. **Communicating the Change Vision:** Promote understanding and commitment in order to accept vision and strategy;
5. **Empowering Broad-Based Action:** Enable people to act and overcome barriers by changing the structures that could undermine the vision and strategy;
6. **Generating Short-Term Wins:** There is a need for visible achievements in order to motivate and ensure further support;
7. **Consolidating Gains and Produce More Change:** Introducing broad range of changes related to systems, structures and policies and maintains change momentum; and
8. **Anchoring New Approaches in the Culture:** Articulate and communicate new values, attitudes and behaviors.

However, besides many change management models developed SMEs have to reckon with a number of forces that resist change and support the status quo within an organization. These forces are both internal and external. An effective change management tries to minimize these forces by satisfying following criteria which affect stakeholders (French, Rayner, Rees, & Rumbles, 2015):

- **Benefit:** The change should have a clear relative advantage for the individuals being asked to change;
- **Compatibility:** The change should be as compatible as possible with the existing values and experiences of the people being asked to change;
- **Complexity:** The change should be no more complex than necessary. It must be as easy as possible to understand and use; and
- **Triability:** The change should be something that people can try on a step-by-step basis and make adjustments throughout the process.

In order to produce changes in people's and SME behavior as well, SME management needs to be very active during change phases in order to implement reinforcement tactics. Verbal reinforcement is a cost free method. For implementation it is important to articulate and permanently repeat desirable behavior that fits into organizational change. As a result, old behaviors become "extinguished" over time, and will be replaced with new (desired) behavior. It is crucial that some sort of reinforcement is necessary for change to take root in the individual. Beside the verbal method, an SME can change its structure and policy in a very formal way, by writing new rules and procedures. In this case, without informal reinforcement, employees in most cases are not going to change quite as easily. So, in organizational changes reinforcement comes in as a vital part of it (Pryor, Taneja, Humphreys, Anderson & Singleton, 2008).

Main Components and Characteristics of Change Management in Sustainable SME Development

As it has been stated previously, SMEs operate in very dynamic environment where uncertainty and particularly changes are among the most important characteristics. Hence, for this chapter, *change* and its management in an SME development is the main theme.

But, how do changes really affect an SME and organization in general? Any change generates one or more problems because permanent structures and (organizational) system need new parameters and actions in order to keep outputs within planned framework. So, managing changes is a permanent/daily activity performed by managers, and can potentially be one of the business functions in an SME as well as in any other organization. If it is not the case, SME will be negatively affected both externally and internally.

The main components of change management could be identified as: change, problems and solutions/management/problem solving (Adizes, 1992). It is very important to mention that these components create a cycle, so, every solution creates another change, and so on. But, it is very important to underline that initial change, marked as "$change_1$" which created "$problem_1$" after reaction of "$management_1$" causes "$change_2$" which differs from "$change_1$", as it presented below:

$change_1 \rightarrow problem(s)_1 \rightarrow solutions_1/management_1/(problem\ solving_1) \rightarrow change_2 \rightarrow \ldots$

Change can be minor or major problem, affecting less or more important components and structures in an SME. Depending on this, the problem(s) created by change will be minor or complex one(s). Minor problems could be solved and managed immediately, while complex ones need much more time, effort and knowledge.

Conflicts are side products of organizational changes and have to be taken into consideration in the change management process. There are both constructive and destructive conflicts, depending on mutual trust and respect between relevant people within an organization. If, there is mutual respect then people may learn from their differences. Also, people then may work together and decisions made will be better and their implementation will be less questionable (Adizes, 1992).

The next very important issue in change management is timing. In order to find out the most appropriate time to implement changes within an SME, so called push and pull factors have to be taken into consideration. If push factors are dominant then SME is forced to be changed in an opposite way it wants to change. It could be crucial to determine why an SME is implementing changes within it. The reason for change determines the strategy that will be implemented.

In case of huge, strategic changes a strategic plan should be developed or at least a new vision or mission statement. There are some proposed factors which may be considered as triggers for change (Rees & Hall, 2013):

- Challenges of growth, especially global markets;
- Challenges of economic downturns and tougher trading conditions;
- Changes in strategy;
- Technological changes;
- Competitive pressures, including mergers and acquisitions;
- Customer pressure, particularly shifting markets;
- Pressures to learn new organization behavior and skills; and
- Government legislation/initiatives.

Additionally, it is crucial to recognize the signs suggesting that some organizational structure isn't working and needs to be changed. When inertia starts to dominate in majority of business processes of an organization or SME, it is time for structural changes. If management neglects this sign and keeps the current structure operational, negative impact on the organization performaces will cause its „death" over time. According to the approach mentioned before (Sisney, 2013) there are five rules of organizational structure in change management process:

1. When the strategy changes, change the structure;
2. Don't allow functions focused on efficiency to control functions focused on effectiveness;
3. Don't allow functions focused on short-range results to control functions focused on long-range development;
4. Balance the need for autonomy with the need for control; and
5. Put the right people in the right functions.

Some authors suggest that change management process could be considered as a project with four parts of its implementation plan (Burton, Obel & Håkonsson, 2015). They are:

1. *What*. The tasks as projects and sub-projects have to be defined;
2. *Who*. There is a need to allocate human resources in order to assign responsibilities;
3. *When*. Decide on milestones and deadlines;
4. *Follow up on milestones*.

More recent researches suggest that organizational infrastructure may have a profound impact on organizational behavior. In process of changing organizational routines there will be influence of not only people's sense-making and behavior but also of the technology and infrastructure with which people interact (Langstrand, 2016).

However, bearing in mind the objective of this chapter which emphasizes sustainable SME development the most desirable change which should be implemented is related to the SME structures and its functions that develop new organizational systems ensuring a higher level and rate of SME survival.

A CONCEPTUAL FRAMEWORK OF STRUCTURAL CHANGE MANAGEMENT IN SUSTAINABLE SME DEVELOPMENT

From the very beginning of its operation every organizational system seems to be forever in reorganization (Drucker & Maciaiello, 2008). There are many reasons for that but spontaneous development of disorder, frictions and underperformance are most important ones that force an organization to change its own structure. It has to be noted that good organization structure doesn't guarantee good performance. But, in opposite, an inappropriate organization structure impedes performance and seriously affects survival of an organization and an SME as well.

As noted earlier, SMEs, like living beings, have their life cycle with characteristic phases they pass through. All of these phases have their own behavioral patterns and need a specific structure for SME operations. Therefore, structures in different phases differ and need to change as SME behavior as well. However, unlike living beings, SMEs, at least theoretically, could escape death as a final step in their life cycle. This is achievable only if the SME management practices proper approach to changes (Adižes, 1994).

By researching SME life cycle phases, it is possible to find some answers relating to SME survival and development. For example, emotional potential and other personal characteristics of the entrepreneurs are very important internal factors in early stage of structuring SMEs. But, good personal characteristics are not enough as a potential to structure successful SMEs in later stages of their development. Obviously, some needs in the market environment could be met by satisfying the demand for SME goods and services. If the needs are more intensive and offer is more attractive then SMEs have a higher potential to create added value. This potential is based on active intelectual potential and knowledge potential as one of the resources in SMEs. Knowledge potential is a multidimensional one because it is related to the product (knowing WHAT to produce), to the technology (HOW to produce) and finally to the organizational system (production management). But, this chapter will show that this resource is still not enough to create a functional SME capable for sustainable development. In addition, several other resources can't be neglected in optimizing the SME management system: machinery, supplies, finance, human capital are among the most important. Maintaining the potential of all of these resources at a certain potential level is a precondition of SMEs *short–term survival.* These resources can't be renewed by themselves, so „somebody" should take care of that. In early stage it is an entrepreneur but later on when the SME spreads out he/she is forced to involve others into the SME, and then its structure optimization actually starts (Černiček & Molnar, 1998). When the first cycle of an SME is completed it has three group of functions:

1. Functions for continuity assurance;
2. Functions for operative management; and
3. The servicing functions.

The SME which incorporated in itself all of these group of functions could be marked as S_{s-t}. All of these groups have some specific functions inside of them and have certain relations between themselves.

An SME evolution, from short-term to long-term survival is followed by some significant changes in its functional structure. This change, as a result, causes other ones, like quality and quantity of resources managed, and particularly SME behaviors expressed internally and externally as well. Relevant for all structures is that every of them is flexible, adaptable and changable at the beggining, but with time they become more inflexible and rigid. So, there is also a moment when none of the structures can respond to changes which happened externally and are not able to perform their main role. This is the final moment when SME has to implement change management in order to create appropriate changes within it. If it is not case, then the SME starts to get organizationally old. There are numerous factors which contribute to SME aging such as changes in individual and social needs, life cycle of the needs, development of science and technology, scarcity of natural resources and growing competition.

During its existance an SME permanently solves problems caused by changes and creates some effects. Figure 1 visualizes effects of an SME growth in order to achieve its sustainable development.

The curves presented in Figure 1 are a synthesis of all internal and external factors that affect an SME. Only one curve, the thickest one, represents successful change management. So, very few SMEs are capable to create the thickest curve in their life cycle. Most of them, turn aside of it with unsucsessful tries in order to realize own potential which is due to poor change management. However, as Figure 1 also suggests, innovation lies behind all successful SME development, not just in products or services but also in structures and management as well[1].

Figure 1. SME growth in the function of sustainable development
Source: *Adapted by author from Černiček & Molnar, 2000, p.232*

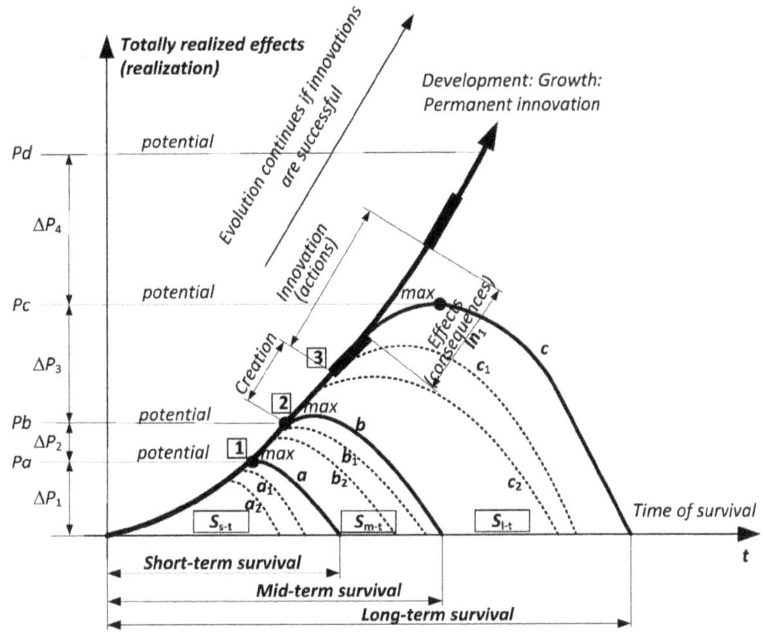

The most dominant feature of SME life cycle is its functional structure development. Every level of it, allows effectiveness up to a certain level (potential). When an SME evolves untill the level of short-term survival, then realized effects permenently increse until point 1 and after that SME starts to decrese its realized effectiveness. Maximum potential of this structure is denoted as ΔP_1, which is same as total potential denoted as P_a. When SME exploits available potential then it reaches a curve marked as „*a*", and if it is not the case then curves „a_1" and „a_2" are reached and are obviously behind curve „*a*" (See Figure 1).

There are four basic functions which assure SMEs continuity/short-term survival, such as: financial, sales, manufacturing (technical) and procurement functions (Černiček & Molnar, 1998). These functions properly interrelated should enable the reproduction cycle in an SME. If some of them fail then SME existance is seriously endangered. However, a question arises about the period of time in which this functional structure is able to ensure SME survival. It strongly depends on potential generated between the SME supply and demand created on the market. Finally, it is limited by the product life cycle which could be reached only by most successful firms. All potentials created this way definitelly vanish at the end of the product life cycle in any case. But, if an SME didn't create new potential it vanishes too.

When an SME reaches potential for longer survival, then realized effects increse until point 2 and after that SME starts to decrese its realized effects. Maximum potential of this structure is P_b, which is higher then P_a for ΔP_2, that originates from newly evolved part of the structure. This part of the structure should focus mostly on new products and/or services, meaning their quality and quantity in general and their market penetration. Additionaly, permanent development of products and/or services should be assured as well, which implies changes in management structure too. If SME exploits available potential then this is observable on a curve marked as „*b*", if it is not the case then curves „b_1" and „b_2" have been created behind curve „*b*".

For those SMEs untouched by negative statistics associated with „life expectancy", during the first „survival" phase, new potential can be created. This is only possible if SME develops new specialised functions, namely, the *SME passes through new structural changes*. Besides financial resources needed for creating the new system it is necessary to choose a certain segment of market environment with new potential. This is followed by creating attractive offers, development of SME processes which will realize the offer and finally the organizational system development with potential to realize all these factors. The research suggests that this system could be a system for *integrative management and development* and it has its own interrelated functions as well as functions previously developed (Zgodavova & Bober, 2012). As such it assures *mid-term survival* of an SME. The SME which successfully incorporates in itself all of these functions is an optimal mid–term SME and could be denoted as $S_{m\text{-}t}$. However, numerous researches on business failures suggests that SME managers neglect or just not able to recognize importance of structural changes of organization they run (Hagen, 2009; Molnar, 2010; Wagner, 2013).

In a situation of strategic development, continuous and accelerated growth could be experienced by an SME due to approaching point 3. Maximum potential of this structure is P_c, which is higher then P_b for the difference denoted as ΔP_3, that originates from newly evolved part of the organizational structure. If SME exploits available potential that is observed on a curve marked as „*c*", and if it is not the case then curves „c_1" and „c_2", have been created which are behind curve „*c*".

In the third phase of SME structure development there are numerous sub-systems incorporated which realize SME potential on the market as well as numerous sub-systems which develop new potential. All of them are in different stages of their life cycle or in different stages of future development of their potential. The main challenge for the SME new functional structure is to measure its own potential and choose the optimal option(s) in combining already developed systems. The new functions developed in this phase are: *strategic management function* and *strategic management servicing function*. The SME which incorporated in itself these functions as well as those developed in previous phases have sustainability potential and could be marked as $S_{l\text{-}t}$. Dominant characteristic of this kind of SMEs is its ***long-term existence***. When SME achieved its highest, strategic level, then it is able to successfully fight against „aging" in order to stay flexible, vital and „young", applying real-time replacememt of its inadequate sub-systems, components and structures with new ones. At this level of survival, an SME has very strong ties with its environment, particularly with the market, and it is quite well integrated with it. On the other hand, this SME developed its own structure in way that it is capable to innovate itself. If an SME (or any other organization) is capable to innovate itself then new potential has been created, such as ΔP_3 and so on[2]. All of these innovations suggest that in first step it is necessary to innovate the system for strategic management in which an SME realizes innovations later on. Every successfuly realized systemic innovation creates new potential for an SME.

After all of these statements, at least one question arises: How long can an SME be vital, flexible and „young"? The answer is very simple, but hard to implement in practice and it is: until SME change management is capable successfuly solving the strategic tasks.

The concept presented above leans on the basic princple of organizational structure that form follows function. Aditionally, there is an approach that describes what really gives an organization its „shape" and controls how it performs (Sisney, 2013). Those are the three main aspects:

1. The functions it performs;
2. The location of each function; and
3. The authority of each function within its domain.

First two of three aspects are completely in line with the concept presented by the author in Figure 1. The "authority" aspect isn't embraced by this research.

This section of the chapter conveys to the readers that every SME or organization in general, is a unique organizational system with its systemic characteristics such as: environment, inputs-outputs, components, processes/functions, structures and behaviors. Because of that, every SME has to have its own strategy of structural change management in its sustainable development which is addressed in the next section.

OPPORTUNITIES AND STRATEGIES OF STRUCTURAL CHANGE MANAGEMENT IN SUSTAINABLE SME DEVELOPMENT

Once opportunities within SME environment have been taken into the consideration then a situational analysis should be implemented. One of the most often used instruments of the situational analysis is PEST analysis which deals with political, economic, social and technological factors from the environment or the so called external drivers. There are also variations, or broader PEST analyses which include other external factors that affect an SME, such as legal and environmental ones when PEST is transformed into PESTLE analysis, as well as STEEPLED where ethical and demographic factors should be added (Rothaermel, 2012). Commonly, PEST analysis starts with a 2x2 matrix as illustrated in Table 1. In case, a PESTLE or a STEEPLED analysis is being performed, then the matrix changes its structure according to the numbers of external factors analysed.

After fulfilling the PEST matrix, or some of its numerous variations, it is necessary to carry out an evaluation of specific factors. This can be carried out by using an instrument illustrated by Table 2.

There are differences among the identified factors or drivers in their impact on an SME. Because of that, they could be ranked on the basis of their importance. This can be done in several ways, such as by introducing a three-star scale. In this way, an economic factor, for example, "Levels of government spending" (which is estimated to have a major impact on the subject of analysis in Table 1), could be marked as "Levels of government spending ***". On the other hand, some factor with a lower impact could be marked as for example "Intellectual property *".

Table 1. Toolbox for a PEST Matrix

Political factors	Economic factors
• Factor 1 • Factor 2 ...	• Factor 1 • Factor 2 ...
Social factors	**Technological factors**
• Factor 1 • Factor 2 ...	• Factor 1 • Factor 2 ...

Source: Compiled by author on the grounds of own research.

Table 2. Toolbox for evaluation of PEST factors

PEST	Factor	Opportunity (Intensity)	Threat (Intensity)
Political factors	• Factor 1 • Factor 2 ...		
Economic factors	• Factor 1 • Factor 2 ...		
Social factors	• Factor 1 • Factor 2 ...		
Technological factors	• Factor 1 • Factor 2 ...		

Source: Compiled by author on the grounds of own research.

In the SME *short-term survival period* external factors analysed in PEST matrix have major influence on its future potential. For example, if "Intellectual property" issue has been neglected it surely will negatively affect an SME in the future, when large companies "borrow" its idea or innovation and start a large scale production under own name.

In order to prepare a relevant strategy for optimizing structural change management in sustainable SME development it should be implemented with another instrument of strategic planning which is SWOT analysis. This instrument combines external factors which affect an SME with internal ones. Internal factors are treated as *Strengths* and *Weaknesses* (Fine, 2009). Analysing internal factors, mostly resources of an SME, are evaluated. This can be done by analyzing each factor individually (expert approach) or using some of the methods such as: Brainstorming, Brainwalking, surveys and others. These approaches could be implemented in PEST analysis as well. The external factors/drivers identified and evaluated in PEST analysis could be used in SWOT analysis as *Opportunities* and *Threats*, depending on the character of their influence.

In the process of SWOT analysis, numerous visions are being considered. One should be selected which best illustrates the real situation. In the case of only one vision being considered in the process of SWOT analysis, it is being confronted with reality and may be redefined.

Commonly, SWOT analysis starts with a matrix 2x2 as it is illustrated by Table 3.

Strengths are all SME resources which create positive external and internal influences.

Weaknesses are primarily all SME resources that can't achieve the intended SME goals, because of their quality, but also because of their quantity.

Opportunities are all external resources that are identified in the environment that can be used in order to improve an SME structure or behaviour. The quality and quantity of these resources can't be influenced by an SME, but they can take advantage of them.

Threats for an SME are all systems, phenomena and events in the environment that exert or may exert a negative influence.

It is important to note that SWOT analysis does not boil down to a fulfilled matrix as shown by Table 3, but also the activities which follow it. After completion of the SWOT matrix it is necessary to identify key factors among those listed in *strengths* and *weaknesses* and propose actions which could improve an SME resource as well as its position in the environment. The proposed actions should lean on *opportunities* from environment and should also take into consideration *threats* that could negatively interfere in actions as well as on SME as an organisational system. For example, if "Lack of skills" has been identified as a weakness of an SME, and there is an opportunity in the environment identified as

Table 3. Toolbox for a SWOT matrix

Strengths	**Weaknesses**
• Factor 1 • Factor 2 …	• Factor 1 • Factor 2 …
Opportunities	**Threats**
• Factor 1 • Factor 2 …	• Factor 1 • Factor 2 …

Source: Compiled by author on the grounds of own research.

"Government subsidies for capacity building in business sector" then SME management could propose an action such as "Documentation submission for governmental subsidies in the business sector". Because of the lack of resources that an SME manages it is crucial to identify actions with their costs which will produce maximum positive effects for an SME as a whole. This process of selection of key factors with actions proposal is the quest for a strategy creation.

SWOT analysis becomes more powerful when after a certain time it is being repeated and actions proposed and undertaken made change resources and the SME as a whole. Then, the effects of the actions undertaken can be measured in more realistic way. For example, if proposed action "Documentation submission for governmental subsidies in business sector" has been undertaken, in next SWOT matrix a weakness previously identified as "Lack of skills" could simply vanish. This means that undertaken action resulted positively on the weakness identified previously. More preferably, this weakness could be converted into the strength "Highly skilled employees" if the action undertaken had deeper impact on weaker resource.

It should be noted that, in addition to internal factors which were the subject of actions, there are also those which could be neglected during time and in this way are being converted from strengths to weaknesses. This is often the case with equipment as well as with skilled people. Also, environmental factors could significantly change over a period of time due to some broader social or natural processes, and can be analyzed. Their impact on internal factors can be more or less significant.

Unlike "ordinary" SWOT analysis, the so-called "advanced" SWOT analysis evaluates identified factors listed in the SWOT matrix and prioritizes them in relation to their impact on the subject of SWOT analysis as well as SME (David, 2009). For this purpose, previously presented SWOT matrix has been modified, while the numbers in following Table 4 being are fictitious for easier explanation.

Table 4. Toolbox for an advanced SWOT matrix

	Importance	**Rank**	**Result**
Strengths			
• Factor 1 • Factor 2 …	0,04 0,30 …	1 2 …	0,04 <u>0,60</u> …
Weaknesses			
• Factor 1 • Factor 2 …	0,06 0,01 …	2 1 …	<u>0,12</u> 0,01 …
	Importance	**Possibility**	**Result**
Opportunities			
• Factor 1 • Factor 2 …	0,03 0,01 …	1 2 …	<u>0,03</u> 0,02 …
Threats			
• Factor 1 • Factor 2 …	0,03 0,15 …	1 2 …	0,03 <u>0,30</u> …

Source: Compiled by author on the grounds of own research.

In the Table 4, where internal characteristics have been analysed, the reader should pay attention to three columns on the right side.

- In the column "Importance", the *significance* of the specific internal factor has been assessed. Every strength and weakness has been assessed by numbers in range of 0,01 (insignificant) to 1,00 (most significant). The sum of all assessed factors has to be equal to 1,00.
- Column "Rank" is determined in the range of 1 (minor) to 3 (main) in order to indicate the strength or weakness for the subject of analysis, and SME as well.
- Column "Result" represents the product of two columns "Importance" and "Rank". It allows prioritization of strengths and weaknesses. An SME has to rely on the most important strengths (the highest score) and should try to convert weaknesses into strengths or to defend its weak points.

In the lower part of Table 4, the external factors should be assessed, with a little modification compared to the internal ones.

- Column "Importance" shows the extent to which external factors can influence the object of analysis. Every opportunity and threat has been assessed by numbers in range of 0,01 (insignificant) to 1,00 (most significant). The sum of all assessed factors has to be equal to 1,00.
- Column "Probability" is determined by the likelihood of outside influence and is expressed by whole numbers in certain range (e.g. 1 to 3, which corresponds to the values of "small", "medium" and "large").
- Column "Result" represents the product of two columns "Importance" and "Probability". It allows the prioritization of opportunities and threats. It must rely on the most important external factors (the highest score), and ignore those who are unlikely to affect SME.

SWOT analysis could show its major relevance in the later period of an SME survival, particularly in the *mid-term survival period*. Actually, at the beginning of this period, when first signs of inertia in business processes have been recognized, it is high time to investigate internal characteristics of SME and put them in correlation with the external ones.

Unlike to PEST and SWOT analysis where the main task is to identify external and internal factors that affect an organisational system such as SME, the TOWS analysis deals with 4 different strategies that the subject of analysis could implement. Methodology of TOWS analysis suggests that external factors should be identified first, and after that internal ones. The results given then should be compared among themselves in order to find the best strategy for the subject of analysis and SME.

The TOWS analysis tries to give answers to the following questions:

- **Comparing Strengths and Opportunities**: How can existing strengths assist in capitalizing opportunities?
- **Comparing Strengths and Threats**: How can existing strengths assist in identifying and avoidance of existing and potential threats?
- **Comparing Weaknesses and Chances**: How existing weaknesses could be overcome using opportunities from the environment?
- **Comparing Weaknesses and Threats**: How could be reduced these weaknesses and to avoided existing and potential threats on the best way?

Answers to these questions are obtained by direct comparison of the factors identified in the TOWS matrix and are the concrete strategy for the subject of analysis, namely (Weihrich, 1982):

1. **MAXI-MAXI. Strategy of Strengths and Opportunities (SO):** Every organization would like to be in position to implement this strategy. However, just some of them are in a position to choose this strategy. For some organisations the option could be, previously convert main weaknesses into the strengths and after that implement this strategy, using the opportunities that come from the environment.
2. **MAXI-mini. Strategy of Strengths and Threats (ST):** In this case, it is relatively easy to counteract threats from the environment, because SME has significant strengths. However, depending on the specific situation, care should be taken about how to use own strengths, which should be usually used with a dose of restraint and discretion.
3. **mini -MAXI. Strategy of Weaknesses and Opportunities (WO):** This strategy seeks to minimize the weaknesses and maximize opportunities. It hides potentially attractive options. In case, that there have been identified some attractive opportunities in the environment but weaknesses are to significant, then should be worked on overcoming own weaknesses and then take advantage from the opportunities.
4. **mini-mini. Strategy Weaknesses and Threats (WT):** This strategy bears potentially high risks. Weaknesses from inside and threats from the outside should be minimized. The situation is unenviable and leads to a real battle for survival.

Table 5 shows position of each main strategy in TOWS matrix.

There are numerous challenges how to maximize strengths, utilize available opportunities, and address threats and weaknesses by turning them into opportunities.

Table 5. Example of a TOWS matrix

Subject of Analysis (SME)	**Strengths** • Factor 1 (Experienced engineers' staff) • Factor 2 (Well equipped R&D sector) •...	**Weaknesses** • Factor 1 (Declining profitability) • Factor 2 (Basic production equipment is out of date) •...
Opportunities • Factor 1 (Government subsidies related to SME sector) • Factor 2 (Growing market for SME products) •...	**MAXI-MAXI (SO)**	**mini –MAXI (WO)**
Threats • Factor 1 (Rising transportation costs) • Factor 2 (Increasing foreign competition) •...	**MAXI-mini (ST)**	**mini-mini (WT)**

Source: Adapted by author from Weihrich (1982, p. 63.)

The TOWS analysis could show its major relevance particularly at the beginning of *long-term survival period*, when some strategic decisions have to made. If strategy, roughly articulated by the TOWS analysis becomes right, SME could count on survival in a long term perspective.

It is relevant to mention that organizational structure follows strategy, not oppositely. Good organizational structure is just a means for achieving the objectives and goals of an SME (Drucker & Maciaiello, 2008).

However, even the best strategy developed for structural change management in SME sustainable development will face[3] with resistance to change. Only the strategy developed that foresees resistance to change, individual and/or group ones, which is implemented by managers who are able to overcome these internal processes could be characterized as promising

FUTURE RESEARCH DIRECTIONS

Many authors, theorists and practitioners too, have developed their own or upgraded existing models which deal with change management. But the fact is that majority of them have limited space of implementation, mostly because of SME complexity, dynamic and turbulent changes in environment.

Future researches in this field may focus on quantification of parameters given by the author in Figure 1, particularly on measurement of SME potential needed for its transition from short-term to mid-term survival as well as from mid-term to long term survival which may happen later on if an SME is a successful one. After these quantifications and mathematical computations the next level in developing the concept presented in this chapter could be a computer simulation which will show, based on quantified parameters, where an SME is in its life cycle depicted in Figure 1.

Change management as a function within SME needs highly educated and skilled staff and managers. Their education and training, in order to be capable in the future to deal with change management issues, should be one of the subjects of future researches too.

In addition, empirical researches should be undertaken, particularly in the field of human resource management aimed at recruitment, training and conditions for employees within an SME who will ultimately be its driving change force as innovators and managers.

CONCLUSION

This chapter dealt with structural changes in SME emphasizing change management as a tool for implementing changes within an organizational system.

In summarizing this chapter it can be pointed out that there are three distinct phases from high entrepreneurial drive until a fully fledged developed SME. In the first one, continuity of SME survival has to be achieved. In the second one, this existence potentially lasts and finally in the third phase long-term existence conditions have to be assured. Each of these phases has been completed by certain functional structure with clearly defined functions. From start-up to maturity, in each phase, the management system of an SME is being developed as a driving force of its development.

The conclusions based on presented material could be summarized as bellow:

- SMEs without strategic and development capabilities are able to survive for only a few years;

- There are three different levels of functional management and other structures in the SME life cycle. Each of them supports a certain kind of behavior and capabilities;
- SMEs with short-term existence potential could only develop the first level of functional structure;
- SMEs with mid-term existence potential are able to develop the first and the second level of functional structure;
- SMEs with all three levels of functional structures have potential to achieve long-term existence.

SMEs which strive to achieve long-term survival must have innovative potential in every phase of their growth. Otherwise they become ordinary economic entities with lesser chances for long-term survival.

Consequently, SME managers who react rapidly on both external and internal factors could be successful. Indeed, those SME managers who are able to anticipate the future and prepare their SMEs for new circumstances are really successful. It may also be concluded that SMEs which do not survive are those who fail to anticipate the future or even adapt to changes that affect them both externally and internally.

Change management should lean on results of environmental assesment implementing instrruments of strategic planing such as PEST, SWOT and TOWS analyses. After that, an SME oriented for sustainable development will be able to produce a viable strategy, taking into consideration all resources it manages. The strategy should include certain criteria for measuring success of an SME which are related to its sustainable development.

Finally, every SME can enhance its chance for a long business life involving both survival and development if it develops and optimizes its own strategic capabilities. It can be reasoned out that structural change management in sustainable SME development results is in the function of long-term survival as a strategic goal for an SME.

REFERENCES

Ackoff, R. L. (1999). *Ackoff's Best: His Classic Writings on Management*. New York, NY: John Wiley & Sons Inc.

Ackoff, R. L., & Emery, F. E. (1972). *On Purposeful Systems*. Chicago: Aldine-Atherton.

Adizes, I. (1992). *Mastering Change: The Power of Mutual Trust and Respect*. Santa Barbara, CA: Adizes Institute.

Adižes, I. (1994). *Životni ciklusi preduzeća*. Beograd, Serbia: Agora – Prometej.

Australian Government: Department of Industry Innovation, Science, Research and Tertiary Education. (2013). *Australian Small Business: Key Statistics and Analysis – December 2012*. Retrieved May 14th 2016, from Australian Treasure web site: http://www.treasury.gov.au/~/media/Treasury/Publications%20and%20Media/Publications/2012/Australian%20Small%20Business%20-%20Key%20Statistics%20and%20Analysis/downloads/PDF/AustralianSmallBusinessKeyStatisticsAndAnalysis.ashx

Birch, D. L. (1981). Who Creates Jobs? *The Public Interest*, *65*, 3–14.

Boulding, K. E. (1956). General Systems Theory - The Skeleton of Science. *Management Science*, *2*(3), 197–208. doi:10.1287/mnsc.2.3.197

Burton, R. M., Obel, B., & Håkonsson, D. D. (2015). *Organizational Design: A Step-by-Step Approach* (3rd ed.). Cambridge University Press.

Černiček, I., & Molnar, R. (1988). Razvoj profitnih potencijala preduzetničkih organizacija. In I. Černiček & R. Molnar (Eds.), Razvoj preduzetništva - šansa za progress (pp. 25-43). Zrenjanin, Serbia: Tehnički fakultet "Mihajlo Pupin".

Černiček, I., & Molnar, R. (2000). Razvoj organizacije. In I. Černiček & B. Marić (Eds.), *Reinženjering proizvodno-poslovnih procesa* (pp. 203-256). Belgrade: Fakultet za preduzetni menadžment Novi Sad, Univerziteta "Braća Karić".

David, F. R. (2009). *Strategic Management: Concepts and Cases*. Upper Saddle River, NJ: FT Prentice Hall.

Drucker, P. F. (1991). *Inovacije i preduzetništvo: Praksa i principi*. Beograd, Serbia: Privredni pregled.

Drucker, P. F., & Maciaiello, J. A. (2008). *Management* (Revised ed.). HarperCollins e-books. Retrieved August 1st 2016, from Youth Portal web site: http://youth-portal.com/wp-content/uploads/2014/10/Peter-F-Drucker-Management-Rev-Ed.pdf

EUROSTAT. (2016). *Business demography statistics - Enterprise Survival Rate*. Retrieved May 13th 2016, from Eurostat web site: http://ec.europa.eu/eurostat/statistics-explained/index.php/ Business_ demography _statistics

Fine, G. L. (2009). *The SWOT Analysis: Using your Strength to overcome Weaknesses, Using Opportunities to overcome Threats*. Kick It. LLC. Retrieved February 15th 2013, from Lawrence Fine web site: http://lawrencefine.com/ downloads/SWOT%20-%20PDF.pdf

French, R., Rayner, Ch., Rees, G., & Rumbles, S. (2015). *Organizational Behaviour* (3rd ed.). New York, NY: John Wiley & Sons.

Hagen, J. (2009). Top 12 Reasons Why Businesses Fail. In E. Koester (Ed.), *What Every Engineer Should Know About Starting a High-Tech Business Venture*. Boca Raton, FL: Taylor & Francis Group.

Kotter, J. (2012). *Leading Change*. Boston, MA: Harvard Business Review Press.

Langstrand, J. (2016). The missing link in systems thinking: The impact of infrastructure on organizational change. *International Journal of Quality and Service Sciences*, *8*(2), 197–208. doi:10.1108/IJQSS-09-2015-0061

Lewin, K. Z. (1947). Frontiers in Group Dynamics: Concept, Method and Reality in Social Science; Social Equilibria and Social Change. *Human Relations*, *1*(1), 36. doi:10.1177/001872674700100103

Mc Bourda, F. (2013). *Change Management Theories and Methodologies* (White paper). Retrieved May 15th 2016, from Tata Consultancy Services web site: http://www.tcs.com/ SiteCollectionDocuments/White%20Papers/EntSol-Whitepaper-Change-Management-Theories-Methodologies-0213-1.pdf

Molnar, R. (2010). *Razvoj sistemske nefinansijske podrške malim i srednjim preduzećima i preduzetništvu u Srbiji* (Unpublished doctoral dissertation). Available from Library at the University of Novi Sad. Serbia.

Penezić, N. & Jakopin, E. (2002). *Analiza razvoja malih i srednjih preduzeća i preduzetništva u Srbiji.* Beograd: Republički zavod za razvoj i Republička agencija za razvoj malih i srednjih preduzeća i preduzetništva.

Pryor, M. G., Taneja, S., Humphreys, J., Anderson, D., & Singleton, L. (2008). Challenges Facing Change Management: Theories and Research. *Delhi Business Review*, *9*(1), 1–20.

Rees, G., & Hall, D. (2013). Managing Change. In G. Rees & R. French (Eds.), *Leading, Managing and Developing People* (pp. 101–126). CIPD.

Rothaermel, F. T. (2012). *Strategic Management: Concepts and Cases.* New York, NY: McGraw-Hill/ Irwin.

SBA. (2014). *Frequently Asked Questions: What is the Survival Rate for New Businesses?* Retrieved May 13th 2016, from U.S. Small Business Administration web site: https://www.sba.gov/sites/default/ files/ FAQ_March_ 2014_0.pdf

Sengi, P. (2003). *Peta disciplina: Umeće i praksa organizacije koja uči.* Novi Sad, Serbia: Adizes MC.

Sisney, L. (2013). *Organizational Physics: The Science of Growing a Business.* Raleigh, NC: Lulu Press.

Stanković, F. (1998). Potencijali i upravljanje rastom u MSP. In I. Černiček & R. Molnar (Eds.), Razvoj preduzetništva - šansa za progress (pp. 44-52). Zrenjanin, Serbia. Tehnički fakultet "Mihajlo Pupin".

Wagner, E. T. (2013). *Five Reasons 8 Out Of 10 Businesses Fail.* Retrieved July 24th 2016 from Forbes web site: http://www.forbes.com/sites/ericwagner/2013/09/12/five-reasons-8-out-of-10-businesses-fail/#741626b05e3c

Weihrich, H. (1982). The TOWS matrix: A tool for situational analysis. *Long Range Planning*, *15*(2), 54–66. doi:10.1016/0024-6301(82)90120-0

Zgodavova, K., & Bober, P. (2012). An Innovative Approach To the Integrated Management System Development: SIMPRO-IMS Web Based Environment. Quality Innovation Prosperity, 16(2), 59-70.

Zhang, X. (2016). Co-evolution between institutional environments and organizational change: The mediating effects of managers uncertainty. *Journal of Organizational Change Management*, *29*(3), 381–403. doi:10.1108/JOCM-12-2014-0218

ADDITIONAL READING

Basile, A., & Faraci, R. (2015). Aligning management model and business model in the management innovation perspective: The role of managerial dynamic capabilities in the organizational change. *Journal of Organizational Change Management*, *28*(1), 43–58. doi:10.1108/JOCM-10-2013-0199

Brunninge, O., Nordqvist, M., & Wiklund, J. (2007). Corporate Governance and Strategic Change in SMEs: The Effects of Ownership, Board Composition and Top Management Teams. *Small Business Economics*, *29*(3), 295–308. doi:10.1007/s11187-006-9021-2

Černiček, I., Nadasdi F., Chaezy, S. & Molnar, R. (2003). *TVM - Total Value Management: Integralno upravljanje vrednostima*. Novi Sad, Serbia: Domla-Publishing.

Cole, G. A. (1994). *Strategic Management: Theory and practice*. London: DP Publications.

Grant, R. M. (2010). *Contemporary Strategy Analysis*. New York, NY: John Wiley & Sons.

Hammer, M., & Champy, J. (1993). *Reengineering the Corporation: A Manifesto for Business Revolution*. New York: Harper Collins.

Hisrich, R. D., Peters, M., & Shepherd, D. (2009). *Entrepreneurship* (8th ed.). New York, NY: McGraw-Hill/Irwin.

Lindgren, M., & Bandhold, H. (2003). *Scenario Planning – The link between future and strategy*. New York: Palgrave Macmillian.

Lowy, A. (2015). The six dilemmas of strategy execution. *Strategy and Leadership*, *43*(6), 18–24. doi:10.1108/SL-07-2015-0062

Luo, W., Song, L. J., Gebert, D. R., Zhang, K., & Feng, Y. (2016). How does leader communication style promote employees commitment at times of change? *Journal of Organizational Change Management*, *29*(2), 242–262. doi:10.1108/JOCM-11-2014-0204

Mellert, L. D., Scherbaum, Ch., Oliveira, J., & Wilke, B. (2015). Examining the relationship between organizational change and financial loss. *Journal of Organizational Change Management*, *28*(1), 59–71. doi:10.1108/JOCM-11-2013-0236

Mintzberg, H. (1987). The Strategy Concept I: Five Ps For Strategy. *California Management Review, 30*(1), Fall/1987, p. 11-24. doi: 10.2307/41165263

Nadasdi, F., & Csernicsek, I. (2000). The Total Value Management Approach. *Value World*, *23*(2), 11–13.

Pollack, J., & Algeo, C. (2016). Project managers and change managers contribution to success. *International Journal of Managing Projects in Business*, *9*(2), 451–465. doi:10.1108/IJMPB-09-2015-0085

Porter, M. (1979). How Competitive Forces Shape Strategy. *Harvard Business Review*, *57*(2), 137–145. PMID:18271320

Schoemaker, P. J. H. (1995, Winter). Scenario Planning: A Tool for Strategic Thinking. *Sloan Management Review*, *36*(2), 25–40.

Trondal, J. (2015). Ambiguities in organizations and the routines of behavior and change. *The International Journal of Organizational Analysis*, *23*(1), 123–141. doi:10.1108/IJOA-04-2013-0654

Udo-Imeh, P. T., Edet, W. E., & Anani, R. B. (2012). Portfolio Analysis Models: A Review. *European Journal of Business and Management*, *4*(18), 101–121.

Valentin, E. K. (2001). SWOT analysis from a resource-based view. *Journal of Marketing Theory and Practice*, *9*(2), 54–68. doi:10.1080/10696679.2001.11501891

Yeo, R. K., & Marquardt, M. J. (2015). Think before you act: Organizing structures of action in technology-induced change. *Journal of Organizational Change Management*, *28*(4), 511–528. doi:10.1108/JOCM-12-2013-0247

KEY TERMS AND DEFINITIONS

Change Management: A systematic approach to dealing with change both from the perspective of an organization and the individual as well.

Entrepreneur: An active subject in the entrepreneurship process. He/she creates something new that differs from other goods available on the market creating new value.

Entrepreneurship: A social phenomenon where somebody (entrepreneur) moves available resources from low productivity sectors to the high productivity ones, bearing risk and expecting a profit for those efforts.

Innovation: A means that enriches a resource with new capacities for wealth creation.

Management: A business function which deals with processes of planning, organizing, commanding, coordinating and controlling.

SME (Small and Medium Sized Enterprise): An economic entity defined by legal documents which differ from country to country. The three main criteria used to define an SME are number of employees, annual turnover and average value of assets.

SME Development: Every SME as an organizational system has its life cycle and passing through it SME changes its elements, structures and behavior.

Systems Theory: An independent scientific discipline that contributes to the development of scientific research and science in general. Instead of analytical and linear methods, its focus is on the totality, on global characteristics, structures and behavior of the researched system.

ENDNOTES

1 Editors' note: See chapter 17, and compare with the concept and principles of open innovation.

2 Editors' note: For the more mathematically minded, the rate of change denoted as ΔP_n is a process of finding the derived function. The derivative of the function, $\Delta P/\Delta t$, is the limit (if it exists) of this expression as Δt approaches zero.

3 Editors' note: A further and comprehensive discussion on sustainability is given in chapter 13.

Chapter 17
Integrating Sustainable Development into Project Portfolio Management through Application of Open Innovation

Hosein Daneshpour
University of Vaasa, Finland

ABSTRACT

In recent years, the idea of integration of sustainability into project management has been put forth, which requires a more comprehensive and holistic approach to project management. Integration of sustainability increases the complexity of project management. Therefore, project management organizations require a framework capable of opening up the traditional organization of the project business and providing competitive advantages. This study argues that the open innovation model presents the requisite capability to approach this issue. This research, through a structured review of the literature and an overview of the open innovation and project management knowledge areas, explains how open innovation can provide a fruitful contribution to the integration of sustainable development into project portfolio management and why it should be favoured. The findings show that open innovation can facilitate the environmental, social and economic sustainability of projects, while rendering the project organization more agile.

INTRODUCTION

Over recent years sustainability has become one of the most significant challenges facing organizations, and companies are thus striving to implement sustainability into their planning processes and strategies. Development of societies is associated with projects that are instruments for change, and that act as stepping stones for the infrastructural changes; therefore project management (PM) skills must be developed in order to lead these projects in a successful and sustainable way. The transition from traditional PM to sustainable PM is a necessity and should be fulfilled by project-oriented organizations in distinct stages.

DOI: 10.4018/978-1-5225-1949-2.ch017

The concepts of sustainability and PM have recently been linked together to form an emerging field of research (Marcelino-Sadab, González-Jaen, & Pérez-Ezcurdia, 2015).

Consideration of sustainability in projects moves PM towards the title of a true profession, and expands the system boundary of PM. Integration of sustainability into PM shifts the traditional PM paradigm, and calls for a new approach to handle this complex issue. This new paradigm incorporates a variety of values of partner's (stakeholder's) to create shared values (Silvius, Schipper, Planko, Brink, & Köhler, 2012).

The open innovation (OI) model presents a new paradigm which shifts firms from a closed to an open model, thereby advancing the business. Primarily, the logic of OI was introduced to the world by Chesbrough (2003) as a new way to cope with the innovation process, having observed that some newcomers in the field of high-tech industries were, surprisingly, managing to compete with large established companies such as IBM and AT&T. Subsequently, in 2006 the OI strategy was furthered by the introduction of open service innovation as a business model, and not just a way of doing R&D.

However, the link between OI, different management disciplines and the economic context has not yet been discussed (West, Salter, Vanhaverbeke, & Chesbroughd, 2014). Consequently, the OI literature has not adequately addressed a number of themes, such as sustainability and PM. Therefore, this research explores the PM potentials of OI, tries to link OI to the project portfolio management context and provides a perspective for future research. In addition, until now, the attention of PM researchers has mainly been focused on integrating one dimension of sustainability (triple bottom line) into PM. This study presents a more comprehensive understanding of this case and incorporates other dimensions by means of OI.

This research is based on a sound review of the (Scopus indexed) literature and is presented as follows: first, the link between OI and the concept of sustainability is evaluated. Next, the gap between sustainability and PM is discussed. Finally, the connection between portfolio management and OI is explored.

OPEN INNOVATION BACKGROUND

The OI model shifts firms from a closed model to an open strategy (Chesbrough, Vanhaverbeke, & West, 2006). Chesbrough (2003) coined the term "open innovation", while highlighting that, in a world of widely-distributed knowledge, firms cannot rely entirely upon internal ideas and should utilize valuable ideas generated outside the company made by customers, suppliers and other stakeholders. Therefore, OI emphasizes that "firms can and should use both external and internal (equal importance) ideas and paths to advance their business" (Chesbrough, 2006a, p. xxiv). OI can be defined as "both a set of practices for profiting from innovation and also a cognitive mode, for creating, interpreting and researching those practices" (Chesbrough et al., 2006, p. 286). Subsequently, Chesbrough (2011) developed the concept of the OI model, broke new ground with the "open service innovation" model that considered the business model from an open service perspective, both for product and service delivery businesses, and concluded that OI is not just a method for approaching R&D.

OI is depicted as a porous funnel that benefits the environment of relevant companies via the use of ideas generated outside an individual company (the outside-in process; acquiring externally-sourced knowledge) and the simultaneous transfer of unused ideas to other companies (the inside-out process; transferring internal ideas to the market), as well as coupled processes that combine outside-in and inside-out processes (Lichtenthaler, 2011). A vital feature of this model is that projects can be initiated based on external resources and progress to market at various stages and in a variety of ways, thereby generating additional value, such as licensing revenue. This process should occur through the explicit and "purposive" incorporation of an open "business model" to "create value" and obtain a portion of that generated value. It is this point that differentiates the OI concept from the previous ordinary utilization of customers' ideas within the company (Chesbrough & Appleyard, 2007). In contrast, the closed model advocates limited interaction with the outside environment, which is similar to a funnel that is surrounded by the limits and borders of the company, while the ideas (projects) generated inside the company progress through internal channels. In the closed model, the science and technology bases of the firm are the main originators of projects, which then progress through internal channels – some are scrapped and others go to market through a sole exit route (Lichtenthaler, 2011).

OI business models can be categorized according to their level of co-creation: market-based innovation strategy (with minimal levels of co-creation and complexity), crowd-based innovation strategy, collaborative innovation strategy and network-based innovation strategy (with maximal levels of co-creation) (Saebi & Foss, 2015). The OI business model acts as a mediator between the technical and economic domains, which takes technological characteristics as inputs and transforms them into economic outputs (Chesbrough & Rosenbloom, 2002). In addition, one of the main features of this model is the creation of value with stakeholders, which helps firms to escape the "commodity trap (Chesbrough, 2006b). Furthermore, OI notes the role of the "Front End" (or "Fuzzy Front End") in the business model, and considers Front End management as a key factor in the success of a portfolio, as a high degree of uncertainty exists in such organizations. The success of the Front End directly depends on stakeholder management, standardized tools and systems and the organization's ideation culture (Heising, 2012).

Nevertheless, what OI means in practice is still up for debate (Giannopoulou, Yström, & Ollila, 2011). Previous studies have also revealed that there are barriers to the adoption of OI approaches (Savitskaya, Salmi, & Torkkeli, 2010) (such as cultural issues, not invented here syndrome, not sold here syndrome, measurement errors, etc.), and the majority of firms are still operating on the basis of the closed model (Lichtenthaler, 2008).

Sustainable Innovation: Incorporating Sustainability into Innovation

The challenge of sustainable development (SD) demands long-term structural changes and innovation in different sectors of society (Geels, 2004). In the year 1972 the "Club of Rome" challenged the concept of development and growth (Turner, 2008). Generally, the concept of sustainability is shaped based on the balance between people, planet and profit or "triple bottom line" (TBL) (Borowy, 2013). Nonetheless, four main dimensions of sustainability can be considered:

1. TBL emphasis
2. Business level;
3. Sustainability oriented innovation;
4. Organizational focus (Amini & Bienstock, 2014).

The concept of "sustainable innovation" (or eco-innovation) merges aspects of SD into innovation (Geels, Hekkert, & Jacobsson, 2008). The interplay between sustainability and innovation is an undeniable fact, as both concepts are critical parts of the corporate agenda (Geels, 2004). Traditionally, the motivation for innovation has been based on cutting the price of production or creating a better product than competitors, while sustainable innovation offers an alternative strategy based on the creation of products or processes with desirable characteristics and shared value. Porter and Kramer implicitly define the concept of sustainable innovation on the basis of "shared value creation" (Porter & Kramer, 2011).

(Hautamäki, 2010) describes the concept of "sustainable innovation" as innovation activities that are in accordance with ethical, social, economic and environmental sustainability. Therefore, while innovation supports "sustainable development" it should be "participative" to involve a variety of stakeholders and also have the ability to keep the renewal process going "continually". Additionally, the "global innovation" approach challenges companies to open their internal innovation processes to the world, while "innovative management" requires company management to motivate innovation. Nevertheless, the concept of "sustainable innovation" is an emerging research field and, as yet, lacks conceptual consensus (Boons & Lüdeke-Freund, 2013).

OPEN INNOVATION ADDS VALUE TO SUSTAINABLE DEVELOPMENT

The relationship of OI to different disciplines of SD has not yet been adequately presented in the literature as it is an emerging field of research (West et al., 2014). Table 1 presents the current contributions of OI to the dimension of SD. Overall, it can be observed that the link between OI and dimensions of sustainability, except at the business level, have not yet been adequately discussed.

As noted in Table 1, the link between OI and the TBL concept requires further research; for instance (Holmes & Smart, 2009) present the link between open innovation and corporate social responsibility. Based on existing scholarly research, the core idea of TBL can be said to be based on the concept of co-creation (or cooperation), which can be related to the OI philosophy (Hamdouch & Zuindeau, 2010). However, further research is required to present the clear contribution of OI to environmental, social and economic pillars. Similarly, the link between OI and the organizational level (supply chain) requires further attention.

Regarding the relationship of OI to sustainable innovation, as discussed in the previous section, and similar to the sustainable innovation notion, the literature has described OI by using some of the same terms, such as "co-creation", that can be regarded as connecting OI and sustainable innovation (Arnold, 2015). Nevertheless, concerning OI and the business level, the proper potential of OI as a business model has been presented and mainly relies on value optimization using a co-creation model. It is still

Table 1. Main contributions of OI to SD

Sustainability Dimensions	OI	
	Paper Indexed	Key Contributions
triple bottom line ("corporate social responsibility")	6	(Holmes & Smart, 2009)
organizational focus ("supply chain management")	15	(Erzurumlu, 2010)
"sustainability innovation"	5	(Arnold, 2015)
business level ("business model")	203	(Chesbrough & Appleyard, 2007)

Source: Own presentation of author

an emerging field of science and other business models can offer supporting elements and advantages (Mutka & Aaltonen, 2013).

This section has mainly illustrated the relationship between sustainability and OI; the next section focuses on the interaction of sustainability and PM.

SUSTAINABILITY IN PROJECT MANAGEMENT

The concepts of sustainable project management (SPM) and traditional project management (TPM) contrast with one another: for instance, SPM focuses on people, planet and profit, utilizing a life cycle approach, while TPM is based on the iron triangle of PM (time, scope and cost) and is less complex (Silvius et al., 2012). The value of incorporating sustainability into PM can be better highlighted by taking into account the fact that sustainability is about both long-term and short-term objectives, and SPM respects both current and future generations with a global PM scale, including all necessary stakeholders (not only shareholders) (Eskerod & Huemann, 2013).

The concept of sustainability in PM is closely related to CSR. From the business point of view, CSR can provide competitive advantages if it provides (direct and indirect) financial benefits based on a win-win relationship with stakeholders (Carroll & Shabana, 2010). CSR strategies can be categorized according to three types: philanthropy, corporate responsibility integration and corporate responsibility innovation. Philanthropy is external to a firm's core business and can only provide benefits such as improvement of the firm reputation (e.g., donation to charity groups). Corporate responsibility integration is closer to the existing core business and can improve the firm's social and environmental aspect (e.g., ISO 14001 certification). Corporate responsibility innovation, with its emphasis on a new business model, can offer the greatest potential benefits while alleviating social and environmental problems. Innovative corporate responsibility presents the highest financial performance potential among the three CSR strategies, and may provide business benefits such as entrance to new market segments (Halme & Laurila, 2009). Therefore, an appropriate business model is a key driver of the realization of sustainability within a business.

On the other hand, the link between CSR and business operations can be illustrated on the basis of four main CSR maturity levels. Project management (project-oriented CSR) is the most immature form of these four as the temporary scale of the project is in contrast with the permanent social system. Quality management (quality-oriented CSR) cannot offer competitive advantages and differentiate companies from competitors, as argued by Chesbrough (2011) regarding techniques such as Six Sigma and TQM. Strategic management (strategic-oriented CSR) can, by contrast, facilitate the innovation process in the strategy of the company, as it is based on the creation of "shared value". Finally, organizational learning (transformational-oriented CSR) is the best strategy, as it is based on the integration of "external stakeholders" and business model transformation, which has rarely been implemented in practice (Martinuzzi & Krumay, 2013).

Implementation of business models (such as business excellence models, or BEMs) in the context of PM is a contentious field of research, and BEMs' capacity to address the need for integration of sustainability into PM is a controversial issue (Westerveld, 2003). Basically, BEMs are frameworks for the management of performance excellence in any organization. Mann, Adebanjo, & Tickle (2010, p. 606) define business excellence as "excellence in strategies, business practices, and stakeholder-related performance results that have been validated by assessments using proven business excellence models".

Nowadays there are many BEMs, such as that of the European Foundation for Quality Management (EFQM), and Silvius et al. (2012) have presented some linkages between EFQM and PM. Nonetheless, there is, thus far, limited discussion concerning the application of BEMs in PM (Qureshi, Warraich, & Hijazi, 2009).

The value of PM can be evaluated on the basis of the level of satisfaction of stakeholders, while the creation (or transformation) of a business model can be the main step for value creation (Mir & Pinnington, 2014). Accordingly, innovation is recognized as a critical element of a business model for creating sustainable advantages (Boons & Lüdeke-Freund, 2013). Table 2 presents some indicators for the main pillars of sustainability.

Navigating Complexity in Sustainable Project Management

Integrating sustainability into management of a project increases the complexity of PM. Furthermore, globalization, competitive markets and the demand for innovation have made organizations' strategies more complex (Whitty & Maylor, 2009). The main feature of complexity in the project is "multiple stakeholders" (Kerzner & Belack, 2010).

According to the Project Management Institute (PMI), there are important reasons why companies should be careful regarding this complexity: the level of complexity increases gradually and the budget for complex project is higher, on average. Therefore, there is more money at risk, but if the challenge can be navigated it can provide sustainable competitive advantages (Project Management Institute, 2013). Based on research conducted by PMI, "multiple stakeholders" and "ambiguity" have been recognized as the main characteristics of complexity in projects, while the main facilitators for the success of project portfolio complexity management are: effective communication, standardized project portfolio practices and talent for leadership, accompanied by agility and effective change management (Project Management Institute, 2013). In addition, the main elements for standardized project management are PM tools and processes and leadership (Milosevic & Patanakul, 2005).

Meanwhile, organizational agility is a crucial requirement of the turbulent project business environment (Serrador & Pinto, 2015). The agility of an organization not only helps the organization to meet its business objectives, but to exceed financial objectives and also to provide competitive advantage (Stettina & Hörz, 2015). To improve agility in project organization the implementation of three competencies is necessary: change management, risk management and standardize portfolio management (Project Management Institute, 2013). Successful change management is the main indicator of the agility of an organization, and can be measured according to the level of interaction with the external environment, formal and standard change management and PM office performance (Harraf, Wanasika, Tate, & Talbott, 2015). Table 3 is a summary of the main contributions and gaps.

Table 2. Examples of sustainability indicators in projects

Economic	Environmental	Social
Direct/Indirect Economic Performance Strategic value Market Presence Business Agility	Energy, water, emissions, waste and etc.	Product Responsibility Human Rights Community support Market communication

Source: (Silvius & Tharp, 2013)

Table 3. Incorporation of sustainability dimensions into project management

Dimensions of Sustainability	Key Contributions and Gaps
Economical/social/environmental (TBL) & organizational level	So far the attention of researchers has primarily been applied to this dimension, though yet more research is required (Marcelino-Sadab et al., 2015)
Business level	PM business models and value creation require further analysis (Wikström, Artto, Kujala, & Söderlund, 2010)
Sustainable innovation	More case studies are required (Brook & Pagnanelli, 2014)

Source: Own presentation of author

Organization of Project Management: The Change

The successful management of change is a critical factor for the success of project organization in the present competitive business environment. Approximately 30 percent of worldwide gross domestic product (GDP) is spent on projects (Turner, Huemann, Anbari, & Bredillet, 2010). However, organizational change management suffers from a lack of a proper framework to manage change (Todnem, 2005). PMI describes change management as "a comprehensive cycle and structured approach for transitioning individual groups and organizations from a current state to a future state with intended business benefits" (Project Management Institute, 2013, p.7). Moreover, the issue of organizational change management in the context of PM has not been adequately discussed in the literature; Hornstein (2015) has highlighted the failures of PM guidelines in coping with organizational change management. To cover this traditional gap, PMI, in 2013, introduced a practical guide for managing change in organizations, formulated as: "define the pacing of change in collaboration with the recipient's representative, in order to be able to mobilize the stakeholders and sustain the change" (Project Management Institute, 2013, p. 69). This is still an emerging field. The required elements for sustainable change mainly comprise: leadership, strategic management, paradigm shift, communication, interaction with drivers of change, better planning processes and portfolio approach (Harrington, Voehl, & Voehl, 2014).

Despite the lack of effective guidelines for change management in the literature, there are some practical guidelines for change management within organizations, such as Kotter's eight-step change process (Kotter, 1996)[1], Luecke's seven-steps model (Luecke, 2003) or Kanter's six-steps approach (Kanter, Stein, & Jick, 1992). Chesbrough & Rosenbloom (2002) explain the transition path from a closed to an open model in four steps: first, think of your organization as a service; second, experiment with a new sources of revenue and change the role of stakeholders from passive to active in order to "co-create" with them; third, based on these successful experiments, draw up a new business model – projects can go through internal channels if they fit this business model, otherwise they can be handled through external channels; forth, scale up (transform your model) and assign the model higher volume across the company and stakeholders.

Using maturity models for change management involves practical tools to translate this complex concept into practice (Crawford, 2006). In other words, the level of maturity of organizations impacts its ability to change (Project Management Institute, 2013). Leading PM organizations have developed both standards of PM and maturity models for the assessment of organizations (e.g., OPM3 and P3M3). These maturity models provide a hierarchical structure that includes a number of capabilities and the relevant indicators for the assessment of three organizational levels; projects, programmes and portfolios and can translate complex project concepts into organizational capabilities (Project Management

Table 4. Contribution of OI to PM

Keywords	Paper Indexed	Remarks	Potentially Relevant
Portfolio management & OI	6	No paper in the well-known journals of project management	(Katzya, Turgut, Holzmann, & Sailer, 2013)
Project management & OI	46	No paper in the well-known journals of project management	(Chiaroni, Chiesa, Frattini, & Terruzzi, 2015)
Project management & Co-creation	13	Three papers in the well-known journals of project management	(Chang, Chih, Chew, & Pisarskia, 2013)

Source: Own presentation of author

Institute, 2013). For instance, the PfM3 (Portfolio Maturity Model) defines seven process perspectives through five maturity levels (AXELOS, 2013). Table 4 presents the contribution of the OI literature to PM. Overall, a lack of sufficient research in this area is apparent.

Perspective of Portfolio Management Standards Regarding SD and OI: A Critical Review

Portfolio management is a crucial element of standardized PM performance. The main objective of portfolio management is to achieve a business's strategic objectives by maximizing the value of the portfolio and striking the right balance between projects (Cooper, Edgetta, & Kleinschmidta, 1999). Internationally and globally employed standards (e.g., the MoP or PMI) are considered to be best practices that fundamentally influence project and portfolio management. These standards reflect the best, common understanding of experts based on their experience in a variety of industrial organizations and projects, and reflect what a project manager should do in order to manage a project in a successful way. Due to their worldwide influence, it is of great importance for this chapter that I also assess these standards from the perspective of SD and open model potential (change aspect).

The management of portfolios (MoP) defines portfolio management as a "coordinated collection of strategic processes and decisions that together enable a more effective balance of organizational change and business as usual" (AXELOS, 2014, p. ix)[2]. MoP refers to the stakeholder in the portfolio delivery cycle in terms of "stakeholder engagement" to meet the needs of both external and internal stakeholders. There is no explicit consideration of environmental sustainability, although the "portfolio definition cycle" (understands) raises the point of environmental analysis and mentions some tools, such as PESTLE analysis (AXELOS, 2014). Additionally, financial management is considered to be a principle of the portfolio delivery cycle without mentioning economic sustainability indicators. It can be concluded that the MoP provides a limited approach to sustainability (TBL) in portfolio management.

The standard for portfolio management (3rd ed. PMI) defines portfolio management as "the centralized management of one or more portfolios to achieve strategic objectives" (Project Management Institute, 2013, p. 29). A portfolio, in turn, is defined as "a component collection of programs, projects, or operations managed as a group to achieve strategic objectives", while it is highlighted that this standard is more a guide than a methodology, and it is therefore advised that project managers may prefer a methodology-based standard, such as MoP (Project Management Institute, 2013, p. 3). Similar to the PMBOK, the method and logic that has been implemented in this standard is based on "Process Group" and "Knowledge Area" mapping, and defining input/output tools and techniques. Although no explicit consideration is given to environmental concerns, Enterprise Environmental Factors (EEFs) are applied to all of these process groups, though no specific tools or techniques are available. In addition, "cost-benefit analysis" is offered as a general economic analysis tool to define the "portfolio road map", which is a key deliverable for "portfolio management processes". Nevertheless, a lack of sufficient techniques for economic sustainability is observed. It can be noted that the standards of PM (Prince2, PMBOK 5th and ICB) also fail to meet the needs of stakeholders (Eskerod & Huemann, 2013).

The key aim of the OI model is to change organizations' business models (opening up). Management of organizational changes is an important part of portfolio management and the standards require portfolio managers to monitor changes continuously (in both internal and external environments). Hence Table 5 also illustrates the change management boundary and potential of the mentioned standards. In PMI, "manage strategic change" is a portfolio management process that allows the portfolio manager to manage changes, encompassing three tools. However, "portfolio communication management" tools as channels of communication can be noted as implicit contributors. MoP has clear, defined roles and responsibilities for the management of change – "change delivery committee" (CDC) and "business change director" – but it does not involve any specific tools (AXELOS, 2014). MoP takes the strategic objective of portfolio management to be to "change the business" and, through the "strategy alignment principle", to align the change initiative with the organization's overall strategy, and argues that "energized change culture" can pave the way. It can be concluded that MoP presents a relatively better and more comprehensive pathway to management of organizational change and change management tasks.

As highlighted in the Table 5, this section has highlighted that the best practices of portfolio management, which act as decision-making business models for project portfolio offices, have not explicitly considered sustainability dimensions in the PM processes and, further, are not capable of fostering the open model perspective. As already discussed, based on OI strategy, the organization should utilize both internal and external resources for the commercialization of products and services, through both external and internal routes (of equal importance), and projects that fit the current strategy of the firm should go through internal channels while those that do not fit should go to the market through external channels. For instance, in the mentioned standards, there is no clear sign that shows how to cope with the projects that at first seem to lack promise, but turn out to be amazingly valuable.

Generally, the study of PM standards (that reflects a shared understanding of PM) has highlighted that the traditional portfolio project management approach is similarly based on the closed model system, and does not provide a fruitful contribution for the PM office and organization towards openness unless some potentials can be exploited.

INCORPORATION OF OPEN INNOVATION INTO SUSTAINABLE PROJECT PORTFOLIO MANAGEMENT

OI can favourably meet the needs of sustainable PM criteria. OI contributes to economic sustainability in PM primarily through its ability to approve market presence. OI attacks the cost side of the organization, both by reducing internal and external development costs and decreasing innovation costs, exploiting economies of scale and scope and providing a new stream of revenue as it addresses a wider range of markets, while also saving time. Therefore, OI supports economic sustainability through market pres-

Table 5. Perspective of portfolio management best practices regarding SD and OI

	The Standard for Portfolio Management-3th Ed. (PMI)			Management of Portfolios (MoP)		
Sustainable Development Pillars	People (stakeholders)	Planet (environmental)	Economic sustainability	People (stakeholders)	Planet (environmental)	Economic sustainability
Relevance	- Manage strategic change - Develop portfolio communication management plan	EEFs	portfolio roadmap	Portfolio delivery cycle: stakeholder engagement	No explicit consideration (Portfolio definition cycle: understand)	Portfolio delivery cycle: Financial management
Tools & Techniques	stakeholder analysis, gap analysis, readiness assessment, elicitation techniques, communication requirements analysis	No explicit consideration	No explicit consideration	Main elements (no explicit tools): managerial support, consistent communication with stakeholders, collaboration with portfolio office, stakeholder engagement plan, key stakeholder identification, active participation of key stakeholders in portfolio cycle	PESTLE, SWOT	No explicit tools and techniques (general portfolio financial management elements)
Potential for OI	Manage strategic change			Energized change culture		

Source: Own presentation of author

ence and return on investment. In addition, the environmental and social pillars of SPM can be very well addressed by OI. OI can benefit SPM through the channel of innovative corporate responsibility or transformational-oriented CSR. Table 6 and Figure 1 summarize the main findings.

Managerial Implications of Open Innovation Model For Project Management

This research primarily underlines the logic behind the proposed scope of the research; however, this section illustrates a perspective with a more practical approach. As discussed above, TPM is similar to the closed approach in organizations and the main PM guidelines have, in relative terms, failed to meet the needs of organizations in regard to both sustainability and change management. Therefore, a transition to an open model can be seen as a necessary solution, and it is thus of great importance that a method for moving towards an open model is made available. This research has highlighted that the main area of cooperation between the OI model and (sustainable) PM is in stakeholder management

Table 6. Summary of the contribution of OI to sustainable project management

	Sustainable Project Management	Contribution of Open Innovation
Indicators	**economic sustainability**	Market presence Direct and indirect economic performance(e.g. economy of scale and scope) Strategic value Flexibility
	social and environmental sustainability	e.g. Corporate responsibility innovation (CR innovation type), transformational oriented CSR
Framework	**portfolio Fuzzy decision making** **project delivery** **paradigm shift**	Fuzzy front end funnel Channels of communication (porous funnel) New paradigm

Source: Own presentation of author

Figure 1. Main channels of contribution of OI to sustainable project management
Source: *Own presentation of author*

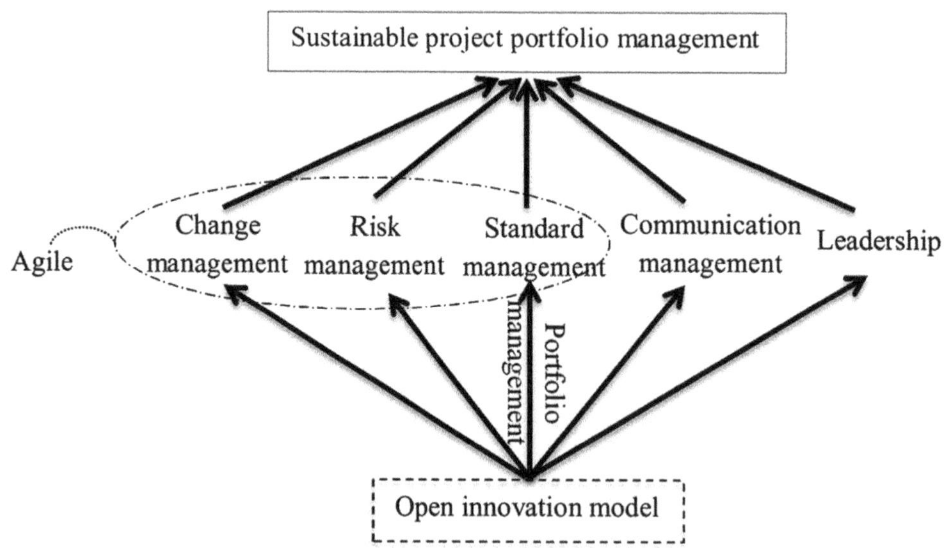

and engagement, and hence the overlapping area of maturity models of OI and PM can help to assess and improve PM practices (Jia, Chen, Xue, Chen, Cao, & Tang, 2011). Table 7 presents a framework based on the portfolio management maturity and OI maturity models. It should be noted that, in this table, the first groups of "undifferentiated" organizations that have been caught in the "commodity trap" have been excluded.

FUTURE RESEARCH DIRECTIONS

This chapter primarily outlines the theory and logic of the application of OI to the project portfolio management of an organization as it favourably impacts SD. Future research should carry out (quantitative) case studies to illustrate how a specific project organization copes with the OI model. Consequently, the influence of OI on the economic side of the project portfolio (quantitative research from the perspective of financial pay-offs) demands further discussion as, from the managerial perspective, it may appear especially salient. Accordingly, further development of the standards of PM and descriptions of OI modelled on the language of PM best practices and methodologies is required.

A general limitation of this research has been the focus of analysis on certain portfolio management guidelines, where different organizations may prefer to implement other guidelines. An emerging field within sustainability and SPM is the improvement of key performance indicators and, in cases of implementation of OI in project organizations, the development of relevant indicators is also required. Furthermore, development of business models for SPM is an important area for future research.

Table 7. Translation of OI based on the portfolio management maturity model

OI Maturity Model				Portfolio Management - Stakeholder Engagement Maturity Model
Platform Player	Business models of company and stakeholders are interconnected and, furthermore, innovation in the organization's business model is an important criterion at this level. Additionally, management of innovation is the responsibility of every unit of the organization. Intellectual property (IP) is a "strategic asset".	Communication is a central facet of the organization's culture, facilitating full engagement of stakeholders and continuous improvement.	**Level 5 Optimized**	
Integrated	External and internal innovations are integrated. The company shares its innovation road map, and aligns its business model with those of stakeholders, while focusing on new business as well as current business. IP is managed as a "financial asset".	Advanced and quantitative techniques used for evaluation of portfolio stakeholders.	**Level 4 Managed**	
Externally Aware	Awareness of external innovations, definition of stakeholders' roles and consideration of IP management as an "enabling asset".	"A centrally managed and consistent approach to stakeholder engagement" exists in the portfolio organization.	**Level 3 Defined**	
Segmented	There is a plan for innovation which is taken into consideration for future investment. The company can choose projects based on its segmented market. Someone is responsible for innovation management and this process is planned.	Stakeholder engagement is based on the "personal initiative of portfolio managers" to a greater extent.	**Level 2 Repeatable**	
Differentiated	Some innovative work is done that provides some differentiations in products or services. The innovation environment is based on ad hoc activities and IP management is unplanned.	"Stakeholder engagement and communication is rarely used".	**Level 1 Awareness**	

Source: Own presentation of author

In this context, discussion regarding small and medium sized enterprises (SMEs) is a matter of great importance. If experiments are performed with SMEs by integrating SD into portfolio management through the application of OI, the capability of OI to overcome organizational and cultural barriers in SMEs, by modifying their organization and enhancing their agility, will be demonstrated (Vrandea, Jong, Vanhaverbeke, & Rochemont, 2009). Furthermore, if costs and benefits are computed on the basis of the OI approach for SMEs, the implementation of OI for the firms' decision makers will be facilitated (Gassmann, Enkel, & Chesbrough, 2010).

CONCLUSION

This research, based on a structured review of the existing literature, has revealed both the potential and weaknesses of the existing research and methodologies surrounding OI and SPM. This study has highlighted the need for a more comprehensive understanding of SPM, and claims that an appropriate portfolio management strategy is the key to the realization of SPM. Accordingly, this research evaluates the link between OI and sustainability disciplines, and shows that OI can facilitate integration of sustainability into PM. The current PM methodologies are not explicitly capable of fostering openness in the organization of PM and lack sufficient attention to SD. The close relationship between PM and change management issues has been highlighted as a main obstacle, while OI can favourably change the business organization. If change is made through the application of OI, the consequence will be an agile organization.

This study argues that, as long as the business and organization of PM is based on the closed model, successful integration of SD into projects is not possible or, at least, cannot provide competitive advantages. The overlapping areas of OI and PM can be addressed by reference to stakeholder management in order to better interact with stakeholders to create shared value. In addition, this research highlights the requirement for further evaluation and quantitative analysis of large, small and medium-sized firms. Although OI offers opportunities to cope with this issue, it is still an emerging field of science and requires more in-depth analysis.

REFERENCES

Amini, M., & Bienstock, C. (2014). Corporate sustainability: An integrative definition and framework to evaluate corporate practice and guide academic research. *Journal of Cleaner Production*, *76*, 12–19. doi:10.1016/j.jclepro.2014.02.016

Arnold, M. (2015). (Article in Press). Fostering sustainability by linking co-creation and relationship management concepts. *Journal of Cleaner Production*. doi:10.1016/j.jclepro.2015.03.059

AXELOS. (2013). *Portfolio, programme and project management maturity model (P3M3®)- P3M3 v 2.1*. AXELOS.

AXELOS. (2014). *Management of portfolios*. London: The Stationery Office.

Boons, F., & Lüdeke-Freund, F. (2013). Business models for sustainable innovation: State-of-the-art and steps towards a research agenda. *Journal of Cleaner Production*, *45*, 9–19. doi:10.1016/j.jclepro.2012.07.007

Borowy, I. (2013). *Defining sustainable development for our common future: a history of the world commission on environment and development (Brundtland commission)*. Taylor and Francis.

Brook, J., & Pagnanelli, F. (2014). Integrating sustainability into innovation project portfolio management - A strategic perspective. *Journal of Engineering and Technology Management*, *34*, 46–62. doi:10.1016/j.jengtecman.2013.11.004

Carroll, A., & Shabana, K. (2010). The business case for corporate social responsibility: A review of concepts research and practice. *International Journal of Management Reviews*, *12*(1), 85–105. doi:10.1111/j.1468-2370.2009.00275.x

Chang, A., Chih, Y., Chew, E., & Pisarskia, A. (2013). Reconceptualising mega project success in Australian Defence: Recognising the importance of value co-creation. *International Journal of Project Management*, *31*(8), 1139–1153. doi:10.1016/j.ijproman.2012.12.005

Chesbrough, H., & Appleyard, M. (2007). Open Innovation and Strategy. *California Management Review*, *50*(1), 57–76. doi:10.2307/41166416

Chesbrough, H., & Rosenbloom, R. (2002). The role of the business model in capturing value from innovation: Evidence from Xerox Corporations technology spin-off companies. *Industrial and Corporate Change*, *11*(3), 529–555. doi:10.1093/icc/11.3.529

Chesbrough, H. (2003). The Era of Open Innovation. *MIT. Sloan Management Review*, *44*(3), 35–41.

Chesbrough, H. (2006a). *Open innovation: the new imperative for creating and profiting from technology*. Boston: Harvard Business school Press.

Chesbrough, H. (2006b). *Open business models: how to thrive in the new innovation landscape*. Harvard Business School.

Chesbrough, H. (2011). *Open services innovation: Rethinking your business to grow and compete in a new era*. San Francisco: Wiley. doi:10.1007/978-88-470-1980-5

Chesbrough, H., Vanhaverbeke, W., & West, J. (2006). *Open innovation, researching a new paradigm*. New York: Oxford University Press.

Chiaroni, D., Chiesa, V., Frattini, F., & Terruzzi, R. (2015). Implementing open innovation: A case study in the renewable energy industry. *International Journal of Technology Intelligence and Planning*, *10*(10), 195–221. doi:10.1504/IJTIP.2015.070847

Crawford, K. (2006). The project management maturity model. *Information Systems Management*, *23*(4), 50–58. doi:10.1201/1078.10580530/46352.23.4.20060901/95113.7

Erzurumlu, S. (2010). Collaborative product development with competitors to stimulate downstream innovation. *International Journal of Innovation Management*, *14*(4), 573–602. doi:10.1142/S1363919610002787

Eskerod, P., & Huemann, M. (2013). Sustainable development and project stakeholder management: What standards say. *International Journal of Managing Projects in Business*, *6*(1), 36–50. doi:10.1108/17538371311291017

Cooper, G. (1999). New product portfolio management: Practices and performance. *Journal of Product Innovation Management*, *16*(4), 333–351. doi:10.1016/S0737-6782(99)00005-3

Gassmann, O., Enkel, E., & Chesbrough, H. (2010). The future of open innovation. *R & D Management*, *40*(3), 213–221. doi:10.1111/j.1467-9310.2010.00605.x

Geels, F. (2004). From sectoral systems of innovation to socio-technical systems From sectoral systems of innovation to socio-technical systems From sectoral systems of innovation to socio-technical systems. *Research Policy*, *33*(6-7), 897–920. doi:10.1016/j.respol.2004.01.015

Geels, F., Hekkert, M., & Jacobsson, S. (2008). The dynamics of sustainable innovation journeys. *Technology Analysis and Strategic Management*, *20*(5), 521–536. doi:10.1080/09537320802292982

Giannopoulou, E., Yström, A., & Ollila, S. (2011). Turning open innovation into practice: Open innovation research through the lens of managers. *International Journal of Innovation Management*, *15*(3), 505–524. doi:10.1142/S1363919611003465

Halme, M., & Laurila, J. (2009). Philanthropy, integration or innovation? Exploring the financial and societal outcomes of different types of corporate responsibility. *Journal of Business Ethics*, *84*(3), 325–339. doi:10.1007/s10551-008-9712-5

Hamdouch, A., & Zuindeau, B. (2010). Sustainable development, 20 years on: Methodological innovations, practices and open issues. *Journal of Environmental Planning and Management*, *53*(4), 427–438. doi:10.1080/09640561003694286

Harraf, A., Wanasika, I., Tate, K., & Talbott, K. (2015). Organizational agility. *Journal of Applied Business Research*, *31*(2), 675–686. doi:10.19030/jabr.v31i2.9160

Harrington, J., Voehl, F., & Voehl, C. (2014). *Model for sustainable change*. Project Management Institute.

Hautamäki, A. (2010). *Sustainable innovation, a new age of Innovation and Finland's innovation policy*. Helsinki: sitra.

Heising, W. (2012). The integration of ideation and project portfolio management-A key factor sustainable success. *International Journal of Project Management*, *30*(5), 582–595. doi:10.1016/j.ijproman.2012.01.014

Holmes, S., & Smart, P. (2009). Exploring open innovation practice in firm-nonprofit engagements: A corporate social responsibility perspective. *R & D Management*, *39*(4), 394–409. doi:10.1111/j.1467-9310.2009.00569.x

Hornstein, H. (2015). The integration of project management and organizational change management is now a necessity. *International Journal of Project Management*, *33*(2), 291–298. doi:10.1016/j.ijproman.2014.08.005

Jia, G., Chen, Y., Xue, X., Chen, J., Cao, J., & Tang, K. (2011). Program management organization maturity integrated model for mega construction programs in China. *International Journal of Project Management, 29*(7), 834–845. doi:10.1016/j.ijproman.2011.03.003

Kanter, R., Stein, B., & Jick, T. (1992). *The challenge of organizational change*. New York: The Free Press.

Katzya, B., Turgut, E., Holzmann, T., & Sailer, K. (2013). Innovation intermediaries: A process view on open innovation coordination. *Technology Analysis and Strategic Management, 25*(3), 295–309. doi:10.1080/09537325.2013.764982

Kerzner, A., & Belack, C. (2010). *Managing complex projects*. John Wiley. doi:10.1002/9780470927977

Kotter, J. (1996). *Leading change*. Boston: Harvard Business School Press.

Lichtenthaler, U. (2008). Open Innovation in Practice: An analysis of strategic approaches to technology transactions. *IEEE Transactions on Engineering Management, 55*(1), 148–157. doi:10.1109/TEM.2007.912932

Lichtenthaler, U. (2011). Open innovation: Past research, current debates, and future directions. *The Academy of Management Perspectives, 25*(1), 75–93. doi:10.5465/AMP.2011.59198451

Luecke, R. (2003). *Managing change and transition*. Harvard Business School Press.

Mann, R., Adebanjo, D., & Tickle, M. (2010). Deployment of business excellence in Asia: An exploratory study. *International Journal of Quality & Reliability Management, 28*(6), 604–627. doi:10.1108/02656711111141184

Marcelino-Sadab, S., González-Jaen, L., & Pérez-Ezcurdia, A. (2015). Using project management as a way to sustainability. From a comprehensive review to a framework definition. *Journal of Cleaner Production, 99*, 1–16. doi:10.1016/j.jclepro.2015.03.020

Martinuzzi, A., & Krumay, B. (2013). The good, the bad, and the successful how corporate social responsibility leads to competitive advantage and organizational transformation. *Journal of Change Management, 13*(4), 24–443. doi:10.1080/14697017.2013.851953

Milosevic, D., & Patanakul, P. (2005). Standardized project management may increase development projects success. *International Journal of Project Management, 23*(3), 181–192. doi:10.1016/j.ijproman.2004.11.002

Mir, F., & Pinnington, A. (2014). Exploring the value of project management: Linking project management performance and project success. *International Journal of Project Management, 25*(2), 202–217. doi:10.1016/j.ijproman.2013.05.012

Mutka, S., & Aaltonen, P. (2013). The impact of a delivery projects business model in a project-based firm. *International Journal of Project Management, 31*(2), 166–176. doi:10.1016/j.ijproman.2012.07.006

Porter, M., & Kramer, M. (2011). Creating shared value. *Harvard Business Review, 89*(1/2), 62–77.

Project Management Institute. (2013a). *Managing change in organizations: a practice guide*. Pennsylvania: PMI.

Project Management Institute. (2013b). *PMI's pulse of the profession in-depth report: navigating complexity*. PMI.

Project Management Institute. (2013c). *The standard for portfolio management* (3rd ed.). PMI.

Qureshi, T., Warraich, A., & Hijazi, T. (2009). Significance of project management performance assessment (PMPA) model. *International Journal of Project Management*, *27*(4), 378–388. doi:10.1016/j.ijproman.2008.05.001

Saebi, T., & Foss, N. J. (2015). Business models for open innovation: Matching heterogeneous open innovation strategies with business model dimensions. *European Management Journal*, *33*(3), 201–213. doi:10.1016/j.emj.2014.11.002

Savitskaya, I., Salmi, P., & Torkkeli, T. (2010). Barriers to open innovation: Case China. *Journal of Technology Management & Innovation*, *5*(4), 10–21. doi:10.4067/S0718-27242010000400002

Serrador, P., & Pinto, J. (2015). Does agile work? A quantitative analysis of agile project success. *International Journal of Project Management*, *33*(5), 1040–1051. doi:10.1016/j.ijproman.2015.01.006

Silvius, G., & Tharp, J. (2013). *Sustainability integration for effective project management*. IGI. doi:10.4018/978-1-4666-4177-8

Silvius, G., Schipper, R., Planko, J., Brink, J., & Köhler, A. (2012). *Sustainability in project management*. Surrey: Gower.

Stettina, C., & Hörz, J. (2015). Agile portfolio management: An empirical perspective on the practice in use. *International Journal of Project Management*, *33*(1), 140–152. doi:10.1016/j.ijproman.2014.03.008

Todnem, R. (2005). Organisational change management: A critical review. *Journal of Change Management*, *5*(4), 369–380. doi:10.1080/14697010500359250

Turner, G. (2008). A comparison of the limits to growth with 30 years of reality. *Global Environmental Change*, *18*(3), 397–411. doi:10.1016/j.gloenvcha.2008.05.001

Turner, R., Huemann, M., Anbari, F., & Bredillet, C. (2010). *Perspectives on projects*. Routledge.

Vrandea, V., Jong, J., Vanhaverbeke, W., & Rochemont, M. (2009). Open innovation in SMEs: Trends, motives and management challenges. *Technovation*, *29*(6-7), 423–437. doi:10.1016/j.technovation.2008.10.001

West, J., Salter, A., Vanhaverbeke, W., & Chesbrough, H. (2014). Open innovation: The next decade. *Research Policy*, *43*(5), 805–811. doi:10.1016/j.respol.2014.03.001

Westerveld, E. (2003). The project excellence model: Linking success criteria and critical success factors. *International Journal of Project Management*, *21*(6), 411–418. doi:10.1016/S0263-7863(02)00112-6

Whitty, S., & Maylor, H. (2009). And then came complex project management. *International Journal of Project Management*, *27*(3), 304–310. doi:10.1016/j.ijproman.2008.03.004

Wikström, K., Artto, K., Kujala, J., & Söderlund, J. (2010). Business models in project business. *International Journal of Project Management*, *28*(8), 832–841. doi:10.1016/j.ijproman.2010.07.001

KEY TERMS AND DEFINITIONS

Corporate Social Responsibility (CSR): CSR is a policy that remarks that firms have a responsibility toward society, and companies can take some strategies in order to voluntarily participate in making a cleaner environment and society.

Open Innovation (OI): A term introduced by Professor Henry Chesbrough in 2003, which proposes that firms should exploit both external and internal ideas and innovations. OI has also been put forth as a business model.

Portfolio Management: Management of one or more portfolios in order to achieve strategic objectives of the organization.

Research and Development (R&D): A set of structured activities to provide innovative products and services with higher quality, lower cost, and new users and customers.

Sustainable Development (SD): Development that is based on three pillars: environmental, social and economic sustainability.

Sustainable Project Management (SPM): Project management that is in accordance with sustainable development.

Traditional Project Management (TPM): Project management methodology that mainly focuses on time, cost, scope and quality management.

ENDNOTES

1. Editors' note: For a detailed presentation of Kotter's eight-step change process, see chapter 16, on structural change management in SMEs.
2. Editors' note: In chapter 14 a concept was already introduced that, at SME level, particularly among medium size enterprises, a more effective process management can take place, while the causal relationships are empirically analyzed and described in terms of barriers, enablers and benefits.

Compilation of References

ABI. (2016). Retrieved August 21st, 2016 from Association of British Insurers web site: https://www.abi.org.uk/

Abubakar, A., & Ilkan, M. (2015). Impact of online WOM on destination travel and intention to travel: A medical tourism perspective. *Journal of Destination Marketing & Management*. 10.1016/j.jdmin.2015.12.005

Ackoff, R. L. (1999). *Ackoff's Best: His Classic Writings on Management*. New York, NY: John Wiley & Sons Inc.

Ackoff, R. L., & Emery, F. E. (1972). *On Purposeful Systems*. Chicago: Aldine-Atherton.

Adižes, I. (1994). *Životni ciklusi preduzeća*. Beograd, Serbia: Agora – Prometej.

Adizes, I. (1992). *Mastering Change: The Power of Mutual Trust and Respect*. Santa Barbara, CA: Adizes Institute.

Aghion, P., & Howitt, P. (1998). *Endogenous growth theory*. Cambridge, MA: The MIT Press.

Alagumalai, S., Curtis, D. D., & Hnugi, N. (2005). *Applied Rasch Measurement: A Book of Exemplars*. Springer.

Al-Ghamdi, S. M. (1998). Obstacles to successful implementation of strategic decisions: The British experience. *European Business Review*, *98*(6), 322–327. doi:10.1108/09555349810241590

Alhojailan, M. I. (2012). Thematic analysis: A critical review of its process and evaluation. *West East Journal of Social Sciences*, *1*(1), 39–47.

Almli, V., Verbeke, W., Vanhonacker, F., Naes, T., & Hersleth, M. (2011). General image and attribute perceptions of traditional food in six European countries. *Food Quality and Preference*, *22*(1), 129–138. doi:10.1016/j.foodqual.2010.08.008

Altenburg, T., & Drachenfels, C. (2006). *Business environment reforms: Why it is necessary to rethink priorities and Strategies*. German Development Institute. Retrieved July 27th, 2016 from German Development Institute web site: https://www.die-gdi.de/uploads/media/Altenburg_Drachenfels_2008_Business_environment_reforms.pdf

Altkorn, J. (2001). Orientacje Marketingowe Polskich Przedsiębiorstw. In *VII Kongres Ekonomistów Polskich, Sesja IV*. Warszawa: Polskie Towarzystwo Ekonomiczne.

Altomonte, C., Navaretti, G. B., di Mauro, F., & Ottaviano, G. (2011). Assessing competitiveness: how firm-level data can help. *Bruegel Policy Contribution, 16*.

Amaro, S., & Duarte, P. (2015). An integrative model of consumers intentions to purchase travel online. *Tourism Management*, *46*, 64–79. doi:10.1016/j.tourman.2014.06.006

Amini, M., & Bienstock, C. (2014). Corporate sustainability: An integrative definition and framework to evaluate corporate practice and guide academic research. *Journal of Cleaner Production*, *76*, 12–19. doi:10.1016/j.jclepro.2014.02.016

Andersen, O. (1997). Internationalization and Market Entry Mode: A Review of Theories and Conceptual Frameworks. *Management International Review, 37*, 27-42.

Andersson, F. O. (2011). Organisational capacity and entrepreneurial behaviour in nonprofit organisations: An empirical exploration. *Voluntary Sector Review, 2*(1), 43–56. doi:10.1332/204080511X560611

Andrijašević, S., & Petranović, V. (1999). *Ekonomika osiguranja*. Zagreb: Alfa d.d.

Annoni, P., & Dijkstra, L. (2013). *EU regional competitiveness index*. Luxembourg: Europe.

Aragón-Correa, J. A., & Sharma, S. (2008). The influence of stakeholders on the environmental strategy of service firms: The moderating effects of complexity uncertainty and munificence. *British Journal of Management, 19*(2), 185–203. doi:10.1111/j.1467-8551.2007.00538.x

Archibugi, D., Filippetti, A., & Frenz, M. (2013). Economic crisis and innovation: Is destruction prevailing over accumulation? *Research Policy, 42*(2), 303–314. doi:10.1016/j.respol.2012.07.002

Arendt, C., Landis, R. & Meister, T. (1995). The human side of change - part 4. *IEE Solutions*, 22-27.

Armistead, C. G., & Rowland, P. (1996). *Managing business processes: BPR and beyond*. New York, NY: Wiley.

Armistead, C., & Pritchard, J. P. (1999). Business process management - lessons from European business. *Business Process Management Journal, 5*(1), 10–35. doi:10.1108/14637159910249144

Armstrong, G., & Kotler, Ph. (2011). *Marketing. An Introduction*. Upper Saddle River, NJ: Pearson.

Arnold, M. (2015). (Article in Press). Fostering sustainability by linking co-creation and relationship management concepts. *Journal of Cleaner Production*. doi:10.1016/j.jclepro.2015.03.059

Artz, K. W., Norman, P. M., Hatfield, D. E., & Cardinal, L. B. (2010). A longitudinal study of the impact of R&D, patents, and product innovation on firm performance. *Journal of Product Innovation Management, 27*(5), 725–740. doi:10.1111/j.1540-5885.2010.00747.x

Atanassova, I., & Clark, L. (2015). Social media practices in SME marketing activities: A theoretical framework and research agenda. *Journal of Customer Behaviour, 14*(2), 163-183.

Audretsch, A. B., & Beckmann, A. M. (2007). From Small Business to Entrepreneurship Policy. In D. B. Audretsch & I. Grilo (Eds.), *Handbook of Research on Entrepreneurship Policy* (pp. 45–47). London: Edward Elgar Pub. doi:10.4337/9781847206794.00007

Auerswald, P. E., & Branscomb, L. M. (2003). Start-ups and Spin-offs: Collective Entrepreneurship Between Invention and Innovation. In D. M. Hart (Ed.), *The Emergence of Entrepreneurship Policy: Governance, Start-Ups, and Growth in the U.S. Knowledge Economy* (pp. 61–65). Cambridge, UK: Cambridge University Press. doi:10.1017/CBO9780511610134.004

Australian Government: Department of Industry Innovation, Science, Research and Tertiary Education. (2013). *Australian Small Business: Key Statistics and Analysis – December 2012*. Retrieved May 14th 2016, from Australian Treasure web site: http://www.treasury.gov.au/~/media/Treasury/Publications%20and%20Media/Publications/2012/Australian%20Small%20Business%20-%20Key%20Statistics%20and%20Analysis/downloads/PDF/AustralianSmallBusinessKeyStatisticsAndAnalysis.ashx

AXA. (2016). Retrieved March 25, 2016, Axa Insurance web site: from https://www.axa.rs/

AXELOS. (2013). *Portfolio, programme and project management maturity model (P3M3®)- P3M3 v 2.1*. AXELOS.

AXELOS. (2014). *Management of portfolios*. London: The Stationery Office.

Ayyagari, M., Demirgüç-Kunt, A., & Maksimovic, V. (2006). *How Important Are Financing Constraints? The Role of Finance in the Business Environment.* World Bank Policy Research Working Paper, No. 3820.

Azevedo, C. L., Almeida, J. P. A., van Sinderen, M., & Pires, L. F. (2015). Towards capturing strategic planning in EA. In *Enterprise Distributed Object Computing Conference (EDOC), 2015 IEEE 19th International*. doi:10.1109/EDOC.2015.31

Balogun, J., & Johnson, G. (2004). Organizational restructuring and middle manager sensemaking. *Academy of Management Journal*, *47*(4), 523–549. doi:10.2307/20159600

Banterle, A., Cavaliere, A., De Marchi, E. (2016). The Italian food industry in the era of the TTIP negotiate. *British Food Journal*, *118*(8), 1930 – 1945.

Banterle, A., Carraresi, L., & Stranieri, S. (2010). Small Business Marketing Capability in the Food Sector:The Cases of Belgium, Hungary and Italy. *International Journal on Food System Dynamics*, *2*, 94–102.

Banterle, A., Cavalieri, A., Carraresi, L., & Stranieri, S. (2014). Food SMEs Face Increasing Competition in the EU Market: Marketing Management Capability Is a Tool for Becoming a Price Maker. *Agribusiness*, *30*(2), 113–131. doi:10.1002/agr.21354

Barcus, S. W. W., & Wilkinson, J. W. (1995). *Handbook of management consulting services*. McGraw-Hill.

Barney, J. (1991). Firm resources and sustained competitive advantage. *Journal of Management*, *17*(1), 99–120. doi:10.1177/014920639101700108

Barney, J. (2001a). Is the resource-based "view" a useful perspective for strategic management research? Yes. *Academy of Management Review*, *26*(1), 41–56.

Barney, J. (2001b). Resource-based theories of competitive advantage: A ten-year retrospective on the resource-based view. *Journal of Management*, *27*(6), 643–650. doi:10.1177/014920630102700602

Bartik, T. (1989). Small business start-ups in the United States: Estaimates of the effects of characteristics of states. *Southern Economic Journal*, *55*(4), 34–56. doi:10.2307/1059479

Bashein, B., Markus, M., & Riley, P. (1994). Precondition for BPR success and how to prevent failures. *Information Systems Management*, *11*(2), 7–13. doi:10.1080/10580539408964630

Baumol, W., Panzar, J., & Willig, R. (1982). *Contestable markets and the theory of industrial structure*. New York: Harcourt Brace Jovanovich, Inc.

Beck, T. H. L., Demirgüc-Kunt, A., & Maksimovic, V. (2008). Financing patterns around the world: Are small firms different? *Journal of Financial Economics*, *89*(3), 467–487. doi:10.1016/j.jfineco.2007.10.005

Beer, M., & Eisenstat, R. A. (2000). The silent killers of strategy implementation and learning. *Sloan Management Review*, *41*(4), 29–40.

Beinhocker, E. D., & Kaplan, S. (2002). Tired of strategic planning. *The McKinsey Quarterly*, *2*, 48–57.

Belanić, L. (2014). Povijesni razvoj osiguranja pravne zaštite. *Evropska revija za pravo osiguranja*, *2*, 26-34.

Belás, J., Demjan, V., Habánik, J., Hudáková, M., & Sipko, J. (2015). The business environment of small and medium-sized enterprises in selected regions of the Czech Republic and Slovakia. *E+ M Ekonomie a Management*, (1), 95.

Belletti, G., Brunori, G., Marescotti, A., Pacciani, A., & Rossi, A. (2010). *Guide to the Valorisation of Typical Agri-food Products. Concepts, Methods and Tools*. Regional Agency for Development and Innovation in Agriculture and Forestry.

Beraha, I. (2016). *Strategija unapređenja poslovnih performansi malih i srednjih preduzeća u lancu snabdevanja automobilske industrije* (Doctoral Dissertation). University "Union-Nikola Tesla", Faculty of Entrepreneurship. Retrieved July 26, 2016, from: http://unionnikolatesla.edu.rs/doktorati/isidoraberaha/disertacija.pdf

Berger, J. (2014). Word of mouth and interpersonal communication: A review and directions for future research. *Journal of Consumer Psychology*, *24*(4), 586–607. doi:10.1016/j.jcps.2014.05.002

Bergey, J., Smith, D., Tiley, S., Weiderman, N. & Woods, S. (1999). Why reengineering projects fail. *Carnegiee Mellon Software Engineering Institute - Product Line Practice Initiative,1*, 1-30.

Berman, S. (2010). *Capitalizing on complexity*. IBM Global Business Services.

Bernoulli, D. (1954). Exposition of a new theory on the measurement of risk. *Econometrica*, *22*(1), 23–36. doi:10.2307/1909829

Berrington, C., & Oblich, R. (1995). Translating reengineering into bottom-line results. *Industrial Engineering (American Institute of Industrial Engineers)*, 24–27.

Bešić, C., Nikolić, M. & Damnjanović, A. (2010). *Strategijski menadžment*. Čačak: Fakultet tehničkih nauka.

Bešić, C., Sajfert, Z. & Damnjanović, A. (2009). *Menadžment promena*. Čačak: Fakultet tehničkih nauka.

Bialková Šimová, D. (2014). *Efficiency of amount incurred to support SMEs from EU funds*. Bratislava, Slovakia: TU SjF Košice.

Birch, D. L. (1981). Who Creates Jobs? *The Public Interest*, *65*, 3–14.

Bjurstrom, E., & Roberts, H. (2007). *The Principle of Connectivity: Networked Asets, Strategic Capabilities and Bundled Outcomes. In Intellectual Capital Revisited: Paradoxes in the Knowledge Intensive Organization* (pp. 45–60). Cheltenham, UK: Edvard Elgar.

Blom, R. J. (2004). *Handboek Midden- en Kleinbedrijf*. Graydon Nederland BV.

Bocquet, R., Bas, C., Mothe, C., & Poussing, N. (2013). Are firms with different CSR profiles equally innovative? Empirical analysis with survey data. *European Management Journal*, *31*(6), 642–654. doi:10.1016/j.emj.2012.07.001

Bodie, Z., Kane, A., & Marcus, A. (2007). *Essentials of investments*. New York, NY: McGraw-Hill Higher Education.

Bodnar, G. M., & Marston, R. C. (2002). Exchange rate exposure: A simple model. *International Finance Review*, *3*, 107–115. doi:10.1016/S1569-3767(02)03010-8

Bond, F. G., & Fox, C. M. (2007). *Applying the Rasch Model. Fundamental Measurement in the Human Sciences* (2nd ed.). Lawrence Erlbaum Associates, Inc.

Bontis, N., Keow, W. C. C., & Richardson, S. (2000). Intellectual capital and business performance in Malaysian industries. *Journal of Intellectual Capital*, *1*(1), 85–100. doi:10.1108/14691930010324188

Boons, F., & Lüdeke-Freund, F. (2013). Business models for sustainable innovation: State-of-the-art and steps towards a research agenda. *Journal of Cleaner Production*, *45*, 9–19. doi:10.1016/j.jclepro.2012.07.007

Bora, S., & Di Ciccia, A. (2008). *Statistica: metodologie per le scienze economiche e sociali*. Milan: Italy McGraw-Hill Education.

Borowy, I. (2013). *Defining sustainable development for our common future: a history of the world commission on environment and development (Brundtland commission)*. Taylor and Francis.

Borselli, A. (2012). Cognosceat emptor: On the Insurer's Duty to Inform the Prospective Policyholder in Europe. *Revija za pravo osiguranja*, *2*, 45-54.

Bosworth, D., & Rogers, M. (1998). *Research and development, intangible qssets and the performance of large Australian companies*. Melbourne: Melbourne Institute.

Botonaki, A., Polymeros, K., Tsakiridou, E., & Mattas, K. (2006). The role of food quality certification on consumers food choices. *British Food Journal*, *108*(2), 77–90. doi:10.1108/00070700610644906

Bottani, E. (2010). Profile and enablers of agile companies: An empirical investigation. *International Journal of Production Economics*, *125*(2), 251–261. doi:10.1016/j.ijpe.2010.02.016

Boulding, K. E. (1956). General Systems Theory - The Skeleton of Science. *Management Science*, *2*(3), 197–208. doi:10.1287/mnsc.2.3.197

Boutellier, R., Gassmann, O., & Von Zedtwitz, M. (2013). *Managing global innovation: uncovering the secrets of future competitiveness*. Springer Science & Business Media.

Brandenburger, A. M., & Nalebuff, B. J. (1995, July-August). The Right Game: Use Game Theory to Shape Strategy. *Harvard Business Review*, 57–74.

Braun, V., & Clarke, V. (2006). Using thematic analysis in psychology. *Qualitative Research in Psychology*, *3*(2), 77–101. doi:10.1191/1478088706qp063oa

Brenes, E. R., Mena, M., & Molina, G. E. (2008). Key success factors for strategy implementation in Latin America. *Journal of Business Research*, *61*(6), 590–598. doi:10.1016/j.jbusres.2007.06.033

Brezinski, H. (1998). The investment behavior of SMEs in East Germany in the 90s, some insights from a survey. In L. Csaba (Ed.), *The Hungarian SME sector development in comparative perspective*. CIPE.

Brigham, E., & Houston, J. (2001). *Fundamentals of financial management*. San Diego, CA: Harcourt. Inc.

Brooking, A. (1996). *Intellectual Capital*. London: Cengage Learning.

Brook, J., & Pagnanelli, F. (2014). Integrating sustainability into innovation project portfolio management - A strategic perspective. *Journal of Engineering and Technology Management*, *34*, 46–62. doi:10.1016/j.jengtecman.2013.11.004

Browning, V., & Sparks, B. (2011). The impact of online reviews on hotel booking intentions and perceptions of trust. *Tourism Management*, *32*(6), 1310–1323. doi:10.1016/j.tourman.2010.12.011

Bryson, J. M. (1988). A strategic planning process for public and non-profit organizations. *Long Range Planning*, *21*(1), 73–81. doi:10.1016/0024-6301(88)90061-1

Buckman Laboratories. (2016). Retrieved August 16th, 2016, from Buckman web site: http://www.buckman.com/en/about-us/history

Bullita, C. (2006). Geschäftsprozessmanagement bei Siemens Medical Solutions. In A. Geschäftsgebiet, H. J. Schmelzer, & W. Sesselmann (Eds.), *Geschäftsprozessmanagement in der Praxis* (pp. 475–489). Munich: Hanser.

Burton, R. M., Obel, B., & Håkonsson, D. D. (2015). *Organizational Design: A Step-by-Step Approach* (3rd ed.). Cambridge University Press.

Business Development Bank of Canada. (2015). *High-Impact Firms: Accelerating Canadian Competitiveness*. Retrieved May 17, 2016 from BDC web site https://www.bdc.ca/EN/Documents/analysis_research/high-impact-firms-accelerating-canadian-competitiveness.pdf

Cachon, G. P., & Netessine, S. (2004). Game theory in supply chain analysis. In D. Smichi-Levi, S. D. Wu, & Z. Shen (Eds.), *Handbook of Quantitative Supply Chain Analysis: Modeling in the eBusiness Era* (pp. 13–59). Boston: Kluwer. doi:10.1007/978-1-4020-7953-5_2

Camera Di Commercio Italo-Cinese. (2014). *Commercio estero: i trend cinese del prime semestre 2014*. Retrieved August 16th from Camera Di Commercio Italo-Cinese web site: http://www.chinaitaly.com/it/commercio-estero-i-trend-cinesi-del-primo-semestre-2014

Canales, R., & Nanda, R. (2012). A darker side to decentralized banks: Market power and credit rationing in SME lending. *Journal of Financial Economics*, *105*(2), 353–366. doi:10.1016/j.jfineco.2012.03.006

Cândido, C. J., & Santos, S. P. (2015). Strategy implementation: What is the failure rate? *Journal of Management & Organization*, *21*(02), 237–262. doi:10.1017/jmo.2014.77

Carayannis, L., & Provance, M. (2010). Measuring firm innovativeness. Towards composite innovation index built on firm innovative posture, propensity and performance attributes. *International Journal of Innovation and Regional Development*, *1*(1), 90–107. doi:10.1504/IJIRD.2008.016861

Carleton, W. T., & Lerner, M. (1966). *A theory of financial analysis*. San Deigo, CA: Harcourt, Brace & World, Inc.

Caron, J. R., Jarvenpaa, S. L., & Stodard, D. B. (1994). Business reengineering at CIGNA Corporation: Experiences and lessons learned from the first five years. *Management Information Systems Quarterly*, *18*(3), 233–250. doi:10.2307/249617

Carroll, A., & Shabana, K. (2010). The business case for corporate social responsibility: A review of concepts research and practice. *International Journal of Management Reviews*, *12*(1), 85–105. doi:10.1111/j.1468-2370.2009.00275.x

Carroll, C., Kaltenthaler, E., FitzGerald, P., Boland, A., & Dickson, R. (2011). A thematic analysis of the strengths and weaknesses of manufacturers submissions to the NICE Single Technology Assessment (STA) process. *Health Policy (Amsterdam)*, *102*(2), 136–144. doi:10.1016/j.healthpol.2011.06.002 PMID:21763025

Casaló, L. V., Flavián, C., & Guinalíu, M. (2011). Understanding the intention to follow the advice obtained in an online travel community. *Computers in Human Behavior*, *27*(2), 622–633. doi:10.1016/j.chb.2010.04.013

Catasus, B. & Chaminade. (2007). Intellectual capital: Paradoxes and expansion. In *Intellectual capital revisited: paradoxes in the knowledge intensive organization*. Edvard Elgar.

Černiček, I., & Molnar, R. (1988). Razvoj profitnih potencijala preduzetničkih organizacija. In I. Černiček & R. Molnar (Eds.), Razvoj preduzetništva - šansa za progress (pp. 25-43). Zrenjanin, Serbia: Tehnički fakultet "Mihajlo Pupin".

Černiček, I., & Molnar, R. (2000). Razvoj organizacije. In I. Černiček & B. Marić (Eds.), *Reinženjering proizvodno-poslovnih procesa* (pp. 203-256). Belgrade: Fakultet za preduzetni menadžment Novi Sad, Univerziteta "Braća Karić".

CFO. (2013). CFO/REL Working Capital Scoreboard. *CFO Magazine*. Retrieved July 2, 2014, from CFO web site: http://www3.cfo.com/working-capital/2013-06

CFO. (2014). CFO/REL Working Capital Scoreboard. *CFO Magazine*. Retrieved October 2, 2014, from CFO web site: http://ww2.cfo.com/cash-flow/2014/07/running-in-place-working-capital/attachment/14jul_workingcap_charts/

Chamber of Commerce and Industry of Serbia. (2016). Retrieved August 21, 2016, from http://www.pks.rs/PrivredaSrbije.aspx?id=20

Chang, A., Chih, Y., Chew, E., & Pisarskia, A. (2013). Reconceptualising mega project success in Australian Defence: Recognising the importance of value co-creation. *International Journal of Project Management*, *31*(8), 1139–1153. doi:10.1016/j.ijproman.2012.12.005

Chao, C. N., Corus, C., & Li, T. (2012). Balancing traditional media and online advertising strategy. *International Journal of Business, Marketing, and Decision Sciences*, *5*(1), 12–24.

Charreton, R., & Bourdaire, J. M. (1985). *La décision économique*. Presses universitaires de France.

Chaston, I. (1990). *Managing for Marketing Excellence*. New York, NY: McGraw-Hill.

Chen, M. C., Cheng, S. J., & Hwang, Y. (2005). An empirical investigation of the relationship between intellectual capital and firms market value and financial performance. *Journal of Intellectual Capital*, *6*(2), 159–176. doi:10.1108/14691930510592771

Chesbrough, H. (2003). The Era of Open Innovation. *MIT. Sloan Management Review*, *44*(3), 35–41.

Chesbrough, H. (2006a). *Open innovation: the new imperative for creating and profiting from technology*. Boston: Harvard Business school Press.

Chesbrough, H. (2006b). *Open business models: how to thrive in the new innovation landscape*. Harvard Business School.

Chesbrough, H. (2011). *Open services innovation: Rethinking your business to grow and compete in a new era*. San Francisco: Wiley. doi:10.1007/978-88-470-1980-5

Chesbrough, H., & Appleyard, M. (2007). Open Innovation and Strategy. *California Management Review*, *50*(1), 57–76. doi:10.2307/41166416

Chesbrough, H., & Rosenbloom, R. (2002). The role of the business model in capturing value from innovation: Evidence from Xerox Corporations technology spin-off companies. *Industrial and Corporate Change*, *11*(3), 529–555. doi:10.1093/icc/11.3.529

Chesbrough, H., Vanhaverbeke, W., & West, J. (2006). *Open innovation, researching a new paradigm*. New York: Oxford University Press.

Chiaroni, D., Chiesa, V., Frattini, F., & Terruzzi, R. (2015). Implementing open innovation: A case study in the renewable energy industry. *International Journal of Technology Intelligence and Planning*, *10*(10), 195–221. doi:10.1504/IJTIP.2015.070847

Cho, D. W., Lee, Y. H., Hwa, A., & Hwang, M. K. (2012). A framework for measuring the performance of service supply chain management. *Computers & Industrial Engineering*, *62*(3), 801–818. doi:10.1016/j.cie.2011.11.014

Ciccone, A., & Papaioannou, E. (2006). *Red tape an delayed entry*. London: CEPR.

Clarkson, P., Li, Y., Richardson, G., & Vasivari, F. (2011). Does it really pay to be green? Determinants and consequences of proactive environmental strategies. *Journal of Accounting and Public Policy*, *30*(2), 122–144. doi:10.1016/j.jaccpubpol.2010.09.013

Classe, D., & Mundle, J. (1997). Geschäftsprozessgestatung im Rahmen des CIP-Prozasses bei Bosch. In S.-P. Riekhof (Ed.), *H.C* (pp. 211–224). Stuttgart: Beschleunigung von Geschäftsprozessen.

Clemmer, J. (1994). Process re-engineering and process improvement - not an either/or choice. *CMA Magazine*, 36-39.

Clifford, J. (2014). *Impact evaluation by social enterprises: measuring the un-measurable?* Retrieved April 20, 2016, from http://www.oecd.org/cfe/leed/10th-fplg-meeting.htm

Ćoćkalo, D., Đorđević, D., Sajfert, Z., & Bogetić, S. (2011). SMEs in the Republic of Serbia: The developing capacities. *Journal of Applied Engineering Science*, *9*(4), 449–456. doi:10.5937/jaes9-1131

Coe, D. T., Helpman, E., & Hoffmaister, W. A. (1995). *North-south R&D spillovers.* NBER Working Paper, No. 5048. Cambridge, MA: National Bureau of Economic Research.

Cohen, B., Smith, B., & Mitchell, R. (2008). Toward a sustainable conceptualization of dependent variables in entrepreneurship research. *Business Strategy and the Environment*, *17*(2), 107–119. doi:10.1002/bse.505

Cohen, W. M., & Klepper, S. (1992). The anatomy of R&D intensity distributions. *The American Economic Review*, *82*(4), 773–799.

Coldiretti. (2015). *Expo: Coldiretti, + 300% export cibo Made in Italy in Cina da inizio crisi.* Retrieved August 16th from Coldiretti web site: http://www.coldiretti.it/News/Pagine/384---9-Giugno-2015.aspx

Cooper, G. (1999). New product portfolio management: Practices and performance. *Journal of Product Innovation Management*, *16*(4), 333–351. doi:10.1016/S0737-6782(99)00005-3

Cooper, R. (1995). *When lean enterprises collide: competing through confrontation.* Harvard Business Press.

Cooper, R., & Markus, M. L. (1995). Human Reengineering. *Sloan Management Review*, *36*(4), 39. PMID:10139766

Corallo, A., Margherita, A., Scalvenzi, M., & Storelli, D. (2010). Building a Process-Based Organization: The Design Roadmap at Superjet International. *Knowledge and Process Management*, *17*(2), 49–61. doi:10.1002/kpm.340

COSME. (2016). *Europe's programme for small and medium-sized enterprises.* Retrieved July 26, 2016, from EU web site: http://ec.europa.eu/growth/smes/cosme/index_en.htm

Crawford, K. (2006). The project management maturity model. *Information Systems Management*, *23*(4), 50–58. doi:10.1201/1078.10580530/46352.23.4.20060901/95113.7

Crichton, D. (2009). *Coping with Climate Change: Risks and Opportunities for Insurers. Chapter 8: Small and medium enterprises.* Retrieved March 25, 2016, from http://www.cii.co.uk/media/4043823/ch8_smes.pdf

Crook Russell, T. (2011). Does Human Capital Matter? Strategic Human Resource Management. *Human Resource Management Journal*, *21*, 93–104.

Dahlmann, F., & Brammer, S. (2011). Exploring and Explaining Patterns of Adaptation and Selection in Corporate Environmental Strategy in the USA. *Organization Studies*, *32*(4), 527–553. doi:10.1177/0170840611403670

Damnjanović, A., & Milisavljević, V. (2009). *Ciljni marketing.*Paper presented at 12th Internatinal Conference Quality and Reliability Management(ICDQM '09). Čačak.

Danske Bank & Ernst & Young. (2009). *Working Capital Management: A Survey of Nordic Companies.* Copenhagen, Denmark: Danske Bank A/S & Ernst & Young.

Daszkiewicz, N. (2004). *Internacjonalizacja Małych i Średnich Przedsiębiorstw we Współczesnej Gospodarce.* Gdańsk: Scientific Publishing Group.

David, F. R. (2009). *Strategic Management: Concepts and Cases.* Upper Saddle River, NJ: FT Prentice Hall.

Dawe, R. L. (1996). Systems are people too. *Transportation & Distribution*, *37*(1), 86.

De Palma, R., & Dobes, V. (2010). An integrated approach towards sustainable entrepreneurship. Experience from the TEST project in transitional economies. *Journal of Cleaner Production*, *18*(18), 1807–1821. doi:10.1016/j.jclepro.2010.07.025

Dees, J. G. (2009). *Social ventures as learning laboratories.* Retrieved March 15, 2016, from https://centers.fuqua.duke.edu/case/wp-content/uploads/sites/7/2015/02/Article_Dees_SocialVenturesasLearningLaboratories_2009.pdf

Dencheva, S. (2014). Electronic business models, adopted by the hospitality industry in Bulgaria. *IIBA conference,Istanbul Conference proceedings*, (pp. 311-317).

DG Enterprise and Industry. (2014). *SBA Fact Sheet: Finland.* Retrieved August 22nd 2016 from EU web site: http://ec.europa.eu/DocsRoom/documents/16121/attachments/11/translations/en/renditions/native

DG GROW. (2015). *SBA Fact Sheet: Finland.* Retrieved August 22nd 2016 from EU web site: http://ec.europa.eu/DocsRoom/documents/16344/attachments/10/translations/en/renditions/native

Dickinson, G. (2001). Enterprise risk management: Its origins and conceptual foundation. *The Geneva Papers on Risk and Insurance. Issues and Practice*, *26*(3), 360–366. doi:10.1111/1468-0440.00121

Dierkes, M., Erner, C., Langer, T., & Norden, L. (2013). Business credit information sharing and default risk of private firms. *Journal of Banking & Finance*, *37*(8), 2867–2878. doi:10.1016/j.jbankfin.2013.03.018

Dixit, A. K., & Nalebuff, B. J. (1993). *Thinking Strategically: The Competitive Edge in Business, Politics, and Everyday Life* (pp. 59–79). New York, NY: W. W. Norton & Company.

Djankov, S., Glanser, T., McLiesh, C., Ramalho, R., & Schleifer, A. (2008). *The effect of corporate taxes on investment and entrepreneurship. NBER WP 13765.* Cambridge: NBER. doi:10.3386/w13756

Doove, S., Kwaak, T., & Span, T. (2015). *Survey on the access to finance of enterprises (SAFE).* European Commission. Retrieved May 27, 2016, from EC web site: http://europa.eu/DocsRoom/documents/14321/attachments/1/translations/en/.../native

Downes, J., & Goodman, J. E. (2003). *Finance and Investment Handbook.* Barron's Educational Series.

Drnevich, P. L., & Kriauciunas, A. P. (2011). Clarifying the conditions and limits of the contributions of ordinary and dynamic capabilities to relative firm performance. *Strategic Management Journal*, *32*(3), 254–279. doi:10.1002/smj.882

Drucker, P. F. (1991). *Inovacije i preduzetništvo: Praksa i principi.* Beograd, Serbia: Privredni pregled.

Drucker, P. F., & Maciaiello, J. A. (2008). *Management* (Revised ed.). HarperCollins e-books. Retrieved August 1st 2016, from Youth Portal web site: http://youth-portal.com/wp-content/uploads/2014/10/Peter-F-Drucker-Management-Rev-Ed.pdf

Drucker, P. F. (1954). *The practice of management: A study of the most important function in American society.* New York, NY: Harper & Brothers.

Đulić, K., & Kuzman, T. (2012). Protection of minority shareholders' rights: legal framework and implementation. *Quarterly Monitor of Economic Trends and Policies in Serbia, 30*, 86-95.

Duliniec, E. (2009). *Marketing międzynarodowy.* Warszawa: Polskie Wydawnictwo Ekonomiczne.

Dumay, J., & Cuganesan, S. (2011). Making sense of intellectual capital complexity: Measuring through narrative. *Journal of Human Resource Costing & Accounting*, *15*(1), 24–49. doi:10.1108/14013381111125305

Dunav. (2016). Retrieved August 15, 2016, from *Dunav Insurance Company* web site: http://www.dunav.com/

Dye, R., & Sibony, O. (2007). How to improve strategic planning. *The McKinsey Quarterly*, *3*, 40–49.

Dzida, L., Orzechowska, U., Piotrowska, J., Pokorska, A., Rozkrut, D., & Wegner, M. (2014). *Działalnosc innowacyjna przedsiebiorstw w latach 2011-2013.* Główny Urzad Statystyczny, Urzad Statystyczny w Szczecinie. Retrieved July 7, 2016 from Central Statistical Office of Poland web site: http://stat.gov.pl/obszary-tematyczne/nauka-i-technika-spoleczenstwo-informacyjne/nauka-i-technika/dzialalnosc-innowacyjna-przedsiebiorstw-w-latach-2011-2013,2,8.html

Eales, A. B. (1995). *Financial risk management.* McGraw-Hill.

Ebers, M. (2004). Information and advising requirements in the financial services sector: Principles and Peculiarities in EC Law. *Electronic Journal of Comparative Law, 8*(2), 1-16. Retrieved March 15, 2016, from http://www.ejcl.org

Eccles, R. (1991). Analyzing cognitive maps to help structure issues or problems. *European Journal of Operational Research, 159*, 673–686.

Edvinson, L. (2002). *Corporate Longitude: What you need to know to navigate the knowledge economy.* Financial Times Prentice Hall.

Eggers, F., Kraus, S., Hughes, M., Laraway, S., & Snycerski, S. (2013). Implications of customer and entrepreneurial orientations for SME growth. *Management Decision, 51*(3), 524–546. doi:10.1108/00251741311309643

Ejermo, O. (2009). Regional innovation measured by patent data – does quality matter? *Industry and Innovation, 16*(2), 141–165. doi:10.1080/13662710902764246

Elzinga, D.J., Horak, T., Lee, C-Y., & Bruner, C. (1995). Business process management: survey and methodology. *Engineering Management, IEEE Transactions on,42*(2), 119-128.

Emerson, J. (2003). The blended value proposition. *California Management Review, 45*(4), 35–51. doi:10.2307/41166187

Emerson, J., Wachowicz, J., & Chun, S. (2000). *Social return on investment: exploring aspects of value creation in the non-profit sector.* San Francisco: The Roberts Foundation.

Emery, F. E., & Trist, E. L. (1965). The causal texture of organizational environments. *Human Relations, 18*(1), 21–32. doi:10.1177/001872676501800103

Epstein, M. J., & McFarlan, F. W. (2011). Measuring the efficiency and effectiveness of a nonprofit's performance. *Strategic Finance, 93*(4), 27–34.

Erikson, T. (2002). Entrepreneurial capital: The emerging ventures most important asset and competitive advantage. *Journal of Business Venturing, 17*(3), 275–290. doi:10.1016/S0883-9026(00)00062-8

Ernst & Young. (2013). *Working capital management report.* Retrieved July 2, 2014, from Ernst & Young web site: http://www.ey.com/Publication/vwLUAssets/Working_capital_management_2013:_all_tied_up/$FILE/All_Tied_Up_2013_DE0430.pdf

Ernst & Young. (2015). *Working capital management report.* Retrieved May 27 2016, from Ernst & Young web site: http://www.ey.com/Publication/vwLUAssets/EY-all-tied-up-working-capital-management-2015/$FILE/EY-all-tied-up-working-capital-management-2015.pdf

Erzurumlu, S. (2010). Collaborative product development with competitors to stimulate downstream innovation. *International Journal of Innovation Management, 14*(4), 573–602. doi:10.1142/S1363919610002787

Eskerod, P., & Huemann, M. (2013). Sustainable development and project stakeholder management: What standards say. *International Journal of Managing Projects in Business, 6*(1), 36–50. doi:10.1108/17538371311291017

Esser, K., Hillebrand, W., Messner, D., & Meyer-Stamer, J. (2013). *Systemic competitiveness: new governance patterns for industrial development.* Routledge.

Eur-Lex. (2016). Retrieved July 28, 2016, from Eur-Lex, Access to European Union law web site: http://eur-lex.europa.eu/legal-content/EN/TXT/?uri=CELEX:52004DC0064

European Commission. (2004). Retrieved July 27, 2016, from EC web site: http://ec.europa.eu/regional_policy/en/projects/finland/e-commerce-for-smes

European Commission. (2007). *Mezzanine Finance: Final Report, 2007.* Retrieved May 29th 2016 from the University of Mannheim European Documentation Center web site: http://edz.bib.uni-mannheim.de/daten/edz-h/gdb/07/round-table_mezzanine_2007.pdf

European Commission. (2010). *Europe 2020. A European strategy for smart, sustainable and inclusive growth.* Retrieved August 16, 2016, from http://ec.europa.eu/eu2020/pdf/COMPLET%20EN%20BARROSO%20%20%20007%20-%20Europe%202020%20-%20EN%20version.pdf

European Commission. (2010). *EUROPE 2020: A strategy for smart, sustainable and inclusive growth.* Retrieved July 27, 2016 from http://publications.europa.eu/resource/cellar/6a915e39-0aab-491c-8881-147ec91fe88a.0004.02/DOC_1

European Commission. (2012). *Entrepreneurship and innovation programme.* EIP performance report. Retrieved August 1, 2016, from http://ec.europa.eu/cip/files/docs/eip_performance_report_january_2012_en.pdf

European Commission. (2014). *Proposed approach to social impact measurement in European Commission legislation and in practice relating to EuSEFs and the EuSI.* Brussels: GECES.

European Commission. (2014). *Research and innovation performance in the EU. Innovation union progress at country level.* Retrieved July 26, 2016, from http://ec.europa.eu/research/innovation-union/pdf/state-of-the-union/2014/iuc_progress_report_2014.pdf

European Commission. (2015a). *Cost of the Cumulative Effects of Compliance with EU Law for SMEs.* Executive summary. Publications Office of the European Union. Retrieved July 26, 2016, from EU web site: ec.europa.eu/DocsRoom/documents/16321/attachments/1/translations/en/renditions/native

European Commission. (2015a). *User guide to the SME Definition.* Retrieved May 27, 2016, from EC web site: http://ec.europa.eu/growth/smes/business-friendly-environment/sme-definition/index_en.htm

European Commission. (2015b). *2015 SBA Fact Sheet-Serbia.* Retrieved July 26, 2016, from EU web site: http://ec.europa.eu/growth/smes/business-friendly-environment/performance-review/

European Commission. (2015b). *Innovation Union Scoreboard.* Retrieved May 27, 2016, from EC web site: http://ec.europa.eu/growth/industry/innovation/facts-figures/scoreboards/files/ius-annex-h_en.pdf

European Commission. (2015c). *Serbia 2015 Report.* Retrieved July 26, 2016, from EU web site: http://ec.europa.eu/enlargement/pdf/key_documents/2015/20151110_report_serbia.pdf

European Commission. (2016). *Entrepreneurship and Small and medium-sized enterprises (SMEs). Business-friendly environment. What is an SME?* Retrieved 5th August, 2016 from EU web site: http://ec.europa.eu/growth/smes/business-friendly-environment/sme-definition/

European Commission. (2016). *The importance of the digital economy.* Retrieved July 27, 2016, from EC web site: https://ec.europa.eu/growth/sectors/digital-economy/importance_en

European Commission. (2016a). *Enterprise.* Retrieved July 26, 2016, from EU web site: https://europa.eu/european-union/topics/enterprise_en

European Commission. (2016b). *The European Company – Your business opportunity?* Retrieved July 26, 2016, from EU web site: http://ec.europa.eu/internal_market/company/societas-europaea/index_en.htm

European Commission. (2016c). *The European Commission on Serbia's progress in 2015.* Retrieved July 26, 2016, EU web site: http://ec.europa.eu/enlargement/pdf/key_documents/2015/20151110_report_serbia.pdf

European Funds and Slovakia. (2014). *The experience of the past for future development I.* NSRR of SR for programing period 2007-2013.

European Integration Office. (2014). *National Program for Adoption Acquis European Union.* Retrieved July 26, 2016, from European Integration Office web site: http://www.seio.gov.rs/upload/documents/nacionalna_dokumenta/npaa/npaa_2014_2018.pdf

EUROSTAT. (2016). *Business demography statistics - Enterprise Survival Rate.* Retrieved May 13th 2016, from Eurostat web site: http://ec.europa.eu/eurostat/statistics-explained/index.php/ Business_ demography _statistics

Eurostat. (2016). *Scambi internazionali di merci.* Retrieved August 16th from Eurostat web site: http://ec.europa.eu/eurostat/statistics-explained/index.php/International_trade_in_goods/it

Evans, D., & Leighton, S. (1989). Some empirical aspects of entrepreneurship. *The American Economic Review, 79*(3), 519–535.

Executive Agency for SME Encouragement in Bulgaria. (2011). *Analysis of the state and the factors for the development of SMEs in Bulgaria.* Sofia: Executive Agency for SME Encouragement in Bulgaria.

Farkas, A. (2005). Benchmark of Working Capital Performance. In *Proceedings-3rd International Conference on Management, Enterprise and Benchmarking (MEB 2005)* (pp. 65-78). Óbuda University, Keleti Faculty of Business and Management.

Farrell, M. A., & Oczkowski, E. (1997). An analysis of the MKTOR and MARKOR measures of market orientation: An Australian perspective. *Marketing Bulletin, 8*, 30–40.

Feldman, M. P. (2003). Entrepreneurship and American Research Universities: Evolution in Technology Transfer. In D. M. Hart (Ed.), *The Emergence of Entrepreneurship Policy: Governance, Start-Ups, and Growth in the U.S. Knowledge Economy* (pp. 91–96). Cambridge, UK: Cambridge University Press. doi:10.1017/CBO9780511610134.005

Feltes, P., & Karuppan, C. (1995). Reengineering: Getting down to the business of doing business. *Industrial Management (Des Plaines), 37*(4), 3–12.

Feurer, R., & Chaharbaghi, K. (1995). Strategy development: Past, present and future. *Management Decision, 33*(6), 11–21. doi:10.1108/00251749510087614

Figueres, C. (2016). *Sustaining momentum on climate action.* Insurance Europa. Retrieved May 15, 2016, from http://www.insuranceeurope.eu/annual-report-2015-2016

Financial Conduct Authority. (2015). *Handling of insurance claims for Small and Medium-sized Enterprises (SMEs), Thematic Review TR15/6.* Retrieved May 19, 2016. from https://www.fca.org.uk/static/documents/thematic-reviews/tr15-06.pdf

Financial Conduct Authority. (2016). *SMEs and insurance brokers.* Retrieved March 10, 2016, from: https://www.the-fca.org.uk/consumers/smes-insurance-brokers

Fine, G. L. (2009). *The SWOT Analysis: Using your Strength to overcome Weaknesses, Using Opportunities to overcome Threats.* Kick It. LLC. Retrieved February 15th 2013, from Lawrence Fine web site: http://lawrencefine.com/ downloads/SWOT%20-%20PDF.pdf

Firoozmand, S., Haxel, P., Jung, E., & Suominen, K. (2015). *State of SME Finance in the United States in 2015*. TradeUp Capital Fund and Nextrade Group LLC. Retrieved May 23, 2016 from TradeUp web site: http://www.tradeupfund.com/uploads/2/6/0/4/26048023/state_of_sme_finance_in_the_united_states_2015.pdf

Fitzsimmons, M. (2006). The problem of uncertainty in strategic planning. *Survival*, *48*(4), 131–146. doi:10.1080/00396330601062808

Fonfara, K. (2009). Istota i zakres procesu internacjonalizacji firmy. In K. Fonfara (Ed.), Zachowanie przedsiębiorstwa w procesie internacjonalizacji – podejście sieciowe (pp. 11-25). Warszawa: Polskie Wydawnictwo Ekonomiczne.

Fonfara, K. (2014). Marketing partnerski na rynku przedsiębiorstw (3rd ed.). Warszawa: Polskie Wydawnictwo Ekonomiczne.

Fonfara, K. (2012). *The Development of Business Networks in the Company Internationalization Process*. Poznań: Poznań University of Economics Press.

Fonseca, R., Lopez-Garcia, P., & Pissarides, C. (2001). Entrepreneurship, start-up costs and employment. *European Economic Review*, *45*(4-6), 692–705. doi:10.1016/S0014-2921(01)00131-3

Franco, M., Santos, M. D. F., Ramalho, I., & Nunes, C. (2014). An exploratory study of entrepreneurial marketing in SMEs. *Journal of Small Business and Enterprise Development*, *21*(2), 265–283. doi:10.1108/JSBED-10-2012-0112

Franken, A., Edwards, C., & Lambert, R. (2009). Executing strategic change: Understanding the critical management elements that lead to success. *California Management Review*, *51*(3), 49–73. doi:10.2307/41166493

Frantzen, D. (2000). R&D, human capital and international technology spillovers: A cross country analysis. *The Scandinavian Journal of Economics*, *102*(1), 57–75. doi:10.1111/1467-9442.00184

French, R., Rayner, Ch., Rees, G., & Rumbles, S. (2015). *Organizational Behaviour* (3rd ed.). New York, NY: John Wiley & Sons.

Friedl, P. (2008). *Model izbire metod evolucijskega načina spreminjanja organizacij: primer gradbene panoge* (Unpublished doctoral dissertation). Univerza na primorskem, Fakulteta za management, Koper, Slovenia.

Froehlich, E. A., Hawranek, P. M., Lettmayr, C. F., & Pichler, J. H. (1994). *Manual for small industrial businesses: Project design and appraisal*. UNIDO General Studies Series.

Frontier Economics. (2013). *Exploring the impact of private equity on economic growth in Europe*. London: Frontier Economics Ltd.

Frost, F. A. (2003). The use of strategic tools by small and medium-sized enterprises: An Australasian study. *Strategic Change*, *12*(1), 49–62. doi:10.1002/jsc.607

Fukuda-Parr, S., & Kumar, A. S. (2003). *Readings in Human Development: Concepts, measures and policies for a development paradigm*. New Delhi: Oxford University Press.

G8. (2014). *Measuring impact*. Retrieved February 20, 2016, from http://www.socialimpactinvestment.org/reports/Measuring%20Impact%20WG%20paper%20FINAL.pdf

Gassmann, O., Enkel, E., & Chesbrough, H. (2010). The future of open innovation. *R & D Management*, *40*(3), 213–221. doi:10.1111/j.1467-9310.2010.00605.x

Gates, L. P. (2010). *Strategic planning with critical success factors and future scenarios: An integrated strategic planning framework* (Technical Report No. CMU/SEI-2010-TR-037). Carnegie Mellon University. Retrieved September 1, 2016 from Software Engineering Institute web site: http://resources.sei.cmu.edu/library/asset-view.cfm?assetID=9673

Geels, F. (2004). From sectoral systems of innovation to socio-technical systems From sectoral systems of innovation to socio-technical systems From sectoral systems of innovation to socio-technical systems. *Research Policy*, *33*(6-7), 897–920. doi:10.1016/j.respol.2004.01.015

Geels, F., Hekkert, M., & Jacobsson, S. (2008). The dynamics of sustainable innovation journeys. *Technology Analysis and Strategic Management*, *20*(5), 521–536. doi:10.1080/09537320802292982

Gellynck, X., Banterle, A., Kühne, B., Carraresi, L., & Stranieri, S. (2010). Market orientation and marketing management of traditional food producers in the EU. *British Food Journal*, *114*(4), 481–499. doi:10.1108/00070701211219513

GEM Consortium. (2009). *Global entrepreneurship monitor 2009. Global report*. Retrieved May 10, 2016, from http://www.babson.edu/Academics/centers/blank-center/global-research/gem/Documents/gem-2009-global-report.pdf

GEM Consortium. (2016). *Global entrepreneurship monitor 2015/2016. Global report*. Retrieved August 4, 2016, from http://www.gemconsortium.org/report/49480

Geroski, P. A. (1989). Entry, innovation, and productivity growth. *The Review of Economics and Statistics*, *71*(4), 572–578. doi:10.2307/1928098

Geroski, P., Machin, S., & Van Reenen, J. (1993). The profitability of innovating firms. *The Rand Journal of Economics*, *24*(2), 198–211. doi:10.2307/2555757

Ghisetti, C., Mazzanti, M., Mancinelli, S., & Zoli, M. (2015). *Do financial constraints make the environment worse off? Understanding the effects of financial barriers on environmental innovations*. SEEDS Working Paper Series.

Giannopoulou, E., Yström, A., & Ollila, S. (2011). Turning open innovation into practice: Open innovation research through the lens of managers. *International Journal of Innovation Management*, *15*(3), 505–524. doi:10.1142/S1363919611003465

Gibbs, D. (2009). Sustainability entrepreneurs, ecopreneurs and the development of a sustainable economy. *Greener Management International*, *55*, 63–78.

Glaeser, E., & Kerr, W. (2009). Local industrial conditions and entrepreneurship: How much of the spatial distribution can we explain? *Journal of Economics & Management Strategy*, *18*(3), 623–663. doi:10.1111/j.1530-9134.2009.00225.x

Glaister, K. W., & Falshaw, J. R. (1999). Strategic planning: Still going strong? *Long Range Planning*, *32*(1), 107–116. doi:10.1016/S0024-6301(98)00131-9

González Montesinos, M. J. (2008). *El Análisis de Reactivos con el Modelo Rasch*. Manual Técnico A. Instituto Nacional para la Evaluación de la Educación, Universidad de Sonora.

Gorynia, M. (2000). Luka konkurencyjna w przedsiębiorstwach a przystąpienie Polski do Unii Europejskiej. *Gospodarka Narodowa*, *10*, 48–67.

Grant, R. M. (1991). The resource-based theory of competitive advantage: Implications for strategy formulation. *California Management Review*, *33*(3), 114–135. doi:10.2307/41166664

Grasenick, K., & Low, J. (2004). Shaken, not stirred: Defining and connecting indicators for the measurement and valuation of intangibles. *Journal of Intellectual Capital*, *5*(2), 268–281. doi:10.1108/14691930410533696

Grawe. (2016). Retrieved August 15, 2016, from Graw Insurance Group web site: http://www.grawe.rs/rs/index.htm

Green, A., & Ryan, J. C. H. (2005). A framework of intangible valuation areas (FIVA): Aligning business strategy and intangible assets. *Journal of Intellectual Capital*, *6*(1), 43–52. doi:10.1108/14691930510574654

Greenspan, A. (2008). *The age of turbulence: Adventures in a new world*. London, UK: Penguin Books.

Griffith, R., Redding, S., & Van Reenen, J. (2004). Mapping the two faces of R&D productivity growth in a panel of OECD industries. *The Review of Economics and Statistics*, *86*(4), 883–895. doi:10.1162/0034653043125194

Guevara, D., & Bounfour, A. (2013). Monitoring intellectual capital: A case study of a large company during the recent economic crisis. *Knowledge Management Research & Practice*, *11*(2), 196–207. doi:10.1057/kmrp.2013.12

Gulbahara, M., & Yilidirima, F. (2015). Marketing Efforts Related to Social Media Channels and Mobile Application Usage in Tourism: Case Study in Istanbul. *Procedia: Social and Behavioral Sciences*, *195*, 453–462. doi:10.1016/j.sbspro.2015.06.489

Guthrie, J. T., Klauda, S. L., & Morrison, D. A. (2012). Motivation, achievement and classroom contexts for information book reading. Retrieved from http//www.corilearning.com/researchpublications/

Hagen, J. (2009). Top 12 Reasons Why Businesses Fail. In E. Koester (Ed.), *What Every Engineer Should Know About Starting a High-Tech Business Venture*. Boca Raton, FL: Taylor & Francis Group.

Hall, J. K., Daneke, G. A., & Lenox, M. J. (2010). Sustainable development and entrepreneurship: Past contributions and future directions. *Journal of Business Venturing*, *25*(5), 439–448. doi:10.1016/j.jbusvent.2010.01.002

Halme, M., & Laurila, J. (2009). Philanthropy, integration or innovation? Exploring the financial and societal outcomes of different types of corporate responsibility. *Journal of Business Ethics*, *84*(3), 325–339. doi:10.1007/s10551-008-9712-5

Hamdouch, A., & Zuindeau, B. (2010). Sustainable development, 20 years on: Methodological innovations, practices and open issues. *Journal of Environmental Planning and Management*, *53*(4), 427–438. doi:10.1080/09640561003694286

Hammer, M. (2007). The Process Audit. *Harvard Business Review*, *85*(4), 111. PMID:17432158

Hammer, M., & Stanton, S. (1995). *The reengineering revolution*. New York: HarperCollins.

Hansson, L., Wrangmo, A., & Søilen, K. S. (2013). Optimal ways for companies to use Facebook as a marketing channel. *Journal of Information. Communication and Ethics in Society*, *11*(2), 112–126. doi:10.1108/JICES-12-2012-0024

Harlock, J. (2013). *Impact measurement practice in the UK third sector: a review of emerging evidence*. Birmingham, UK: University of Birmingham.

Harraf, A., Wanasika, I., Tate, K., & Talbott, K. (2015). Organizational agility. *Journal of Applied Business Research*, *31*(2), 675–686. doi:10.19030/jabr.v31i2.9160

Harrington, J., Voehl, F., & Voehl, C. (2014). *Model for sustainable change*. Project Management Institute.

Harris, J., & Deacon, J. (2011). Marketing in context – the marketing authenticity of owner/entrepreneurs of small firms: Case evidence from Welsh (UK) SME food and drink producers and retailers. *Small Enterprise Research*, *18*(1), 33–50. doi:10.5172/ser.18.1.33

Hart, O. (1995). *Firms, contracts, and financial structure*. Oxford, UK: Clarendon Press. doi:10.1093/0198288816.001.0001

Hassan,, S.B., Hamid, M.S.B., & Bohairy, H. A. (2010). Perception of destination branding measures: A case study of Alexandria destination marketing organizations. *International Journal of Euro-Mediterranean Studies*, *3*(2), 271–288.

Hautamäki, A. (2010). *Sustainable innovation, a new age of Innovation and Finland's innovation policy*. Helsinki: sitra.

Heising, W. (2012). The integration of ideation and project portfolio management-A key factor sustainable success. *International Journal of Project Management*, *30*(5), 582–595. doi:10.1016/j.ijproman.2012.01.014

Hennet, J.-C. (2008). SMEs in supply chain networks: cooperation and competition. In A. Villa (Ed.), *Road Map to the Development of European SME Networks Towards Collaborative Innovation* (pp. 86–96). Berlin: Springer-Verlag.

Hennet, J.-C., & Yesemin, A. (2008). Supply chain coordination: A game-theory approach. *Engineering Applications of Artificial Intelligence*, *21*(3), 399–405. doi:10.1016/j.engappai.2007.10.003

Herold, K., Tarkiainen, A., & Sundqvist, S. (2016). How the source of word-of-mouth influences information processing in the formation of brand attitudes. *Journal of Marketing for Higher Education*, *26*(1), 64–85. doi:10.1080/08841241.2016.1146387

Hertz, S., Johansson, J. K., & de Jager, F. (2001). Customer-oriented cost cutting: Process management at Volvo. *Supply Chain Management*, *6*(3/4), 128–142. doi:10.1108/13598540110399174

Higgs, M., & Rowland, D. (2005). All changes great and small: Exploring approaches to change and its leadership. *Journal of Change Management*, *5*(2), 121–151. doi:10.1080/14697010500082902

Highfield, R., & Smiley, R. (1987). New business start-ups and economic activity: An empirical investigation. *International Journal of Industrial Organization*, *5*, 51–66. doi:10.1016/0167-7187(87)90006-3

Hillary, R. (2000). *Small and Medium Sized Enterprises and the Environment*: *Business Imperatives*. Greenleaf Publishing. Retrieved May 12th, 2016 from web site: http://reports.weforum.org/global-competitiveness-report-2015-2016/

Hill, F. M., & Collins, L. K. (1998). The positioning of BPR and TQM in long-term organisational change strategies. *TQM Journal*, *10*(6), 438–446. doi:10.1108/09544789810239182

Hinterhuber, H. (1995). Business process management: The European Approach. *Business Change and Re-engineering*, *2*(4), 63–73.

Hirzel, M. (2008). Erfolgsfaktor Prozessmanagement. In M. Hirzel, F. Kühn, & I. Gaida (Eds.), *Prozessmanagement in der Praxis* (pp. 11–22). Gabler.

Hisrich, R. D., & Peters, M. P. (2002). *Entrepreneurship*. New York, NY: McGraw-Hill/Irwin.

Hoffman, D., & Fodor, M. (2010). Can you measure the ROI of your social media marketing. *MIT Sloan Management Review*, *1*(52), 323–343.

Holland, D., & Kumar, S. (1995). Getting past the obstacles to successful reengineering. *Business Horizons*, *38*(3), 79–85. doi:10.1007/978-3-8349-9436-3_1

Hollanders, H., Es-Sadki, N., & Kanerva, M. (2015). *Innovation Union Scoreboard 2015* (pp. 7–25). Brussels: European Commission, Directorate-General for Internal Market, Industry, Entrepreneurship and SMEs.

Holmes, S., & Smart, P. (2009). Exploring open innovation practice in firm-nonprofit engagements: A corporate social responsibility perspective. *R & D Management*, *39*(4), 394–409. doi:10.1111/j.1467-9310.2009.00569.x

Hooley, G., Cox, T., Fahy, J., Shipley, D., Beracs, J., Fonfara, K., & Snoj, B. (2000). Market Orientation in the Transition Economies of Central Europe. *Journal of Business Research*, *50*(3), 273–285. doi:10.1016/S0148-2963(99)00105-8

Hornstein, H. (2015). The integration of project management and organizational change management is now a necessity. *International Journal of Project Management*, *33*(2), 291–298. doi:10.1016/j.ijproman.2014.08.005

Houy, C., Fettke, P., & Loos, P. (2010). Empirical research in business process management. Analysis of an emerging field. *Business Process Management Journal*, *16*(4), 619–661. doi:10.1108/14637151011065946

Howells, J. (1995). A socio-cognitive approach to innovation. *Research Policy*, *24*(6), 883–894. doi:10.1016/0048-7333(94)00804-3

Hoying, T., Agarwal, R., Rapsch, S., & Yasushi, S. (2014). *Mining the Untapped Gold in SME Commercial Insurance.* Retrieved March 10, 2016, from https://www.bcgperspectives.com/content/articles/insurance_growth_mining_untapped_gold_sme_commercial_insurance/

Huang, C. J., & Liu, C. J. (2005). Exploration for the relationship between innovation, IT and performance. *Journal of Intellectual Capital*, *6*(2), 237–252. doi:10.1108/14691930510592825

Hubbard, R. G. (2008). *Nondestructive creation. Entrepreneurship and management research in the study of growth.* Retrieved August 11, 2016, from http://www.sciencedirect.com/science/article/pii/S0161893808000355

Hubbard, R., & Armstrong, J. S. (1994). Replications and extensions in marketing: Rarely published but quite contrary. *International Journal of Research in Marketing*, *11*(3), 233–248. doi:10.1016/0167-8116(94)90003-5

Hughes, M., Hughes, P., & Morgan, R. (2007). Exploitative Learning and Entrepreneurial Orientation Alignment in Emerging Young Firms: Implications for Market and Response Performance. *British Journal of Management*, *18*(4), 359–375. doi:10.1111/j.1467-8551.2007.00519.x

Hung, R. Y. Y. (2006). Business Process Management as competitive advantage: A review and empirical study. *Total Quality Management & Business Excellence*, *17*(1), 21–40. doi:10.1080/14783360500249836

Hu, X., Li, L., Xie, C., & Zhou, J. (2008). The effects of country-of-origin on Chinese consumers wine purchasing behavior. *Journal of Technology Management in China*, *3*(3), 292–306. doi:10.1108/17468770810916195

IBM Global CEO Study Team. (2006). *Expanding the Innovation Horizon: The Global CEO Study 2006.* IBM Global Services.

Ilkić, Z. (2015). Osiguranje mašina od loma, upravljanje rizicima i preventiva u Srbiji. *Evropska revija za pravo osiguranja*, *14*(2), 45-51.

InCaS. (2008). *Intellectual Capital Statement for Europe. A new approach to make SMEs successful.* Retrieved August 16th, 2016, from InCaS web site: http://www.psych.lse.ac.uk/incas/page99/page99.html

INE. (2016). *Instituto Nacional de Estadística (Spanish Statistical Institute).* Retrieved 5th August, 2016 from INE web site: http://www.ine.es/

Insurance Council of Australia. (2008). *Non-Insurance in the Small to Medium sized Enterprise sector, December 2008.* Retrieved March 10, 2016, from http://www.insurancecouncil.com.au/assets/report/non-insurance%20in%20the%20small%20to%20medium%20sized%20enterprise%20sector.pdf

International Finance Corporation. (2008). *Corporate Governance – Manual* (updated edition). Retrieved July 26, 2016, from International Finance Corporation web site: http://www.ifc.org/wps/wcm/connect/topics_ext_content/ifc_external_corporate_site/corporate+governance/publications/toolkits+and+manuals/ifc+corporate+governance+manual

Ittner, C. D., & Larcker, D. F. (1997). The performance effects of process management techniques. *Management Science*, *43*(4), 522–534. doi:10.1287/mnsc.43.4.522

Ivančević, K. (2006). Upravljanje rizikom u osiguranju. In Krizni menadžment: Prevencija krize, (hrestomatija), (pp. 413-438). Beograd: Univerzitet u Beogradu, Fakultet bezbednosti.

Ivančević, K. (2010). *Pravna zaštita potrošača korisnika usluge osiguranja i bankarske usluge* (Unpublished doctoral dissertacion). Univerzitet u Beogradu, Pravni fakultet.

Ivančević, K. (2014). Posredovanje u osiguranju. In Uslužni poslovi (pp. 143-166). Pravni fakultet Univerziteta u Kragujevcu.

Ivančević, K. (2015). Finansijsko obrazovanje potrošača i delatnost osiguranja. In Reforme i novi izazovi u pravu osiguranja (pp. 230-252). Udruženje za pravo osiguranja Srbije, Palić.

Ivanjko, Š. (2005). Osiguranje pravne zaštite. In Aktuelnosti građanskog i trgovačkog zakonodavstva i pravne prakse (pp. 39-49). Neum: BiH, Pravni fakultet u Mostaru.

Jackson, K. M., & Trochim, W. M. (2002). Concept mapping as an alternative approach for the analysis of open-ended survey responses. *Organizational Research Methods*, *5*(4), 307–336. doi:10.1177/109442802237114

Jackson, M. O. (2010). *Social and Economic Networks*. Princeton, NJ: Princeton University Press.

Jackson, N. (1997). Business Process Re-engineering '96. *Management Services*, *41*(2), 34.

Jeanmart, N. (2016). *Cross-border insurance, Insurance Europa, Annual Report 2015-2016*. Retrieved June 10, 2016, from http://www.insuranceeurope.eu/annual-report-2015-2016

Jeremić, Lj., & Živić, M. (2011). Uloga malih i srednjih preduzeća u ekonomskom razvoju i zaposlenosti u Evropskoj uniji. *Ekonomski vidici*, *91(1)*, 77–85.

Jevremović-Petrović, T. (2005). European business units. In Guide through the EU law (pp. 9-11). Belgrade: Institute of International Politics and Economics.

Jevremović-Petrović, T. (2011). Creditor protection in company law. *Annals of the Faculty of Law*, *59*(1), 223–254.

Jevtić, B., Damnjanović, A., Matić, R., & Grozdanić, R. (2011). Izazovi industrijske politike u IKT sektoru. *Ekonomski vidici*,*16*(3), 572-588.

Jia, G., Chen, Y., Xue, X., Chen, J., Cao, J., & Tang, K. (2011). Program management organization maturity integrated model for mega construction programs in China. *International Journal of Project Management*, *29*(7), 834–845. doi:10.1016/j.ijproman.2011.03.003

Jimenez, J. D., & Sanz-Valle, R. (2011). Innovation, organizational learning and performance. *Journal of Business Research*, *64*(4), 408–417. doi:10.1016/j.jbusres.2010.09.010

Jokić, D. (2010). *Preduzetništvo. Užice: Naučno istraživački centar (NIC)*. Užice.

Jones, R., Alford, P., & Wolfenden, S. (2015). *Entrepreneurial Marketing in the Digital Age: A study of the SME tourism industry*. Retrieved August 7, 2016 from Bournemouth University web site: http://eprints.bournemouth.ac.uk/22932/1/Entrepreneurial%20Marketing%20in%20the%20Digital%20Age%20A%

Jüttner, U., Christopher, M., & Godsell, J. (2010). A strategic framework for integrating marketing and supply chain strategies. *The International Journal of Logistics Management*, *21*(1), 104–126. doi:10.1108/09574091011042205

Kaczmarek, A. (2009). Strategie działania małych i średnich przedsiębiorstw. *Zeszyty Naukowe Małopolskiej Wyższej Szkoły Ekonomicznej w Tarnowie*, *2*(13), 260-271.

Kannebley, S., Sekkel, J. V., & Araujo, B. C. (2010). Economic performance of Brazilian manufacturing firms: A counterfactual analysis of innovation impacts. *Small Business Economics*, *34*(3), 339–354. doi:10.1007/s11187-008-9118-x

Kanter, R. M. (2006). Innovation: The classic traps. *Harvard Business Review*, *84*(11), 72–83. PMID:17131564

Kanter, R., Stein, B., & Jick, T. (1992). *The challenge of organizational change*. New York: The Free Press.

Kantorovich, L. V. (1959). *Ekonomicheskii raschet nailuchshego ispolzovania resursov* [The Best Use of Economic Resources]. Acad. of SC., USSR, Centar za kulturnu djelatnost Zagreb.

Kaplan, R. S., & Norton, D. P. (2004). *Strategy maps: Converting intangible assets into tangible outcomes*. Harvard Business Press.

Kariuki, A. K., Obonyo, P., & Ogutu, M. (2015). Intellectual Capital and Performance of Firms Listed on Nairobi Securities Exchange. *1st DBA-Africa Management Review International Conference* (pp. 118-142).

Katzya, B., Turgut, E., Holzmann, T., & Sailer, K. (2013). Innovation intermediaries: A process view on open innovation coordination. *Technology Analysis and Strategic Management*, *25*(3), 295–309. doi:10.1080/09537325.2013.764982

Kaur, G., Sharma, R. D., & Seli, N. (2013). A Balanced Approach towards Market Orientation. *Vikalpa*, *38*(3), 51–65.

Kavoura, A., & Stavrianea, A. (2011). Following and belonging to an online travel community in social media, its shared characteristics and gender differences. *Procedia: Social and Behavioral Sciences*, *175*, 515–521. doi:10.1016/j.sbspro.2015.01.1231

Keilbach, M., & Audretsch, D. (2005). *The knowledge spillover theory of entrepreneurship*. Retrieved August 10, 2016, from http://web.mit.edu/iandeseminar/Papers/Fall2005/audretschkeilbach.pdf

Kerzner, A., & Belack, C. (2010). *Managing complex projects*. John Wiley. doi:10.1002/9780470927977

Kianto, A. (2007). What do we really mean by dynamic intellectual capital*? International Journal of Learning and Intellectual Capital*, *4*(4), 342–356. doi:10.1504/IJLIC.2007.016332

Kim, Ch. (2006). *E-tourism: An innovative approach for the small and medium sized tourism enterprises (SMTES) in Korea*. Retrieved August 16, 2016 form OECD web-site: http://www.oecd.org/industry/tourism/34268048.pdf

Kiralova, A., & Pavliceka, A. (2015). Development of social media strategies in tourism destination. *Procedia: Social and Behavioral Sciences*, *175*, 358–366. doi:10.1016/j.sbspro.2015.01.1211

Kirchmer, M. (2009). *High Performance through Process Excellence*. Berlin: Springer.

Klapper, L. F., Sarria-Allende, V., & Sulla, V. (2002). *Small and medium-size enterprise financing in Eastern Europe* (Working Paper No.2933). Retrieved June 15, 2013, from World Bank website: http://info.worldbank.org/etools/docs/library/28751/SME%20financing%20Eastern%20Europe.pdf

Kleindl, B. (1999). A game theoretic perspective on market-oriented versus innovative strategic choice. *Journal of Strategic Marketing*, *7*(4), 265–274. doi:10.1080/096525499346378

Kočović, J., & Šulejić, P. (2002). *Osiguranje*. Beograd, Srbija: Ekonomski fakultet Beograd.

Kohlbacher, M. (2010). The effects of process orientation: A literature review. *Business Process Management Journal*, *16*(1), 135–152. doi:10.1108/14637151011017985

Kohli, A. K., & Jaworski, B. J. (1990). Market Orientation: The Construct, Research Propositions, and Managerial Implications. *Journal of Marketing*, *54*(April), 1–18. doi:10.2307/1251866

Kohli, A. K., Jaworski, B. J., & Kumar, A. (1993). MARKOR: A Measure of Market Orientation. *JMR, Journal of Marketing Research*, *30*(November), 467–477. doi:10.2307/3172691

Korderl, Z. (2015). Strategie sektora MSP w Polsce. *Logistyka*, *3*, 98–103.

Kornai, J. (1971). *Anti-equilibrium: On economic systems theory and the tasks of research*. Amsterdam: North-Holland.

Koschate-Fischer, N., Diamantopoulos, A., & Oldenkotte, K. (2012). Are Consumers Really Willing to Pay More for a Favorable Country Image? A Study of Country-of-Origin Effects on Willingness to Pay. *Journal of International Marketing, American Marketing Association*, *20*(1), 19–41.

Kotler, P. (1988). *Marketing Management: Analysis, Planning, and Control.* Zagreb: Informator.

Kotter, J. (1996). *Leading change.* Boston: Harvard Business School Press.

Kotter, J. (2012). *Leading Change.* Boston, MA: Harvard Business Review Press.

Kováč, M., Malega, P., & Spišáková, E. (2013). *SMEs.* Košice, Slovakia: TU, SjF.

Kozioł-Nadolna, K. (2013). *Internacjonalizacja Działalności Badawczo-rozwojowej w Kształtowaniu Procesów Innowacyjnych Przedsiębiorstw w Polsce.* Warszawa: CeDeWu.

KPMG. (2005). *Working Capital Management Survey: How European companies manage their working capital.* Zurich: KPMG Europe LLP & Mitglied des KPMG – Netzwerks unabhängiger Mitgliedsfirmen, die KPMG International Cooperative.

KPMG. (2009). *Singapore Cash Management Survey.* Singapore: KPMG Advisory Services Pte. Ltd.

KPMG. (2012). *Annual Working Capital Survey.* Retrieved July 2, 2014, from KPMG web site: http://www.kpmg.com/DE/de/Documents/Annual-working-capital-survey-2012-KPMG.pdf

Kraus, S., Rigtering, J. C., Hughes, M., & Hosman, V. (2012). Entrepreneurial orientation and the business performance of SMEs: A quantitative study from the Netherlands. *Review of Managerial Science, 6*(2), 161–182. doi:10.1007/s11846-011-0062-9

Krejcie, R., & Morgan, D. (1970). Determining sample size for research activities. *Educational and Psychological Measurement.*

Kremp, E., & Mairesse, J. (2004). *Knowledge Management, Innovation and Productivity: A Firm Level Exploration Based on French Manufacturing CIS3 Data.* NBER Working Paper No. 10237. Retrieved from http://www.nber.org/papers/w10237.pdf

Kristandl, G., & Bontis, N. (2007). Constructing a definition for intangibles using resourced based view of the firm. *Management Decision, 45*(9), 1510–1524. doi:10.1108/00251740710828744

Kritikos, A. (2010). *Entrepreneurs and their impact on jobs and economic growth.* Retrieved August 1, 2016, from http://wol.iza.org/articles/entrepreneurs-and-their-impact-on-jobs-and-economic-growth.pdf

Kukoleča, S. (1980). *Osnovi teorije organizacionih sistema. In Privredno finansijski vodič.* Oeconomica.

Kunc, M. & Bhandari R. (2011). Strategic development processes during economic and financial crisis. *Management Decision, 49*(8), 1343 – 1353.

Küng, P., & Hagen, C. (2007). The fruits of Business Process Management: An experience report from a Swiss bank. *Business Process Management Journal, 13*(4), 477–487. doi:10.1108/14637150710763522

Lange, O. (1958). *Wstęp do ekonometrii.* Panstwowe Wydawnictwo Naukowe.

Langstrand, J. (2016). The missing link in systems thinking: The impact of infrastructure on organizational change. *International Journal of Quality and Service Sciences, 8*(2), 197–208. doi:10.1108/IJQSS-09-2015-0061

Lasagni, A. (2012). How can external relationships enhance innovation in SMEs? New evidence for Europe. *Journal of Small Business Management, 50*(2), 310–339. doi:10.1111/j.1540-627X.2012.00355.x

Laville, J. (2010). *The social enterprise: towards a theoretical socio-economic approach.* Retrieved January 12, 2016, from http://www.ssc.wisc.edu/~wright/Social%20Economy%20PDFs/THE%20EMERGENCE%20OF%20SOCIAL%20ENTERPRISE/TEOSE-17.pdf

Lee, R. G., & Dale, B. G. (1998). Business process management: A review and evaluation. *Business Process Management Journal*, *4*(3), 214–225. doi:10.1108/14637159810224322

Lee, S. S.-Y. (2008). Drivers for the participation of small and medium-sized suppliers in green supply chain initiatives. *Supply Chain Management: An International Journal*, *13*(3), 185–198. doi:10.1108/13598540810871235

Lee, S.-Y., & Klassen, R. D. (2008). Drivers and Enablers That Foster Environmental Management Capabilities in Small- and Medium-Sized Suppliers in Supply Chains. *Production and Operations Management*, *17*(6), 573–586. doi:10.3401/poms.1080.0063

Lev, B. (2001). *Intangibles: Management, measurement, and reporting*. Washington, DC: Brookings Institution Press.

Leviner, N., Churchfield, L., & Wells, D. (2012). *Understanding the impact of social entrepreneurs: Ashoka's answer to the challenge of measuring effectiveness*. Retrieved January 16, 2016, from http://www.nj.gov/state/programs/pdf/faith-based-social-understanding-impact.pdf

Lewin, K. Z. (1947). Frontiers in Group Dynamics: Concept, Method and Reality in Social Science; Social Equilibria and Social Change. *Human Relations*, *1*(1), 36. doi:10.1177/001872674700100103

Leyton-Brown, K., & Shoham, Y. (2008). *Essentials of Game Theory: A Concise, Multidisciplinary Introduction*. New York, NY: Morgan Claypool Publishers.

Lichtenthaler, U. (2008). Open Innovation in Practice: An analysis of strategic approaches to technology transactions. *IEEE Transactions on Engineering Management*, *55*(1), 148–157. doi:10.1109/TEM.2007.912932

Lichtenthaler, U. (2011). Open innovation: Past research, current debates, and future directions. *The Academy of Management Perspectives*, *25*(1), 75–93. doi:10.5465/AMP.2011.59198451

Linacre, J. M. (2012). Winsteps® Rasch measurement computer program User's Guide. Beaverton, OR: Winsteps.com.

Linacre, J. M. (2016). Winsteps® Rasch measurement computer program. Beaverton, OR: Winsteps.com.

Liu, Zh., & Park, S. (2015). What makes a useful online review? Implication for travel product websites. *Tourism Management*, *47*, 140–151. doi:10.1016/j.tourman.2014.09.020

Loefsten, H. (2014). Product innovation processes and the trade-off between product innovation performance and business performance. *European Journal of Innovation Management*, *17*(1), 61–84. doi:10.1108/EJIM-04-2013-0034

Luecke, R. (2003). *Managing change and transition*. Harvard Business School Press.

Lueckgen, I. (2004). *Nascent entrepreneurs in German regions. Evidence from the Regional Entrepreneurship Monitor Germany*. Retrieved August 2, 2016, from http://repec.iza.org/dp1394.pdf

Luo,Y. (2009). Analysis of Culture and Buyer Behavior in Chinese Market. *Asian Culture and History*, *1*(1), 25-30.

Mahoney, J. T., & Pandian, J. R. (1992). The resource-based view within the conversation of strategic management. *Strategic Management Journal*, *13*(5), 363–380. doi:10.1002/smj.4250130505

Májková-Sobeková, M. (2011). Analyse of barriers and factors of SMEs funding in Slovak republic. *Economic Magazine*, (10), 1033-1049.

Malega, P. (2010). External business environment searching as the important tool for decision-making. *Managerial Forum*, *8*(2), 79-81.

Malega, P. (2012). Competitiveness of Slovak enterprises and possibilities of the comparison. *Managerial Forum*, *12*(1), 17-22.

Malega, P. (2014). Entry strategy into foreign markets of SMEs. *Management of Companies.*, *4*(2), 89–93.

Manchanda, P., Packard, G., & Pattabhiramaiah, A. (2014). The Economic Impact of Customer Participation in a Firm-Sponsored Online Customer Community. *Marketing Science*, *34*(3), 367–387. doi:10.1287/mksc.2014.0890

Mankins, M. C., & Steele, R. (2005). Turning great strategy into great performance. *Harvard Business Review*, *83*(7/8), 64–72. PMID:16028817

Mann, R., Adebanjo, D., & Tickle, M. (2010). Deployment of business excellence in Asia: An exploratory study. *International Journal of Quality & Reliability Management*, *28*(6), 604–627. doi:10.1108/02656711111141184

Marcelino-Sadab, S., González-Jaen, L., & Pérez-Ezcurdia, A. (2015). Using project management as a way to sustainability. From a comprehensive review to a framework definition. *Journal of Cleaner Production*, *99*, 1–16. doi:10.1016/j.jclepro.2015.03.020

Mariani, M., Felice, M., & Mura, M. (2015). Facebook as a destination marketing tool: Evidence from Italian regional Destination Management Organizations. *Tourism Management*, *54*, 321–343. doi:10.1016/j.tourman.2015.12.008

Markowitz, H. (1952). Portfolio selection. *The Journal of Finance*, *7*(1), 77–91.

Martínez-Torres, M. R. (2014). Identification of intangible assets in knowledge based organizations using concept mapping techniques. *R & D Management*, *44*(1), 42–52. doi:10.1111/radm.12037

Martinuzzi, A., & Krumay, B. (2013). The good, the bad, and the successful how corporate social responsibility leads to competitive advantage and organizational transformation. *Journal of Change Management*, *13*(4), 24–443. doi:10.1080/14697017.2013.851953

Mašić, B. (2010). *Menadžment*. Beograd: Univerzitet Singidunum.

Mc Bourda, F. (2013). *Change Management Theories and Methodologies* (White paper). Retrieved May 15th 2016, from Tata Consultancy Services web site: http://www.tcs.com/ SiteCollectionDocuments/White%20Papers/EntSol-Whitepaper-Change-Management-Theories-Methodologies-0213-1.pdf

McAdam, R., & McCormack, D. (2001). Integrating business processes for global alignment and supply chain management. *Business Process Management Journal*, *7*(2), 113–130. doi:10.1108/14637150110389696

McGrath, R. G., & MacMillan, I. C. (2000). *The entrepreneurial mindset: Strategies for continuously creating opportunity in an age of uncertainty* (Vol. 284). Harvard Business Press.

Melan, E. H. (1989). Process Management: A Unifying Framework For Improvement. *National Productivity Review*, *8*(4), 395–406. doi:10.1002/npr.4040080407

Menger, C. (1996). *Investigations into the Method of the Social Sciences*. Ludwig von Mises Institute.

Meredith, G. G. (1987). Management services for small enterprises. In Small Enterprise Development: Policies and programmes (pp. 125-139). Geneva: International Labor Organization.

Micco, A., & Pages, C. (2006). *The economic effects of employment protection: evidence from international industry-level data.* Discussion paper 2633. Bonn: IZA.

Michaelis, M., Woisetschlager, D. M., Backhaus, C., & Ahlert, D. (2008). The effects of country of origin and corporate reputation on initial trust. *International Marketing Review*, *25*(4), 404–422. doi:10.1108/02651330810887468

Middel, R., Weegh, S. O. D., & Gieskes, J. (2007). Continuous improvement in The Netherlands: A survey-based study into current practices. *International Journal of Technology Management*, *37*(3-4), 259–271. doi:10.1504/IJTM.2007.012262

Miles, M. B., & Huberman, A. M. (1994). *Qualitative data analysis: An expanded sourcebook*. SAGE Publications.

Milosavljević, M., Veinović, M., & Grubor, G. (2013). *Informatika*. Beograd: Univerzitet Singidunum.

Milošević, D. (2014). *Strategija rasta i razvoja preduzeća*. Beograd: Visoka škola za poslovnu ekonomiju i preduzetništvo.

Milosevic, D., & Patanakul, P. (2005). Standardized project management may increase development projects success. *International Journal of Project Management*, *23*(3), 181–192. doi:10.1016/j.ijproman.2004.11.002

Ministry of Economy and Regional Development, Republic Development Bureau, Republic Agency for Development of SME and Entrepreneurship. (2009). *Report on small and medium-sized enterprises and entrepreneurship 2008*. Belgrade: Ministry of Economy and Regional Development Republic Development Bureau Republic Agency for Development of SME and Entrepreneurship.

Ministry of Economy of Poland. (2015, September). *Entrepreneurship in Poland*. Retrieved July 26th, 2016, from Ministry of Economy of Poland website: https://www.mr.gov.pl/media/15459/Entrepreneurship_in_Poland_2015.pdf

Mintzberg, H., Ahlstrand, B., & Lampel, J. B. (2009). *Strategy safari: Your complete guide through the wilds of strategic management*. Edinburg Gate, UK: Pearson Education Limited.

Mintzberg, H., & Waters, J. A. (1985). Of strategies, deliberate and emergent. *Strategic Management Journal*, *6*(3), 257–272. doi:10.1002/smj.4250060306

Mir, A., Mir, R., & Mosca, J. B. (2002). The new age employee: An exploration of changing employee-organization relations. *Public Personnel Management*, *31*(2), 187–200. doi:10.1177/009102600203100205

Mir, F., & Pinnington, A. (2014). Exploring the value of project management: Linking project management performance and project success. *International Journal of Project Management*, *25*(2), 202–217. doi:10.1016/j.ijproman.2013.05.012

Mittermaier, G., & Braun, M. (2004). Geschäftsprozessmanagement bei Infineon. In H. Ellringmann & H. J. Schmelzer (Eds.), *Geschäftsprozessmanagement inside* (pp. 1–40). Munich: Hanser.

Modigliani, F., & Miller, M. H. (1958). The cost of capital, corporation finance, and the theory of investment. *The American Economic Review*, *48*(3), 261–297.

Mohnen, A., & Nasev, J. (2009). *Growth of Small and Medium Sized firms in Germany. In Betriebswirtschaftliche Forschung und Praxis [Business Research and Practice]* (Vol. 5). Academic Press.

Molnar, R. (2010). *Razvoj sistemske nefinansijske podrške malim i srednjim preduzećima i preduzetništvu u Srbiji* (Unpublished doctoral dissertation). Available from Library at the University of Novi Sad. Serbia.

Molodchik, M., Shakina, E., & Bykova, A. (2012). Intellectual capital transformation evaluating model. *Journal of Intellectual Capital*, *13*(4), 444–461. doi:10.1108/14691931211276089

Monostori, L., Valckenaers, P., Dolgui, A., Panetto, H., Brdys, M., & Csanád Csáj, B. (2015). Cooperative Control in Production and Logistics. Report prepared by the IFAC Coordinating Committee on Manufacturing and Logistics Systems.

Montemari, M., & Nielsen, C. (2012). Value creation maps. In Business Models: Networking, Innovating and Globalizing. BookBoon.com.

Moon, H., & Ruona, W. (2015). Towards a deeper understanding of strategic learning. *Leadership and Organization Development Journal*, *36*(6), 657–674. doi:10.1108/LODJ-10-2013-0139

Mortensen, P. S., & Bloch, C. W. (2005). *Oslo Manual-Guidelines for collecting and interpreting innovation data*. Organisation for Economic Cooporation and Development, OECD.

Mouritsen, J. (2004). Measuring and intervening: How do we theorize intellectual capital management? *Journal of Intellectual Capital*, *5*(2), 257–267. doi:10.1108/14691930410533687

Mouritsen, J. (2009). Classification, measurement and the ontology of intellectual capital entities. *Journal of Human Resource Costing & Accounting*, *13*(2), 154–162. doi:10.1108/14013380910968665

Mrkšić, D., Petrović, Z., & Ivančević, K. (2014). *Pravo osiguranja*. Beograd, Srbija: Pravni fakultet Univerziteta Union.

Mrkšić, D., Petrović, Z., & Ivančević, K. (2007). *Neživotna osiguranja*. Beograd: Intermex.

Mulgan, G. (2010). *Measuring social value*. Retrieved May 12, 2016, from http://www.ssireview.org/articles/entry/measuring_social_value

Muller, P., Caliandro, C., Peycheva, V., Gagliardi, D., Marzocchi, C., Ramlogan, R., & Cox, D. (2015, November). *SME Performance Review 2014/2015 (pp. 4-12)*. Retrieved July 26th 2016, from European Commission, Directorate-General for Internal Market, Industry, Entrepreneurship and SMEs website: http://ec.europa.eu/growth/smes/business-friendly-environment/performance-review/index_en.htm

Müller-Stewens, G., & Lechner, C. (2005). *Strategisches Management: Wie strategische Initiativen zum Wandel führen, 3., aktualisierte Auflage*. Stuttgart: Schäffer-Poeschel.

Munar, M., & Jacobsen, J. (2014). Motivations for sharing tourism experiences through social media. *Tourism Management*, *43*, 46–54. doi:10.1016/j.tourman.2014.01.012

Murphy, K., Schleifer, A., & Vishny, R. (1999). The allocation of talent: The implications for growth. *The Quarterly Journal of Economics*, *106*(2), 503–530. doi:10.2307/2937945

Mutka, S., & Aaltonen, P. (2013). The impact of a delivery projects business model in a project-based firm. *International Journal of Project Management*, *31*(2), 166–176. doi:10.1016/j.ijproman.2012.07.006

Narver, J. C., & Slater, S. F. (1990). The Effect of a Market Orientation on Business Profitability. *Journal of Marketing*, *54*(4), 20–35. doi:10.2307/1251757

National Bank of Serbia. (2016). *Insurance Supervision Department Report for 2015*. Retrieved March 10, 2016, from National Bank of Serbia web site: http://nbs.rs/internet/english/60/index.html

Naujoks, H. (2010). Strategic management in turbulent times. In L. Stadtler, A. Schmitt, P. Klarner, & T. Straub (Eds.), *More than Bricks in the Wall: Organizational Perspectives for Sustainable Success* (pp. 104–111). Springer Gabler. doi:10.1007/978-3-8349-8945-1_11

Nazemi, S., Asadi, S. T., & Asadi, S. T. (2015). Barriers to strategic planning implementation; case of: Mashhad electricity distribution company. *Procedia: Social and Behavioral Sciences*, *207*, 2–9. doi:10.1016/j.sbspro.2015.10.142

Nelson, D. (2003). *The Penguin dictionary of mathematics*. London, UK: Penguin UK.

Neubauer, T. (2009). An empirical study about the status of business process management. *Business Process Management Journal*, *15*(2), 166–183. doi:10.1108/14637150910949434

Neuberger, D., & Räthke, S. (2009). Microenterprises and multiple bank relationships: The case of professionals. *Small Business Economics*, *32*(2), 207–229. doi:10.1007/s11187-007-9076-8

Nevens, T., Summer, G. L., & Uttal, B. (1990). Commercializing technology: What the best companies do. *Harvard Business Review*, (May–June), 154–163.

Nguyen, K., & Coudounaris, D. (2015). The mechanism of online review management: A qualitative study. *Tourism Management Perspectives*, *16*, 163–175. doi:10.1016/j.tmp.2015.08.002

Nicholson, W. (1997). *Intermediate microeconomics and its application*. The Dryden Press.

Nickerson, J., & Silverman, B. (1998). Intellectual Capital Management Strategy: The Foundation of Successful New Business Generation. *Journal of Knowledge Management*, *1*(4), 320–331. doi:10.1108/EUM0000000004603

Nielsen, C. & Dane-Nielsen, H. (2010). The emergent properties of intellectual capital: A conceptual offering. *Journal of Human Resource Costing &Accounting*, *14*.

Niestrój, R. (2003). Istota i znaczenie marketingu. In J. Altkorn (Ed.), Podstawy marketingu (pp. 13-44). Kraków: Instytut Marketingu.

Njegomir, V. (2011). *Osiguranje*. Novi Sad, Srbija: Ortomedics book.

Njeguš A. (2010). *Informacioni sistemi u turističkom poslovanju*. Beograd: Fakultet za turistički i hotelijerski menadžment, Univerzitet Singidunum.

Njeguš, A. (2009). *Poslovni informacioni sistemi*. Beograd: Univerzitet Singidunum.

Noble, Ch. H., Sinha, R. K., & Kumar, A. (2002). Market Orientation and Alternative Strategic Orientations: A Longitudinal Assessment of Performance Implications. *Journal of Marketing*, *66*(4), 25–39. doi:10.1509/jmkg.66.4.25.18513

Nooteboom, B. (1994). Innovation and diffusion in small firms: Theory and evidence. *Small Business Economics*, *6*(5), 327–347. doi:10.1007/BF01065137

NUI Galaway & Whitaker University. (2015). *Macro and firm level innovation measurements and indicators*. Retrieved May 25, 2016, from http://galwaydashboard.ie/publications/white-paper-measurement-2015.pdf

Nummela, N. (2011). *International Growth of SMEs*. New York: Routledge.

Nyssens, M. (2006). *Social enterprise at the crossroads of market, public and civil society*. London: Routledge.

O'Reilly, C. A., & Tushman, M. L. (2004). The ambidextrous organization. *Harvard Business Review*, *82*(4), 74–83. PMID:15077368

Oakland, J. S., & Sohal, A. (1996). *Total Quality Management: Text with Cases*. Melbourne: Butterworth-Heinemann.

Obradović, S., Jevtić, B., Grozdanić, R., Damnjanović, A., & Vulović, R. (2011). *Uloga IKT u koordinaciji intermodalnog transporta.*Paper presented at14thInternational Conference Quality and Reliability Management (ICDQM '11), Čačak.

ODonnell, D., Henriksen, L. B., & Voelpel, S. C. (2006). Guest editorial: Becoming critical on intellectual capital. *Journal of Intellectual Capital*, *7*(1), 5–11. doi:10.1108/14691930610639732

OECD. (2004). *ICT, E-business and SMEs*. Retrieved July 27th, 2016 from OECD web site: http://www.oecd.org/internet/ieconomy/34228733.pdf

OECD. (2004). *Networks, Partnerships, Clusters and Intellectual Property Rights: Opportunities and Challenges for Innovative SMEs in a Global Economy*. Retrieved July 26, 2016, from OECD web site: http://www.oecd.org/cfe/smes/31919590.pdf

OECD. (2005). *Oslo Manual*. Retrieved May 28, 2016, from https://www.oecd.org/sti/inno/2367580.pdf

OECD. (2012). *SME Policy Index: Eastern Partner Countries 2012: Progress in the Implementation of the Small Business Act for Europe*. OECD Publishing. Retrieved July 26, from: http://www.keepeek.com/Digital-Asset-Management/oecd/finance-and-investment/sme-policy-index-eastern-partner-countries-2012_9789264178847-en#page1

OECD. (2015). *Building the evidence base. Social impact investment. Preliminary version*. Retrieved May 16, 2016, from http://www.oecd.org/sti/ind/social-impact-investment.pdf

OECD. (2016). *SME Policy Index: Western Balkans and Turkey 2016: Assessing the Implementation of the Small Business Act for Europe*. OECD Publishing. Retrieved July 26, 2016 from: http://www.keepeek.com/Digital-Asset-Management/oecd/development/sme-policy-index-western-balkans-and-turkey-2016_9789264254473-en#page313

Ohmae, K. (1982). *The mind of the strategist: The art of Japanese business*. New York, NY: McGraw-Hill.

ONeill, P., & Sohal, A. S. (1999). Business Process Reengineering A review of recent literature. *Technovation*, *19*(9), 571–581. doi:10.1016/S0166-4972(99)00059-0

Ongaro, E. (2004). Process management in the public sector: The experience of one-stop shops in Italy. *International Journal of Public Sector Management*, *17*(1), 81–107. doi:10.1108/09513550410515592

Onkelinx J. & Sleuwaegen L. (2010, October). Internationalization strategy and performance of small and medium sized enterprises. *NBB Working Paper, 197*, 1-45.

ORegan, N., & Ghobadian, A. (2002). Effective strategic planning in small and medium sized firms. *Management Decision*, *40*(7), 663–671. doi:10.1108/00251740210438490

Orłowski, W., Pasternak, R., Flaht, K., & Szubert, D. (2010). *Procesy inwestycyjne i strategie przedsiębiorstw w czasach kryzysu*. Warsaw: PARP.

Ožegović, L., & Sajfert Z. (2009). *Preduzetništvo*. Novi Sad: Fakultet za ekonomiju i inženjerski menadžment.

Ožegović, L., & Pavlović, N. (2012). Management of small and medium size enterprises as a carrier of economic growth. *School of Business*, *1*, 74–84. doi:10.5937/skolbiz1201074O

Öztamur, D., & Karakadılar, İ. (2014). Exploring the Role of Social Media for SMEs: As a New Marketing Strategy Tool for the Firm Performance Perspective. *Procedia: Social and Behavioral Sciences*, *150*, 511–520. doi:10.1016/j.sbspro.2014.09.067

Paces, A., & Griliches, Z. (1984). Patents and R&D at the firm level: a first look. In Z. Griliches (Ed.), *R&D, patents, and productivity* (pp. 55–72). Chicago: University of Chicago Press.

Pacheco, D., Dean, T., & Payne, D. (2010). Escaping the green prison: Entrepreneurship and the creation of opportunities for sustainable development. *Journal of Business Venturing*, *25*(5), 464–480. doi:10.1016/j.jbusvent.2009.07.006

Paciolla, R., & Mai, L. W. (2011). The Impact of Italianate on Consumers' Brand Perceptions of Luxury Brands. *European Advances in Consumer Research*, *9*, 360–366.

Palmberg, K. (2010). Experiences of implementing process management: A multiple-case study. *Business Process Management Journal*, *16*(1), 93–113. doi:10.1108/14637151011017967

Paper, D., & Rodger, J. A. (2001). Managing radical transformation: a holistic paradigm. Academic Press.

Paper, D., Rodger, J.A., & Pendharkar, P.C. (2000). *Development and initial testing of a theoretical model of transformation*. Academic Press.

Paper, D. (1998). BPR: Creating the conditions for success. *Long Range Planning*, *31*(3), 426–435. doi:10.1016/S0024-6301(98)80009-5

Paper, D. (1999). The enterprise transformation paradigm: The case of Honeywells industrial automation and control unit. *Journal of Information Technology Cases and Applications*, *1*(1), 4–23. doi:10.1080/15228053.1999.10855927

Paper, D., & Chang, R. (2005). The state of business process reengineering: A search for success factors. *Total Quality Management & Business Excellence*, *16*(1), 121–133. doi:10.1080/1478336042000309907

Paper, D., & Dickinson, S. (1997). A comprehensive process improvement methodology: experiences at Caterpillar's Mossville engine centre. In M. Khosrowpour & J. Liebowitz (Eds.), *Cases on Information Technology Management in Modern Organizations*.

Paper, D., & Mok, W. (2003). The relation between BPR and ERP systems: A failed project. *Annals of Cases on Information Technology*, *5*, 45–62. doi:10.4018/978-1-59140-061-5.ch003

Paper, D., Rodger, J. A., & Pendharkar, P. C. (2001). A BPR case study at Honeywell. *Business Process Management Journal*, *7*(2), 85–99. doi:10.1108/14637150110389416

Parker, S. R., Haytko, D. L., & Hermans, C. M. (2011). Ethnocentrism and Its Effect on the Chinese Consumer: A Threat to Foreign Goods? *Journal of Global Marketing*, *24*(1), 4–17. doi:10.1080/08911762.2011.545716

PARP. (2015, April). *Report on the condition of small and medium-sized enterprise sector in Poland in 2012-2013*. Retrieved July 26th, 2016, from Polish Agency for Enterprise Development website https://badania.parp.gov.pl/images/ReportonSMESectorinPoland_2014.pdf

Parrish, D. (2010). Sustainability-driven entrepreneurship: Principles of organization design. *Journal of Business Venturing*, *25*(5), 510–523. doi:10.1016/j.jbusvent.2009.05.005

Patel, N., & Narain, U. (2008). *Can intangible assets act as a cushion in a downturn?* Retrieved December 16, 2015, from Watson Wyatt web site: http://www.watsonwyatt.com/pubs/directions/media/2009_EU_12783_ Directions_CP05_Intangibles.pdf

Pazio, N. M. (2007). *Podstawy marketingu*. Warszawa: Oficyna Wydawnicza Politechniki Warszawskiej.

Peach International. (2012). *The Chinese food market*. Beijing: Author.

Pelham, A. M., & Wilson, D. T. (1996). A Longitudinal Study of the Impact of Market Structure, Firm Structure, Strategy, and Market Orientation Culture on Dimensions of Small-Firm Performance. *Journal of the Academy of Marketing Science*, *24*(1), 27–43. doi:10.1007/BF02893935

Penezić, N. & Jakopin, E. (2002). *Analiza razvoja malih i srednjih preduzeća i preduzetništva u Srbiji*. Beograd: Republički zavod za razvoj i Republička agencija za razvoj malih i srednjih preduzeća i preduzetništva.

Penrose, E. (1959). *The theory of the growth of the firm*. Oxford, UK: Oxford University Press.

Pérez Fernández de Velasco, J. A. (2004). *Gestión por procesos: cómo utilizar ISO 9001:2000 para mejorar la gestión de la organización*. Madrid: ESIC.

Peteraf, M. A. (1993). The cornerstones of competitive advantage: A resource-based view. *Strategic Management Journal*, *14*(3), 179–191. doi:10.1002/smj.4250140303

Peteraf, M. A., & Bergen, M. E. (2003). Scanning dynamic competitive landscapes: A market-based and resource-based framework. *Strategic Management Journal*, *24*(10), 1027–1041. doi:10.1002/smj.325

Petrović, G., & Grujović, M. (2016). *Ekonomske štete od elementarnih nepogoda u Srbiji i šumadijskom okrugu*. Retrieved March 10, 2016, from scindeks-clanci.ceon.rs/.../1452-44571502099P.pdf

Popping, R. (2015). Analyzing open-ended questions by means of text analysis procedures. *Bulletin of Sociological Methodology. Bulletin de Methodologie Sociologique*, *128*(1), 23–39. doi:10.1177/0759106315597389

Portela, M. (2013). *Entrepreneurship and economic growth: macroeconomic analysis and effects of social capital in EU*. Retrieved August 1, 2016, from http://www.intechopen.com/books/entrepreneurship-born-made-and-educated/entrepreneurship-and-economic-growth-macroeconomic-analysis-and-effects-of-social-capital-in-the-eu

Porter, M. E. (1985). *Competitive advantage: Creating and sustaining superior performance*. New York, NY: FreePress.

Porter, M. E. (1998). *Competitive advantage: creating and sustaining superior performance*. The Free Press. doi:10.1007/978-1-349-14865-3

Porter, M., & Kramer, M. (2011). Creating shared value. *Harvard Business Review*, *89*(1/2), 62–77.

Potter, J., & Proto, A. (2010). OECD Review of SME and Entrepreneurship Issues and Policies at National and Local Levels in Poland). Paris: OECD.

Preece, J. (2001). Sociability and usability: Twenty years of chatting online. *Behavior and Information Technology Journal*, *20*(5), 347–356. doi:10.1080/01449290110084683

Project Management Institute. (2013a). *Managing change in organizations: a practice guide*. Pennsylvania: PMI.

Project Management Institute. (2013b). *PMI's pulse of the profession in-depth report: navigating complexity*. PMI.

Project Management Institute. (2013c). *The standard for portfolio management* (3rd ed.). PMI.

Pryor, M. G., Taneja, S., Humphreys, J., Anderson, D., & Singleton, L. (2008). Challenges Facing Change Management: Theories and Research. *Delhi Business Review*, *9*(1), 1–20.

Pulic, A. (2000). *VAIC – an accounting tool for IC management*. Retrieved 11, March, 2016, from VAIC web site: http://www.vaic-on.net/start.html

PWC. (2013). *European Working Capital Annual Review*. Retrieved July 2, 2015, from PWC web site: http://www.pwc.be/en/publications/2013/capital_market_survey_2013.pdf

PWC. (2014). *Cash for Growth. Annual Global Working Capital Survey*. Retrieved May 27, 2016, from PWC web site: http://www.pwc.com/gx/en/financial-services/publications/assets/working-capital-2014.pdf

PWC. (2015). *Annual Global Working Capital Survey*. Retrieved May 27, 2016, from PWC web site: http://www.pwc.com/gx/en/business-recovery-restructuring-services/working-capital-management/working-capital-survey/2015/assets/global-working-capital-survey-2015-report.pdf

Qu, H., & Lee, H. (2011). Travelers social identification and membership behaviors in online travel community. *Tourism Management*, *32*(6), 1262–1270. doi:10.1016/j.tourman.2010.12.002

Qureshi, S., & Kratzer, J. (2011). An Investigation of Antecedents and Outcomes of Marketing Capabilities in Entrepreneurial Firms: An Empirical Study of Small Technology-Based Firms in Germany. *Journal of Small Business and Entrepreneurship*, *24*(1), 49–66. doi:10.1080/08276331.2011.10593525

Qureshi, T., Warraich, A., & Hijazi, T. (2009). Significance of project management performance assessment (PMPA) model. *International Journal of Project Management*, *27*(4), 378–388. doi:10.1016/j.ijproman.2008.05.001

Radomska, J. (2014). Linking the main obstacles to the strategy implementation with the companys performance. *Procedia: Social and Behavioral Sciences*, *150*, 263–270. doi:10.1016/j.sbspro.2014.09.054

Raduan, C. R., Jegak, U., Haslinda, A., & Alimin, I. I. (2009). Management, strategic management theories and the linkage with organizational competitive advantage from the resource-based view. *European Journal of Soil Science*, *11*(3), 402–418.

Ralston, P. M., Blackhurst, J., Cantor, D. E., & Crum, M. R. (2015). A structure–conduct–performance perspective of how strategic supply chain integration affects firm performance. *Journal of Supply Chain Management*, *51*(2), 47–64. doi:10.1111/jscm.12064

Ranđić, D., & Lekić, S. (2009). *Menadžment*. Beograd: Beogradska poslovna škola.

Rao, P. (2007). Greening of the supply chain: An empirical study for SMES in the Philippine context. *Journal of Asia Business Studies*.

Raps, A. (2005). Strategy implementation – an insurmountable obstacle? In Handbook of Business Strategy. Emerald Group Publishing Limited.

Rasch, G. (1960). *Probabilistic models for some intelligence and attainment tests*. Chicago: The University of Chicado Press.

Rastogi, P. (1994). Nature, significance and rationale of business process reengineering. *Productivity*, *35*(3), 467–476.

Rauch, A., Wiklund, J., Lumpkin, G. T., & Freese, M. (2009). Entrepreneurial orientation and business performance: An assessment of past research and suggestion for the future. *Entrepreneurship. Theory into Practice*, *33*(3), 761–781.

Ravesteyn, P., & Batenburg, R. (2010). Surveying the critical success factors of BPM-systems implementation. *Business Process Management Journal*, *16*(3), 492–506. doi:10.1108/14637151011049467

Recklies, O. (2008). Problems and barriers to strategic planning. *Economics and Organization of Future Enterprise*, *1*(1), 3–11. doi:10.2478/v10061-008-0001-2

Rees, G., & Hall, D. (2013). Managing Change. In G. Rees & R. French (Eds.), *Leading, Managing and Developing People* (pp. 101–126). CIPD.

Reijers, H. A. (2006). Implementing BPM systems: The role of process orientation. *Business Process Management Journal*, *12*(4), 389–409. doi:10.1108/14637150610678041

Reijonen, H., & Komppula, R. (2010). The adoption of market orientation in SMEs: Required capabilities and relation to success. *Journal of Strategic Marketing*, *18*(1), 19–37. doi:10.1080/09652540903511282

Resnick, S. M., Cheng, R., Simpson, M., & Lourenço, F. (2016). Marketing in SMEs: A 4Ps self-branding model. *International Journal of Entrepreneurial Behavior & Research*, *22*(1), 1–20. doi:10.1108/IJEBR-07-2014-0139

Revell, A., & Blackburn, R. (2007). The business case for sustainability? An examination of small firms in the UKs construction and restaurant sectors. *Business Strategy and the Environment*, *16*(6), 404–420. doi:10.1002/bse.499

Reynolds, P. (1992). Predicting new-firm births: interactions of organizational and human populations. In D. Sexton & J. Kasarda (Eds.), *Entrepreneurship in the 1990s* (pp. 59–71). Boston: PWS-Kent Publishing.

Reynolds, P. D. (2000). National Panel Study of U.S. Business Start-ups: Background and Methodology. In J. A. Katz (Ed.), *Databases for the Study of Entrepreneurship* (pp. 153–227). Amsterdam: JAI Press. doi:10.1016/S1074-7540(00)04006-X

Rimmel, G. (2004). *Danish Guidelines on Intellectual Capital Reporting*. Retrieved May 6, 2016, from Gothenburg University web site: http://gupea.ub.gu.se/bitstream/2077/3058/1/SNEE2004_article_The_Danish_Guidelines_on_Intellectual_Capital_Reporting1.PDF

Roberts, H. J. E. (1999). The Control of Intangibles in the Knowledge-intensive Firm.*22nd Annual Congress of the European Accounting Association*, Bordeaux, France.

Roberts, M. E., Stewart, B. M., Tingley, D., Lucas, C., Leder-Luis, J., Gadarian, S. K., & Rand, D. G. (2014). Structural topic models for open-ended survey responses. *American Journal of Political Science*, *58*(4), 1064–1082. doi:10.1111/ajps.12103

Roberts, S., Lawson, R., & Nicholls, J. (2006). Generating regional-scale improvements in SME corporate responsibility performance: Lessons from responsibility Northwest. *Journal of Business Ethics*, *67*(3), 275–286. doi:10.1007/s10551-006-9184-4

Rodgers, M. (1998). *The definition and measurement of innovation*. Working paper 10. Melbourne: Melbourne Institute of Applied Economic and Social Research.

Rodriguez, M., Peterson, R. M., & Krishnan, V. (2012). Social Medias Influence on Business-to-Business Sales Performance. *Journal of Personal Selling & Sales Management*, *32*(3), 365–378. doi:10.2753/PSS0885-3134320306

Rogers, S., Lukens, S., Lin, S., & Jon, E. (2007). *Balancing Risk and Performance with an Integrated Finance Organization: The Global CFO Study 2008*. IBM Corporation in cooperation with The Wharton School and Economist Intelligence Unit.

Rogers, E. (1962). *Diffusion of Innovations*. Free Press of Glencoe.

Romer, M. (1986). Increasing returns and long run growth. *Journal of Political Economy*, *94*(5), 1002–1037. doi:10.1086/261420

Rosenbusch, N., Brinckmann, J., & Bausch, A. (2011). Is innovation always beneficial? A meta-analysis of the relationship between innovation and performance in SMEs. *Journal of Business Venturing*, *26*(4), 441–457. doi:10.1016/j.jbusvent.2009.12.002

Rosłon, J. (2011). Mikro i małe przedsiębiorstwa w Polsce – wybrane aspekty funkcjonowania i rozwoju. *Zeszyty Naukowe Uniwersytetu Szczecińskiego*, *81*, 139–147.

Roth, A. (2015). *Who Gets What and Why: The New Economics of Matchmaking and Market Design*. Denver, CO: Houghton Mifflin Harcourt.

Rubikom Strategy Consultants. (2011). *Kondycja Marketingu w przedsiębiorstwach sektora MSP*. Warsaw: Rubikom Strategy Consultants.

Rugman, A. M., Oh, C. H., & Lim, D. S. (2012). The regional and global competitiveness of multinational firms. *Journal of the Academy of Marketing Science*, *40*(2), 218–235. doi:10.1007/s11747-011-0270-5

Rymarczyk, J. (2004). *Internacjonalizacja i Globalizacja Przedsiębiorstwa*. Warszawa: Polskie Wydawnictwo Ekonomiczne.

Saebi, T., & Foss, N. J. (2015). Business models for open innovation: Matching heterogeneous open innovation strategies with business model dimensions. *European Management Journal*, *33*(3), 201–213. doi:10.1016/j.emj.2014.11.002

Saković, L. (2015). Osiguranje od elementarnih nepogoda: Ruku na oči, prognoze pod tepih. *Biznis & finansije*, *118/119*. Retrieved March 10, 2016, from: http://bif.rs/2015/09/osiguranje-od-elementarnih-nepogoda-ruku-na-oci-prognoze-pod-tepih/

Salgueiro, A. (1999). *Cómo Mejorar los Procesos y la Productividad*. Madrid: AENOR.

Sanchez-Franco, M., & Cataluna, R. (2010). Virtual travel communities and customer loyalty: Customer purchase involvement and web site design. *Electronic Commerce Research and Applications, 9*, 171-182. Retrieved August 26th, 2016 from Research Gate web site: https://www.researchgate.net/publication/220066649_Virtual_travel_communities_and_customer_loyalty_Customer_purchase_involvement_and_web_site_design

Sanchez, L., & Blanco, B. (2014). La gestión por procesos. Un campo por explorar. *Dirección y Organización, 54*, 54–71.

Sandhu, M., & Gunasekaran, A. (2004). Business Process Development in project-based industry. *Business Process Management Journal, 10*(6), 673–693. doi:10.1108/14637150410567875

Šarac, M., Jevremović, A., & Radovanović, D. (2015). *Internet marketing*. Beograd: Fakultet za informatiku i računarstvo, Univerzitet Singidunum.

Savage, L. J. (1954). *The foundations of statistics*. John Wiley & Sons.

Savignac, F. (2008). *Impact of financial constraints on innovation: what can be learned from a direct measure?* Econ. Innov. New Techn.

Savitskaya, I., Salmi, P., & Torkkeli, T. (2010). Barriers to open innovation: Case China. *Journal of Technology Management & Innovation, 5*(4), 10–21. doi:10.4067/S0718-27242010000400002

Savvas, A. (2005, March). Cultural resistance main cause of BPM project failure. *Computer Weekly*.

Savvides, A., & Zachariadis, M. (2003). *International technology diffusion and TFP growth*. Oklahoma State University.

SBA. (2014). *Frequently Asked Questions: What is the Survival Rate for New Businesses?* Retrieved May 13th 2016, from U.S. Small Business Administration web site: https://www.sba.gov/sites/default/files/ FAQ_March_ 2014_0.pdf

Scherer, F. M. (1982). Inter-industry technology flows and productivity growth. *The Review of Economics and Statistics, 64*(4), 627–634. doi:10.2307/1923947

Schima, M. (2004). Geschäftsprozessmanagement bei der Kurz Gruppe. In H. Ellringmann & H. J. Schmelzer (Eds.), *Geschäftsprozessmanagement inside* (pp. 1–40). Munich: Hanser.

Schmelzer, H. J., & Sesselmann, W. (2006). *Geschäftsprozessmanagement in der Praxis*. Munich: Carl Hanser.

Schoemaker, P. J., & Krupp, S. (2015). Overcoming barriers to integrating strategy and leadership. *Strategy and Leadership, 43*(2), 23–32. doi:10.1108/SL-01-2015-0001

Scholes, K., & Johnson, G. (2005). *Exploring corporate strategy*. Edinburg Gate, UK: Prentice Hall International.

Schumpeter, J. A. (1934). *The theory of economic development. An inquiry into profits, capital, credit, interest, and the business cycle*. Cambridge, MA: Harvard University Press.

Schwarcz, P. (2011). *European Union programs and their funding*. Nitra, Slovakia: Slovak University of Agriculture in Nitra.

Seelos, C., & Mair, J. (2005a). Social entrepreneurship: Creating new business models to serve the poor. *Business Horizons, 48*(3), 241–246. doi:10.1016/j.bushor.2004.11.006

Sengi, P. (2003). *Peta disciplina: Umeće i praksa organizacije koja uči*. Novi Sad, Serbia: Adizes MC.

Sentanin, O. F., Almada Santos, F. C., & Chiappetta Jabbour, C. J. (2008). Business process management in a Brazilian public research centre. *Business Process Management Journal, 14*(4), 483–496. doi:10.1108/14637150810888037

Serbia Accounting Law. (2013). Official Gazette of the Republic of Serbia no. No.62/2013.

Serbia Bankruptcy Law. (2014). Official Gazette of RS, No. 104/2009, 99/2011§ other Law, 71/2012 § decision of Constitutional Court and 83/2014.

Serbia Code of Business Conduct. (2006). Official Gazette of the Republic of Serbia no. 1/2006.

Serbia Company Law. (2015). Official Gazette of the Republic of Serbia nos. 36/2011, 99/2011, 83/2014 - § other Law and 5/2015.

Serbia General Collective Agreement. (2008). Official Gazette of RS, No. 50/2008, 104/2008 - Annex I and 8/2009 - Annex II.

Serbia Insurance Law. (2014). Official Gazette of Serbia, no. 139/2014.

Serbia Labor Law. (2014). Official Gazette of the Republic of Serbia nos. 24/2005, 61/2005, 54/2009, 32/2013 & 75/2014.

Serbia Law on Civil Procedure. (2013). Official Gazette of Serbia, no. 49/2013; decisions of Constitutional Court, § 74/2013; § 55/2014.

Serbia Law on Contracts and Torts. (2003). Official Gazette of the SFRY nos. 29/78, 39/85, 45/89, decision of Constitutional Court 57/89, Official Gazette of the FRY 31/93, Official Gazette of Serbia and Montenegro no. 1/2003.

Serbia Law on Public Services. (2014). Official Gazette of the Republic of Serbia nos. No.42/91, 71/94, 79/2005 § other law, 81/2005 § correction, 83/2005 - correction i 83/2014 § other law.

Serbia Law on Safety and Health at Work. (2015). Official Gazette of Serbia, no. 101/2005, 91/2015.

Serbia Law on the Procedure of Registration with the Business Registers Agency. (2014). Official Gazette of the Republic of Serbia nos. 99/2011 and 83/2014.

Serbia Public Procurement Law. (2015). Official Gazette of the Republic of Serbia nos.124/2012, 14/2015 and 68/2015.

Serbia Strategy for Supporting the Development of SMEs, Entrepreneurship and Competitiveness. (2014). Official Gazette of the Republic of Serbia, No. 35/2015.

Serrador, P., & Pinto, J. (2015). Does agile work? A quantitative analysis of agile project success. *International Journal of Project Management*, *33*(5), 1040–1051. doi:10.1016/j.ijproman.2015.01.006

Setti, C., & Stückl, R. (2006). Geschäftsprozessmanagement in der Praxis. In H. J. Schmelzer & W. Sesselmann (Eds.), *Geschäftprozessmanagement in der Praxis* (pp. 461–474). Munich: Hanser.

Shakina, E., & Barajas, A. (2015). An intangible-intensive profile of companies: Protection during the economic crisis 2008-2009. *Management WP BRP*, *33*, 1–22.

Shapiro, B. P. (1988). What the Hell is 'Market Oriented'? *Harvard Business Review*, *66*, 119–125.

Shiu, H. (2006). The Application of the Value Added Intellectual Coefficient to Measure Corporate Performance: Evidence from Technological Firms. *International Journal of Management*, *23*(2), 356–365.

Shukla, A. C., Deshmukh, S. G., & Kanda, A. (2010). Flexibility and Sustainability of supply chains: Are they together? *Global Journal of Flexible Systems Management*, *11*(1/2), 25.

Sidorova, A., & Isik, O. (2010). Business process research: A cross-disciplinary review. *Business Process Management Journal*, *16*(4), 566–597. doi:10.1108/14637151011065928

Sierpińska, M., & Niedbała, B. (2003). *Controlling operacyjny w przedsiębiorstwie. Centra odpowiedzialności w teorii i praktyce*. Warszawa: PWN.

Silva, I. (2012). *Entrepreneurial and innovative behavior in Spanish SMEs. Essay on profitability, growth and export intensity* (Doctoral dissertation). Retrieved August 12, 2016, from https://ddd.uab.cat/pub/tesis/2013/hdl_10803_117454/imds1de1.pdf

Silvestro, R., & Westley, C. (2002). Challenging the paradigm of the process enterprise: A case-study analysis of BPR implementation. *Omega*, *30*(3), 215–225. doi:10.1016/S0305-0483(02)00028-2

Silvius, G., Schipper, R., Planko, J., Brink, J., & Köhler, A. (2012). *Sustainability in project management*. Surrey: Gower.

Silvius, G., & Tharp, J. (2013). *Sustainability integration for effective project management*. IGI. doi:10.4018/978-1-4666-4177-8

Simons, R. (1998). How risky is your company? *Harvard Business Review*, *77*(3), 85–94. PMID:10387580

Simpson, M., Taylor, N., & Barker, K. (2004). Environmental responsibility in SMEs: Does it deliver competitive advantage? *Business Strategy and the Environment*, *13*(3), 156–171. doi:10.1002/bse.398

Şipoş, C. (2009). Optimization models for managerial decisions. *Megatrend Review*, *6*(2).

Sirmon, D. G., Hitt, M. A., & Ireland, R. D. (2007). Managing firm resources in dynamic environments to create value: Looking inside the black box. *Academy of Management Review*, *32*(1), 273–292. doi:10.5465/AMR.2007.23466005

Sisney, L. (2013). *Organizational Physics: The Science of Growing a Business*. Raleigh, NC: Lulu Press.

Skoko, H. (1986). *Optimalne kombinacije faktora čovjek-stroj, istraživanje na modelu tekstilne industrije u Hrvatskoj*. Informator.

Skrinjar, R., Bosilj-Vuksic, V., & Indihar-Stemberger, M. (2008). The impact of business process management orientation on financial and non-financial performance. *Business Process Management Journal*, *14*(5), 738–754. doi:10.1108/14637150810903084

Smit, A. J. (2010). The competitive advantage of nations: Is Porter's Diamond Framework a new theory that explains the international competitiveness of countries. *Southern African Business Review*, *14*(1), 105–130.

Smith, M., & Culbertson, S. (2010). *Enhancing entrepreneurship: The role of goal orientation and self-efficacy*. Retrieved August 8, 2016, from http://jca.sagepub.com/content/early/2011/01/13/1069072710385543.abstract

Smith, H., & Fingar, P. (2007). *Business Process Management: The Third Wave*. Meghan-Kiffer Press.

Smolarek, M. (2012). Strategie konkurencyjne małych przedsiębiorstw. *Zeszyty Naukowe Wyższej Szkoły Humanitas*, *1*, 53–68.

Sneyd, M. R. (2003). *Insurance. Professional Reading Skills*. Pearson Education Limited, Longman.

Snyman, H. A., Kennon, D., Schutte, C. S. L., & Von Leipzig, K. (2013). Formulating a strategic framework to promote SME development. In *SAIIE25 Proceedings*.

Sobczyk, G. (2006). *Strategie konkurencji małych i średnich przedsiębiorstw*. Wydawnictwo Uniwersytetu Marii Curie-Skłodowskiej.

Soininen, J., Martikainen, M., Puumalainen, K., & Kyläheiko, K. (2012). Entrepreneurial orientation: Growth and profitability of Finnish small-and medium-sized enterprises. *International Journal of Production Economics*, *140*(2), 614–621. doi:10.1016/j.ijpe.2011.05.029

Soniewicki, M. (2015). *The company's international competitive advantage – the role of knowledge*. Warsaw: Difin.

Spanos, Y. E., & Lioukas, S. (2001). An examination into the causal logic of rent generation: Contrasting Porters competitive strategy framework and the resource-based perspective. *Strategic Management Journal*, *22*(10), 907–934. doi:10.1002/smj.174

Stacey, R. D. (1996). *Complexity and creativity in organizations*. Oakland, CA: Berrett-Koehler Publishers.

Stam, E., & Hartog, C., Stel van, A., & Thurik, R. (2009). *Ambitious entrepreneurship, highgrowth firms and macroeconomic growth*. EIM Research Reports. Retrieved July 13, 2016, from Research Gate web site: https://www.researchgate.net/publication/46455754_Ambitious_Entrepreneurship_High-Growth_Firms_and_Macroeconomic_Growth

Stamatović, M., & Vemić, M. (2009). Elaboration of the Profound Effects of the Global Economic Crisis on Small and Medium Enterprises in Serbia. In *Proceedings of the 3rd Central European Conference in Regional Science (CERS 2009)*. Retrieved July 26, 2016, from Faculty of Economics, Tehnical University in Košice web site: http://www3.ekf.tuke.sk/cers/cers2009/PDF/02_66_Vemic_Stamatovic.pdf

Stam, E., & Wennberg, K. (2009). The roles of R&D in new firm growth. *Small Business Economics*, *33*(1), 77–89. doi:10.1007/s11187-009-9183-9

Standing, C., Holzweber, M., & Mattson, J. (2016). Exploring emotional expressions in word-of-mouth from online communities. *Information Processing & Management*, *000*, 1–12.

Stanković, F. (1998). Potencijali i upravljanje rastom u MSP. In I. Černiček & R. Molnar (Eds.), Razvoj preduzetništva - šansa za progress (pp. 44-52). Zrenjanin, Serbia. Tehnički fakultet "Mihajlo Pupin".

Stanton, S., Hammer, M., & Power, B. (1993). Reengineering: Getting everyone on board. *I.T. Magazine*, *25*(4), 22.

Stefanova, J. (2014). *Venture capital in the EU: prospects of Bulgarian companies to access new forms of financing*. Veliko Turnovo: Faber Publishing House.

Stefanović, Z. (2011). *European Union Law*. Belgrade: Faculty of Law Union University and Official Gazette.

Stephen, A. T., & Lehman, D. R. (2016). *How Word-Of-Mouth Transmission Encouragement Affect Consumers' Transmission Decisions, Receiver Selection, and Diffusion Speed*. Oxford, UK: Saïd Business School.

Stępień, B. (2011). Międzynarodowa kooperacja gospodarcza – przyczyny i uwarunkowania instytucjonalne i kulturowe. In B. Stępień (Ed.), Międzynarodowa kooperacja gospodarcza z polskiej perspektywy (pp. 90-130). Warszawa: Polskie Wydawnictwo Ekonomiczne.

Stettina, C., & Hörz, J. (2015). Agile portfolio management: An empirical perspective on the practice in use. *International Journal of Project Management*, *33*(1), 140–152. doi:10.1016/j.ijproman.2014.03.008

Stewart, T. A. (1997). *Intellectual capital: The New Wealth of Organizations*. New York: Doubleday.

Stokes, G. (1900). The Theory of Mathematical Inference. *The American Mathematical Monthly*, *7*(1), 1-8. Retrieved July 7, 2016, from http://www.jstor.org/stable/2971489 doi:1

Strychalska-Rudzewicz, A. & Wysocki, J. (2013). Strategie małych i średnich przedsiębiorstw w działalności innowacyjnej. *Organizacja i kierowanie*, *2*, 67-78.

Stuart, T. E. (2000). Inter-organizational alliances and the performance of firms: A study of growth and innovation rates in a high technology industry. *Strategic Management Journal*, *21*(8), 791–811. doi:10.1002/1097-0266(200008)21:8<791::AID-SMJ121>3.0.CO;2-K

Subramaniam, M., & Youndt, M. A. (2005). The influence of intellectual capital on the types of innovative capabilities. *Academy of Management Journal*, *48*(3), 450–463. doi:10.5465/AMJ.2005.17407911

Sucky, E. (2007). A model for dynamic strategic vendor selection. *Computers & Operations Research, 34*(12), 3638–3651. doi:10.1016/j.cor.2006.01.006

Šulejić, P. (2005). *Pravo osiguranja*. Beograd: Dosije.

Sveiby, K. E. (2007). *Methods for Measuring Intangible Assets*. Retrieved 10, March, 2016, from Sveiby web site: http://www.sveiby.com/articles/ IntangibleMethods.html

Swedish Entrepreneurship Forum. (2016). *The Entrepreneurial challenge. A comparative study of entrepreneurial dynamics in China, Europe and the US.* Retrieved August 4, 2016, from http://entreprenorskapsforum.se/wp-content/uploads/2016/06/GEM_Internationell_Rapport_2016_Webb.pdf

Swiss Re. Sigma. (2014). *Natural catastrophes and man-made disasters in 2013: large losses from floods and hail; Haiyan hits the Philippines*. Retrieved March 10, 2016, from http://www.swissre.com/sigma/?year=2014

Sykes, W. (1990). Validity and reliability in qualitative market research: A review of the literature. *Journal of the Market Research Society, 32*(3), 289–328.

Sykes, W. (1991). Taking stock: Issues from the literature on validity and reliability in qualitative research. *Journal of the Market Research Society, 33*(1), 3–12.

Szerb, L., Zoltan, A., Komlosi, E., & Argiles, R. (2015). *Measuring entrepreneurial systems: The regional entrepreneurship and development index*. Retrieved August 1, 2016, from http://ec.europa.eu/regional_policy/sources/docgener/studies/pdf/regional_entrepreneurship_development_index.pdf

Tajeddini, K., Trueman, M., & Larsen, G. (2006). Examining the Effect of Market Orientation on Innovativeness. *Journal of Marketing Management, 22*(5-6), 529–551. doi:10.1362/026725706777978640

Takala, J., & Usitalo, T. (2012). Resilient and proactive utilization of opportunities and uncertainties in service business. *Proceedings of the University ofVaasa, Finland.* University of Vaasa.

Taleb, N. N., Goldstein, D. G., & Spitznagel, M. W. (2009). The six mistakes executives make in risk management. *Harvard Business Review, 87*(10), 78–81. PMID:19968059

Talwar, R. (1993). Business re-engineering—a strategy-driven approach. *Long Range Planning, 26*(6), 22–40. doi:10.1016/0024-6301(93)90204-S

Thomas, J., Harden, A., Oakley, A., Oliver, S., Sutcliffe, K., Rees, R., & Kavanagh, J. (2004). Integrating qualitative research with trials in systematic reviews. *BMJ (Clinical Research Ed.), 328*(7446), 1010–1012. doi:10.1136/bmj.328.7446.1010 PMID:15105329

Timmers, P. (1998). Business Models for Electronic Markets. *Electronic Markets, 8*. Retrieved August 26th, 2016 from University of Utrecht web site: http://www.cs.uu.nl/docs/vakken/ec/Timmers_BMem.pdf

Tirole, J. (1988). *The Theory of Industrial Organization*. Cambridge, MA: MIT Press.

TNS Political & Social. (2012a). *SMEs, Resource Efficiency and Green Markets*. Retrieved August 22nd 2016 from EU web site: http://ec.europa.eu/public_opinion/flash/fl_342_sum_en.pdf

TNS Political & Social. (2012b). *SMEs, Resource Efficiency and Green Markets*. Retrieved August 22nd 2016 from EU web site: http://ec.europa.eu/public_opinion/flash/fl_342_en.pdf

Todnem, R. (2005). Organisational change management: A critical review. *Journal of Change Management, 5*(4), 369–380. doi:10.1080/14697010500359250

Todorović, Z., Bunić, S., & Todorović, I. (2013). Transparency in corporate governance. *Acta Economica, 11*(9), 9–39. doi:10.7251/ACE1319009T

Towers, S. (1996). Re-engineering: Middle managers are the key asset. *Management Services, 40*(12), 17.

Tranter, C. J. (1975). *Advanced level pure mathematics*. Hodder and Stoughton Educational.

Tseng, C., & Goo, Y. (2005). Intellectual capital and corporate value in an emerging economy: Empirical study of Taiwanese manufacturers. *R & D Management, 35*(2), 187–201. doi:10.1111/j.1467-9310.2005.00382.x

Turner, G. (2008). A comparison of the limits to growth with 30 years of reality. *Global Environmental Change, 18*(3), 397–411. doi:10.1016/j.gloenvcha.2008.05.001

Turner, R., Huemann, M., Anbari, F., & Bredillet, C. (2010). *Perspectives on projects*. Routledge.

Udovičić, A., & Kadlec, Ž. (2013). Analiza rizika upravljanja preduzećem. *Praktični menadžment, 4*(1), 50-60.

Ulku, H. (2004). *R&D, innovation and economic growth. An emprical evidence.* IMF Working Papers 04/185. Washington, DC: IMF.

Unger, J. M., Rauch, A., Frese, M., & Rosenbusch, N. (2011). Human capital and entrepreneurial success: Ameta-analytical review. *Journal of Business Venturing, 26*(3), 341–358. doi:10.1016/j.jbusvent.2009.09.004

United Nations. (2000). *Millenium declaration.* Retrieved August, 16, 2016, from http://www.un.org/millennium/declaration/ares552e.htm

United Nations. (2015). *Transforming our world: the 2030 agenda for sustainable development.* Retrieved August, 16, 2016, from http://www.un.org/ga/search/view_doc.asp?symbol=A/RES/70/1&Lang=E

USAID Business Enabling Project. (2015a). *Business enabling project qualitative research-Results of focus group discussions and in-depth interviews*. IPOS. Retrieved July 26, 2016, from site: http://www.policycafe.rs/documents/business/survey/Focus%20Groups%20Report%202015%20final.pdf

USAID Business Enabling Project. (2015b). *Business survey Serbia 2015: Executive Summary*. Retrieved July 26, 2016, from site: http://www.policycafe.rs/Anketa2015/BEP%20Survey%202015%20Executive%20Summary%20-%20english.pdf

Vaismoradi, M., Turunen, H., & Bondas, T. (2013). Content analysis and thematic analysis: Implications for conducting a qualitative descriptive study. *Nursing & Health Sciences, 15*(3), 398–405. doi:10.1111/nhs.12048 PMID:23480423

Van der Aalst, W. M. P. (2004). Business Process Management: A personal view. *Business Process Management Journal, 10*(2), 135. doi:10.1108/bpmj.2004.15710baa.001

Vanhonacker, F., Verbeke, W., Guerrero, L., Claret, A., Sulmont-Rossé, C., & Raude, J. (2010). How European consumers define the concept of traditional food: Evidence from a survey in six countries. *Agribusiness, 26*(4), 453–476. doi:10.1002/agr.20241

Vaughan, E., & Vaughan, T. M. (2000).*Osnove osiguranja, upravljanje rizicima*. Zagreb: Mate d.o.o.

Veale, R., Quester, P., & Karunaratna, A. (2006). *The role of intrinsic (sensory) cues and the extrinsic cues of country of origin and price on food product evaluation.* Paper presented at the 3rd International Wine Business & Marketing Research Conference, Montpellier.

Veljović, A., Vulović, R. & Damnjanović, A. (2009). *Informaciono komunikacione tehnologije u menadžmentu*. Čačak: Tehnički fakultet.

Vemic, M. B. (2015). Strategic Data-Based Optimization of Working Capital Management in Medium-Sized Firms: A Real-World Case Study. In Strategic Data-Based Wisdom in the Big Data Era (pp. 15-50). Hershey, PA: IGI Global. doi:10.4018/978-1-4666-8122-4.ch002

Vemić, M., (2008). *Nauka o privrednom sistemu i ekonomskoj politici u tranziciji.* Novi Sad: Cekom books.

Vemic, M. (2013). Evaluation of knowledge-based high-growth enterprises in Serbia's economic development. *Online Journal of Applied Knowledge Management*, *1*(2), 1–14.

Venohr, B., Fear, J., & Witt, A. (2015). *Best of German Mittelstand-The World Market Leaders.* Academic Press.

Vidaković, M. (2010). Smanjenje obima štete kroz obuku osiguranika. In Osiguranje u CEFTA regionu, (pp. 273-287). Zlatibor: Privredna komora Srbije.

Villa, A., & Antonelli, D. (2009). *A Road Map to the Development of European SME Networks.* Berlin: Springer. doi:10.1007/978-1-84800-342-2

Von Davier, M., & Carstensen, C. H. (2007). Multivariate and Mixture Distribution Rasch Models. Extensions and Applications. New York: Springer Science + Business Media, LLC. doi:10.1007/978-0-387-49839-3

Von Neumann, J., & Morgenstern, O. (1944). *Theory of games and economic behavior.* Princeton University Press.

Vos, J.-P. (2004). Developing strategic self-descriptions of SMEs. *Technovation*, *25*(9), 989–999. doi:10.1016/j.technovation.2004.02.014

Vossen, R. (2010). *Combining small and large firm advantages in innovation: theory and evidence.* Retrieved August 1, 2016, from http://www.rug.nl/research/portal/files/3183201/98b21.pdf

Vrandea, V., Jong, J., Vanhaverbeke, W., & Rochemont, M. (2009). Open innovation in SMEs: Trends, motives and management challenges. *Technovation*, *29*(6-7), 423–437. doi:10.1016/j.technovation.2008.10.001

Wach, K. (2012). *Impact of the regional business environment on the development of SMEs in Southern Poland.* Retrieved August 3, 2016, from http://www.businessenvironment.org/dyn/be/docs/272/The%20Impact%20of%20Business%20Environment%20on%20Small%20and%20.pdf

Wagner, E. T. (2013). *Five Reasons 8 Out Of 10 Businesses Fail.* Retrieved July 24th 2016 from Forbes web site: http://www.forbes.com/sites/ericwagner/2013/09/12/five-reasons-8-out-of-10-businesses-fail/#741626b05e3c

Wahlich, S. M. (2004). Prozessorientatierte Organisation bei Vaillant Hepworth. In H. Ellringmann & H. J. Schmelzer (Eds.), *Geschäftsprozessmanagement inside* (pp. 1–40). Munich: Hanser, Munich.

Wainwright, S. (2002). *Measuring impact: A guide to resources.* London: NCVO.

Wald, A. (1950). *Statistical decision functions.* John Wiley & Sons.

Walton, S. V., Handfield, R. B., & Melnyk, S. A. (1998). The green supply chain: integrating suppliers into environmental management processes. *International Journal of Purchasing and Materials Management, 34*(1), 2-11.

Wang, L., & Xu, Y. (2010). Game Theory Analysis on SME Financing Under FTW. *Orient Academic Forum,* (pp. 434-438).

Wang, S. (2016). *Enterprise Management and Performance Evaluation of E-business in SMEs: Based on Human Resource Management Perspective.* Retrieved July 27th, 2016 from SERSC web site: http://www.sersc.org/journals/IJUNESST/vol9_no3/31.pdf

Wang, O., Gellynck, X., & Verbeke, W. (2015). General image of and beliefs about European food in two mainland Chinese cities. *British Food Journal*, *117*(5), 1581–1595. doi:10.1108/BFJ-07-2014-0232

Wang, W. Y., & Chang, C. (2006). Intellectual capital and performance in causal models: Evidence from the information technology industry in Taiwan. *Journal of Intellectual Capital*, *6*(2), 222–236. doi:10.1108/14691930510592816

Wang, Y., Shuang, S., & Li, D. (2015). Customer participation in virtual brand communities: The self-construal perspective. *Information & Management*, *52*(5), 577–587. doi:10.1016/j.im.2015.04.003

Waterman, R. H. Jr, Peters, T. J., & Phillips, J. R. (1980). Structure is not organization. *Business Horizons*, *23*(3), 14–26. doi:10.1016/0007-6813(80)90027-0

Weatherill, S. (2001). Jusifying the Limits to Party Autonomy in the Internal Market - EC Legislation in the Field of Consumer Protection. In S. Grundmann, W. Kerber, & S. Weatherill (Eds.), *Party autonomy and the role of information in the internal market* (pp. 173–196). Berlin: Walter de Gruyter. doi:10.1515/9783110873030.173

Weihrich, H. (1982). The TOWS matrix: A tool for situational analysis. *Long Range Planning*, *15*(2), 54–66. doi:10.1016/0024-6301(82)90120-0

Wernerfelt, B. (1984). A resource-based view of the firm. *Strategic Management Journal*, *5*(2), 171–180. doi:10.1002/smj.4250050207

Wernerfelt, B. (1995). The resource-based view of the firm: Ten years after. *Strategic Management Journal*, *16*(3), 171–174. doi:10.1002/smj.4250160303

Westerveld, E. (2003). The project excellence model: Linking success criteria and critical success factors. *International Journal of Project Management*, *21*(6), 411–418. doi:10.1016/S0263-7863(02)00112-6

West, J., Salter, A., Vanhaverbeke, W., & Chesbrough, H. (2014). Open innovation: The next decade. *Research Policy*, *43*(5), 805–811. doi:10.1016/j.respol.2014.03.001

White, G. I., Sondhi, A. C., & Fried, D. (2003). *The analysis and use of financial statements* (Vol. 1). John Wiley & Sons.

Whitty, S., & Maylor, H. (2009). And then came complex project management. *International Journal of Project Management*, *27*(3), 304–310. doi:10.1016/j.ijproman.2008.03.004

Wiener. (2016). Retrieved March 15, 2016, from Wiener Stadtishe Insurance Stadtishe web site: http://en.wiener.co.rs/

Wikström, K., Artto, K., Kujala, J., & Söderlund, J. (2010). Business models in project business. *International Journal of Project Management*, *28*(8), 832–841. doi:10.1016/j.ijproman.2010.07.001

Williams, C., & Lee, S. H. (2009). Exploring the internal and external venturing of large R&D-intensive firms. *R & D Management*, *39*(3), 231–246. doi:10.1111/j.1467-9310.2009.00553.x

Williams, C., & Nones, B. (2009). R&D subsidiary isolation in knowledge intensive industries: Evidence from Austria. *R & D Management*, *39*(2), 111–123. doi:10.1111/j.1467-9310.2009.00544.x

Williamson, O. (1971). The verical integration of production: Market failure considerations. *The American Economic Review*, *61*(2), 112–123.

Wilson, K. E., Silva, F., & Richardson, D. (2015). *Social impact investment: building the evidence base.* Retrieved December 14, 2015, from http://www.oecd.org/sti/ind/social-impact-investment.pdf

Wilson, J. W., & Eilertsen, S. (2010). How Did Strategic Planning Help During The Economic Crisis? *Strategy and Leadership*, *38*(2), 1087–8572. doi:10.1108/10878571011029000

Winklhofer, H. (2002). Information Systems project management during organizational change. *Engineering Management Journal*, *14*(2), 33–37. doi:10.1080/10429247.2002.11415160

Wolfe, R. (1994). Organizational innovation: Review, critique and suggested research directions. *Journal of Management Studies, 31*(3), 405–431. doi:10.1111/j.1467-6486.1994.tb00624.x

World Bank Centre for Financial Reporting Reform (CFRR) and the Austrian Institute for SME Research. (2013). A Survey of SME Accounting and Reporting Practices in Austria. Vienna: World Bank Centre for Financial Reporting Reform (CFRR) and the Austrian Institute for SME Research.

World Bank. (2016). *Global Economic Prospects June 2016: Divergences and Risks*. Washington, DC: World Bank. Washington, DC: World Bank. Retrieved September 7th, 2016 from World Bank web site: http://www.worldbank.org/en/publication/global-economic-prospects#data

World Commission on Environment and Development ("The Brundtland Commission"). (1997). *Our Common Future*. Retrieved May 30, 2016, from http://www.un-documents.net/our-common-future.pdf

World Economic Forum. (2014). *The Europe 2020 competitiveness report. Insight report. Building a more competitive Europe*. Retrieved July 25, 2016, from http://www.weforum.org/Europe2020

World Economic Forum. (2016). *Global Competitiveness Report*. Retrieved July 27th, 2016 from World Economic Forum web site: https://www.weforum.org/reports/global-competitiveness-report-2014-2015

World Economic Forum. (2016). Retrieved August 17th, 2016 from WEF web site: http://reports.weforum.org/global-competitiveness-report-2015-2016/

World Economic Forum. (2016). *The Global Competitiveness Report 2015-2016*. Retrieved July 26, 2016, from World Economic Forum web site: http://reports.weforum.org/global-competitiveness-report-2015-2016/

Wrenn, B. (1997). The Market Orientation Construct: Measurement and Scaling Issues. *Journal of Marketing Theory and Practice, 5*(3), 31-54.

Yueliang, S., & Nan, L. (2015). Simulation of Game Model for Supply Chain Finance Credit Risk Based on Multi-Agent. *Open Journal of Social Sciences, 3*(01), 31–36. doi:10.4236/jss.2015.31004

Zachariadis, M. (2003). R&D, innovation, and technological progress: A test of the Schumpeterian framework without scale effects. *The Canadian Journal of Economics. Revue Canadienne dEconomique, 36*(3), 566–686. doi:10.1111/1540-5982.t01-2-00003

Zairi, M. A. M. (1999). BPR implementation process: An analysis of key success and failure factors. *Business Process Management Journal, 5*(1), 87–112. doi:10.1108/14637159910249108

Zairi, M., & Sinclair, D. (1995). Business process re-engineering and process management A survey of current practice and future trends in integrated management. *Business Process Re-Engineering & Management Journal, 1*(1), 8–30. doi:10.1108/14637159510798248

Zajko, M., Chodasová, Z., Jemala, Ľ., & Materák, M. (2010). *Management of SMEs*. Bratislava, Slovakia: STU in Bratislava.

Zgodavova, K., & Bober, P. (2012). An Innovative Approach To the Integrated Management System Development: SIMPRO-IMS Web Based Environment. Quality Innovation Prosperity, 16(2), 59-70.

Zhang, X. (2016). Co-evolution between institutional environments and organizational change: The mediating effects of managers uncertainty. *Journal of Organizational Change Management, 29*(3), 381–403. doi:10.1108/JOCM-12-2014-0218

Zhong, D., & Luo, Q. (2015). Using social network analysis to explain communication characteristics of travel-related electronic word-of-mouth on social networking sites. *Tourism Management, 46*, 274–282. doi:10.1016/j.tourman.2014.07.007

Zhu, Q., Geng, Y., Fujita, T., & Hashimoto, S. (2010). Green supply chain management in leading manufacturers. *Management Research Review*, *33*(4), 380–392. doi:10.1108/01409171011030471

Zhu, Q., Sarkis, J., & Geng, Y. (2005). Green supply chain management in China: Pressures, practices and performance. *International Journal of Operations & Production Management*, *25*(5), 449–468. doi:10.1108/01443570510593148

Živadinović, J., (2010). *Poslovna informatika*. Beograd: Visoka škola za poslovnu ekonomiju i preduzetništvo.

Байбурина Э.Р. Головко, Т.В. (2008). Эмпирическое исследование интеллектуальной стоимости крупных российских компаний и факторов ее роста. Корпоративные финансы. N° 2 (6). (pp. 5–19).

About the Contributors

Milan B. Vemić, Ph.D., is an enterprise development consultant and chairman of council at Higher Education Institution "DOSITEJ", Serbia. As associate professor of economics and finance he teaches graduate and postgraduate courses in Serbia, Kazakhstan and other countries. Milan has published five books: *Fundamental Principles of Financial Management, The Science of Economic System and Economic Policy in Transition, The Science of Financial Management, Internal audit and Practicum in Finance.* His recent publications are: (1) Strategic data based optimization of working capital management in medium-sized firms, IGI Global Publishing, 2015; (2) Development of some specific methodology elements for supporting innovative business development in West Balkans and Black Sea Region, 2015. He is a partner in Matapan Consulting, a firm dedicated to enterprise development, and worked as an expert on several USAID and EU development projects. His research focuses on business economics, finance and management, financial and non-financial performance of SMEs. An active keynote speaker at industry forums and conferences Milan is a member of Central Eurasian Studies Society and of Cluster House Serbia.

* * *

Olja M. Arsenijević, Ph.D., was born in 1967 in Novi Sad. She graduated at the University of Novi Sad, Faculty of Philosophy, at the Department of Serbo-Croatian language and Yugoslav literature. She earned a Master's degree at the Faculty of Management in 2003 with the thesis "Organizational culture of higher education institutions, which studied management". She received her PhD from the Faculty of Management in 2007 with the thesis "Modeling of Management Education Evaluation". Scientific fields of interest: change management, business communication, knowledge management. Author of more than 100 scientific papers in these areas, which were published in the proceedings of scientific conferences, scientific journals, national and international monographs and monograph "Business and collective bargaining." Lecturer at the Faculty of Business Studies and Law, courses: Knowledge management and Knowledge economics. She was editor of the scientific journal "Leader - director". Editor of 7 collection of works from the International Scientific Conference "Towards the age of knowledge". Editor of 3 monographs and leader of the project "Innovative culture as a factor in the formation and development of innovation systems." Project team member on the project "Media technology and socio-educational change", funded by the Ministry of Education and Science of the Republic of Serbia and the editor of 2 monographs "Knowledge, education, the media" arising as a result of the work on that project.

Aleksandar M. Damnjanović, Ph.D., is a specialist in entrepreneurial management, born on February 5, 1977 in Belgrade. Completed postgraduate studies at the Technical Faculty Cacak, Department of Industrial Management, field of Entrepreneurial Management in 2006. He specialized on the topic: "Improving the financial business environment for small and medium-sized enterprises in Serbia" and acquired the title of Specialist in entrepreneurial management. From 2005 he served as a teaching assistant at the Cacak Technical Faculty in the Department of Industrial Management. During his studies he published and announced several scientific papers and also participated in several professional conferences. Completed master studies at the Technical Faculty, Department of Industrial Management, in 2007. In 2014 he defended his doctoral dissertation at the Faculty of Trade and Banking, ALFA University - Belgrade, titled: "The impact of modern technology on the strategic competitiveness of small and medium-sized enterprises", and then gained a Ph.D in economics. He participated in scientific symposiums and conferences of national and international importance. The broader field of scientific and professional work of Aleksandar Damnjanović Ph.D, includes strategic management and management information systems and statistical data processing, entrepreneurial management (SMEs), management, finance.

Hosein Daneshpour is a doctoral candidate at the University of Vaasa, Finland, at the industrial management department. He fulfilled the master of sciences program in the field of environmental energy technology at Lappeenranta University of Technology in Finland in 2014. Additionally, he carried out his bachelor's degree studies in industrial engineering from 2001 to 2005. He has several years of working experience as a project management officer in the field of EPC projects.

Katarina Ivančević, Ph.D., is a full professor of Law at the Law School of Union University Belgrade, with teaching and research experience in Insurance Law, Contract Law, Torts and Consumer Protection Law. She graduated from the Belgrade University Faculty of Law passed the bar exam and received her doctoral degree on Legal Protection of Insurance and Banking Service Clients from Faculty of Law, Union University. She started her carrier in Beobanka legal department and advanced as manager of insurance affairs at Beobanka Insurance Company and Triglav Pension Fund with parallel engagement as professor at the Law Faculty. She is an author of numerous scientific and professional papers on insurance, civil rights, consumer protection and EU Law published in professional journals and proceedings from technical meetings, conferences and symposia. Katarina authored a university text book Consumer law and co-authored a university text book Insurance Law both of which are used at the Faculty of Law Union University and at Academy of Commerce in Novi Sad. She co-authored the monograph Non-Life Insurance. Katarina is a member of the National Council for Consumer Protection, Administrative Boards for Insurance Law of Serbia and Association for Damage Law of Serbia.

Edita Kastratović, Ph.D., graduated in 1986 at the Faculty of Economics in Podgorica, Montenegro. She obtained her PhD in Economics in 1995 at the Ukrainian Academy of Agrarian Sciences in Kiev – Institute for Agrarian Economy. During the nineties she was engaged on international education projects of BK Company in Moscow, with the title of Director of Education for CIS countries. She has participated in establishing the International University for Business and Management in Moscow with MSU Lomonosov, where she worked as a Director and then also as a Vice-rector of the University. As coordinator and co-author of numerous international projects, she has worked as Deputy Manager of Center for International Relations of the Institute of Sociology of the Russian Academy of Sciences. In

Serbia she has worked as a General Manager of the SME Institute, BK TV, BK University, and also as a Dean of Faculty of Management in Sports. She is author of over a hundred scientific and professional papers, textbooks and monographs in the field of management, entrepreneurship and sport. She has participated in the implementation of over seventy scientific, research and developmental projects, more than two hundred scientific studies and projects in the field of system solutions for the development of SME, investments, education, privatization, quality, strategies of local economic development. She is a member of editorial boards, editor and reviewer of national and international journals in the field of economics, sport, legislation, science, politics and entrepreneurship. She is a recipient of various awards, letters of commendation and plaques. She is a sports pilot and enjoys sailing.

Mariusz Maciejczak, Ph.D., is an economist who graduated from Faculty of Economic Sciences of Warsaw University of Life Sciences - SGGW in Poland and at CAH University of Applied Science in The Netherlands. Since 2005 he is working as an assistant professor at the Faculty of Economic Sciences of WULS-SGGW. His work focuses on issues related to economics and management. He does research on institutions and institutional governance, knowlegde-based bioeconomy and the economic issues of the co-existence between products coming from different production systems in the value supply chains. Furthermore, he works on matters related to technologies and innovations' economics and management.

Peter Malega, Ph.D., is an Assistant Professor of Industrial Engineering and Management at Technical University of Košice, Faculty of Mechanical Engineering. He received his doctoral degree in Engineering Technologies and Materials. Peter teaches undergraduate and graduate classes in the area of industrial engineering and economics and management of small and medium enterprises. He is a professional member of the Junior Chamber International, Slovakia. His research interests include strategic management, production management, economics and management of small and medium enterprises and work studies. Peter has published three books and a number of articles in domestic and foreign journals and also in conferences. Peter actively cooperates with a number of companies in Slovakia where he realizes both his research and development activity.

Olgica Milošević, PhD candidate, is a specialist in European Law, born on July 5, 1978 in Belgrade. Graduated from Faculty of Law University in Belgrade in 2008 and the Union Law School in 2010. Completed master's studies in European Law at the Union Law School in 2014 with a Master's thesis: "A wide margin of appreciation doctrine in the European Court of Human Rights' decisions regarding bioethical issues" and acquired the title Master in European Law. Currently she is a PhD candidate at the same faculty with the title of doctoral dissertation approved: "Legal protection of people's reproductive health as their personal right". Employed since 2011 at the University Business Academy in Novi Sad, Faculty of Applied Management, Economy and Finance, Belgrade on a position of Assistant Lecturer in Legal field and Secretary of Faculty. She published and announced several scientific papers and also participated in scientific symposiums and conferences of national and international importance. The broader field of scientific and professional work of Olgica Milošević includes European Law, Company Law, Labor Law, Entrepreneurial Management (SMEs), Environmental Law, Bioethics, Medical Law.

Robert M. Molnar, Ph.D., is a specialist in entrepreneurship and strategic management from Serbia. He obtained his PhD in the area of Industrial Engineering and Engineering Management at University of Novi Sad in 2010. The title of his dissertation is "Development of a system of non-financial support to small and medium-sized enterprises and entrepreneurship in Serbia". He has been engaged in higher and university education in Serbia from 1994, at the beginning as a teaching assistant and later on as a professor. So far, he published more than 50 relevant articles as well as 16 books on strategic management, SME sector development, education and innovation. From the year 2000 he is also active as a freelance consultant and has shared his knowledge and experience through technical assistance to SMEs, industry experts, academic institutions, international organizations and public administration. He actively participates in different projects in his field of expertise. One of his most relevant projects is "Establishment of the Regional Center for Development of SMEs and Entrepreneurship – BANAT, Ltd., Zrenjanin", when he served as general manager for 4 years (2002-2006), and later assisted with transformation into the Regional Agency for Socio-Economic Development in Banat, Ltd., Zrenjanin.

Aida Nurul is a Ph.D. candidate working under the guidance of Prof. Josu Takala at University of Vaasa, Finland. Her interest lies broadly in the area of environmental sustainability network. Aida graduated from Kolej Universiti Tun Hussein Onn (KUITTHO) Malaysia of Bachelor of Management of Technology and received her Master Degree in Master of Science (Quantitative Science) from Universiti Teknologi MARA (UiTM), Malaysia. She is currently employed by University Tun Hussein Onn Malaysia (UTHM), Malaysia.

Drago Orčić, M.Sci., is an expert for small and medium-sized enterprises, born 10.06.1966 in Subotica. He is entrepreneur since 1991 when he opened his first private company. Since 2003 he lead several EU projects dedicated to the development of small and medium-sized enterprises in the region. In 2006 he has been certified as consultant for innovations by ''CENTRIM'' University of Brighton, UK. Since 2001 he has been consulting SMEs. In 2012 he has been certified as a consultant for SMEs as per EU standards by the Ministry of Economy, Regional Development and the National Agency for Regional Development. He graduated in 2011 at the Faculty of Management in Novi Sad, with the subject "Comparative analysis of resistance to change in SMEs". He obtained a master's degree in 2012 defending the theme "Measuring the value of intellectual capital in SMEs". In 2012 he enrolled to doctoral studies in Techical Engineering, University of Belgrade, where he does research on intellectual capital and intangible resources in SMEs. Drago Orčić was an active participant in many international scientific conferences. He has substantial experience in the profit and non-profit sector as well as in working with SMEs. His field of scientific expertise is intellectual capital, human capital, identification, measurement and launching of the intangible resources and potential.

Beatriz Blanco Rojo holds a PhD in Economics and Management from the Universidad Autónoma of Madrid. She has been a Senior Lecturer in the Operations Management and Logistics area from the University of Cantabria since 1997. Throughout her long career, she has participated in over 24 research projects and she is coauthor of over 40 publications (18 of which are indexed). She has also supervised PhD theses defended in 2010, 2014 and two in 2016. Her areas of research focus on logistics, maritime sector and operations research (Lean Management).

Lidia Sánchez Ruiz, Ph.D., has been an Assistant Professor at the University of Cantabria for 6 years. Despite her short career, she has participated in several R&D competitive projects and she is coauthor of several book chapters and research articles, some of them in international journals. She has also participated in numerous conferences and scientific events, both national and international. Her areas of research focus on process management, continuous improvement, Lean Management philosophy, Rasch Measurement Theory and logistics (mainly port innovation).

Adrian Słodki, M.Sci., is an economist who graduated from Faculty of Economic Sciences of Warsaw University of Life Sciences (WULS-SGGW). His main point of interest during the master studies (2013-2014) was Game Theory and its implementation in small enterprises. Since 2014 he is working as Anti-Money Laundering and Counter Terrorist Financing (AML/CTF) Analyst at Royal Bank of Scotland. His work focuses on delivering Customer Due Diligence (Know Your Customer) based on the bank policy which adheres to the Financial Conduct Authority regulations, European Commission AML/CTF directives and Financial Action Task Force recommendations.

Marcin Soniewicki, Ph.D., in Economics, Assistant Professor at Poznan University of Economics and Business, Faculty of International Business and Economics, International Marketing Department. He is University of London graduate and conducts research in the area of knowledge management – technological and organizational aspects; strategic management; competitiveness of international enterprises; marketing and market orientation. He specializes in the role of market and technical knowledge in creating innovation as well as competitive advantage of enterprises.

Julia Stefanova, Ph.D., is a Senior Assistant Professor at the Economic Research Institute at the Bulgarian Academy of Sciences. Her academic endeavors cover the integration of the EU capital markets, regional cooperation and consolidations among stock exchanges in Central and Eastern Europe, the Danube region countries, Latin American capital markets development, etc. She has completed research projects on venture capital in EU and the possibilities for access of Bulgarian companies to venture capitals and on the capital markets of South-Eastern Europe countries (Serbia, Albania and FYR Macedonia) in the course of their preparation for EU accession. Currently she has been involved in academic research on opportunities and challenges in development and social impact investing in EU in close collaboration with her American Fulbright colleague Mr. Zachary Wenner. Dr. Julia Stefanova has PhD degree in International Economic Relations and World Economy from the Economic Research Institute (The Bulgarian Academy of Sciences) and Master's degrees in International Economic Relations, Finance and English Linguistics. Besides her academic experience she has over ten years of banking background working as bank branch manager, credit officer, and relationship manager in several commercial banks in Bulgaria and as an expert in payments and settlements systems' oversight in the Central Bank of Bulgaria. She speaks English, French and Greek languages.

Josu Takala, Ph.D., has vast experience in both industry and academia. From 1979-1992 he worked for Asea Brown Boveri Group (ABB Strömberg) in R&D and Quality Assurance divisions as researcher and manager involving automation tasks. With 30 independent companies in a broad area of electrical technology ABB Strömberg achieves 8% turnover in R&D (3000 Million USD/1995). Josu graduated at Tampere University of Technology M.Sc (El.Eng) 1980 and received his doctoral degree in electrical engineering and business studies in the University of Vaasa in 1988 as well as Doctorates (HC) from the Technical University of Košice (2009) and from Universiti Tun Hussein Onn Malaysia (2015) in Technology Management and Business. Currently (1988-) he is Professor in Industrial Management at the University of Vaasa, and cooperates with multiple universities in Finland, Thailand, Malaysia, China,

Slovakia, Slovenia etc. His field of interest is technology management within sustainable competitive strategies of private-public organizations of manufacturing networks through generic utilization of quality management, (technology) strategy, new product development, production management, logistics and environmental management. Josu published more than 500 scientific articles. Co-Editor, Special Issue Editor, Member of Scientific Board in Journals (like MPER, MNG), key note speaker, Chairman or Honored Chairman in international conferences.

Binod Timilsina has received the M.Sc. degree in Industrial management in 2013 from the University of Vaasa, Finland. Currently he is enrolled as a doctoral student at department of production, University of Vaasa, Finland. He is also the member of editorial board of Acta Logistica (International scientific journal about logistics). His research interests are in the area of strategic management, operations strategy, and operational excellence. His researches are mostly related to small and medium size enterprises. Previously, he has published research papers in peer reviewed international journals, and has worked as a reviewer for the book entitled "Supply chain strategies and the engineering-to-order approach" published by IGI Global.

Maria Giovanna Tongiani is a researcher in Economy and Management of Companies, Department of Econony and Management of University of Pisa. Her primary teaching areas are undergraduate economy and management of retailing and wholesaling firms, graduate economy, management and strategy of retailing firms. Her research interests include small firms and distribution channels, retailing and wholesaling management and strategy, business process management and the business value and adoption of e-business, shopper marketing and in-store marketing.

Zachary Wenner is a currently residing in Sofia, Bulgaria, where he is conducting research related to development finance and impact investing pursuant to a Fulbright Scholarship. Previously, Mr. Wenner was a Research Scholar at the University of Pennsylvania's Wharton School where he researched impact investing private equity as part of the Wharton Social Impact Initiative. Mr. Wenner led the Aspen Institute's Impact Economy Initiative, a policy program designed to create an enabling environment for impact investing and social enterprise development in collaboration with the Gates Foundation and the Omidyar Network. During his tenure at the Aspen Institute, Mr. Wenner was a Fellow with Social Finance, a Social Impact Bond intermediary, and part of the Grameen Banks "Bankers without Borders" program. Prior to working at the Aspen Institute, Zachary was a financial analyst at Holliday Fenoglio and Fowler. Mr. Wenner graduated from Dartmouth College, where he was a Presidential Scholar.

Silvena Dencheva Yordanova graduated at University of Economics-Varna, Bulgaria, in 1999 with bachelor's degree in Marketing and management. Her MBA degree from the 2000 year is from University of Economics-Varna. Her Ph.D. degree is with a dissertation "Electronic Commerce in Bulgaria" also from University of Economics-Varna, Bulgaria.

Sheng Zhao is living in Nanjing (China). She received a B.A.in Economics and commerce from University of Pisa in 2013 and a M.A. in Marketing and Market Research from University of Pisa in 2016 with a scholarship of the Italian Region Tuscany. Her graduate dissertation and actual research interests are on Italian traditional food and Chinese consumer.

Index

F

G

H

I

J

K

L

M

www.ingramcontent.com/pod-product-compliance
Lightning Source LLC
LaVergne TN
LVHW060356300926
849162LV00006B/1284